T0202400

Current Trends in Atomic Physics

Lecture Notes of the Les Houches Summer School:

Volume 107, 4–29 July 2016

Current Trends in Atomic Physics

Edited by

Antoine Browaeys, Thierry Lahaye, Trey Porto,
Charles S. Adams, Matthias Weidemüller, and
Leticia F. Cugliandolo

OXFORD
UNIVERSITY PRESS

UNIVERSITY PRESS

Great Clarendon Street, Oxford, OX2 6DP,
United Kingdom

Oxford University Press is a department of the University of Oxford.
It furthers the University's objective of excellence in research, scholarship,
and education by publishing worldwide. Oxford is a registered trade mark of
Oxford University Press in the UK and in certain other countries

First Edition published in 2019

Impression: 1

Published in the United States of America by Oxford University Press
198 Madison Avenue, New York, NY 10016, United States of America

British Library Cataloguing in Publication Data
Data available

Library of Congress Control Number: 2018963932

ISBN 978–0–19–883719–0

DOI: 10.1093/oso/9780198837190.001.0001

Printed and bound by
CPI Group (UK) Ltd, Croydon, CR0 4YY

École de Physique des Houches

Service inter-universitaire commun
à l'Université Joseph Fourier de Grenoble
et à l'Institut National Polytechnique de Grenoble

Subventionné par l'Université Joseph Fourier de Grenoble,
le Centre National de la Recherche Scientifique,
le Commissariat à l'Énergie Atomique

Previous sessions

Publishers

- Session VIII: Dunod, Wiley, Methuen
- Sessions IX and X: Herman, Wiley
- Session XI: Gordon and Breach, Presses Universitaires
- Sessions XII–XXV: Gordon and Breach
- Sessions XXVI–LXVIII: North Holland
- Session LXIX–LXXVIII: EDP Sciences, Springer
- Session LXXIX–LXXXVIII: Elsevier
- Session LXXXIX– : Oxford University Press

Preface

Following the laser cooling revolution of the 1980s, a Les Houches session devoted to atomic physics was held in 1990. Some of the speakers in this school went on to win Nobel Prizes and many of the students went on to become Professors. Subsequently, more specialized sessions followed such as "Coherent atomic matter waves" (1999), "Ultra-cold atoms and quantum information" (2009), "Many-body physics with ultracold gases" (2010), "Quantum machines: measurement and control of engineered quantum systems" (2011). A quarter of a century from the classic 1990 school, we felt it was time to organize a summer school that revisited the general theme of atomic physics.

One of the most notable evolutions over the last twenty-five years is the extent to which atomic physics has spread into the interfaces between different areas of physics. The precision and control of few-body atomic systems manipulated by classical or quantum electromagnetic fields allows atomic physicists to address fundamental questions on open problems in a wide variety of areas. There have been two general trends in this direction. The first is the increasing precision of measurements performed on isolated individual atomic systems. Along this line, tests of fundamental symmetries, of fundamental laws (the equivalence principle, the quest for new forces), and tests of QED now complement those performed in high-energy physics. The second trend is connected to the unprecedented control gained over laser-cooled atoms and ions, as well as more recently over artificial atoms or hybrid systems. In this context, the interactions between the atoms can be tailored, offering model systems to explore open questions in condensed-matter physics (quantum many-body problems) or chemistry. At the interface between these two trends, atomic physics also provides wonderful systems to explore the foundations of quantum physics, with possible applications in quantum optics and quantum information processing.

The goal of this session was to illustrate these many facets of atomic physics through a combination of long tutorials and more specialized seminars. We did not aim at covering a particular topic in full depth (there are many schools held to fill this need) but rather to give a broad perspective on the field and see how atomic physics interconnects with other areas of physics. We grouped the lectures into five broad themes. The first theme highlighted the connections between atomic, molecular, and optical (AMO) physics and solid-state systems that are designed and studied using the same tools as in AMO physics. We had three lectures and one seminar to address how atomic physics meets artificial atoms. Mikhail D. Lukin lectured on quantum optics with atom-like systems (see Chapter 1 on page 3 of this volume); Benoît Deveaud gave an introduction to the physics of composite bosons in solids; Steven M. Girvin presented lectures on circuit QED (see Chapter 11 on page 404); and Tristan Briant gave an introduction to opto-mechanics in a seminar.

A second theme dealt with high-precision measurements. David DeMille described how AMO experiments can be used, complementary to high-energy physics, for searching new particles (see Chapter 2 on page 33); Thomas Udem lectured on frequency combs and high-precision spectroscopy of hydrogen (see Chapter 4 on page 145); Mark A. Kasevich lectured on atom interferometry (see Chapter 6 on page 224); and finally Thorsten Schumm gave a seminar on atomic and nuclear clocks.

Another theme focused on the connections between atomic physics and chemistry, and comprised four lectures on cold molecules and ultrafast processes. Frédéric Merkt gave a series of lectures on cold chemistry and the manipulation of cold molecules (see Chapter 3 on page 85); Matthias Weidemüller lectured on ultra-cold chemistry, Anne L'Huillier gave an introduction to the physics of ultra-fast processes (Chapter 8 on page 330), and finally Alexander Kuleff illustrated how ultrafast electron dynamics allow for the exploration of chemical processes (Chapter 9 on page 345).

A fourth set of lectures dealt with quantum simulation of condensed-matter systems using the tools of atomic physics: Thierry Giamarchi gave a general introduction on quantum simulation from a condensed-matter point of view (see Chapter 5 on page 220); Chris Monroe lectured on quantum simulation with trapped ions; Immanuel Bloch on quantum simulation with cold atoms; and Andreas Läuchli gave some insights on computational aspects of quantum simulation.

The final theme dealt with the foundations of quantum physics, which have traditionally been studied using atomic physics tools. Wojciech Hubert Zureck gave a series of lectures on decoherence and beyond (Chapter 7 on page 296); Ivan Deutsch gave an introduction on quantum control, measurement, and tomography; Alain Aspect lectured on how land-mark quantum optics experiments can now be addressed with atoms (Chapter 12 on page 430); and Markus Arndt gave a seminar on matter wave physics with "large" objects (Chapter 10 on page 369).

At a time where many tutorials, lecture notes, and review articles are available online, we are particularly grateful to the lecturers for having taken the burden of writing these original notes for this volume. We have no doubt that, beyond illustrating the status of the field in 2016, many of them will not be outdated for a long time.

Finally, we would like to warmly thank the students of the school, who showed great enthusiasm both for the lectures and for the rest of this one-month experience in Les Houches. We also thank the various funding agencies for their generous support: the German-French University, the Air Force Office of Scientific Research (USA), the Intercan European network, the iXcore foundation (France), the Physics Frontier Center (USA), and the CNRS Groupement de Recherche on "Quantum Information: from foundations to applications" (France). We also warmly thank Murielle Gardette, Isabelle Lelièvre, and Flora Gheno for their beautiful efficiency, their kindness, and permanent good mood. They made our life as organizers shamingly easy, and we could enjoy the school as much as the students did! Special thanks also for the cook and the staff at the restaurant for treating us so well and to Alain Aspect for giving the traditional public lecture in Les Houches!

A. Browaeys, T. Lahaye, T. Porto, C. S. Adams,
M. Weidemüller, and L. F. Cugliandolo
July 2018

Contents

List of participants

ORGANIZERS

ANTOINE BROWAEYS
IOGS and CNRS, Palaiseau, France

THIERRY LAHAYE
IOGS and CNRS, Palaiseau, France

TREY PORTO
NIST and University of Maryland, USA

CHARLES S. ADAMS
Durham University, UK

MATTHIAS WEIDEMÜLLER
University of Heidelberg, Germany

LECTURERS

MARKUS ARNDT
University of Vienna, Austria

ALAIN ASPECT
Institut d'Optique, Palaiseau, France

IMMANUEL BLOCH
Max Planck Institute, Garching, Germany

TRISTAN BRIANT
Laboratoire Kastler-Brossel, Paris, France

DAVID DEMILLE
Yale University, New Haven, USA

IVAN DEUTSCH
University of New Mexico, USA

BENOÎT DEVEAUD
EPLF, Switzerland

THIERRY GIAMARCHI
University of Geneva, Switzerland

STEVEN M. GIRVIN
Yale University, New Haven, USA

MARK A. KASEVICH
Stanford University, USA

ALEXANDER KULEFF
Heidelberg University, Germany

ANDREAS LÄUCHLI
University of Innsbruck, Austria

ANNE L'HUILLIER
University of Lund, Sweden

MIKHAIL D. LUKIN
Harvard University, Cambridge, USA

FRÉDÉRIC MERKT
ETH, Zurich, Switzerland

CHRISTOPHER MONROE
University of Maryland, College Park, USA

THORSTEN SCHUMM
University of Vienna, Austria

THOMAS UDEM
Max Planck Institute, Garching, Germany

WOJCIECH HUBERT ZURECK
Los Alamos National Laboratory, USA

STUDENTS

MR. ANDREA AMICO
University of Florence, Italy

MR. CHITRAM BANERJEE
University of Paris-Sud, Orsay, France

MR. BRYNLE BARRETT
iXBlue, Saint Germain en Laye, France

MR. MARK BROWN
University of Colorado, USA

MR. TOBIAS BRÜNNER
University of Freiburg, Germany

MR. LORENZO CARDARELLI
University of Hannover, Germany

MR. KRZYSZTOF CHABUDA
University of Warsaw, Poland

MR. STEVEN CONNELL
Griffith University, Brisbane, Australia

MR. FLORENT COTTIER
University of São Paulo, São Carlos, Brazil

MR. SYLVAIN DE LÉSÉLEUC
Institut d'Optique, Palaiseau, France

MR. FRANCESCO DELFINO
University of Pisa, Italy

MRS. EMMA DOWD
University of Berkeley, USA

MR. PIERRE DOYEUX
Laboratoire Charles Coulomb, Montpellier, France

MR. DAVIDE DREON
Laboratoire Kastler-Brossel, Paris, France

MR. MAXIME FAVIER
SYRTE, Paris Observatory, France

MR. DARIO ALESSANDRO FIORETTO
University of Innsbruck, Austria

MR. OON FONG EN
University of Singapore

MR. TUVIA GEFEN
The Hebrew University, Jerusalem, Israel

MR. MANUEL GERKEN
University of Heidelberg, Germany

MR. PRASOON GUPTA
University of Maryland, College Park, USA

MR. HAN JINGSHAN
University of Singapore

MR. ANDRE HEINZ
Max Planck Institute, Garching Germany

MR. MENGZI HUANG
Laboratoire Kastler-Brossel, Paris, France

MR. MARTIN IBRÜGGER
Max Planck Institute, Garching, Germany

MR. YONGGUAN KE
Sun Yat-Sen University, Guangzhou, China

MR. ALEXANDER KEESLING
Harvard University, Cambridge, USA

MR. BUMSUK KO
Seoul National University, Republic of Korea

MRS. MARIE LABEYE
Université Paris 6, Paris, France

MRS. KATHRIN LUKSCH
Clarendon Laboratory, Oxford, UK

MR. MACIEJ MALINOWSKI
ETH, Zurich, Switzerland

MR. MICHITERU MIZOGUCHI
IMS, Okazaki, Japan

MR. MUSAWWADAH MUKHTAR
Institut d'Optique, Palaiseau, France

MR. BRUNO NAYLOR
Université Paris 13, Villetaneuse, France

MRS. NIKOLETT NEMET
University of Auckland, New Zealand

MRS. MARÍA AUXILIADORA PADRÓN BRITO
ICFO, Castelldefels, Spain

MR. CHRISTOPHER PARMEE
Cavendish Laboratory, Cambridge, UK

MR. GREGOR PIEPLOW
University of Madrid, Spain

MR. MATTHIEU PIERCE
Laboratoire Kastler-Brossel, Paris, France

MR. MARCIN PLODZIEN
University of Eindhoven, The Netherlands

MRS. APICHAYAPORN RATKATA
Durham University, UK

MR. DAVID RAVENTÓS
ICFO, Castelldefels, Spain

MR. ANTONIO RUBIO-ABADAL
Max Planck Institute, Garching, Germany

MR. ILAN SHLESINGER
Institut d'Optique, Palaiseau France

MR. PABLO SOLANO
University of Maryland, College Park, USA

MR. NICHOLAS SPONG
Durham University, UK

MRS. KRISHNAPRIYA SUBRAMONIAN RAJASREE
Okinawa Institute, Japan

MR. OLIVER THOMAS
University of Kaiserslautern, Germany

MR. KONSTANTIN TIUREV
Aalto University, Finland

MR. KAI VOGES
University of Hannover, Germany

MR. SHENGTAO WANG
University of Michigan, Ann Arbor, USA

MR. FRANZ FERDINAND WIESER
University of Vienna, Austria

MR. DAVID WONG-CAMPOS
University of Maryland, College Park, USA

MR. LUYAO
University of Singapore

MR. YICONG YU
University of Wuhan, China

1

Quantum optics with diamond color centers coupled to nanophotonic devices

ALP SIPAHIGIL AND MIKHAIL D. LUKIN

Department of Physics, Harvard University
Cambridge, MA 02138 USA

Sipahigil, A. and Lukin, M. D. "Quantum optics with diamond color centers coupled to nanophotonic devices." In *Current Trends in Atomic Physics*. Edited by Antoine Browaeys, Thierry Lahaye, Trey Porto, Charles S. Adams, Matthias Weidemüller, and Leticia F. Cugliandolo. Oxford University Press (2019). © Oxford University Press.
DOI: 10.1093/oso/9780198837190.003.0001

Chapter Contents

1.1   Introduction: quantum optics with solid-state systems

In the past two decades major advances in the isolation and coherent control of individual quantum systems ranging from neutral atoms, ions, and single photons to Josephson junction based circuits and electron spins in solids (Wineland, 2013; Haroche, 2013) have been made. The ability to control single quantum systems has opened up new possibilities ranging from quantum information processing and simulation (Cirac and Zoller, 1995; Jaksch et al., 1998), to quantum-enhanced sensing (Maze et al., 2008; Rondin et al., 2014) and quantum communication (Cirac et al., 1997; Kimble, 2008). Many of these potential applications require the realization of controllable interactions between multiple qubits. Several techniques have been developed to create such interactions between qubits over short distances at the single device level. Examples include dipole-dipole interactions between neutral atoms (Jaksch et al., 2000; Urban et al., 2009), phonon-mediated interactions in trapped ion crystals (Cirac and Zoller, 1995; Monroe et al., 1995) and photon-mediated interactions in cavity and circuit quantum electrodynamics (QED) (Pellizzari et al., 1995; Neuzner et al., 2016; Majer et al., 2007). Recent experiments have used these interactions to entangle up to a few tens of qubits (Monz et al., 2011; Barends et al., 2014) and control quantum dynamics of more than fifty qubits (Bernien et al., 2017; Zhang et al., 2017). The current challenge involves scaling these systems up to a large number of connected, controlled qubits.

One approach to scalability involves distributing quantum information between remote nodes that each contain multiple qubits (Kimble, 2008). This approach has motivated the development of quantum networks where stationary qubits that can store quantum information are interfaced with optical photons for distributing quantum information (Monroe et al., 2014, 2016; Andrews et al., 2014). Such nodes also form the basis of quantum repeater architectures for long-distance quantum communication, where stationary qubits are used as quantum memories for optical photons (Briegel et al., 1998).

Optical interfaces for different qubit architectures are now being actively explored. Microwave-to-optical frequency conversion using mechanical transducers is being investigated to interface microwave qubits based on Josephson junctions with optical photons (Andrews et al., 2014). For optical emitters such as neutral atoms and trapped ions, cavity QED techniques have been developed to interface hyperfine qubits with optical photons (Cirac et al., 1997; Ritter et al., 2012). In this chapter, we will discuss recent advances using color centers in diamond that demonstrate key elements required to realize a solid-state quantum network node.

Color centers in diamond are part of a growing list of solid-state optical emitters (Aharonovich et al., 2016) that include quantum dots (Michler et al., 2000), rare-earth ions (Lvovsky et al., 2009), single molecules (Orrit and Bernard, 1990), and point defects in crystals (Gruber et al., 1997). The recent interest in solid-state emitters for quantum applications builds upon a long history of laser-science, optoelectronics, and microscopy research. Since the 1960s, transition-metal and rare-earth ions embedded in solids have found wide use as an optical gain medium (Maiman, 1960) based on their weakly allowed optical transitions (Judd, 1962; Liu and Jacquier, 2006). These studies enabled applications ranging from the development of optical amplifiers based on erbium-doped silica fibers which form the basis of modern telecommunication infrastructure (Mears et al., 1987) to widely tunable solid-state lasers such as Ti:Sapphire lasers.

The early investigations of solid-state emitters were primarily interested in understanding the properties and applications of ensembles of dopants and point defects (Liu and Jacquier, 2006; Stoneham, 1975). The desired properties for quantum applications, however, differ strongly from those of a gain medium. Specifically, the ability to isolate and control single emitters is critical for realizing controllable qubits. Moreover, in order to achieve coherent atom-photon interactions based on single solid-state emitters, it is necessary to identify systems with strong transition dipole moments (to enable high radiative decay rates), weak static dipole moments (to minimize optical dephasing due to environmental noise), and weak vibronic coupling (to minimize phonon broadening). In addition, the presence of metastable spin sublevels is necessary to create a long-lived memory.

By the early 1990s, advances in optical spectroscopy and microscopy enabled the first major step in this direction, the optical detection of single molecules inside solid-state matrices (Moerner and Kador, 1989; Orrit and Bernard, 1990). In the past two decades, these techniques have been extended to studies of quantum dots and nitrogen-vacancy (NV) color centers in diamond. These systems have shown promising spin and optical properties which made them leading candidates for use in solid-state quantum network nodes (Gao et al., 2015).

Optically active quantum dots are mesoscopic semiconductor structures where electrons and holes are confined to result in a discrete, atom-like optical spectra (Michler et al., 2000; Santori et al., 2002). Self-assembled InGaAs quantum dots embedded in a GaAs matrix have strong optical dipole transitions with excited state lifetimes in the range of ~ 1 ns. They can be integrated into nanophotonic structures (Lodahl et al., 2015) to achieve strong light-matter interactions at high bandwidths (Hennessy et al., 2007; Englund et al., 2007). However, the spatial and spectral positions of these emitters are non-deterministic due to the self-assembly process (Lodahl et al., 2015).

Charged quantum dots also have spin degrees of freedom which can be used to store quantum information. The coherence of the optical quantum dot spin is limited by the high density nuclear spin bath in the host crystal which leads to an inhomogeneous spin dephasing timescale of T_2^* of ~ 2 ns and a coherence time (T_2) of 1-3μs using dynamical decoupling sequences (Greilich, 2006; Press et al., 2010; Bechtold et al., 2015; Stockill et al., 2016). The few microsecond long coherence time limits the potential use of InGaAs quantum dots as quantum memories for long-distance quantum communication applications.

The nitrogen-vacancy (NV) color center in diamond consists of a substitutional nitrogen atom with a neighboring vacancy (Gruber et al., 1997; Doherty et al., 2013). The NV center has a spin-triplet ground state with a long coherence time of up to 2 ms at room temperature (Balasubramanian et al., 2009). Interestingly, the NV spin can be optically polarized and read out at room temperature using off-resonant excitation (Jelezko et al., 2004). The ability to initialize, coherently control, and read out the NV spin at room temperature using a simple experimental setup resulted in its wide use in nanoscale sensing applications such as nanoscale magnetometery and thermometry (Taylor et al., 2008; Lovchinsky et al., 2016; Kucsko et al., 2013). At cryogenic temperatures, about 4 % of the NV fluorescence is emitted into a narrowband zero-phonon line (ZPL). For NV centers in bulk diamond, the linewidth of this ZPL transition can be close to the lifetime-limited linewidth of 13 MHz where the excited state lifetime is 12 ns (Tamarat et al., 2006). For NV centers that are close to surfaces (e.g. in a nanostructure), electric field noise due to charge fluctuations in the environment results in spectral diffusion of the optical transitions and non-radiatively broadens the optical transition to few GHz (Faraon et al., 2012; Riedel et al., 2017). This poses a major challenge for realizing a nearly deterministic spin-photon interface based on an NV center in a microphotonic device. In the past decade, efforts to overcome these challenges focused on integrating NV centers in photonic crystals (Englund et al., 2010; Faraon et al., 2012; Hausmann et al., 2013) or fiber-based Fabry-Perot cavities (Albrecht et al., 2013; Riedel et al., 2017) for Purcell-enhancing the ZPL emission rate. In addition, methods to minimize the electric field noise originating from surfaces and fabrication-induced damage are being actively explored (Chu et al., 2014).

In this chapter, we discuss recent progress in addressing these challenges using a new family of color centers in diamond coupled to nanophotonic devices. Specifically, we focus on the optical and spin coherence properties of the silicon-vacancy (SiV) center (Neu et al., 2011; Hepp, 2014; Hepp et al., 2014) at cryogenic temperatures that are relevant for applications in quantum science and technology. This solid-state platform, which combines coherent optical and spin transitions, recently emerged as a promising approach for the realization of scalable quantum networks.

This chapter is organized as follows. In Sec. 1.2 we discuss the relevant broadening mechanisms of the optical transitions for solid-state emitters and the figures of merit to achieve a deterministic spin-photon interface. In Sec. 1.3, we show that SiV centers in high-quality diamond crystals have a narrow inhomogeneous distribution and can efficiently generate indistinguishable photons. Sec. 1.4 shows that the optical coherence properties of SiV centers are maintained in nanophotonic structures. Sec. 1.5 discusses experiments where SiV centers are strongly coupled to photonic crystal cavities to achieve a deterministic spin-photon interface and optical nonlinearities at the single-photon level. Entanglement generation between two SiV centers in a single nanophotonic device is described in Sec. 1.6. The spin properties of the SiV center which are important for realizing a long-lived quantum memory are discussed in Sec. 1.7. This section presents a microscopic model of the electron-phonon interactions that result in spin dephasing and recent experiments that obtained 13 ms coherence at dilution fridge temperatures. We conclude with an outlook discussing prospects for realizing quantum repeater nodes based on long-lived SiV spin qubits strongly interacting with optical photons.

1.2 Coherent atom-photon interactions using solid-state emitters

Solid-state quantum emitters are tightly confined in a crystal matrix. This allows the placement of emitters in the close vicinity of dielectric or plasmonic nanophotonic structures without the challenges associated with trapping neutral atoms or ions near surfaces. The ability to position emitters in the near field of strongly confined modes enables strongly enhanced light-matter interactions. Using this approach, Purcell enhancement of optical transitions (Purcell, 1946; Lodahl et al., 2004) has been demonstrated for a wide range of quantum emitters (Aharonovich et al., 2016). Strong coupling has also been achieved using quantum dots embedded in photonic crystal cavities (Senellart et al., 2017).

Purcell-enhanced emission into nanophotonic modes ensures that the emitters interact with a well-defined spatial mode as opposed to 4π emission into free space. However, for most applications in quantum information science, the temporal mode also needs to be well defined. In other words, the emission linewidth (γ) should match the lifetime limit ($2\pi\gamma_{rad} = 1/T_1$) to be able to generate indistinguishable single photons (Lettow et al., 2010; Legero et al., 2003). While this is often true for cold atomic systems with radiatively broadened transitions ($\gamma = \gamma_{rad}$), the complex environment of solid-state emitters can result in homogeneous and inhomogeneous dephasing processes (γ_d) that nonradiatively broaden the optical transitions with $\gamma = \gamma_{rad} + \gamma_d$.

Phonons and strain can provide such additional homogeneous and inhomogeneous broadening mechanisms. Displacements of atoms in the host crystal can affect the optical transitions in two different ways. Static lattice distortions (i.e. strain) may reduce the symmetry of the defect and change the energy splittings (Sternschulte et al., 1994) between electronic orbitals. A variation in local strain contributes to the inhomogeneous distribution of the resonance frequencies. Displacements of the lattice can also give rise to dynamic effects during an optical excitation cycle. A homogeneous dephasing mechanism is caused by elastic and inelastic scattering of bulk acoustic phonons. While this process does not modify the single-photon generation abilities at room temperature, it limits the coherence time of the generated photons typically below a ps timescale. Photons generated at room temperature are therefore not suitable for the observation of quantum interference effects. The dephasing due to acoustic phonons can often be completely suppressed by operating at liquid helium temperatures. A microscopic model of this acoustic phonon broadening mechanism is discussed in Refs. (Fu et al., 2009; Goldman et al., 2015) for NV centers and in Ref. (Jahnke et al., 2015) for SiV centers.

In addition, at both ambient and cryogenic temperatures, the emission of a photon can be accompanied by the spontaneous creation of local, high-frequency (few THz), short-lived (ps decay timescale) molecular vibrations of the color center based on the Franck-Condon principle. Since these high-frequency localized modes decay very rapidly into bulk phonon modes (Huxter et al., 2013), only the zero-phonon line (ZPL) emission has sufficiently narrow linewidths for indistinguishable photon generation. The probability of emission into the ZPL is $\sim 4\%$ for the NV center and $\sim 70\%$ for the SiV center. The high emission probability into the ZPL for the SiV center can be understood based on the following consideration. The SiV optical transitions take place between orbital states of different parity, ^2E_g and ^2E_u, which differ in phase but have similar charge densities (Gali and Maze, 2013). This small change in the electronic charge density results in the strong ZPL, since optical excitations do not couple efficiently to local vibrations.

For NV centers, the dominant dephasing process at cryogenic temperatures is caused by charge fluctuations in the solid-state environment. As shown in Fig. 1.1, the

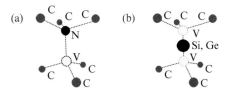

Fig. 1.1 Molecular structure of the NV, SiV, and GeV centers in diamond. (A) The NV center consists of a substitutional nitrogen (N) with a neighboring lattice vacancy (V). (B) For the SiV and GeV centers, the impurity atom (Si or Ge) is centered between two lattice vacancies (V) and constitutes an inversion center.

NV center consists of a nitrogen impurity and a neighboring lattice vacancy each occupying a substitutional site. This structure resembles that of a polar molecule and results in electronic orbitals with a large static dipole moment (~ 1 Debye). The energy of these electronic orbitals therefore shifts with any applied electric field, resulting in a broadening of the optical transition with charge fluctuations in the environment. The level of charge noise and spectral diffusion can be minimal for NV centers in high-quality single crystal bulk crystals (Bernien et al., 2012; Sipahigil et al., 2012). However, this dephasing process becomes more severe inside nanofabricated structures due to increased exposure to surface and defect states (Faraon et al., 2012).

We recently demonstrated that silicon-vacancy (SiV) (Evans et al., 2016; Sipahigil et al., 2016) and germanium-vacancy (GeV) (Bhaskar et al., 2017) centers in diamond nanophotonic structures maintain their optical coherence with a strong supression of the spectral diffusion observed in NV centers. The origin of this spectral stability for SiV and GeVs can be understood as a consequence of their inversion symmetry (Sipahigil et al., 2014). For defects such as SiV and GeV centers (Fig. 1.1) the impurity atom (Si or Ge) is located at an interstitial site between two lattice vacancies. In this geometry, the impurity atom constitutes an inversion center. Owing to the inversion symmetry of the structure, electronic orbitals are symmetric with respect to the origin and have zero static electric dipole moments. The absence of a static electric dipole moment results in a first-order insensitivity to electric fields for SiVs and GeVs[1]. This allows the incorporation of optical emitters with coherent optical transitions into nanophotonic devices which is the key enabling aspect for the experiments discussed in the following sections.

1.3 Indistinguishable photons from separated silicon-vacancy centers in diamond

The symmetry properties discussed in Sec. 1.2 lead to the absence of spectral diffusion (Rogers et al., 2014c) and a narrow inhomogeneous distribution (Sternschulte et al., 1994) for SiV centers in bulk diamond. In Ref. (Sipahigil et al., 2014), we used

---

[1] We note that this property parallels a similar idea based on a charge-insensitive superconducting qubit design (i.e. the transmon qubit) which lead to major advances in circuit quantum electrodynamics in the past decade (Koch et al., 2007).

SiV centers incorporated in a high-quality single-crystal diamond sample to generate indistinguishable photons from separate emitters. These SiV centers were incorporated in the high-quality crystals during the growth process and displayed a narrow inhomogeneous distribution of $\gamma_{inh}/2\pi \sim 1$ GHz. For comparison, the natural linewidth of an SiV center is $\gamma_{nat}/2\pi = 94$ MHz corresponding to an excited state lifetime of 1.73 ns. To our knowledge, this ratio of the ensemble inhomogeneous distribution and the natural linewidth $\gamma_{inh}/\gamma_{nat}$ is the smallest observed among solid-state quantum emitters (Aharonovich et al., 2016). This narrow distribution makes SiV centers suitable for indistinguishable photon generation from separated emitters.

Figs. 1.2(a) and (b) illustrate the narrow inhomogeneous distribution of the SiV centers by showing a comparison of the density of SiV centers at the center frequency (ν_0) of the inhomogeneous distribution and at a detuning of 1.5 GHz. When we excite the sample at the center frequency of the ensemble ν_0, a high density of emitters are visible with a large backround due to emitters in a different focal plane. Detuning the excitation laser only by 1.5 GHz from this frequency results in a drastic reduction in the density of visible emitters.

To generate single photons from isolated, single SiV centers and minimize background from other emitters (Moerner and Kador, 1989), the laser was tuned to the edge of the inhomogeneous distribution ($\nu = \nu_0 + 1.5$ GHz) where nearly resonant, single SiV centers can be spatially resolved as shown in Figs. 1.3(c,d). Photoluminescence excitation (PLE) spectra of the emitters, SiV_I and SiV_{II}, reveal transitions separated by 52.1 MHz with a linewidth of 136 and 135 MHz respectively. For comparison, the lifetime of the excited states was measured to be 1.73 ± 0.05 ns at temperatures below 50 K corresponding to a transform limited linewidth of 94 MHz. The narrow linewidths of the optical transitions are close to the lifetime-limited linewidth, indicating coherent single-photon generation suitable for efficient indistinguishable photon generation.

To test the indistinguishability of the generated photons, single photons from SiV_I and SiV_{II} (Fig. 1.2c, d) were directed to the input ports 1 and 2 of a beam splitter respectively. Figure 1.2 shows two-photon interference (Hong et al., 1987) measurements where the degree of indistinguishability of single photons is varied by changing the photon polarization. The two datasets show the second-order intensity correlation function, $g^2(\tau)$, measured for indistinguishable (II) and distinguishable (I) photon states. For identically polarized indistinguishable photons, we find $g_{\parallel}^2(0) = 0.26 \pm 0.05$ where the error bars denote shot noise estimates. After rotating the fluorescence polarization of SiV_{II} by $90°$ to make the photon sources distinguishable, $g_{\perp}^2(0) = 0.66 \pm 0.08$ was observed. These results clearly demonstrate two-photon interference corresponding to a measured two-photon interference visibility of $\eta = 0.72 \pm 0.05$ that is limited by detector timing resolution and background photons (Sipahigil et al., 2014).

These observations established the SiV center as an excellent source of indistinguishable single photons with the combination of a strong ZPL transition, narrow inhomogeneous distribution, and spectrally stable optical transitions. These results also suggested the possibility of integrating SiV centers inside nanophotonic cavities to obtain strong coupling (Burek et al., 2012; Hausmann et al., 2013; Riedrich-Möller et al., 2012; Lee et al., 2012; Faraon et al., 2012) while maintaining their spectral

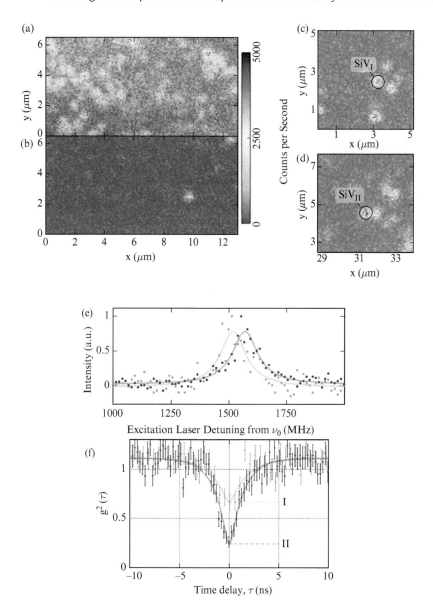

Fig. 1.2 (a) The probe laser frequency was fixed to the ensemble average of $\nu_0 = 406.7001$ THz while spatially scanning the sample and monitoring fluorescence. A high density of resonant emitters is visible with a large background. (b) Scan of the same region with the laser tuned to $\nu_1 = \nu_0 + 1.5$ GHz. Due to the narrow inhomogeneous distribution, only few resonant emitters are visible and the background level is reduced. (c) and (d) show the two emitters, SiV_I and SiV_{II}, used for the two-photon interference experiment at frequency $\nu \sim \nu_1$. (e) PLE spectrum of the two SiVs: transition linewidth is 135 ± 2 MHz for each emitter and the detuning is 52 MHz. (f) The single photons from the two SiVs are interfered on a beam splitter. The resulting second-order intensity correlation function $g^2(\tau)$ at the output ports is plotted for two cases: (I) Single photons from the two emitters are chosen to be orthogonally polarized and hence distinguishable, $g_\perp^2(0) = 0.66 \pm 0.08$. (II) For indistinguishable single photons with identical polarizations, $g_\parallel^2(0) = 0.26 \pm 0.05$. Figure adapted from (Sipahigil et al., 2014).

stability. The optical properties of SiV centers in nanophotonic devices will be discussed in Sec. 1.4 and 1.5.

1.4   Narrow-linewidth optical emitters in diamond nanostructures via silicon ion implantation

One major advantage of building quantum devices with solid-state emitters rather than trapped atoms or ions is that solid-state systems are typically more easily integrated into nanofabricated electrical and optical structures that enable strong light-matter interactions (Ladd et al., 2010; Vahala, 2003). The scalability of these systems is important for practical realization of even simple quantum optical devices (Li et al., 2015). Motivated by these considerations and the robustness of the SiV optical transitions in bulk crystals as described in Sec. 1.3, in Ref. (Evans et al., 2016), we studied the optical properties of SiV centers created in diamond nanophotonic structures via Si^+ ion implantation.

Silicon-vacancy centers occur only rarely in natural diamond (Lo, 2014), and are typically introduced during CVD growth via deliberate doping with silane (Edmonds et al., 2008; D'Haenens-Johansson et al., 2011) or via silicon contamination (Rogers et al., 2014c; Neu et al., 2013; Clark et al., 1995; Sternschulte et al., 1994; Zhang et al., 2016). While these techniques typically result in a narrow inhomogeneous distribution of SiV fluorescence wavelengths (Rogers et al., 2014c), these samples have a number of disadvantages. For example, the concentration of SiV centers can be difficult to control and localization of SiV centers in three dimensions is not possible. Ion implantation is a commercially available technology that offers a promising solution to these problems. By controlling the energy, quantity, and isotopic purity of the source ions, the depth, concentration, and isotope of the resulting implanted ions can be controlled.

To create SiV centers in bulk diamond, we first implant Si^+ ions (Innovion Corporation) at a dose of 10^{10} ions/cm^2 and an energy of 150 keV resulting in the placement of Si atoms at an estimated depth of 100 ± 20 nm (Ziegler et al., 2010). After implantation, we clean the samples using an an oxidative acid clean (boiling 1 : 1 : 1 perchloric : nitric : sulfuric acid) (Hauf et al., 2011) and then perform a high-vacuum ($< 10^{-6}$ Torr) anneal at a temperature of $1100°$C for two hours. At temperatures above 800 °C, lattice vacancies can diffuse and combine with the implanted Si ions to form SiV centers. At higher temperatures of $1100°$C, undesired defect complexes (e.g. divacancies) that can result in magnetic or electric field noise anneal out (Acosta et al., 2009; Yamamoto et al., 2013) and partial healing of the lattice damage helps to reduce strain (Orwa et al., 2011).

Following these steps, SiV centers created in bulk diamond via ion implantation were characterized with photoluminescence excitation (PLE) spectroscopy at 4K. Based on a comparison of the Si^+ ion implantation density and the measured density of SiV centers in the sample, we estimate our SiV creation yield to be 0.5–1%. The PLE measurements in Ref. (Evans et al., 2016) showed that implanted SiV centers in bulk diamond have narrow optical transitions with linewidths of $\gamma/2\pi = 320 \pm 180$ MHz (mean and standard deviation for N = 13 spatially resolved emitters). Almost all SiV centers had a linewidth within a factor of three of the lifetime limit. For the 13 SiV centers characterized, about half of the optical transitions were in a 15 GHz window.

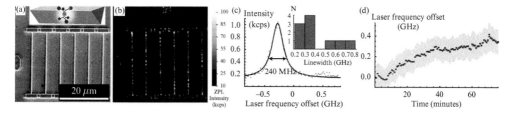

Fig. 1.3 SiV centers in nanostructures. (a) Scanning electron micrograph of six nanobeam waveguides. Inset: schematic of a triangular diamond nanobeam containing an SiV center. (b) Photoluminescence image of the structures in (a). Multiple SiV centers are visible in each waveguide. (c) Linewidth of representative implanted SiV inside a nano-waveguide measured by PLE spectroscopy. Inset: histogram of emitter linewidths in nanostructures. Most emitters have linewidths within a factor of four of the lifetime limit of 94 MHz. (d) Spectral diffusion of the emitter measured in part c. The total spectral diffusion is under 400 MHz even after more than an hour of continuous measurement. This diffusion is quantified by measuring the drift of the fitted center frequency of resonance fluorescence scans as a function of time. Error bars are statistical error on the fitted center position. The lighter outline is the linewidth of the fitted Lorentzian at each time point. Figure adapted from (Evans et al., 2016).

To test the optical coherence of the implanted SiV centers in nanophotonic devices, we fabricated an array of diamond waveguides (Fig. 1.3a) on the sample characterized above using previously reported methods (Burek et al., 2012; Hausmann et al., 2013). Each waveguide (Fig. 1.3a, inset) is 23 μm long with approximately equilateral-triangle cross sections of side length 300–500 nm. After fabrication, we again performed the same 1100° annealing and acid cleaning procedure. This process leads to the creation of many SiV centers as seen in the fluorescence image of the final structures of Fig. 1.3b.

SiV centers in nanostructures display narrow-linewidth optical transitions with a full-width at half-maximum (FWHM) of $\gamma_n/2\pi = 410 \pm 160$MHz (mean and standard deviation for N = 10 emitters; see Fig. 1.3 inset for linewidth histogram), only a factor of 4.4 greater than the lifetime limited linewidth $\gamma/2\pi = 94$MHz. The linewidths measured in nanostructures are comparable to those measured in bulk (unstructured) diamond ($\gamma_b/2\pi = 320 \pm 180$MHz). The ratios γ_n/γ and γ_b/γ are much lower than the values for NV centers, where the current state of the art for typical implanted NV centers in nanostructures (Faraon et al., 2012) and in bulk (Chu et al., 2014) is $\gamma_n/\gamma > 100$–200 and $\gamma_b/\gamma > 10$ ($\gamma/2\pi = 13$MHz for NV centers).

By extracting the center frequency of each individual scan, we also determine the rate of fluctuation of the ZPL frequency and therefore quantify spectral diffusion (Fig. 1.3d). Optical transition frequencies in SiV centers are stable throughout the course of our experiment, with spectral diffusion on the order of the lifetime-limited linewidth even after more than an hour.

The residual broadening of the optical transition can result from a combination of second-order Stark shifts and phonon-induced broadening. The presence of a strong static electric field would result in an induced dipole that linearly couples to charge fluctuations, accounting for the slow diffusion. Determining the precise mechanisms for the residual broadening of the SiV optical linewidths remains an important topic of future study. It should be possible to test the linear and nonlinear susceptibility of the optical transition frequencies to electric fields by applying large electric fields from nearby electrodes (Acosta et al., 2012). Despite the deviation from an ideal behavior by a factor of 4 in linewidth and $10 - 100$ in ensemble inhomogenous distribution, the observed optical properties are already sufficient to observe strong atom-photon interactions in photonic crystal cavities with high cooperativity, a topic that will be discussed in the next section.

1.5   Diamond nanophotonics platform for quantum nonlinear optics

We next review experiments using negatively charged silicon-vacancy (SiV) color centers coupled to diamond nanophotonic devices that demonstrate strong interactions between single photons and a single SiV center. In Ref. (Sipahigil et al., 2016), we created SiV centers inside one-dimensional diamond waveguides and photonic-crystal cavities with small mode volumes ($V \sim \lambda^3$) and large quality factors ($Q \sim 7000$) as illustrated in Fig. 1.4. These nanophotonic devices are fabricated using angled reactive-ion etching to create free-standing single-mode structures starting from bulk diamond (Burek et al., 2012, 2014). A recent review article on different diamond nanophotonic device fabrication approaches can be found in Ref. (Schröder et al., 2016).

To obtain optimal coupling between the emitter and the cavity mode, the emitter needs to be positioned at the field maximum of the cavity mode. This can be achieved with two possible approaches. In Refs. (Sipahigil et al., 2016) and (Schröder et al., 2017), a Si^+ ion beam was focused at the center of the cavities to create SiV centers as illustrated in Fig. 1.4. Using this approach, the emitters can be positioned with close to 40 nm precision which is limited by a combination of ion-beam spot size, alignment errors and straggle in the crystal. An alternative approach is to fabricate lithographically defined small apertures and use commerically available ion beams to implant the Si^+ ions (Schröder et al., 2016). Using these techniques, a large array of coupled emitter-cavity systems can be fabricated on a single sample.

An example spectrum of the coupled SiV-cavity system at 4 K is shown in Fig. 1.4. Each narrow dip within the broad cavity transmission band is caused by a single SiV. On SiV resonance, the emitter results in a strong extinction ($\Delta T/T = 38(3)\%$) of cavity transmission (Sipahigil et al., 2016). Based on the measurements shown in Fig. 1.4f, it is possible to infer a cooperativity of $C = 4g^2/\kappa\gamma = 1.0(1)$ for the SiV-cavity system with cavity QED parameters $\{g, \kappa, \gamma\}/2\pi = \{2.1, 57, 0.30\}$ GHz where g is the single-photon Rabi frequency, κ is the cavity intensity decay rate and γ is the SiV optical transition linewidth, cooperativity C is a parameter that characterizes the ratio of coherent coupling rate (g) with dissipation rates (γ, κ). We note that the SiV optical transition linewidth of 300 MHz includes the sum of free-space decay, non-radiative decay and pure dephasing rates.

The observed cooperativity of $C = 1$ in the experiments marks the onset of nonlinear effects at the single-photon level. Owing to the strong interaction between a single-photon and the SiV center, the SiV is saturated at a flux of single-photon per Purcell enhanced lifetime (~ 300 ps) as shown in Fig. 1.5. In this regime, the

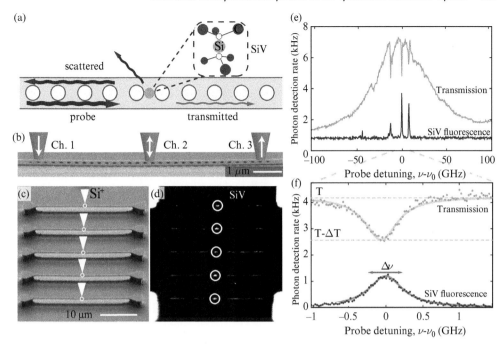

Fig. 1.4 (a) Schematic of an SiV center in a diamond photonic crystal cavity. (b) Scanning electron micrograph (SEM) of a diamond photonic crystal cavity. (c) SEM of five cavities fabricated out of undoped diamond. After fabrication, SiV centers are positioned at the center of each cavity using focused Si^+ ion beam implantation. (d) SiV fluorescence is detected at the center of each nanocavity shown in (c). (e) Measured cavity transmission and SiV fluorescence spectrum. Three SiVs are coupled to the cavity and each results in suppressed transmission at the corresponding frequencies. (f) On resonance, a single SiV results in $\Delta T/T = 38(3)\%$ extinction of cavity transmission. Figure reproduced from (Sipahigil et al., 2016).

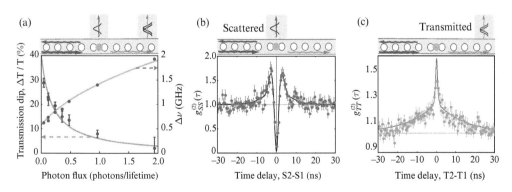

Fig. 1.5 Single-photon nonlinearities. (a) Cavity transmission and SiV transition linewidth measured at different probe intensities. (b, c) Intensity autocorrelations of the scattered (fluorescence) and transmitted fields. The scattered field shows antibunching (b), while the transmitted photons are bunched with an increased contribution from photon pairs (c). Figure reproduced from (Sipahigil et al., 2016).

strongly coupled SiV-cavity system acts as a photon number sorter that scatters single photons while having a high transmission amplitude for photon pairs due to saturated atomic response. This effect can be observed in photon intensity correlation measurements that show antibunching for the scattered field (Fig. 1.5b) and bunching for the transmitted field (Fig. 1.5c) (Sipahigil et al., 2016).

1.6 Two-SiV entanglement in a nanophotonic device

Experiments discussed in the previous section demonstrated coherent single-photon single-emitter interactions. As seen in Fig. 1.4e, the optical transitions of SiV centers in nanostructures have an inhomogeneous distribution of ~ 20 GHz. This inhomogeneity implies that different emitters will emit spectrally distinguishable photons that are not suitable for photon-mediated interactions between multiple emitters. To generate indistinguishable photons from different nodes, it will be necessary to either improve material properties by reducing the strain variations in the crystal or develop active tuning methods.

To generate spectrally tunable photons, one can use Raman transitions between the metastable orbital states of SiV centers (Sipahigil et al., 2016). When a single SiV is excited from the state $|u\rangle$ at a detuning Δ (Fig. 1.6), the emission spectrum includes a spontaneous component at frequency ν_{ec} and a Raman component at frequency $\nu_{ec} - \Delta$ that is tunable by choosing Δ. As shown in Fig. 1.6c, the Raman emission frequency can be tuned by varying the excitation laser detuning Δ. This technique allows frequency tuning of the Raman emission by ± 10 GHz which is comparable to the inhomogeneous distribution of SiV centers in nanophotonic devices.

The Raman tuning approach enables the tuning of multiple spatially resolved emitters in a single nanophotonic device into resonance. To achieve this, we applied a separate Raman control field for two SiV centers in the same diamond waveguide (Fig. 1.6d) and measured intensity correalations of Raman photons emitted into the waveguide mode (Sipahigil et al., 2016). In this experiment, the two SiVs emit Raman photons into the same spatial mode and the frequency indistinguishability is achieved by Raman tuning. If the Raman emission of the two SiVs are not tuned into resonance, the photons are distinguishable, resulting in the measured $g_{dist}^{(2)}(0) = 0.63\,(3)$ (curve II, Fig. 1.6e) close to the conventional limit associated with two distinguishable single-photon emitters $g_{dist}^{(2)}(0) = 0.5$. Alternatively, if the Raman transitions of the two SiVs are tuned instead into resonance with each other, an interference feature is observed in photon correlations around zero time delay with $g_{ind}^{(2)}(0) = 0.98(5)$ (curve III, Fig. 1.6e).

The observed enhancement of the $g_{ind}^{(2)}(0)$ results from the collectively enhanced decay of the two SiV centers. When the Raman transitions of the two SiVs are tuned into resonance with each other, it is not possible to distinguish which of the two emitters produced a waveguide photon. Thus, the emission of an indistinguishable single photon leaves the two SiVs prepared in the entangled state $|B\rangle = (|cu\rangle + e^{i\phi}|uc\rangle)/\sqrt{2}$ (Cabrillo et al., 1999), where ϕ is set by the propagation phase between emitters spaced by ΔL and the relative phase of the Raman control lasers. This state is a two-atom superradiant state with respect to the waveguide mode and scatters

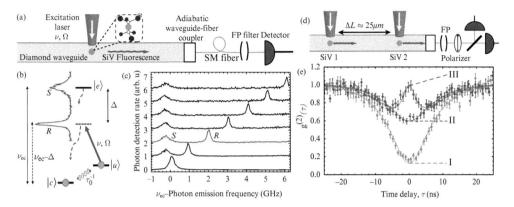

Fig. 1.6 (a-c) Spectrally tunable single-photons using Raman transitions. (a) Photons scattered by a single SiV into a diamond waveguide are coupled to a single-mode (SM) fiber. A scanning Fabry-Perot (FP) cavity measures the emission spectra. (b) Under excitation at a detuning Δ, the emission spectrum contains spontaneous emission (labeled S) at frequency ν_{ec} and narrow Raman emission (R) at frequency $\nu_{ec} - \Delta$. (c) Δ is varied from 0 to 6 GHz in steps of 1 GHz and a corresponding tuning of the Raman emission frequency is observed. (d) Schematic for the two-SiV entanglement generation experiment. Photons scattered by two SiV centers into a diamond waveguide are detected after a polarizer. (e) Intensity autocorrelations for the waveguide photons. Exciting only a single SiV (I) yields $g^{(2)}_{single}(0) = 0.16(3)$ for SiV1, and $g^{(2)}_{single}(0) = 0.16(2)$ for SiV2. $g^{(2)}_{dist}(0) = 0.63(3)$ when both SiVs excited and Raman photons are spectrally distinguishable (II). $g^{(2)}_{ind}(0) = 0.98(5)$ when both SiVs excited and Raman photons are tuned to be indistinguishable (III). The observed contrast between curves II and III at $g^{(2)}(0)$ is due to the collectively enhanced decay of state $|B\rangle$. Figure reproduced from (Sipahigil et al., 2016).

Raman photons at a collectively enhanced rate that is twice the scattering rate of a single emitter. This enhanced emission rate into the waveguide mode results in the experimentally observed interference peak at short time delays (curve III, Fig. 1.6e) and is a signature of entanglement.

In these experiments, the coherence time of the entangled state and control over the SiV orbital states is limited by the occupation of ~ 50 GHz phonons at 4 K, which causes relaxation between the metastable orbital states $|u\rangle$ and $|c\rangle$ and limits their coherence times to less than 50 ns. In the following section, we discuss a microscopic model of this dephasing process and a recent experiment that extended SiV spin coherence to 13 ms by operation at 100 mK.

1.7   SiV spin coherence at low temperatures

In this section, we focus on the dynamics of the spin and orbital degrees of freedom of the SiV center that are relevant for storing quantum information. In Sec. 1.7.1, we first present high temperature (4 – 22 K) measurements of the electronic orbital relaxation dynamics and a microscopic model of the SiV-phonon interactions that limit spin coherence (Jahnke et al., 2015). After a discussion of the predictions of this model for low temperatures, we present recent experiments that demonstrate a long coherence time of 13 ms at 100 mK (Sukachev et al., 2017) in Sec. 1.7.2.

1.7.1   Electron-phonon processes of the silicon-vacancy center in diamond

The symmetry properties of the SiV center results in ground (^2E_g) and excited (^2E_u) electronic orbitals that both have E symmetry and double orbital degeneracy (Hepp et al., 2014). The two degenerate ground state orbitals are occupied by a single hole with $S = 1/2$ resulting in four degenerate ground states with orbital ($\{|e_+\rangle, |e_-\rangle\}$) and spin ($\{|\downarrow\rangle, |\uparrow\rangle\}$) degrees of freedom (D'Haenens-Johansson et al., 2011; Rogers et al., 2014a; Hepp et al., 2014; Gali and Maze, 2013). The fourfold degeneracy of the ground states is partially lifted by the spin-orbit interaction ($\Delta_{GS}S_zL_z$ with $\Delta_{GS} \sim 45$ GHz) which results in two ground spin-orbit branches (UB, LB in Fig. 1.7a). Each spin orbit branch consists of two degenerate states with well defined orbital and spin angular momentum (in Fig. 1.7b) at zero magnetic field (Hepp et al., 2014). Acoustic phonons can also drive spin-conserving transitions between $|e_+\rangle$ and $|e_-\rangle$ orbitals, resulting in population transfer between orbitals at rates $\gamma_{+,-}$ (Ham, 1965; Fischer, 1984).

In Ref. (Jahnke et al., 2015), the relaxation within the ground state doublet, $\gamma_{+, -}$ in Fig. 1.7b, was probed directly using pulsed optical excitation and time-resolved fluorescence measurements. This measurement was repeated for a single SiV center at various temperatures between 4.5 K and 22 K and the relaxation rate was found to scale linearly with temperature (Fig. 1.7c). The slowest rate observed was

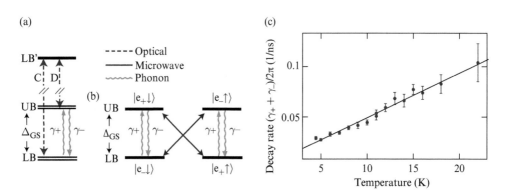

Fig. 1.7 (a) Relevant electronic states of the SiV center at cryogenic temperatures. Optical dipole transitions (labeled C, D) connect the two spin-orbit branches UB and LB to the lower branch LB′ of the excited states. (b) Each spin orbit branch contains two electronic states with orthogonal electronic orbital and spin states (e.g. $|e_-\uparrow\rangle$, $|e_+\downarrow\rangle$ for the UB). Phonon transitions are allowed between different orbital states with the same spin projection. Microwave transitions are allowed between different spin states with the same orbital projection. (c) The measured orbital relaxation rate, $\gamma_+ + \gamma_-$, at different tempratures. Figure adapted from (Jahnke et al., 2015).

$\gamma_+ + \gamma_- = (39\text{ns})^{-1}$ at $T = 5$ K. This fast decay mechanism implies that any coherence created between the states shown in Fig. 1.7 will decay at the 100 ns timescale at 4 K. Recent experiments at 4 K (Pingault et al., 2014; Rogers et al., 2014b; Becker et al., 2016; Pingault et al., 2017) that observed coherence times in this range can therefore be explained by this phonon-induced orbital relaxation process.

We next discuss a microscopic model of the orbital relaxation processes within the ground electronic levels. The electron-phonon processes are consequences of the linear Jahn-Teller interaction between the E-symmetric electronic states and E-symmetric acoustic phonon modes (Fischer, 1984; Maze et al., 2011; Doherty et al., 2011). Since the phonons couple to the orbital degree of freedom, phonon transitions are spin-conserving at zero magnetic field and we can focus on the orbital degree of freedom to understand the electronic dynamics of ground levels. For a given spin state, the effective zero-field orbital Hamiltonian takes the following form

$$H_0 = \pm \frac{1}{2}\hbar \Delta_{\text{GS}} \sigma_z, \tag{1.1}$$

where σ_z is the usual Pauli operator for orbital states in the $\{|e_+\rangle, |e_-\rangle\}$ basis, $\hbar\Delta_{\text{GS}}$ is the magnitude of the spin-orbit splitting, which is $-\hbar\Delta_{\text{GS}}$ for $|\uparrow\rangle$ and $+\hbar\Delta_{\text{GS}}$ for $|\downarrow\rangle$.

The interaction between the orbital states $\{|e_+\rangle, |e_-\rangle\}$ and phonon modes is described most easily if the phonon modes are linearly transformed to be circularly polarized. With this transformation, the phonon Hamiltonian and the linear electron-phonon interaction are

$$\hat{H}_{\text{E}} = \sum_{p,k} \hbar \omega_k a_{p,k}^\dagger a_{p,k} \tag{1.2}$$

$$\hat{V}_{\text{E}} = \sum_{k} \hbar \chi_k [\sigma_+ (a_{-,k} + a_{-,k}^\dagger) + \sigma_- (a_{+,k} + a_{+,k}^\dagger)], \tag{1.3}$$

where χ_k is the coupling strength for a single phonon, σ_+ (σ_-) is the raising (lowering) operator for the orbital states, and $a_{p,k}^\ddagger$ ($a_{p,k}$) is the creation (annihilation) operator for phonons with polarization $p = \{-, +\}$ and wave vector k. The coupling strength and the density of phonon modes are approximately $\overline{|\chi_k(\omega)|^2} \approx \chi\omega$ and $\rho(\omega) = \rho\omega^2$, respectively, where the overbar denotes the average over all modes with frequency $\omega_k = \omega$ and χ and ρ are proportionality constants (Fu et al., 2009; Abtew et al., 2011). Treating $\hat{V}_{\text{E}}$ as a time-dependent perturbation, the first-order transitions between the orbital states involve the absorption or emission of a single phonon whose frequency is resonant with the splitting Δ_{GS} (see Fig. 1.7b). The corresponding transition rates are

$$\gamma_+ = 2\pi \sum_{k} n_{-,k} |\chi_k|^2 \delta(\Delta_{\text{GS}} - \omega_k)$$

$$\gamma_- = 2\pi \sum_{k} (n_{+,k} + 1)|\chi_k|^2 \delta(\Delta_{\text{GS}} - \omega_k), \tag{1.4}$$

where $n_{p,k}$ is the occupation of the phonon mode with polarization p and wave vector k. Assuming acoustic phonons, performing the thermal average over initial states and the sum over all final states leads to

$$\gamma_+ = 2\pi\chi\rho\Delta_{\mathrm{GS}}^3 n(\Delta_{\mathrm{GS}}, T)$$
$$\gamma_- = 2\pi\chi\rho\Delta_{\mathrm{GS}}^3 [n(\Delta_{\mathrm{GS}}, T) + 1]. \tag{1.5}$$

For temperatures $T > \hbar\Delta_{\mathrm{GS}}/k_{\mathrm{B}}$, Eq. (1.5) can be approximated by a single relaxation rate with a linear temperature dependence

$$\gamma_+ \approx \gamma_- \approx \frac{2\pi}{\hbar}\chi\rho\Delta_{\mathrm{GS}}^2 k_{\mathrm{B}} T. \tag{1.6}$$

The measurements presented in Fig. 1.7 demonstrated a linear dependence of decay rates $\gamma_{+,-}$ for temperatures below 20 K, but greater than the spin orbit splitting $(T > \hbar\Delta_{\mathrm{GS}}/k_{\mathrm{B}} \sim 2.4\mathrm{K})$. We therefore conclude that the relaxation mechanisms are dominated by a resonant single phonon process at liquid helium temperatures. For a discussion of higher-order phonon processes that become dominant above 20 K, we refer the reader to Ref. (Jahnke et al., 2015).

We next discuss the implications of the linear electron-phonon interactions for qubit coherence and approaches that could be used to enhance coherence times. At temperatures above the spin-orbit gap $(T > \hbar\Delta_{\mathrm{GS}}/k_{\mathrm{B}} \sim 2.4\mathrm{K})$, any coherence formed between the states shown in 1.7b will decohere quickly at the 100 ns timescale limited by $\gamma_{+,-}$. Recent experiments that probed ground state coherences have reported T_2^* values that are in good agreement with this observation (Pingault et al., 2014; Rogers et al., 2014b; Becker et al., 2016; Pingault et al., 2017). Eq. 1.5 shows that the phonon decay rates that limit coherence are determined by a combination of phonon density of states and occupation $(\gamma_\pm \sim \rho(\Delta_{\mathrm{GS}})(2n(\Delta_{\mathrm{GS}}, T) + 1 \mp 1))$ at the energy of the spin-orbit splitting with $\Delta_{\mathrm{GS}} \sim 45\mathrm{GHz}$. Since the interaction with the phonon bath is a Markovian process, dynamical decoupling sequences cannot be applied to extend coherences. To extend T_2^*, we will therefore focus on approaches that reduce the orbital relaxation rates $\gamma_\pm$. The first two approaches focus on reducing phonon occupation $n(\Delta_{\mathrm{GS}}, T)$ to decrease γ_+. The occupation depends on the ratio, T/Δ_{GS}, of the temperature and the energy splitting between the coupled orbital states. Substantial improvements can be achieved by minimizing this ratio in cooling the sample to lower temperatures $(T \ll \Delta \sim 2.4 \mathrm{~K})$. Based on the fit in Fig. 1.7, the expected orbital relaxation timescale is given by $1/\gamma_+ = 200(e^{2.4\mathrm{K}/T} - 1)$ ns which correspond to 2 μs at 1 K and 2 ms at 0.26 K. A recent experiment by (Sukachev et al., 2017) operating at these low temperatures will be discussed in Sec. 1.7.2. A second approach is to increase Δ_{GS} by using emitters subject to high strain that increases the splitting between the spin-orbit branches. At the limit of $\hbar\Delta_{\mathrm{GS}} \gg kT$, similar reductions in phonon occupation can be used to suppress relaxation rates at 4 K, an effect that was recently observed in Ref. (Sohn et al., 2017). In both cases, only the two lowest energy states constitute a subspace that does not couple to phonons. The lowest two energy states are therefore expected to have long coherence times and could be used as a long-lived spin qubit.

We note that the linear electron-phonon interaction Hamiltonian (Eq. 1.3) and the resulting single-phonon orbital relaxation process are analogous to the Jaynes-Cummings Hamiltonian and Wigner-Weisskopf model of spontaneous emission used in quantum optics. One can therefore use ideas developed in the context of cavity QED to minimize relaxation rates $\gamma_{+,\,-}$ by use of phononic bandgaps (Burek et al., 2016) or strong coupling to well defined nanomechanical modes to realize multi-qubit interactions (Burek et al., 2016; Sohn et al., 2017).

1.7.2 SiV spin at low temperatures: a long-lived quantum memory

Motivated by the prediction of long spin coherence at low temperatures, we probed the coherence properties of single SiV centers at mK temperatures in Ref. (Sukachev et al., 2017). The key idea of this experiment can be understood by considering the energy level diagram of the SiV in Fig. 1.8a. Application of a magnetic field lifts the degeneracy between the spin-orbit states in the LB ($|e_+\uparrow\rangle$ and $|e_-\downarrow\rangle$ abbreviated as $|\uparrow\rangle$ and $|\downarrow\rangle$ from here onwards) with different spin projections. These states, which are defined as qubit states, decay at a rate γ_+ determined by the phonon occupation at frequency Δ_{GS} (Eq. 1.5). By reducing the occupation of phonon modes at Δ_{GS} at lower temperatures, one can suppress the rate γ_+, leaving the spin qubit in a manifold free from phonon-induced decoherence, thereby increasing spin coherence (Jahnke et al., 2015).

In Ref. (Sukachev et al., 2017), we investigated the SiV spin properties below 500 mK using a dilution refrigerator with a free-space confocal microscope and a vector magnet as shown in Fig. 1.8b. We first studied the thermal population of the LB and the UB between 0.1 and 10 K using an ensemble of as-grown SiV centers to probe the effective sample temperature. We probed the relative populations in the LB and the UB (and therefore the ratio γ_+/γ_-) by measuring the absorption

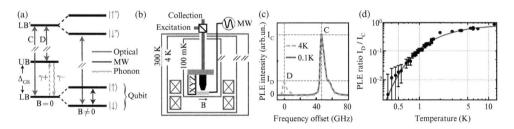

Fig. 1.8 (a) SiV electronic structure. Optical transitions C and D connect the lower (LB) and upper (UB) spin-orbit branches to the lowest-energy optical excited state (LB′). Each branch is split into two spin sublevels in a magnetic field $\vec{B}$. γ_+ and γ_- are phonon-induced decay rates. (b) Schematic of the setup. An objective is mounted on piezo positioners to image the diamond sample using free-space optics. The combined system is attached to the mixing plate of a dilution refrigerator and placed inside a superconducting vector magnet. (c) PLE spectra of an SiV ensemble at $B = 0$ for $T = 4$ K and 0.1 K. The peak intensity I_C (I_D) is proportional to the population in the LB (UB). (d) I_D/I_C (and γ_+/γ_-) is reduced at low temperatures, following $e^{-\hbar\Delta/k_B T}$ with $\Delta_{\mathrm{fit}} = 42 \pm 2$ GHz in agreement with the measured $\Delta_{\mathrm{GS}} = 48$ GHz. Figure adapted from (Sukachev et al., 2017).

spectrum of transitions C and D. Transitions C and D were both visible in PLE at 4 K, which indicates comparable thermal population in the LB and UB [Fig. 1.8c]. As the temperature was lowered [Fig.1.8d], the ratio of the transition D and C peak amplitudes (I_D/I_C) reduces by more than two orders of magnitude and follows $e^{-\hbar\Delta_{GS}/k_B T}$ (Jahnke et al., 2015). These measurements demonstrated an orbital polarization in the LB of $> 99\%$ below 500 mK. At these low temperatures, $\gamma_+ << \gamma_-$ and the qubit states are effectively decoupled from the phonon bath.

We used microwave fields resonant with the qubit transition to coherently control the SiV spin and probe its coherence (Sukachev et al., 2017). In these experiments, single strained SiV centers with $\Delta_{GS} \sim 2\pi 80$ GHz are used. When crystal strain is comparable to spin-orbit coupling (~ 48 GHz), the orbital components of the qubit states are no longer orthogonal (Hepp et al., 2014), leading to an allowed magnetic dipole transition between the qubit states (Pingault et al., 2017). Fig. 1.9 shows measurements where a resonant microwave field drives coherent Rabi oscillations of the spin qubit. In these experiments, a long laser pulse at frequency $f_{\downarrow\downarrow'}$ first initializes the spin in state $|\uparrow\rangle$ via optical pumping. After a microwave pulse of duration τ, a second laser pulse at $f_{\downarrow\downarrow'}$ reads out the population in state $|\downarrow\rangle$. Ramsey interference was used to measure the spin dephasing time T_2^* for SiV centers in two different samples [Fig. 1.9b]. For a sample that contained a low concentration of ^{13}C nuclear spins (Sample-12, circles), we measured a dephasing time in the range of $T_2^* \sim 4\mu s$. For a second sample that contained a natural abundance of ^{13}C nuclear spins (Sample-13, dark gray squares), we measured $T_2^* \approx 300$ ns which is similar to typical values observed with NV centers. These results demonstrate that the dephasing time T_2^* of SiV centers is primarily limited by the nuclear spin bath in the diamond host with a natural abundance of ^{13}C (Childress et al., 2006).

Dephasing due to slowly evolving fluctuations in the environment (e.g. nuclear spins) can be suppressed by using dynamical decoupling techniques (Ryan et al., 2010;

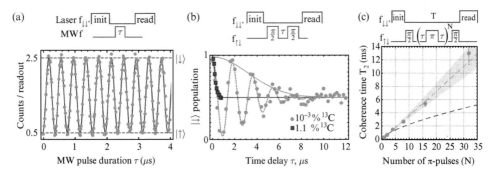

Fig. 1.9 Coherent spin control. (a) Resonant driving at frequency $f_{\uparrow\downarrow}$ results in Rabi oscillations between states $|\uparrow\rangle$ and $|\downarrow\rangle$. (b) Ramsey interference measurement of T_2^* for the two samples (see text for details). MW pulses are detuned by ~ 550 kHz from the $f_{\uparrow\downarrow}$ for the circles. (c) T_2 coherence vs. number of rephasing pulses N for Sample-12. Fitting to $T_2 \propto N^\beta$ gives $\beta = 1.02 \pm 0.05$ (light gray dashed line), the shaded region represents a standard deviation of 0.05. For comparison, the dark gray dashed line shows $N^{2/3}$ scaling. Figure adapted from (Sukachev et al., 2017).

de Lange et al., 2010). We extended the spin coherence time T_2 by implementing Carr–Purcell–Meiboom–Gill (CPMG) sequences with $N = 1, 2, 4, 8, 16$, and 32 rephasing pulses (Meiboom and Gill, 1958) in the isotopically purified sample. Fig. 1.9c shows that the coherence time increases approximately linearly with the number of rephasing π−pulses N. The longest observed coherence time is $T_2 = 13 \pm 1.7$ ms for $N = 32$. Repeating the CPMG sequences for $N = 1, 2$, and 4 with the Sample-13 gave similar coherence times T_2 as for Sample-12. Surprisingly, the observation that the coherence time T_2 in both samples is identical for a given N indicates that the coherence time T_2 is not limited by the nuclear spin bath, but by another noise source. While the origin of the noise source is at present not understood, the linear dependence of T_2 on N suggests that T_2 can potentially be further improved by using additional rephasing pulses (Sukachev et al., 2017).

These observations establish the SiV center as a promising solid-state quantum emitter for the realization of quantum network nodes using integrated diamond nanophotonics (Sipahigil et al., 2016). The demonstrated coherence time of 13 ms is already sufficient to maintain quantum states between quantum repeater nodes separated by 10^3 km (Childress et al., 2006). The quantum memory lifetime could be further extended by implementing robust dynamical decoupling schemes (de Lange et al., 2010) or using coherently coupled nuclear spins as longer-lived memories (Maurer et al., 2012).

1.8 Outlook

In this chapter, we reviewed recent progress on the realization of near-deterministic spin-photon interactions based on a single SiV in a cavity and two-emitter entanglement generation in a single nanophotonic device (Sipahigil et al., 2016). When combined with the recent demonstration of 13ms spin coherence for SiV centers (Sukachev et al., 2017), these advances make SiV centers the first solid-state quantum emitter that has the desired combination of long-lived spin coherence with a deterministic spin-photon interface.

Here we outline several research directions to advance this diamond nanophotonics platform towards the realization of scalable quantum repeater nodes. Nanocavity designs with smaller mode volumes recently led to an improved cooperativity of $C \sim 23$ and enabled the observation of photon-mediated interactions between two SiV centers (Evans et al., 2018). With further advances in diamond nanofabrication and emitter properties, these experiments can be extended to realize deterministic entanglement generation between two SiVs in a single nanocavity at high rates. Current experiments are also investigating using the hyperfine interaction between the SiV electron spin and Carbon-13 nuclear spins in the diamond lattice to realize multi-qubit registers with second scale coherence times. For long-distance quantum communication applications, such nanophotonic devices containing multi-qubit registers are expected to improve the entanglement generation rates by many orders of magnitude compared with the state of the art (Rosenfeld et al., 2017; Hucul et al., 2015; Hensen et al., 2015). We expect that these technical improvements will bring the realization of a multi-node quantum repeater within reach. For long-distance quantum communication, the SiV photons should be downconverted to the telecommunication-band for low-loss photon distribution in a fiber network. The improved collection efficiency from the nanophotonic sturctures is expected to result in a high signal-to-noise ratio quantum frequency conversion to the 1310−1350 nm telecommunication band with reduced spontaneous noise contribution from a pump in the 1620−1685 nm band (Zaske et al., 2012).

Finally, other color centers such as the germanium-vacancy (GeV) (Palyanov et al., 2015) and the tin-vacancy (SnV) (Iwasaki et al., 2017) centers with similar symmetry properties could yield improved optical and spin properties and recently started being actively explored. The GeV center was demonstrated to be superior to SiV centers in terms of its quantum efficiency (Bhaskar et al., 2017) and is expected to show similar spin properties at dilution fridge temperatures (Siyushev et al., 2017). To date, however, there has not been a system that combined good optical and spin properties at higher temperatures of 4 K. In principle, no fundamental reason prevents obtaining spectrally stable optical transitions and long spin coherence times at 4 K. Recent experiments with the neutral silicon-vacancy centers have already started investigating this possibility (Green et al., 2017; Rose et al., 2017). Considering the existence of hundreds of color centers in diamond (Zaitsev, 2001), a systematic experimental survey of different color centers with guidance from density functional theory calculations (Goss et al., 2005) is likely to result in new systems with improved device performance at elevated temperatures, which could be instrumental for realizing practical solid-state quantum repeater nodes.

Acknowledgments

The work described in this chapter highlights the results of a collaborative team effort over the past decade. Most of the experiments presented were carried out by members of the Harvard quantum optics group including Ruffin Evans, Denis Sukachev, Mihir Bhaskar, Christian Nguyen and Alexander Zibrov. Furthermore, we are grateful to Fedor Jelezko and members of the Ulm quantum optics group, including Kay Jahnke, Lachlan Rogers, Mathias Metsch, and Petr Siyushev for a close collaboration on quantum optics with novel color centers; to Marko Loncar, Hongkun Park, and Michael Burek for critical contributions including the development and fabrication of the diamond nanophotonic devices. We additionally acknowledge many discussions and collaborations with Ed Bielejec, Ryan Camacho, and Dirk Englund. This work was supported by the NSF, CUA, AFOSR MURI, ARL CDQI and DURIP Grant No. N00014-15-1-28461234 through ARO.

References

Abtew, T. A. et al. (2011, Sep.). Dynamic jahn-teller effect in the nv⁻ center in diamond. *Phys. Rev. Lett.*, **107**, 146403.

Acosta, V. M. et al. (2009). Diamonds with a high density of nitrogen-vacancy centers for magnetometry applications. *Phys. Rev. B*, **80**(11), 115202.

Acosta, V. M. et al. (2012, May). Dynamic stabilization of the optical resonances of single nitrogen-vacancy centers in diamond. *Phys. Rev. Lett.*, **108**(20), 206401.

Aharonovich, I., Englund, D., and Toth, M. (2016). Solid-state single-photon emitters. *Nat. Photon.*, **10**(10), 631–41.

Albrecht, R. et al. (2013). Coupling of a single nitrogen-vacancy center in diamond to a fiber-based microcavity. *Phys. Rev. Lett.*, **110**(24), 243602.

Andrews, R. W. et al. (2014). Bidirectional and efficient conversion between microwave and optical light. *Nat. Phys.*, **10**(4), 321–6.

Balasubramanian, G. et al. (2009). Ultralong spin coherence time in isotopically engineered diamond. *Nat. Mater.*, **8**(5), 383–7.

Barends, R. et al. (2014). Superconducting quantum circuits at the surface code threshold for fault tolerance. *Nature*, **508**(7497), 500–3.

Bechtold, A. et al. (2015). Three-stage decoherence dynamics of an electron spin qubit in an optically active quantum dot. *Nat. Phys.*, **11**(12), 1005–8.

Becker, J. N. et al. (2016, Nov.). Ultrafast all-optical coherent control of single silicon vacancy colour centres in diamond. *Nat. Commun.*, **7**, 13512.

Becker, J. N. et al. (2017). All-optical control of the silicon-vacancy spin in diamond at millikelvin temperatures. *arXiv preprint arXiv:1708.08263*.

Bernien, H., Childress, L., Robledo, L., Markham, M., Twitchen, D. J., and Hanson, R. (2012, Jan.). Two-photon quantum interference from separate nitrogen vacancy centers in diamond. *Phys. Rev. Lett.*, **108**(4), 1–5.

Bernien, H. et al. (2017, Nov.). Probing many-body dynamics on a 51-atom quantum simulator. *Nature*, **551**(7682), 579–84.

Bhaskar, M. K. et al. (2017, May). Quantum nonlinear optics with a germanium-vacancy color center in a nanoscale diamond waveguide. *Phys. Rev. Lett.*, **118**, 223603.

Briegel, H. et al. (1998, Dec.). Quantum Repeaters: The Role of Imperfect Local Operations in Quantum Communication. *Phys. Rev. Lett.*, **81**(26), 5932–5.

Burek, M. J. et al. (2014). High quality-factor optical nanocavities in bulk single-crystal diamond. *Nat. Commun.*, **5**, 5718.

Burek, M. J. et al. (2016, Dec.). Diamond optomechanical crystals. *Optica*, **3**(12), 1404–11.

Burek, M. J. et al. (2012). Free-standing mechanical and photonic nanostructures in single-crystal diamond. *Nano Lett.*, **12**(12), 6084–9.

Burek, M. J. et al. (2017, Aug.). Fiber-coupled diamond quantum nanophotonic interface. *Phys. Rev. Applied*, **8**, 024026.

Cabrillo, C. et al. (1999). Creation of entangled states of distant atoms by interference. *Phys. Rev. A*, **59**(2), 1025.

Childress, L. et al. (2006, Oct.). Coherent dynamics of coupled electron and nuclear spin qubits in diamond. *Science*, **314**(5797), 281–5.

Chu, Y. et al. (2014). Coherent optical transitions in implanted nitrogen vacancy centers. *Nano Lett.*, **14**(4), 1982–6.

Cirac, J. I. et al. (1997). Quantum state transfer and entanglement distribution among distant nodes in a quantum network. *Phys. Rev. Lett.*, **78**(16), 3221.

Cirac, J. I., and Zoller, P. (1995). Quantum computations with cold trapped ions. *Phys. Rev. Lett.*, **74**(20), 4091.

Clark, C. D. et al. (1995, Jun.). Silicon defects in diamond. *Phys. Rev. B*, **51**, 16681–8.

de Lange, G. et al. (2010, Oct.). Universal dynamical decoupling of a single solid-state spin from a spin bath. *Science*, **330**(6000), 60–3.

D'Haenens-Johansson, U. F. S. et al. (2011, Dec.). Optical properties of the neutral silicon split-vacancy center in diamond. *Phys. Rev. B*, **84**, 245208.

Doherty, M. W. et al. (2011). The negatively charged nitrogen-vacancy centre in diamond: the electronic solution. *New J. Phys.*, **13**(2), 025019.

Doherty, M. W. et al. (2013). The nitrogen-vacancy colour centre in diamond. *Phys. Rep.*, **528**(1), 1–45.

Edmonds, A. M. et al. (2008). Electron paramagnetic resonance studies of silicon-related defects in diamond. *Phys. Rev. B*, **77**(24), 245205.

Englund, D. et al. (2007). Controlling cavity reflectivity with a single quantum dot. *Nature*, **450**(7171), 857–61.

Englund, D. et al. (2010). Deterministic coupling of a single nitrogen vacancy center to a photonic crystal cavity. *Nano Lett.*, **1**, 1–10.

Evans, R. E. et al. (2016). Narrow-linewidth homogeneous optical emitters in diamond nanostructures via silicon ion implantation. *Phys. Rev. Appl.*, **5**(4), 044010.

Evans, R. E. et al. (2018). Photon-mediated interactions between quantum emitters in a diamond nanocavity. *Science*, **362**(6415), 662–5.

Faraon, A. et al. (2012, Jul.). Coupling of nitrogen-vacancy centers to photonic crystal cavities in monocrystalline diamond. *Phys. Rev. Lett.*, 109, 033604.

Fischer, G. (1984). Vibronic coupling: The interaction between the electronic and nuclear motions. Academic Press, London.

Fu, K.-M. C. et al. (2009, Dec.). Observation of the dynamic jahn-teller effect in the excited states of nitrogen-vacancy centers in diamond. *Phys. Rev. Lett.*, 103, 256404.

Gali, A., and Maze, J. R. (2013, Dec.). Ab initio study of the split silicon-vacancy defect in diamond: electronic structure and related properties. *Phys. Rev. B*, 88(23), 235205.

Gao, W. B. et al. (2015). Coherent manipulation, measurement and entanglement of individual solid-state spins using optical fields. *Nat. Photon.*, 9(6), 363–73.

Goldman, M. L. et al. (2015, Apr.). Phonon-induced population dynamics and intersystem crossing in nitrogen-vacancy centers. *Phys. Rev. Lett.*, 114, 145502.

Goss, J. P. et al. (2005). Vacancy-impurity complexes and limitations for implantation doping of diamond. *Phys. Rev. B*, 72(3), 035214.

Green, B. L. et al. (2017, Aug.). Neutral silicon-vacancy center in diamond: spin polarization and lifetimes. *Phys. Rev. Lett.*, 119, 096402.

Greilich, A. (2006, Jul.). Mode locking of electron spin coherences in singly charged quantum dots. *Science*, 313(5785), 341–5.

Gruber, A. et al. (1997). Scanning confocal optical microscopy and magnetic resonance on single defect centers. *Science*, 276(5321), 2012–14.

Ham, F. S. (1965). Dynamical jahn-teller effect in paramagnetic resonance spectra: orbital reduction factors and partial quenching of spin-orbit interaction. *Phys. Rev.*, 138(6A), A1727.

Haroche, S. (2013, Jul.). Nobel lecture: controlling photons in a box and exploring the quantum to classical boundary. *Rev. Mod. Phys.*, 85, 1083–1102.

Hauf, M. V. et al. (2011). Chemical control of the charge state of nitrogen-vacancy centers in diamond. *Phys. Rev. B*, 83(8), 081304.

Hausmann, B. J. M. et al. (2013). Coupling of nv centers to photonic crystal nanobeams in diamond. *Nano Lett.*, 13(12), 5791–6.

Hennessy, K. et al. (2007). Quantum nature of a strongly coupled single quantum dot–cavity system. *Nature*, 445(7130), 896–9.

Hensen, B. et al. (2015). Loophole-free bell inequality violation using electron spins separated by 1.3 kilometres. *Nature*, 526(7575), 682–6.

Hepp, C. (2014). Electronic structure of the silicon vacancy color center in diamond. Ph.D. thesis, Saarbrücken, Universität des Saarlandes, Diss., 2014.

Hepp, C. et al. (2014, Jan.). Electronic structure of the silicon vacancy color center in diamond. *Phys. Rev. Lett.*, 112(3), 036405.

Hong, C. K., Ou, Z. Y., and Mandel, L. (1987). Measurement of subpicosecond time intervals between two photons by interference. *Phys. Rev. Lett.*, 59(18), 2044–6.

Hucul, D. et al. (2015). Modular entanglement of atomic qubits using photons and phonons. *Nat. Phys.*, 11(1), 37–42.

Huxter, V. M. et al. (2013, Sep.). Vibrational and electronic dynamics of nitrogen "vacancy centres in diamond revealed by two-dimensional ultrafast spectroscopy. *Nature Phys.*, 9(11), 744–9.

Iwasaki, T. et al. (2017). Tin-vacancy quantum emitters in diamond. arXiv preprint arXiv:1708.03576.

Jahnke, K. D. et al. (2015). Electron–phonon processes of the silicon-vacancy centre in diamond. *New J. Phys.*, 17(4), 043011.

Jaksch, D. et al. (1998). Cold bosonic atoms in optical lattices. *Phys. Rev. Lett.*, 81(15), 3108.

Jaksch, D. et al. (2000). Fast quantum gates for neutral atoms. *Phys. Rev. Lett.*, 85(10), 2208.

Jelezko, F. et al. (2004). Observation of coherent oscillations in a single electron spin. *Phys. Rev. Lett.*, 92(7), 076401.

Judd, B. R. (1962). Optical absorption intensities of rare-earth ions. *Phys. Rev.*, 127(3), 750.

Kimble, H. J. (2008). The quantum internet. *Nature*, 453(7198), 1023–30.

Koch, J. et al. (2007, Oct.). Charge-insensitive qubit design derived from the cooper pair box. *Phys. Rev. A*, 76, 042319.

Kucsko, G. et al. (2013, Aug.). Nanometre-scale thermometry in a living cell. *Nature* (London), 500(7460), 54–8.

Ladd, T. D. et al. (2010). Quantum computers. *Nature*, 464(7285), 45–53.

Lee, J. C. et al. (2012, Apr.). Coupling of silicon-vacancy centers to a single crystal diamond cavity. *Opt. Express*, 20(8), 8891.

Legero, T. et al. (2003, Dec.). Time-resolved two-photon quantum interference. *Applied Physics B: Lasers and Optics*, 77(8), 797–802.

Lettow, R. et al. (2010, Mar.). Quantum interference of tunably indistinguishable photons from remote organic molecules. *Phys. Rev. Lett.*, 104(12), 123605–.

Li, Y. et al. (2015, Oct.). Resource costs for fault-tolerant linear optical quantum computing. *Phys. Rev. X*, 5, 041007.

Liu, G., and Jacquier, B. (2006). Spectroscopic properties of rare earths in optical materials. Volume 83. Springer Science & Business Media.

Lo, C. (2014). Natural colorless type IaB diamond with silicon-vacancy defect center. *Gems and Gemology*, L, 293.

Lodahl, P., Mahmoodian, S., and Stobbe, S. (2015). Interfacing single photons and single quantum dots with photonic nanostructures. *Rev. Mod. Phys.*, 87(2), 347.

Lodahl, P. et al. (2004). Controlling the dynamics of spontaneous emission from quantum dots by photonic crystals. *Nature*, 430(7000), 654–7.

Lovchinsky, I. et al. (2016). Nuclear magnetic resonance detection and spectroscopy of single proteins using quantum logic. *Science*, 351(6275), 836–41.

Lvovsky, A. I., Sanders, B. C., and Tittel, W. (2009). Optical quantum memory. *Nat. Photon.*, 3(12), 706–14.

Maiman, T. H. (1960). Stimulated optical radiation in ruby. Nature, 187(4736), 493–4.

Majer, et al. (2007). Coupling superconducting qubits via a cavity bus. *Nature*, 449(7161), 443–7.

Maurer, P. C. et al. (2012). Room-temperature quantum bit memory exceeding one second. *Science*, 336(6086), 1283–6.

Maze, J. R. et al. (2008). Nanoscale magnetic sensing with an individual electronic spin in diamond. *Nature*, 455(7213), 644–7.

Maze, J. R. et al. (2011). Properties of nitrogen-vacancy centers in diamond: the group theoretic approach. *New J. Phys.*, 13(2), 025025.

Mears, R. J. et al. (1987). Low-noise erbium-doped fibre amplifier operating at 1.54 μm. Electron. *Lett.*, 23(19), 1026–8.

Meiboom, S., and Gill, D. (1958, Aug.). Modified spin echo method for measuring nuclear relaxation times. *Review of Scientific Instruments*, 29(8), 688–91.

Michler, P. et al. (2000). A quantum dot single-photon turnstile device. *Science*, 290(5500), 2282–5.

Moerner, W. E., and Kador, L. (1989). Optical detection and spectroscopy of single molecules in a solid. *Phys. Rev. Lett.*, 62(21), 2535.

Monroe, Chris, Meekhof, D. M., King, B. E., Itano, Wayne M., and Wineland, David J. (1995). Demonstration of a fundamental quantum logic gate. *Phys. Rev. Lett.*, 75(25), 4714.

Monroe, C., Raussendorf, R., Ruthven, A., Brown, K. R., Maunz, P., Duan, L.-M., and Kim, J. (2014). Large-scale modular quantum-computer architecture with atomic memory and photonic interconnects. *Phys. Rev. A*, 89(2), 022317.

Monroe, Christopher R., Schoelkopf, Robert J., and Lukin, Mikhail D. (2016). Quantum connections. *Sci. Amer.*, 314(5), 50–7.

Monz, Thomas, Schindler, Philipp, Barreiro, Julio T., Chwalla, Michael, Nigg, Daniel, Coish, William a., Harlander, Maximilian, Hänsel, Wolfgang, Hennrich, Markus, and Blatt, Rainer (2011, March). 14-Qubit Entanglement: Creation and Coherence. *Phys. Rev. Lett.*, 106(13), 130506.

Neu, E., Hepp, C., Hauschild, M., Gsell, S., Fischer, M., Sternschulte, H., Steinmüller-Nethl, D., Schreck, M., and Becher, C. (2013, April). Low-temperature investigations of single silicon vacancy colour centres in diamond. *New J. Phys.*, 15(4), 043005.

Neu, E., Steinmetz, D., Riedrich-Möller, J., Gsell, S., Fischer, M., Schreck, M., and Becher, C. (2011). Single photon emission from silicon-vacancy colour centres in chemical vapour deposition nano-diamonds on iridium. *New J. Phys.*, 13(2), 025012.

Neuzner, Andreas, Körber, Matthias, Morin, Olivier, Ritter, Stephan, and Rempe, Gerhard (2016). Interference and dynamics of light from a distance-controlled atom pair in an optical cavity. *Nat. Photon.*, 10(5), 303–6.

Orrit, M. and Bernard, J. (1990). Single pentacene molecules detected by fluorescence excitation in ap-terphenyl crystal. *Phys. Rev. Lett.*, 65(21), 2716–19.

Orwa, J. O., Santori, C., Fu, K. M. C., Gibson, B., Simpson, D., Aharonovich, I., Stacey, A., Cimmino, A., Balog, P., Markham, M. et al. (2011). Engineering of nitrogen-vacancy color centers in high purity diamond by ion implantation and annealing. *J. Appl. Phys.*, 109(8), 083530.

Palyanov, Y. N. et al. (2015). Germanium: a new catalyst for diamond synthesis and a new optically active impurity in diamond. *Sci. Rep.*, 5, 14789.

Pellizzari, T. et al. (1995). Decoherence, continuous observation, and quantum computing: a cavity qed model. *Phys. Rev. Lett.*, 75(21), 3788.

Pingault, B. et al. (2014). All-optical formation of coherent dark states of silicon-vacancy spins in diamond. *Phys. Rev. Lett.*, 113(26), 263601.

Pingault, B. et al. (2017, May). Coherent control of the silicon-vacancy spin in diamond. *Nature Communications*, 8, 15579.

Press, D. et al. (2010). Ultrafast optical spin echo in a single quantum dot. *Nat. Photon.*, 4(6), 367–70.

Purcell, E. M. (1946). Spontaneous emission probabilities at radio frequencies. *Phys. Rev.*, 69, 681.

Riedel, D. et al. (2017, Sep.). Deterministic enhancement of coherent photon generation from a nitrogen-vacancy center in ultrapure diamond. *Physical Review X*, 7(3).

Riedrich-Möller, J. et al. (2012, Jan.). One- and two-dimensional photonic crystal microcavities in single crystal diamond. *Nat. Nanotechnol.*, 7(1), 69–74.

Ritter, S. et al. (2012, Apr.). An elementary quantum network of single atoms in optical cavities. *Nature*, 484(7393), 195–200.

Rogers, L. J. et al. (2014a, Jun.). Electronic structure of the negatively charged silicon-vacancy center in diamond. *Phys. Rev. B*, 89(23), 235101.

Rogers, L. J. et al. (2014b). All-optical initialization, readout, and coherent preparation of single silicon-vacancy spins in diamond. *Phys. Rev. Lett.*, 113(26), 263602.

Rogers, L. J. et al. (2014c, Aug.). Multiple intrinsically identical single-photon emitters in the solid state. *Nat. Commun.*, 5, 4739.

Rondin, L. et al. (2014). Magnetometry with nitrogen-vacancy defects in diamond. *Rep. Prog. Phys.*, 77(5), 056503.

Rose, B. C. et al. (2017). Observation of an environmentally insensitive solid state spin defect in diamond. arXiv preprint arXiv:1706.01555.

Rosenfeld, W. et al. (2017, Jul.). Event-ready bell test using entangled atoms simultaneously closing detection and locality loopholes. *Phys. Rev. Lett.*, 119, 010402.

Ryan, C. A., Hodges, J. S., and Cory, D. G. (2010, Nov.). Robust decoupling techniques to extend quantum coherence in diamond. *Phys. Rev. Lett.*, 105(20), 200402.

Santori, C. et al. (2002). Indistinguishable photons from a single-photon device. *Nature*, 419(6907), 594–7.

Schröder, T. et al. (2016, Apr.). Quantum nanophotonics in diamond. *J. Opt. Soc. Am. B*, 33(4), B65–B83.

Schröder, T. et al. (2017, May). Scalable focused ion beam creation of nearly lifetime-limited single quantum emitters in diamond nanostructures. *Nature Communications*, 8, 15376.

Senellart, P., Solomon, G., and White, A. (2017, Nov). High-performance semiconductor quantum-dot single-photon sources. *Nature Nanotechnology*, 12(11), 1026–39.

Sipahigil, A. et al. (2016, Oct.). An integrated diamond nanophotonics platform for quantum-optical networks. *Science*, 354(6314), 847–50.

Sipahigil, A. et al. (2012, Apr.). Quantum interference of single photons from remote nitrogen-vacancy centers in diamond. *Phys. Rev. Lett.*, 108(14), 143601.

Sipahigil, A. et al. (2014, Sep.). Indistinguishable photons from separated silicon-vacancy centers in diamond. *Phys. Rev. Lett.*, 113, 113602.

Siyushev, P. et al. (2017, Aug.). Optical and microwave control of germanium-vacancy center spins in diamond. Phys. Rev. B, 96, 081201.

Sohn, Y. I. et al. (2017). Engineering a diamond spin-qubit with a nano-electro-mechanical system. arXiv preprint arXiv:1706.03881.

Sternschulte, H. et al. (1994, Nov.). 1.681-eV luminescence center in chemical-vapor-deposited homoepitaxial diamond films. *Phys. Rev. B*, 50(19), 14554–60.

Stockill, R. et al. (2016). Quantum dot spin coherence governed by a strained nuclear environment. *Nat. Commun.*, 7, 12745.

Stoneham, A. M. (1975). Theory of defects in solids. Clarendon Press, Oxford.

Sukachev, D. et al. (2017, Nov.). Silicon-vacancy spin qubit in diamond: a quantum memory exceeding 10 ms with single-shot state readout. *Phys. Rev. Lett.*, 119, 223602.

Tamarat, P. et al. (2006). Stark shift control of single optical centers in diamond. *Phys. Rev. Lett.*, 97(8), 083002.

Taylor, J. M. et al. (2008). High-sensitivity diamond magnetometer with nanoscale resolution. *Nat. Phys.*, 4(10), 810–16.

Urban, E. et al. (2009). Observation of rydberg blockade between two atoms. *Nat. Phys.*, 5(2), 110–14.

Vahala, K. J. (2003). Optical microcavities. *Nature*, 424(6950), 839–46.

Wineland, D. J. (2013, Jul.). Nobel lecture: superposition, entanglement, and raising Schrödinger's cat. *Rev. Mod. Phys.*, 85, 1103–14.

Yamamoto, T. et al. (2013). Strongly coupled diamond spin qubits by molecular nitrogen implantation. *Phys. Rev. B*, 88(20), 201201.

Zaitsev, A. M. (2001). Optical properties of diamond: a data handbook. Springer Science & Business Media.

Zaske, S. et al. (2012, Oct.). Visible-to-telecom quantum frequency conversion of light from a single quantum emitter. *Phys. Rev. Lett.*, 109, 147404.

Zhang, J. et al. (2017, Nov.). Observation of a many-body dynamical phase transition with a 53-qubit quantum simulator. *Nature*, 551(7682), 601–4.

Zhang, J. L. et al. (2016). Hybrid group iv nanophotonic structures incorporating diamond silicon-vacancy color centers. *Nano Lett.*, 16(1), 212–17.

Ziegler, J. F., Ziegler, M. D., and Biersack, J. P. (2010). Srim—the stopping and range of ions in matter (2010). *Nucl. Instrum. Meth. B*, 268(11), 1818–23.

2
Searches for new, massive particles with AMO experiments

David DeMille

Department of Physics, Yale University
New Haven, USA

DeMille, D. "Searches for new, massive particles with AMO experiments." In *Current Trends in Atomic Physics*. Edited by Antoine Browaeys, Thierry Lahaye, Trey Porto, Charles S. Adams, Matthias Weidemüller, and Leticia F. Cugliandolo. Oxford University Press (2019).
© Oxford University Press. DOI: 10.1093/oso/9780198837190.003.0002

Chapter Contents

The goal of these lectures will be to explain how atomic/molecular/optical (AMO) physics measurements can have an impact in the world of particle physics. Part of our task will be to translate from the terminology used in particle physics to language that is commonly used and understood by AMO physicists. Of course, the language of particle physics is relativistic quantum field theory, so we will inevitably need to use both some concepts of field theory and some relativistic notation. However, these lectures will assume no knowledge of field theory beyond what I consider standard in a typical undergraduate plus introductory graduate curriculum. In particular, I will assume familiarity with the quantized electromagnetic (EM) field, and with the relativistically covariant formulations of Maxwell's equations.

With that in mind, we will start with a familiar example: the EM field, and its particle-like excitations known as photons. The photon is of course a particle with spin 1 (a "vector boson") and zero mass. (The vanishing mass means that a photon spin orientation perpendicular to the direction of motion cannot occur physically.)

In standard relativistic 4-vector notation, everything about the EM field can be described in terms of the (antisymmetric) field tensor $F^{\mu\nu} = \partial^\mu A^\nu - \partial^\nu A^\mu$, where A^μ is the 4-potential $(\Phi, \vec{A})$. Here and throughout, I use cgs/Gaussian units for electromagnetism (i.e., no μ_0 or ϵ_0, just c!) and the usual particle physics sign convention for the metric of Minkowski space: $g_{\mu\nu} = (g_{00}, g_{ii}) = (+, -, -, -)$. When there are source charges and currents, encapsulated in the 4-vector $j^\mu = (c\rho, \vec{j})$ (where ρ is the charge density and $\vec{j}$ is the current density), then Maxwell's equations can be written in the compact form $\partial_\nu F^{\mu\nu} = 4\pi j^\mu/c$. With a particular choice of gauge (Lorentz gauge, such that $\partial_\nu A^\nu = 0$), we can also derive a simple version of the EM wave equation in terms of the 4-potential from Maxwell's equations. In particular, we write the wave equation in the form $\partial_\nu \partial^\nu A^\mu = 4\pi j^\mu/c = \Box A^\mu$, where $\Box \equiv \partial_\nu \partial^\nu = \frac{1}{c^2}\frac{\partial^2}{\partial t^2} - \nabla^2$. In the absence of sources, this just says $\Box A^\mu = \partial_\nu \partial^\nu A^\mu = 0$, which is the standard source-free wave equation, written in terms of the potentials rather than the fields. If none of this is familiar, you can find it in any standard intermediate-level or higher textbook on electromagnetism (e.g., the texts by (Griffiths, 2017) or (Jackson, 1998)).

2.1 Electromagnetism with a massive photon

Now, let's generalize this in an unfamiliar way. What we will do is work out how Maxwell's equations would be modified if the photon had a finite mass rather than its usual zero mass. How can we do this? We begin by recognizing that the wave equation for the 4-potential looks just like the relativistic quantum equation of motion for a massless particle. That is: in general, for relativistic particles, $E^2 = p^2 c^2 + m^2 c^4$. Hence for massless particles, $-p^2 c^2 + E^2 = 0$. In quantum mechanics, we are used to the relations $\vec{p} = -i\hbar\vec{\nabla}$ and $E = i\hbar\partial/\partial t$. Because everything we do going forward will involve quantum mechanics, we take $\hbar = 1$ from here on. Hence the relativistic energy-momentum relation for a massless particle with wavefunction ψ can be written as $\left[\nabla^2 c^2 - \frac{\partial^2}{\partial t^2}\right]\psi = 0$; that is, $\left[\nabla^2 - \frac{\partial^2}{c^2 \partial t^2}\right]\psi = 0 = \partial_\nu \partial^\nu \psi$. Thus we see that each component of the 4-vector A_μ satisfies the relativistic, quantum energy-momentum expression for the wavefunction of a massless particle.

This gives us some insight as to how to proceed if the photon had nonzero mass M. Here, we know that $-p^2c^2 + E^2 - M^2c^4 = 0$. So, it is natural to expect the mass-bearing EM field to satisfy the wave equation $c^2\partial_\nu\partial^\nu A^\mu - M^2c^4 A^\mu = 0$, i.e. $\partial_\nu\partial^\nu A^\mu - \mu^2 A^\mu = 0$, where $\mu \equiv Mc$ is the inverse of the Compton wavelength associated with the massive photon. (This is known as the Proca equation; it is the relativistic version of the Schrödinger equation, for a spin-1 particle with mass M.) Working backwards from here, one finds that this wave equation arises if Maxwell's equations in the presence of sources are modified to take the form $\partial_\mu F^{\mu\nu} + \mu^2 A^\nu = 4\pi j^\nu/c$.

What happens if the source of this modified EM field is a static charge point charge at the origin, such that $\rho(\vec{r}) = q\delta^3(\vec{r})$ and $A_\mu = (\Phi,0)$? Here, the source 4-current is $j^\mu = (c\rho,0) = (cq\delta^3(\vec{r}),0)$. Then the $\nu = 0$ component of Maxwell's equations says

$$4\pi\rho = \partial_\mu F^{\mu 0} + \mu^2\Phi = \frac{\partial}{\partial x^\mu}\left(\frac{\partial}{\partial x_\mu}A^0 - \frac{\partial}{\partial x_0}A^\mu\right) + \mu^2\Phi \tag{2.1}$$

$$= \frac{\partial}{\partial x^\mu}\left(\frac{\partial}{\partial x_\mu}\Phi\right) + \mu^2\Phi = -\nabla^2\Phi + \mu^2\Phi. \tag{2.2}$$

So, we find

$$\nabla^2\Phi - \mu^2\Phi = 4\pi\rho. \tag{2.3}$$

For a point charge q at the origin, this equation thus leads to a potential of the form $\Phi(\vec{r}) = \frac{q}{r}e^{-\mu r}$. To see this, recall that in spherical coordinates, for our spherically symmetric situation, $\nabla^2 = \frac{1}{r^2}\frac{\partial}{\partial r}\left(r^2\frac{\partial}{\partial r}\right)$. Hence, except at the origin,

$$\nabla^2\Phi = \frac{1}{r^2}\frac{\partial}{\partial r}\left(r^2\frac{\partial}{\partial r}\right)\frac{1}{r}e^{-\mu r} = \frac{1}{r^2}\frac{\partial}{\partial r}\left(r^2\left[-\frac{1}{r^2}e^{-\mu r} - \mu\frac{1}{r}e^{-\mu r}\right]\right)$$

$$= \frac{1}{r^2}\frac{\partial}{\partial r}\left(-e^{-\mu r} - \mu r e^{-\mu r}\right) = \frac{1}{r^2}\left(\mu e^{-\mu r} - \mu e^{-\mu r} + \mu^2 r e^{-\mu r}\right)$$

$$= \mu^2\frac{1}{r}e^{-\mu r} = \mu^2\Phi. \tag{2.4}$$

As you approach the origin, $\Phi(\vec{r}) \to \frac{q}{r}$, so there

$$\nabla^2\Phi = q\nabla^2\left(\frac{1}{r}\right) = q[-4\pi\delta^3(\vec{r})] = -4\pi\rho, \tag{2.5}$$

as claimed.

The potential $\Phi(r) = \frac{q}{r}e^{-\mu r}$ is known as a Yukawa potential. It is similar to a Coulomb potential, but falls off more rapidly due to the exponential term; its characteristic length scale is $\mu^{-1} = 1/(Mc)$. That is: massive photons would give rise to shorter-range potentials than ordinary massless photons and their Coulomb potential.

2.2 Searching for a new particle: the "hidden photon"

From a variety of experiments—primarily tests for macroscopic or even astronomical-scale deviations from the usual $1/r^2$ scaling of electric field from a point charge—we know that the photon mass M_γ, if it does not exactly vanish, is extraordinarily small [(Goldhaber and Nieto, 2010)]: the upper bound is $M_\gamma < 10^{-18}$ eV! However, it is perfectly conceivable that an entirely different particle exists, which is "like a photon"—in the sense that it also is associated with the excitations of a different 4-vector field, whose components all satisfy the relativistic Proca equation—but which has some mass different from zero. In addition, there is no particular reason why the strength of interaction of familiar particles (electrons, protons, neutrons) with this new field—that is, the analogue of the electric charge for a static particle—needs to be the same as the *actual* electric charge that describes each particle's coupling to the EM field. For simplicity, here we write our equations under the assumption that electrons and protons have equal and opposite effective charge for coupling to this new particle, and that neutrons have zero effective charge. However, it is entirely possible that, for example, the effective charge of a neutron could be the same as the effective charge of a proton, rather than just zero as it is for the neutron in the EM interaction.

From many different types of experiments, we would have seen such a new "heavy photon analogue" particle already, if it did not differ significantly in either mass or coupling strength from the ordinary photon. However, if ordinary matter has a sufficiently weak coupling to such a new particle (i.e., a tiny "effective charge"), its existence might not have been noticed in any experiment. Similarly, if the new particle is sufficiently massive, its potential will have a very short range—and hence its effects on ordinary matter might conceivably have been missed.

Looking for such weakly coupled and/or very heavy photon analogues—generically known as "dark photons" or "hidden photons"—turns out to be a very active area of research in particle physics today [(Collar et al., 2012)]. Additional particles of this type can easily be added to the Standard Model of particle theory without upsetting the internal consistency of the theory. In addition, particles of this type tend to arise naturally in string theory and some versions of supersymmetric (SUSY) theories. There is, however, little guidance as to their mass, or the strength of their couplings, in these theoretical models. Hence there are experimental searches for particles of this type with mass M ranging from TeV to femto-eV scales, and as small a coupling as it is possible to detect. (Note that here and throughout, I use the standard particle physics convention to express mass in units of energy—i.e., in terms of Mc^2.)

It is not hard to understand that, in principle, the existence of such a particle would have an observable effect on atomic spectra—for example, on the level splittings in hydrogen atoms, which can be calculated very precisely and have also been measured with great accuracy. So, let us analyze quantitatively how "hidden photons" would affect the energy levels in H atoms.

It is standard to parameterize the potential between an electron with charge e and a nucleus with charge $+Ze$, in the presence of a hidden photon plus the ordinary Coulomb interaction, in the form

$$V(r) = -\frac{Ze^2}{r}(1+\chi^2 e^{-\mu r}) = -\frac{Z\alpha c}{r}(1+\chi^2 e^{-\mu r}). \tag{2.6}$$

Here the dimensionless parameter χ^2 describes the strength of the electron-proton coupling via the hidden photon, relative to the ordinary EM interaction.[1]

If the hidden photon's mass M is small enough that the range of the potential is large compared to the size of the atom (i.e., $\mu^{-1} \gg a_0$), then for a typical electron orbital, the new potential looks simply like a redefinition of the EM strength α—that is, it is as if α were replaced by the new value $\alpha' = \alpha(1+\chi^2)$. Note, however, that the value of α is itself an experimentally determined quantity, from experiments that use only electrons and/or protons. Hence, the effect of the hidden photon will not be observable from atomic spectroscopy when the mass M is too small. Instead, in this case, calculated values for the expected energy levels in the H atom will depend on the value of α, which would be modified in the same way both in H atoms and in the atom-based experiments that were used to define α in the first place.

However, if the mass M is larger, so that the range of the interaction with the hidden photon is short enough that $\mu^{-1} \lesssim a_0$, then the relative binding energy of different electron orbitals will be noticeably affected. In particular, orbitals with higher electron wavefunction amplitude near the nucleus (i.e., low principal quantum number n and low orbital angular momentum ℓ) will be affected more than those with electrons mostly at larger distances. Of course, if the mass M is very large, this new Yukawa term in the potential will be very short-range, and we would naïvely expect it to have comparatively little effect.

Let's work out explicitly what happens in this large-mass limit. Consider the energy shift in first-order perturbation theory, with the perturbation given by the Yukawa potential $V'(r) = -\chi^2 Z\alpha c \frac{e^{-\mu r}}{r}$, with large μ (and hence short range). The shift is given by the expression

$$\delta E^{(1)} = \langle \psi | H' | \psi \rangle = \int \psi^* V' \psi d^3 \vec{r} = -\chi^2 Z\alpha c \int \psi^* \frac{e^{-\mu r}}{r} \psi d^3 \vec{r}. \tag{2.7}$$

Since we are interested in the case where this is a very short range potential, let's only consider s-state wavefunctions, which are the only ones with nonzero amplitude at the nucleus. Then

$$\delta E^{(1)} = -4\pi \chi^2 Z\alpha c \int_0^\infty R_{ns}(r) \frac{e^{-\mu r}}{r} R_{ns}(r) r^2 dr, \tag{2.8}$$

---

[1] This ignores the possibility of coupling to neutrons, which could be present in general. The relation can be modified easily to account for this possibility, but for simplicity we do not do this here.

where $R_{ns}(r)$ is the radial wavefunction for the $\ell = 0$ state with principal quantum number n. Recall that $R_{ns}(r)$ has the form of a polynomial times an exponential, i.e. $R_{ns}(r) = C(1 + c_1 r/a_0 + c_2 r^2/a_0^2 + c_{n-1} r^{n-1}) e^{-r/(na_0)}$. If M is large enough that $\mu^{-1} \ll a_0$, only the first term in the radial wavefunction will be significant while the exponential in the integrand is still substantial. So, we can replace $R_{ns}(r)$ by $R_{ns}(0)$, and write

$$\delta E^{(1)} = -4\pi \chi^2 Z \alpha c R_{ns}^2(0) \int_0^\infty e^{-\rho} \frac{\rho}{\mu} \frac{d\rho}{\mu}, \tag{2.9}$$

where $\rho \equiv \mu r$. Then

$$\delta E^{(1)} = -4\pi \frac{\chi^2 Z \alpha c}{\mu^2} R_{ns}^2 0 = -\chi^2 \frac{Z \alpha c}{M^2 c^2} |\psi(0)|^2. \tag{2.10}$$

This shows one of main points that will recur in these lectures: for large masses of new particles M, the size of the energy shift in atoms generally scales like $1/M^2$. So, as you might naïvely expect, the effect of any new particle shrinks rapidly as its mass M increases—and it is generically hard to see the effect of very heavy particles in AMO experiments. Nevertheless, as we shall see, in some cases these experiments can detect particles with mass well above that of the heaviest known elementary particles!

Based on this analysis, for large mass M, we can write an effective Hamiltonian in the form

$$H_{\text{eff}}(\vec{r}) = -\frac{\chi^2}{M^2 c^2} Z \alpha c \delta^3(\vec{r}). \tag{2.11}$$

This effective Hamiltonian will have the same matrix elements as the actual Yukawa potential $V'(r)$. That is: for large mass, the term in the Yukawa potential $\frac{e^{-\mu r}}{r}$ has the same effect as a term $\delta^3(\vec{r})/\mu^2$. The latter form is generally more convenient to work with, so we will use it a lot.

Now, let's return to the problem at hand: how would a heavy hidden photon affect the spectrum of a hydrogen atom? To quantify the effect, we need to recall the atomic wavefunction's density at the nucleus. For hydrogenic atoms/ions with nuclear charge Z, this is found in standard textbooks:

$$|\psi_{n\ell m}^{(Z)}(0)|^2 = \frac{1}{4\pi} \frac{4Z^3}{n^3 a_0^3}. \tag{2.12}$$

This makes it clear that the effect will be larger for bigger values of the atomic charge Z. However, most high-precision AMO experiments are done with neutral atoms (or perhaps singly charged ions), not high-Z hydrogenic ions. This raises an issue, which is generically important for experiments that are searching for effects due to exchange of high-mass force-carrying particles (with short-range Yukawa interactions). That is: how does the probability to find the electron in contact with the nucleus, i.e. $|\psi(0)|^2$, change across neutral atoms? We will return to this later. For the moment, let's just think about neutral hydrogen atoms, with $Z = 1$. We have a clear prediction about the size of energy shifts δE in H atoms (in the state with quantum numbers n, ℓ) due to a very massive hidden photon, namely:

$$\delta E_{n\ell}(\mathrm{H}) = -\chi^2 \frac{\alpha c}{M^2 c^2} |\psi(0)|^2 = -\frac{1}{\pi}\chi^2 \frac{\alpha c}{M^2 c^2 n^3 a_0^3} \delta_{\ell,0}. \tag{2.13}$$

We can search for such a shift—and in its absence, set quantitative limits on the combination χ^2/M^2—by comparing measured spectra to predictions based on ordinary electromagnetic interactions.

We have performed this analysis in the large-mass case because the equations are especially simple there, and because it will be useful for further discussions. As we noted, as the mass M grows, for a given experimental precision the sensitivity to the coupling strength χ^2 becomes worse. Recall also that in the zero-mass limit, the effect is not observable. So, it is interesting to ask: for what mass M is this type of experiment most sensitive to a very weak coupling χ^2? From simple dimensional analysis, the best sensitivity to small values χ^2 will occur when the effective range of the Yukawa potential is about one Bohr radius, i.e., when $\mu^{-1} = 1/(Mc) \sim a_0 = 1/(m_e c\alpha)$. This implies that $M \sim \alpha m_e \sim 3$ keV.

Of course, no particle of this type has been found so far, so all current data can be interpreted as setting limits on some combination of the coupling strength χ^2 and the mass M. Plots showing limits from both atomic experiments and other types of experiments are shown in fig. 1 of (Jaeckel and Roy, 2010) and fig. 6.3 of (Collar et al., 2012). The optimal sensitivity range described above is evident in these plots.

As you can see from these plots, the limits from atomic hydrogen spectroscopy are not quite competitive with those from other methods. The reason, fundamentally, is that here we must compare *absolutely* measured and theoretically calculated quantities. Absolute measurements are generically hard! The best absolute energy measurement that can be compared to anything fundamentally calculable is of the Rydberg constant [(Mohr, Newell, and Taylor, 2016)]. The state of the art has fractional uncertainty 6×10^{-12}, i.e. absolute uncertainty of 20 kHz out of 3×10^{15} Hz (though the theoretical value is much less accurately known). The best absolute determination of α comes from measurements of the electron's magnetic anomaly $(g - 2)$, which (together with QED theory) yields α with an fractional uncertainty of ~ 0.3 ppb $= 300 \times 10^{-12}$ [(Gabrielse, 2013)]. These numbers set the natural scale at which atomic spectroscopy can (in the best case) set limits on the coupling strength, χ^2, of a hidden photon to electrons and protons.

As we will see next, it is possible to make measurements with more sensitivity to new particles by looking at signals that are, by definition, zero under ordinary conditions—but which can be nonzero in the presence of the new particle.

2.3   Generalization into the language of particle physics

To further explore ways that AMO experiments can be sensitive to new particles, we will need to generalize our approach and begin using language that is more common in particle physics than in AMO physics. In particular: instead of starting with a potential as we did for the hidden photon, we will start from an intuitive picture of interacting particles: a Feynman diagram. In the Feynman diagram representation, the interaction between electron and nucleus—which we originally described as the electron moving in a potential created by the nucleus—will instead be described as the exchange of a force-carrying particle between the electron and nucleus.

Let's again start with the familiar example of the electromagnetic interaction—in a simple atom, just the Coulomb interaction. Here, the relevant Feynman diagram will correspond to the exchange of a photon between the electron and the nucleus (Fig. 2.1a). In particle physics, it is much more common to think about free particles, and collisions between them, than the bound states we usually deal with in AMO physics. Thus, to translate to the language of particle physics, we should describe our quantized system in a basis of free particle plane waves, i.e. definite-momentum states. It is easy to do this, just by taking Fourier transforms of the usual wavefunctions we know and love in the position representation. Similarly, we can understand the interaction between particles in the momentum-space representation rather than the usual position-space representation, by taking the Fourier transform of the potential that describes the interaction.

Let's do this for the case of the Yukawa interaction. In the position representation we worked out earlier, we can write the potential for the Yukawa interaction in the form $V(\vec{r}) = q_1 q_2 \frac{e^{-\mu r}}{r}$. Here, I have replaced the EM charges ($-e$ for the electron and Ze for the nucleus) with generalized charges q_1 and q_2. The Fourier transform $\tilde{V}(\vec{k})$ is straightforward to evaluate:

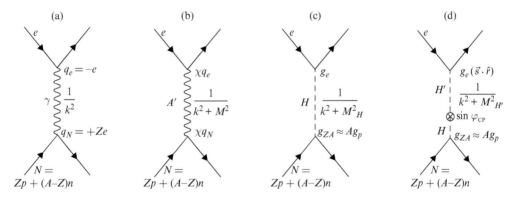

Fig. 2.1 Feynman diagrams describing electron-nucleus interaction. (a) Coulomb interaction. q_e and q_N denote the electron (e) and nuclear (N) electric charges, respectively. The force-mediating photon is described by the propagator $\frac{1}{k^2}$, where k is the momentum transfer. (b) Force mediated by a hidden photon, A', with mass M and effective charges χq_e and χq_N. (c) Force mediated by the Standard Model Higgs boson, H. The effective charges g_e and $g_{ZA} \approx A g_p$ are proportional to the particle masses. (d) T, P-violating force mediated by Higgs bosons, in the case of a more complex Higgs sector. Here we assume an extra neutral psuedoscalar Higgs, H', and CP-violating mixing between H and H' with strength characterized by the phase ϕ_{CP}.

$$\tilde{V}(\vec{k}) \propto q_1 q_2 \int e^{i\vec{k}\cdot\vec{r}} \frac{e^{-\mu r}}{r} d^3\vec{r} \propto q_1 q_2 \iiint e^{ikr\cos\theta} \frac{e^{-\mu r}}{r} r^2 dr d(\cos\theta) d\phi$$

$$\propto q_1 q_2 \int_0^\infty \left\{ \int_{-1}^1 e^{ikr\cos\theta} d(\cos\theta) \right\} \frac{e^{-\mu r}}{r} r^2 dr \propto q_1 q_2 \int_0^\infty \frac{e^{ikr} - e^{-ikr}}{ikr} \frac{e^{-\mu r}}{r} r^2 dr$$

$$\propto \frac{q_1 q_2}{k} \int_0^\infty \left\{ e^{-(\mu+ik)r} - e^{-(\mu-ik)r} \right\} dr \propto \frac{q_1 q_2}{k} \left\{ \frac{1}{\mu+ik} - \frac{1}{\mu-ik} \right\}$$

$$\propto \frac{q_1 q_2}{\mu^2 + k^2}. \tag{2.14}$$

This can be interpreted in terms of a scattering event, in the following way. Particle 1, with initial momentum $\vec{p}_1$, scatters off particle 2 with initial momentum $\vec{p}_2$. After scattering, the final momenta are $\vec{p}_1'$ and $\vec{p}_2'$. The interaction between these particles proceeds via the exchange of a force-carrying particle with momentum $\vec{k}$. From conservation of momentum, $\vec{p}_1 + \vec{p}_2 = \vec{p}_1' + \vec{p}_2'$, and so $\vec{p}_1 - \vec{p}_1' = \vec{p}_2' - \vec{p}_2 \equiv \vec{k}$. That is: here, $\vec{k}$ is the momentum transfer during the collision. This suggests a language to describe the Feynman diagram we drew before, for an electron-nucleus interaction mediated by a massive photon, as a sequence of events. That is: we can think of the interaction $\tilde{V}(\vec{k}) \propto \frac{q_1 q_2}{\mu^2+k^2}$ as the sequence

1. particle 1, with charge q_1, emits the mediating particle
2. the mediating particle, carrying momentum $\vec{k}$, propagates towards particle 2; mathematically, this is described by the "propagator" term $\frac{1}{k^2+\mu^2}$
3. particle 2, with charge q_2, absorbs the mediating particle.

This completes a description of the interaction as a charge-charge coupling, mediated by the force-carrying field as described by the propagator, and with the strength of the interaction between the particle and the force carrier at each vertex in the Feynman diagram described by the generalized charge q_i. An example for a force mediated by a hidden photon, A', is shown in Fig. 2.1b.

We derived the Yukawa potential under the assumption of stationary charges— i.e., as the generalization of the static Coulomb interaction. However, we know that magnetic effects can play a role in the electromagnetic interaction when the charges are moving. So, it is evident that we need to generalize our description to account for the motion of the particles. In the spirit of particle physics, it makes sense to seek a relativistically covariant description of the Yukawa interaction. Our derivation assumed a static charge density, which is the $\mu = 0$ (time-like) component of the current 4-vector j^μ. So, it is natural to expect that we should replace the simple charge for each particle with this 4-current; that is, we replace q_1 and q_2 with j_1^μ and j_2^ν. To get a relativistically invariant interaction, we can connect these currents via a propagator of the form $g_{\mu\nu}/(k^2 + \mu^2)$, where now k is a 4-vector momentum transfer.[2] Then, the total interaction can be written in relativistically covariant form as $\tilde{V}(k) = j_1^\mu \frac{g_{\mu\nu}}{(k^2+m^2)} j_2^\nu$. This is a current-propagator-current description of the interaction, analogous to the earlier charge-propagator-charge description.

---

[2] This is not quite right; when done correctly, the numerator in the propagator for a spin-one boson gets an extra term for nonzero mass, but it will not be important for our discussion here.

2.4 New types of interactions from new types of force-mediating particles

There can be forces between the electron and the nucleus in an atom which are mediated by particles other than the photon (or even massive analogues to the photon). These "more exotic" force-mediating particles can interact with electrons and nucleons in ways that are fundamentally different from how photons interact with them. The classification of the possible types of interactions is a generalization of the electromagnetic 4-current.

Let us enumerate various things that can be different about these interactions, relative to the familiar case of the Coulomb interaction:

1. The strength of the interaction—the "generalized charge" at each vertex—can be quite different from the EM charge strength. We have already discussed this for the case of hidden photons. We will see several different examples going forward.
2. The mass of the force-carrying particle can, in principle, take any value. This is just as we saw for hidden photons. The mass changes the propagator term from $1/k^2$ to $1/(k^2 + \mu^2)$.
3. Most unfamiliar is the fact that different particles can couple to electrons and nucleons in a fundamentally different way. Rather than simply a number associated with the charge (for a static particle), couplings mediated by other types of particles can involve dynamical variables. In particular, a "charge" can be replaced by "charge$'$ $* \vec{s} \cdot \vec{p}$", or "charge$'$ $* \vec{s} \cdot \vec{r}/r$", or even more complicated operators. The standard way to classify the interactions is according to the spin of the force-carrying boson particle, and the symmetry properties of the coupling, as follows:

    (a) Spin-1 particles like the photon or the hidden photon are known as "vector bosons". Their vector (4-)potential can couple to the current associated with the effective charge and motion of an electron and/or nucleus, via an interaction term with the familiar form $V = -\frac{1}{c} j^\mu A_\mu \propto -q\vec{p} \cdot \vec{A} + q\Phi$. Note that here, the 3-vector part $\vec{j} = q\vec{p}$ of the usual EM current is an ordinary vector: it changes sign under a parity transformation.

    (b) Spin-0 particles such as the Higgs boson are knowns as "scalar bosons". Their (scalar) potential couples not to a 4-vector associated with the electron or nucleus, but rather to a scalar quantity associated with these particles. In the end, this quantity acts just like the charge of a static particle. However, instead of having the form 4-current * 4-vector, the interaction looks like a relativistic scalar, independent of particle motion. Scalar "charges" of this type are obviously even under a parity transformation.

    (c) Other spin-1 particles like the Z and W bosons are also "vector bosons". However, their potential can couple not only to the 4-current of a particle, but also to an *axial*-vector analogue of the usual 4-current. What sort of axial vector can be associated with a particle? Of course, there is orbital angular momentum—but that requires nonzero distance (since $\vec{L} = \vec{r} \times \vec{p}$), and we will be interested in short-range forces from exchange of high-mass particles. Beyond that, the only axial vector associated with a particle is its spin, $\vec{s}$. So, we should expect axial-current interactions to be associated with the spin of the particle. Each of the couplings (polar-vector and axial-vector) has its own separate associated "charge". Much like for the vector EM field, for nonrelativistically moving particles, the only important part of the current is the time-like component, j^0, which does not have spatial vector components. This means that we should expect terms like "charge$'$ $* \vec{s} \cdot \vec{p}$", or "charge$'$ $* \vec{s} \cdot \vec{r}/r$" to appear in expressions describing the lowest-order interactions of electrons and nuclei with these types of particles.

(d) Similarly, some other types of spin-0 particles are called "pseudo-scalar bosons". The pion is one well-known particle of this type, though it does not couple directly to electrons. Other "new", hypothetical particles also are of this type. One often-discussed example is the particle called the "axion", which (if it exists) must have very low mass and extremely weak couplings to matter [(Graham, Irastorza, Lamoreaux, Lindner, and van Bibber, 2015)]. Heavier analogues of the Higgs boson can also be pseudoscalar bosons [(Barr and Zee, 1990)]. For interactions with pseudoscalar bosons, the "effective charge" has to be a pseudoscalar quantity created out of dynamical variables associated with the particle. An example would be charge$'$ $* \vec{s} \cdot \vec{p}$.

(e) Just as for Z and W bosons, there can be spin-0 particles that couple *both* to a scalar and to a pseudoscalar associated with the electron and/or the nucleon. We will discuss this case at length, below.

2.5 Example: Higgs boson exchange in atoms

Let's look at a potentially interesting example that has been discussed recently in (Delaunay, Ozeri, Perez, and Soreq, 2017): the effect on atomic spectra due to exchange of a Higgs boson between atomic electrons and the nucleus. In the Standard Model of particle physics, the Higgs is a pure scalar boson. It is very heavy, with mass $m_{\mathrm{H}} = 125$ GeV $\approx 2.5 \times 10^5 m_e$, so we can treat its effect as a very short-range Yukawa interaction. The dimensionless coupling strength of the Higgs to other fundamental particles such as leptons and quarks, in the Standard Model, is proportional to the mass of the particle. In particular, the "effective charge" g of a fundamental particle with mass m is $g = m/\langle v \rangle$; here $\langle v \rangle = 246$ GeV$/c^2$ is the "vacuum expectation value of the Higgs field", a nonzero constant value everywhere in the vacuum. (The interaction with this "background" Higgs field gives ordinary particles their mass, via $m = g\langle v \rangle$.) So, an electron has effective charge $g_e = m_e/\langle v \rangle \approx 2 \times 10^{-6}$ when interacting with the Higgs. For a proton or neutron, the story is more complicated: their mass mainly comes from the energy of interacting gluons, which are themselves massless and do not couple directly to the Higgs. Only a small portion of proton and neutron masses comes from the up and down quarks within them. These quarks couple to the Higgs, but only weakly because they are light (about 2 and 5 MeV for up and down quark, respectively—i.e., not more than one order of magnitude heavier than an electron). Even more complicated contributions come when the gluons convert into virtual pairs of heavy quarks, which can then couple very strongly to the Higgs. When all is said and done, it is crudely estimated that the effective charges of protons and neutrons coupling to the Higgs will be $g_{p[n]} \approx 10^{-3}$. (This is about 10 times smaller than you might have expected if you naïvely assumed the coupling was given by $m_{p[n]}/\langle v \rangle$.)

Note particularly that a neutron has about the same effective charge as a proton; this is strikingly different than for EM interactions, where neutrons have zero charge. The Feynman diagram depicting the electron-nucleus interaction, mediated by the Higgs boson, is shown in Fig. 2.1c.

It is important to note that direct attempts to measure the strength of coupling of the Higgs to electrons, protons, and neutrons at the Large Hadron Collider (LHC) have not yet succeeded. Here, measurement of the braching ratio for decays of Higgs bosons to e, p, or n would determine these coupling strengths, but so far these decays have not been observed because the couplings are so weak [(Delaunay, Ozeri, Perez, and Soreq, 2017)]. Instead, the LHC experiments set an upper limit on these branching ratios, and hence limits on the couplings we are discussing. So far, the LHC can say that $g_e \lesssim 10^{-3}$ and $g_{p,n} \lesssim 10^{-1}$. So, if you could detect a joint coupling strength $g_e g_{p,n} \lesssim 10^{-4}$, you would be improving the current state of the art in knowledge of the Higgs boson's properties. This is a nontrivial statement: it is still not entirely clear that "Higgs particle" seen by LHC is really the same as the Higgs boson predicted within the Standard Model, and measuring branching ratios for its decay to different particles is widely recognized as one of the best ways to distinguish it from other, similar particles.

With this in mind, let's work out the actual interaction between electrons and a nucleus (with atomic number Z and mass number A) due to the Higgs, and its effect on atomic energies. Following our analogy to the EM field, we can consider this as a perturbation to the usual atomic structure, given by the Yukawa potential $V_H(\vec{r}) = c g_{ZA} g_e \frac{e^{-\mu r}}{r}$, for large μ (hence short range). Here, $g_{ZA} = Z g_p + (A - Z) g_n \approx A g_p$ is the coherent coupling to all the nucleons in the nucleus—the total effective charge of the nucleus. By direct analogy with our earlier calculation, this shifts atomic energy levels by the amount $\delta dE = c \frac{g_{ZA} g_e}{M_H^2 c^2} |\psi(0)|^2$.

This means that to understand the energy shifts due to the Higgs exchange, we will need to know the atomic wavefunction density at the nucleus. For hydrogenic atoms/ions, as we said earlier, $|\psi_{n\ell m}^{(Z)}(0)|^2 = \frac{4Z^3}{4\pi n^3 a_0^3}$. For neutral atoms and singly charged ions, of the type typically used in high-precision experiments, this formula cannot be valid for the valence electron. That is because here, unlike in a hydrogenic atom, the outer electrons are shielded from the Coulomb field of the nucleus by all the closed electron shells. Nevertheless, there is a simple formula, worked out decades ago, that accurately describes the wavefunctions of valence electrons in these systems in the vicinity of the nucleus. I will not go through the (somewhat lengthy) derivation of this formula here; it can be found in several textbooks [(Budker, Kimball, and DeMille, 2004; Landau and Lifshitz, 1977)]. Instead, I will just state the result: for a valence electron in a ground-state s orbital of a neutral or singly ionized atom with atomic number Z, the probability density near the nucleus, $|\psi_s(0)|^2$, is given by

$$|\psi_s(0)|^2 \sim \frac{1}{4\pi} \frac{Z}{a_0^3}. \tag{2.15}$$

That is: the electron density near the nucleus grows linearly with atomic number Z for these more widely used cases, rather than as Z^3 for the hydrogenic systems. Note that in this formula, I have left out various numerical factors of order unity;

these are not important for our purposes, but are worked out in detail in the cited references.

With this relation, we can estimate the size of energy shifts due to the Higgs, for an electron in a neutral atom with charge Z. This shift, δE_{Higgs}, is given by

$$\delta E_{\text{Higgs}} \sim \frac{c}{4\pi} \frac{g_Z A g_e}{M_H^2 c^2} \frac{Z}{a_0^3} \sim \frac{c}{4\pi} \frac{A g_p g_e}{M_H^2 c^2} Z(\alpha m_e c)^3 = \frac{1}{4\pi} A g_p g_e \alpha \frac{m_e^2}{M_H^2} Z(\alpha^2 m_e c^2)$$

$$\approx \frac{1}{10} A \cdot (1 \times 10^{-3}) \cdot (2 \times 10^{-6}) \cdot \frac{1}{137} \cdot \left(\frac{0.5 \times 10^{-3}}{125} \right)^2 \cdot Z \cdot \frac{e^2}{a_0} \qquad (2.16)$$

$$\approx ZA \cdot 3 \times 10^{-23} \cdot \frac{e^2}{a_0} \approx ZA \cdot 2\pi \times 0.1 \ \mu\text{Hz}.$$

Here we used the relation $\alpha^2 m_e c^2 = e^2/a_0$, the atomic unit of energy. For hydrogen, $Z = A = 1$, and this is effect is so small that it is hopeless to observe: there are uncertainties in theoretically calculated energies at the level $\sim 2\pi \times 10$ kHz, which is $\gtrsim 10^{10}$ times larger than the shift due to the Higgs! However, consider the case of Yb atoms, where $Z = 70$ and $A \approx 170$. Here, the energy shift due to Higgs is much larger: $\delta E_{\text{Higgs}} \approx 2\pi \times 10$ mHz. This value is similar to the absolute energy stability of modern optical atomic clocks, some of which are based on Yb^+ or Yb. So, this sounds promising!

Can we detect the Higgs this way? Unfortunately, the answer is again no. The issue is that we need something to compare to. Unfortunately, atomic theory for complicated atoms like Yb cannot come close to predicting energies with this type of accuracy. The best available calculations have uncertainties on the level of $\sim.01$ eV $\sim 2\pi \times 1$ THz—again, $\gtrsim 10^{10}$ times larger than the size of the expected effect due to the Higgs.

Let us then consider a different way to potentially detect the Higgs electron-nucleus coupling in atoms. The idea is to use a property that distinguishes the Higgs coupling from the EM coupling. In particular, adding neutrons to the nucleus has relatively little effect on the atomic energies (we will discuss these small effects shortly), but a relatively large effect on Higgs coupling strength. In particular: suppose you measure the transition between two energy levels, but in two isotopes of the same atom, with neutron numbers N and N', with (say) $N - N' = 8$ (which is about the largest difference you can find with stable isotopes). Let's say you do this for a $s - d$ transition (e.g., the narrow quadrupole transition used in optical clocks based on Yb^+). Here, the Higgs energy shift only applies to the lower level of the transition—it vanishes for the d state, since the d electron never touches the nucleus. The difference in energy shifts between the two isotopes will be $\delta E_{\text{Higgs}}(A') - \delta E_{\text{Higgs}}(A) \approx \frac{N'-N}{A} E_{\text{Higgs}}(A) \approx 2\pi \times 0.1$ mHz. So, by using this approach you require somewhat higher sensitivity, but it seems plausible that you could make progress on the question of the coupling of the Higgs to the particles that make up ordinary matter.

However, changing the neutron number also changes the energy, just because of EM interactions. This is called an isotope shift. There are two fundamentally different types of effects. The first is known as the "mass shift", which comes from changing the reduced mass of the electron in the atom with a given nucleus. The mass shift itself has two parts. The first is from the simple change you would expect, related to the electron reduced mass $\mu_A = m_e \frac{1}{1+m_e/m_A}$. Since atomic energies are all proportional to the reduced mass, the fractional difference in atomic energies, between two isotopes, is

$$\frac{E_A - E_{A'}}{E_A} = \frac{\mu_A - \mu_{A'}}{\mu_A} = 1 - \frac{1+m_e/m_A}{1+m_e/m_{A'}} \approx \left(1 + \frac{1}{A}\frac{m_e}{m_p}\right)\left(1 - \frac{1}{A'}\frac{m_e}{m_p}\right)$$

$$\approx 1 + \frac{A'-A}{AA'}\frac{m_e}{m_p} \approx 1 + 1\times 10^{-7}. \tag{2.17}$$

So, the absolute difference in energy in energy due to the change in reduced mass has typical value $\Delta E_A \approx 10^{-7} E_A \approx 3 \times 10^{-7}$ eV ≈ 100 MHz.

Note that these relations assume that the masses are strictly proportional to A, but they are not. You need to know the masses exactly for this method. Fortunately, the relevant masses have typically been measured to within 1 part in 10^{10}. So, this fact alone would only give an uncertainty in the normal mass shift part of isotope shift that is comparable to the expected shift due to the Higgs coupling.

The other part of the mass shift, called the "specific mass shift", is more problematic. This is again related to the reduced mass, but is a more complicated effect that arises in many-electron atoms [(King, 1984)]. That is: the center of mass of any individual electron and the nucleus is no longer the center of mass of the whole system. When center of mass motion of the whole system is removed from the Hamiltonian, there is a nontrivial term due to electron-electron correlations, which is strictly proportional to the reduced mass—but hard to calculate in terms of known or accurately measurable quantities. Uncertainties in the specific mass shift, except for the very simplest atoms, are typically $\gtrsim 2\pi \times 1$ MHz.

In addition, there is another contribution to the isotope shift, known as the field shift. This arises because the volume of the nucleus—and hence its charge distribution—changes when the neutron number changes. This is especially large in heavy atoms (which are of most interest both because the Higgs effect is largest, and because modern clocks based on heavy atoms are the most accurate in use today). Typical values of the field shift are $\sim 2\pi \times 1$ GHz for heavy atoms like Yb. Calculating this effect requires knowledge both of the electron wavefunction amplitude at the nucleus, and the change in nuclear size. Both are known only at the 1% level. So, again, this leads to uncertainties at the level of $\sim 2\pi \times 1$ MHz in the size of the expected shift. Things again look hopeless: uncertainties in isotope shifts are ~ 10 orders of magnitude larger than the expected size of the shift due to Higgs coupling.

So far, it is not clear that there is a way around these problems that will allow AMO experiments to make a significant contribution to the understanding of how the Higgs boson couples to the particles that make up ordinary matter. This is disappointing, but perhaps a good idea can still be found for how to take advantage of this effect. It also shows a general principle: you must be quite clever to devise an AMO experiment that is sensitive to physics happening at the 100 GeV mass scale! However, as we will discuss next, such clever ideas do exist, and are being actively pursued in many labs today.

2.6 Effects in atoms from a generalized Higgs boson sector

In this section, we will consider the effect of "extra" Higgs bosons, in addition to the one present in the Standard Model. Many well-motivated particle theory models include additional Higgs bosons. For example, all models that incorporate supersymmetry (SUSY) have this property. SUSY models have been of tremendous interest for decades because they can repair many of the known deficiencies of the Standard Model. Because of its relevance to this section, a brief aside on SUSY is in order.

2.6.1 A digression on supersymmetry

SUSY is a symmetry in which, for each known particle that is a boson (fermion), there exists a partner particle that is a fermion (boson). Since these partner particles have not been observed, SUSY is said to be a broken symmetry at low energy. However, this symmetry is restored at high energies, if the partner particles have masses large enough that they have so far escaped detection. If the typical SUSY partner mass, M_{SUSY}, is around the scale of electroweak symmetry breaking (i.e., near the mass of the Higgs, $W^{\pm}$, and Z^0 bosons, which are all around 100 GeV), many known problems with the Standard Model can be solved. For example, the Standard Model has no particle with the properties of dark matter, and no mechanism to produce the observed imbalance between matter and antimatter in the aftermath of the Big Bang. By contrast, SUSY models broken at the electroweak scale give plausible explanations for both these observations. Electroweak-scale SUSY also solves certain theoretical issues related to extending the Standard Model, such as difficulty explaining how the Higgs mass can be stabilized at its observed value, and in understanding how the strong, electromagnetic, and weak forces can be unified at some high energy scale. For all these reasons, weak-scale SUSY appears to be a very attractive model—and one that is currently being probed both at the LHC, and in AMO experiments [(Feng, 2013; Pospelov and Ritz, 2005; Engel, Ramsey-Musolf and van Kolck, 2013; Barr, 1993)].

2.6.2 A new effect from Higgs exchange: violation of time-reversal and parity

Now we return to the question at hand: what is the effect of additional Higgs bosons in AMO experiments? In models where there is more than one Higgs boson, the effective couplings of the Higgs to ordinary particles are more complex than the simple scalar couplings we discussed in the last section. In particular, mixing between the different Higgs fields typically has the effect of inducing, in addition to the usual scalar coupling terms we just described, a pseudo-scalar coupling to the electron for to least one of the new Higgs particles [(Barr and Zee, 1990)].

This new, additional coupling gives a new effective term in the Hamiltonian, which we write as

$$H_{\text{H}'} = i \sin\phi_{\text{CP}} \cdot 4\pi c \frac{Ag_p g_e}{m_{\text{H}'}^2 c^2} \frac{1}{m_e c} \left\{ \vec{s} \cdot \vec{p}\delta^3(\vec{r}) + \delta^3(\vec{r})\vec{s} \cdot \vec{p} \right\}. \tag{2.18}$$

This is similar to the effective Hamiltonian due to Higgs exchange with scalar couplings, but with a few key differences. The first is that, due to the pseudoscalar coupling to the electron, we have replaced the number g_e that described the scalar effective charge with the more complicated object, $g_e(\vec{s}\cdot\vec{p})$, that is needed to describe a pseudoscalar coupling. (That this appears in an anticommutator with the δ-function is just a detail—this symmetrization simply ensures that the Hamiltonian is Hermitian.) In addition, to account for the possibly different mass of the new Higgs particle compared to the known Higgs, we have replaced M_H with $M_{H'}$. Finally, we have introduced an overall factor $i\sin\phi_{CP}$. The magnitude of this term accounts for the strength of the psuedoscalar coupling of the Higgs to the electron, relative to the scalar coupling. The Feynman diagram depicting the associated electron-nucleus interaction is shown in Fig. 2.1d.

It is crucial that this overall factor is a pure imaginary number. This has a deep physical consequence: it means that the Hamiltonian $H_{H'}$ *violates time-reversal invariance*, i.e., $T^{-1}H_{H'}T = -H_{H'}$, where T is the time-reversal operator. As you may recall, T is an antiunitary operator: when acting on other operators $f(\vec{r},\vec{s},\vec{p},t)$, it has the effect $Tf(\vec{r},\vec{s},\vec{p},t) = f^*(\vec{r},-\vec{s},-\vec{p},-t)$. By inspection, the factor of i in $H_{H'}$ causes it to change sign under T. In general, in a quantum field theory *any* irreducible complex number is associated with T-violation. Since any complex number can be described as a real amplitude times a complex exponential $e^{i\phi}$, $\sin\phi$ encodes the strength of the T-violation in a given interaction, relative to the T-conserving part.

In addition to violating time-reversal symmetry, the Hamiltonian $H_{H'}$ also *violates spatial inversion symmetry*. That is, $P^{-1}H_{H'}P = -H_{H'}$, where P is the parity operator. You can see this from the parity transformation of the spin operater $\vec{s}$ (an axial vector, even under P) and the momentum operator $\vec{p}$ (a polar vector, odd under P). This means that, in the presence of this new interaction, parity is no longer a good quantum number in atoms. Instead, the energy eigenstates will be mixtures of s and p states. (States with higher ℓ are not affected, since they do not overlap with the nucleus.) Note that this also means that there is no first-order energy shift from $H_{H'}$: it has no diagonal matrix elements between eigenstates of the ordinary EM Hamiltonian, i.e., the Coulomb, spin-orbit, hyperfine, etc. interactions that we usually think of as determining atomic structure.

2.6.3 A digression on *T*- and *CP*-violation

It has long been known that all Lorentz-invariant quantum field theories obey the combined symmetry CPT, where C is charge conjugation (particle↔antiparticle) [(Streater and Wightman, 2000)]. Hence, any violation of T must come along with an equal and opposite CP-violation, so that CPT is conserved—that is, T- and CP-violation are equivalent in quantum field theory. For historical reasons, most particle theorists refer to phenomena that violate both T and CP simply as CP-violation. We adopt this notation, and denote the phase that encodes the strength of CP- (and hence also T-) violation as ϕ_{CP}.

It has been known since the 1950s that the weak interaction violates parity—that is, the strength of processes mediated by the weak interaction is different in right-versus left-handed systems. This difference can be at the level of 100%: for example,

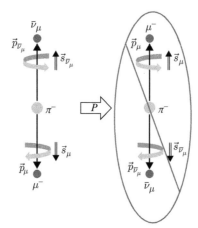

Fig. 2.2 Decay of a negative pion, π^-. The decay into a right-handed negative muon, μ^- and a right-handed muon antineutrino, $\bar{\nu}_\mu$, shown on the left, is allowed. Here, the handedness refers to the relative directions of the momentum $\vec{p}$ and spin $\vec{s}$: if they are aligned as shown here, putting the thumb of your right hand along $\vec{p}$ will make your fingers curl around in the same manner as the rotation associated with the spin. If $\vec{p}$ and $\vec{s}$ are anti-aligned, this can only be done with your left hand. The mirror-image decay into left-handed particles, shown on the right, is forbidden. This means that P symmetry is broken in the decay, which is purely due to the weak interaction.

a negative pion, π^-, can decay only via the weak interaction, resulting (usually) in an negative muon, μ^-, and an antineutrino. For a pion at rest, the muon can only be emitted with its spin $\vec{s}$ along its momentum $\vec{p}$, so that the P-odd quantity $\vec{s} \cdot \vec{p}$ has a net positive value. In other words, the muon produced in this decay has a definite handedness, which clearly violates mirror symmetry (see Fig. 2.2). There is a long history of measuring the effect of the weak interaction in atoms, where the exchange of Z^0 bosons between electrons and the nucleus leads to parity violation.

In the meantime, it has been known since the 1960s that the weak interaction also violates the combined symmetry CP. This has been observed now in many experiments, all of which study the decay of hadronic particles (i.e., particles containing quarks) [(Patrignani and Others, 2016)]. The violation of CP observed in all these experiments is successfully described within the framework of the Standard Model, where a single CP-violating complex phase appears. The process that leads to CP-violation in the Standard Model is intimately linked to processes that change the flavor of quarks: it arises exactly and only due to the quark mixing matrix, known as the CKM matrix. This is why CP-violation so far has been observed only in the flavor-changing decays of certain hadronic mesons. In stable atoms, of course, quarks do not change flavor, so this mechanism is deeply suppressed and negligibly small.[3]

---

[3] It still does occur, but appears only at a high order of perturbation theory. It leads to an electron-nucleus interaction equivalent to the one we are considering here, but with coupling strength many orders of magnitude smaller than what is accessible to present experiments [(Pospelov and Ritz, 2014)].

The study of *CP*-violation turns out to be one of the more compelling questions in particle physics at present. This is because it is known that there *must* be some new, *CP*-violating mechanism beyond the known one in the Standard Model. The reason is very striking. It is an observational fact that the universe today is made almost entirely of matter, with almost no antimatter around. (If significant amounts of antimatter were still present, we would easily see annihilation gamma rays from where it collides with matter.) Yet, as the energy of the Big Bang converted to massive particles, there should have been equal numbers of matter and antimatter particles created. Somehow, since then, the matter "won" in the sense that some matter survives, while essentially no antimatter does. The formation of this matter-antimatter imbalance is known as baryogenesis [(Trodden, 1999)].

In the 1960s, Sakharov described three minimum properties of Nature which are required for baryogenesis to occur. These are: 1. Baryon number violation; 2. Non-equilibrium evolution; and 3. *CP* (and *C*) violation. Remarkably, all of these are present in the Standard Model, when at high temperature and evolving in the time after the Big Bang. However, the actual amount of *CP* violation in the Standard Model is far too small to explain the observed degree of matter-antimatter asymmetry. This is usually described by the baryon-to-photon ratio, $\eta_B \sim 6 \times 10^{-10}$, which roughly quantifies the fraction of matter particles which survived annihilation with antimatter in the aftermath of the Big Bang. This is a very tiny excess, but still too large to be explained by the *CP*-violation in the Standard Model. It turns out that this value of η_B *could* rather naturally be explained, if there are new particles with mass around the electroweak scale (i.e., similar to masses of the *Z*, *W*, and *H* particles), whose interactions involve near-maximally large *CP*-violating phases [(Trodden, 1999)].

In addition, the manner in which *CP*-violation enters the Standard Model is quite peculiar. In almost every theoretical model that adds new particles to the Standard Model, the associated new degrees of freedom allow new *CP*-violating phases to appear in the theory. Since we know from experiment that *CP* is not a good symmetry of nature, there is no good reason why such new phases should be zero. In fact, the single *CP*-violating phase in the Standard Model has a magnitude of about 1 radian [(Patrignani and Others, 2016)]. (It is only because it participates solely in quark flavor-changing processes that its observable effect is usually small.)

So, in summary: the idea that there should be new sources of *CP*-violation, associated with particles beyond those in the Standard Model, seems both quite natural, and necessary [(Ibrahim and Nath, 2008)]. This gives strong motivation to our discussion of possible *CP*-violating phenomena in atoms and molecules. In practice, all of the *CP*-violating effects that are observable in atoms have the property of violating both *T* and *P*. So, we refer to these effects going forward as *T*, *P*-violating effects.

2.6.4 Atomic state mixing due to the *T*, *P*-violating interaction

The natural size of the *T*, *P*- violating effect due to Higgs exchange described above can be readily estimated. The calculations are similar to those we did before for the simpler scalar couplings due to Higgs exchange from a single, Standard Model Higgs field. Since now, with a psuedoscalar coupling, the spin $\vec{s}$ is involved, we should work in a basis of spin-orbit states (with quantum numbers n, ℓ, j, m_j). Because $H_{H'}$ (like any Hamiltonian) is a scalar operator under rotations,[4] it only couples states with the same values of j and m_j. Let's consider alkali atoms to simplify the discussion. Then, the ground $s_{1/2}$ state gets mixed with $p_{1/2}$ states due to the presence of $H_{H'}$. We would like to estimate the size of the $s_{1/2} - p_{1/2}$ mixing amplitude. To do this, we will again need to know how the wavefunction of the valence electron behaves near the nucleus (the only place where $H_{H'}$ gives a substantial contribution, due to the short range of the Yukawa potential from the presumably heavy H' particle). We gave an expression for the square of the wavefunction, near the origin, of an atomic s-orbital earlier; now we also need the wavefunction of the p-orbital. For convenience, we write both these radial wavefunctions in the limit $r \to 0$ here:

$$R_s(r) \sim Z^{1/2} \frac{1}{a_0^{3/2}}$$

$$R_p(r) \sim Z^{1/2} \frac{1}{a_0^{3/2}} \frac{r}{(a_0/Z)} \sim Z^{3/2} \frac{1}{a_0^{5/2}} r. \tag{2.19}$$

In our expression for the matrix element of $H_{H'}$, we will need to operate on the p-state wavefunction with the momentum operator $p \sim -i \, d/dr$, again when $r \to 0$; this yields

$$p R_p(r) \sim -i \frac{\partial}{\partial r} R_p(r) \sim i Z^{3/2} \frac{1}{a_0^{5/2}}. \tag{2.20}$$

Now we can evaluate the matrix element of $\langle n p_{1/2} | H_{H'} | n_{1/2} s \rangle$ and the associated $s_{1/2} - p_{1/2}$ mixing coefficient, $\eta_{H'}$, from first-order perturbation theory. Here, the first term in the curly brackets of Eq. 2.18 gives a nonzero result: the momentum operator acts to the left on the p state wavefunction and the δ-function acts to the right on the s state wavefunction. Notice that the factor of i in $H_{H'}$ is cancelled by the factor of i from the momentum operator, so that the matrix element is a real number. The matrix element then has magnitude

$$\langle n p_{1/2} | H_{H'} | n_{1/2} s \rangle \sim c \sin \phi_{CP} \frac{A g_p g_e}{m_{H'}^2 c^2} \frac{1}{m_e c} \left[\langle p_{1/2} | \vec{s} \cdot \vec{p} \, \right] \left[\delta^3 (\vec{r}) | s_{1/2} \rangle \right]$$

$$\sim c \sin \phi_{CP} \frac{A g_p g_e}{m_{H'}^2 c^2} \frac{1}{m_e c} \times \frac{Z^2}{a_0^4}. \tag{2.21}$$

From ordinary first-order perturbation theory, the mixing coefficient $\eta_{H'}$ is given by

$$\eta_{H'} = \frac{\langle n p_{1/2} | H_{H'} | n_{1/2} s \rangle}{E_s - E_p}. \tag{2.22}$$

---

[4] This is meant in the sense of the Wigner-Eckart theorem.

The energy splitting $E_s - E_p$ has magnitude $E_s - E_p \sim e^2/a_0 = \alpha^2 m_e c^2$. Thus, the mixing coefficient is

$$
\eta_{H'} \sim c \sin\phi_{CP} \frac{A g_p g_e}{m_{H'}^2 c^2} \frac{1}{m_e c} \times \frac{Z^2}{a_0^4} \frac{1}{\alpha^2 m_e c^2}
$$

$$
\sim A Z^2 \sin\phi_{CP} g_p g_e \frac{1}{m_{H'}^2 c^2} \frac{1}{\alpha^2 m_e^2 c^3} (\alpha m_e c)^4 \sim Z^3 \alpha^2 \sin\phi_{CP} g_p g_e \frac{m_e^2}{m_{H'}^2}, \qquad (2.23)
$$

where we used $A \sim Z$ to write a simplified expression just in terms of Z. Numerically, assuming $m_{H'} \sim m_H$, using the Standard Model value $g_p g_e \approx 2 \times 10^{-9}$ mentioned above, and assuming a maximal CP-violating phase such that $\sin\phi_{CP} = 1$, you can work out that for $Z = 80$ and $A = 200$, $\eta_{H'} \sim 10^{-17}$. That seems like a very small mixing, but—as we will see—it is actually possible to detect it!

As an aside, note that the results of experiments looking for this effect are usually cast in terms of a dimensionless quantity called C_S; in our notation, up to some factors of order unity,

$$
C_S \sim \sin\phi_{CP} \frac{g_p g_e}{\alpha} \frac{m_W^2}{m_{H'}^2}, \qquad (2.24)
$$

where m_W is the mass of the W boson, which is a convenient normalizing scale for new particles anticipated to arise near the electroweak scale.

2.6.5 Measurable effect of T, P-violation: atomic electric dipole moment

Next we explore how to look for an effect of this type, experimentally. It turns out that this mixing between s and p states leads to a linear energy shift when you apply an electric field $\vec{\mathcal{E}}$ to the atom. (This is in contrast to the usual Stark effect, which leads only to shifts that are quadratic in $\vec{\mathcal{E}}$.) To see how the linear shift arises, we will work in a toy model that accounts only for the lowest two states of the atom, which we write as $s_{1/2}$ and $p_{1/2}$. In the presence of a pseudoscalar-scalar Higgs coupling, the $s_{1/2}$ ground state—in its spin-up sublevel, with $m = +1/2$—will mix with the $p_{1/2}$ state. The perturbed state can be written as

$$
|\psi\rangle = |s_{1/2}, m = +1/2\rangle + \eta_{H'} |p_{1/2}, m = +1/2\rangle. \qquad (2.25)
$$

Including an electric field $\vec{\mathcal{E}} = \mathcal{E}\hat{z}$ adds the usual Stark term to the atomic Hamiltonian,

$$
H_{St} = e\mathcal{E}z. \qquad (2.26)
$$

To see that the combined effect of the Stark and T, P-violating Higgs interactions gives rise to an energy shift ΔE that is linear in $\mathcal{E}$, we compute the energy shift to first order in perturbation theory, $\Delta E^{(1)}$:

$$\Delta E^{(1)} = \langle \psi | H_{\mathrm{St}} | \psi \rangle$$

$$= e\mathcal{E} \left(\left(\left\langle s_{1/2}, m = \frac{1}{2} \right| + \eta_{\mathrm{H}'} \left\langle p_{1/2}, m = \frac{1}{2} \right| \right) z \left(\left| s_{1/2}, m = \frac{1}{2} \right\rangle + \eta_{\mathrm{H}'} \left| p_{1/2}, m = \frac{1}{2} \right\rangle \right) \right)$$

$$= e\mathcal{E} \left(\eta_{\mathrm{H}'} \left\langle s_{1/2}, m = \frac{1}{2} \right| z \left| p_{1/2}, m = \frac{1}{2} \right\rangle + \eta_{\mathrm{H}'} \left\langle p_{1/2}, m = \frac{1}{2} \right| z \left| s_{1/2}, m = \frac{1}{2} \right\rangle \right)$$

$$= 2\eta_{\mathrm{H}'} e\mathcal{E} \left\langle s_{1/2}, m = \frac{1}{2} \right| z \left| p_{1/2}, m = \frac{1}{2} \right\rangle = 2\eta_{\mathrm{H}'} e\mathcal{E} \langle z \rangle_{sp}. \tag{2.27}$$

Next, let's see what happens to the $m = -1/2$ states. To do so, we use the Wigner-Eckhart theorem, taking into account that the operator z is a rank-1 tensor with projection $q = 0$, and write the two matrix elements of interest in terms of the reduced matrix element $\langle s_{1/2} \| z \| p_{1/2} \rangle$. We see that

$$\left\langle s_{1/2}, m = \frac{1}{2} \right| z \left| p_{1/2}, m = \frac{1}{2} \right\rangle = \left\langle \frac{1}{2}, \frac{1}{2}; 1, 0 \right| \left. \frac{1}{2}, \frac{1}{2} \right\rangle \langle s_{1/2} \| z \| p_{1/2} \rangle$$

$$= -\sqrt{\frac{1}{3}} \langle s_{1/2} \| z \| p_{1/2} \rangle \tag{2.28}$$

and

$$\left\langle s_{1/2}, m = -\frac{1}{2} \right| z \left| p_{1/2}, m = -\frac{1}{2} \right\rangle = \left\langle \frac{1}{2}, -\frac{1}{2}; 1, 0 \right| \left. \frac{1}{2}, -\frac{1}{2} \right\rangle \langle s_{1/2} \| z \| p_{1/2} \rangle$$

$$= +\sqrt{\frac{1}{3}} \langle s_{1/2} \| z \| p_{1/2} \rangle. \tag{2.29}$$

This shows that the energy shift changes sign with the relative direction of the angular momentum $\vec{j}$ and the applied electric field $\vec{\mathcal{E}}$. In other words: the T, P-violating Higgs interaction causes the atom to interact with the $\mathcal{E}$-field in a manner equivalent to an effective Hamiltonian, H_{eff}, with form

$$H_{\mathrm{eff}} = 4\eta_{\mathrm{H}'} e \langle z \rangle_{sp} \vec{\mathcal{E}} \cdot \vec{j} = -\vec{D} \cdot \vec{\mathcal{E}}, \tag{2.30}$$

where $\vec{D} \equiv 4\eta_{\mathrm{H}'} e \langle z \rangle_{sp} \vec{j}$. This means that, due to the T, P-violating Higgs interaction, the atomic state has acquired a permanent electric dipole moment (EDM) $\vec{D} = D\vec{j}$ along its angular momentum (see Fig. 2.3).

Hence, upon application of the $\mathcal{E}$-field, there is an energy shift between spin up and spin down, much like a Zeeman shift. This effective EDM Hamiltonian clearly violates both parity and time-reversal invariance, since the axial vector $\vec{j}$ is odd under T and even under P, while the polar vector $\vec{\mathcal{E}}$ is even under T and odd under P (see Fig. 2.4).

Now, let's estimate the size of the energy shifts ΔE due to the atomic EDM. If $\langle z \rangle_{sp} \approx a_0$ (a typical value), then $\Delta E = 4\eta_{\mathrm{H}'} e \langle z \rangle_{sp} \mathcal{E} \approx 10^{-16} e a_0 \mathcal{E} \approx 2\pi \times 10^{-10} \mathrm{Hz} \cdot$

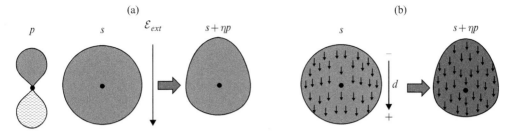

Fig. 2.3 Induced dipole moments in an atom. (a) Polarization of an atom by application of an external $\mathcal{E}$-field. The $\mathcal{E}$-field mixes s and p states, whose wavefunctions are depicted on the left. The different shading in the lower part of the p-state wavefunction indicates a negative wavefunction amplitude. Mixing of the states leads to destructive interference below the nucleus, and constructive interference above it. That shifts the electron probability distribution to above the nucleus, and induces a net dipole $\vec{D}$ in the atom. (b) Polarization of an atom due to an electron EDM. In the $s_{1/2}$ orbital shown here, the electron spin is oriented uniformly throughout the atom. If the spin carries an electric dipole moment $\vec{d}$ as shown, the parts of the wavfunction above the positively charged nucleus will be repelled from it, and the parts below will be attracted. That shifts the electron probability distribution to above the nucleus, and induces a net dipole $\vec{D}$ in the atom. An entirely analogous polarization occurs when the $s_{1/2}$ and $p_{1/2}$ states are mixed due to the T, P-odd electron-nucleon interaction.

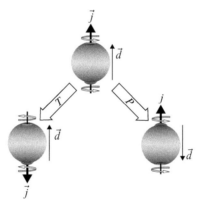

Fig. 2.4 Discrete symmetry-violating properties of an electric dipole moment. Consider a particle with electric dipole moment $\vec{d}$ in the same direction as its angular momentum $\vec{j}$. The time-reversed version of this particle (obtained by operating with T) and the mirror image of the particle (obtained by operating with P) both have $\vec{d}$ antiparallel to $\vec{s}$. Hence, if all particles of a certain type (atoms, electrons, nuclei, etc.) have $\vec{d}$ in the same direction as $\vec{j}$, then the time-reversed and mirror-image versions of the particle do not exist in nature. Then, we would say that both T and P symmetries are violated.

$\mathcal{E}/(V/\text{cm})$. The largest electric fields available in the lab are $\mathcal{E} \sim 100$ kV/cm—above this, electrodes will arc even in a good vacuum. A field of this size gives an energy shift of $\Delta E \approx 2\pi \times 10$ μHz. Remarkably, atomic EDM experiments for atoms with unpaired electrons are able to see shifts of around this size. There is a key difference between the situation here and the case of the T, P-even scalar Higgs interaction:

while for the scalar interaction it was necessary to compare an absolute energy to a calculated value in the absence of the Higgs, here one is looking for a linear Stark shift, which must identically vanish in the absence of T, P-violation. This means that absolute calibrations and ultra-precise calculations are not needed; rather one needs only experimental sensitivity to a small, controllable energy shift.

Experiments with this level of sensitivity have already been able to probe realistic predictions of T, P-violation due to "extra" Higgs boson exchange as described above [(Nakai and Reece, 2017; Jung and Pich, 2014)]. In principle there are other types of particles and interactions that also lead to an equivalent scalar-pseudoscalar nucleon-electron interactions, such as exchange of hypothetical particles known as leptoquarks [(Barr, 1992)]. However, the multi-Higgs scenario is particularly well-motivated: additional Higgs bosons are generally seen as one of the most simple and natural extensions to Standard Model, and are explicitly required in all SUSY models [(Jung and Pich, 2014)]. Hence, experiments looking for EDMs along the angular momentum of atoms are having a significant impact on attempts to build theories that extend the current Standard Model.

2.6.6   How to measure an atomic EDM

Next, we outline the basic idea of experiments to search for an atomic EDM. Conceptually, the following sequence of events is used. (Here, as throughout, we will not be careful about factors of 2.)

1. Polarize the angular momentum $\vec{j}$ in the ground state. This is typically done via optical pumping—for example, by applying circularly polarized light on the $s_{1/2} \to p_{1/2}$ transition.
2. (a) Apply an external electric field $\vec{\mathcal{E}}$ perpendicular to $\vec{j}$. This defines a quantization axis $\hat{z}$, so that the polarized state is in a superposition of $\pm m_j$ states. The EDM, if it exists, causes a relative energy shift between these states of $\hbar\omega_{\mathcal{E}} = D\mathcal{E}$. The energy shift causes the state to evolve in time; the evolution is equivalent to a precession of the expectation value of the angular momentum, $\langle\vec{j}\rangle$, around $\vec{\mathcal{E}}$, at frequency $\omega_{\mathcal{E}} = D\mathcal{E}$, for an atom with EDM $\vec{D} = D\vec{J}$.
   (b) (Optional—but often useful for technical reasons) Apply a magnetic field $\vec{B}$ parallel to $\vec{\mathcal{E}}$. This provides a bias energy shift due to the Zeeman effect (or equivalently, a bias Larmor precession), with associated frequency $\omega_B = \mu B$, where μ is the magnetic moment of the atomic state.
3. Wait for time T. This results in a precession angle $\phi = \omega T = D\mathcal{E}T + \mu BT$. Evidently, a larger value of T will make this effect easier to measure. Equivalently, if you think of this an an energy measurement, detection after evolution time T allows frequency resolution $\Delta\omega = 1/T$ for a measurement on any individual particle.
4. Detect either the spin precession angle $\phi = (\omega_{\mathcal{E}} + \omega_B)T$ or, equivalently, the energy splitting $\Delta E = \hbar(\omega_{\mathcal{E}} + \omega_B)$. The energy splitting can be measured, for example, via a standard Ramsey measurement protocol. The final spin direction can be measured, for example, by applying circularly polarized light along the direction of the original optical pumping beam (perpendicular to $\vec{\mathcal{E}}$) and detecting the

induced fluorescence: it will be zero if $\vec{j}$ is oriented in the original direction, and maximal if $\vec{j}$ is aligned in the opposite direction.

5. Reverse the direction of $\vec{\mathcal{E}}$, and repeat the previous steps. Then, take the difference between the two measurements to extract the purely EDM-induced precession angle or EDM-induced energy shift.

6. Naturally, for better statistical sensitivity you would like to do the same measurement not just on a single atom, but on a large ensemble of N independent atoms. If you think of this as a measurement of the precession angle ϕ, averaging over the N independent atoms will allow a measurement with uncertainty $\delta\phi = 1/\sqrt{N}$, and hence you could detect a minimum EDM $\delta D = \delta\phi/(\mathcal{E}T) = 1/(\mathcal{E}T\sqrt{N})$. If you prefer to think of this as a frequency measurement, then the minimum detectable frequency shift will be $\delta\omega = \Delta\omega/\sqrt{N} = 1/(\sqrt{N}T)$, leading to the same conclusion about the minimum detectable EDM.

A classic example of this type of experiment was developed over nearly two decades in the lab of Eugene Commins at Berkeley, culminating in a 2002 result that stood as the best in the field until recently [(Regan, Commins, Schmidt, and DeMille, 2002)]. This experiment used Tl atoms ($Z = 81$) in a thermal atomic beam, ~ 1 meter long. The brightness of the atomic source was $B_{at} \sim 3 \times 10^{17}/\text{sr/s}$. With a thermal velocity $v \sim 400$ m/s, the interaction time per atom was $T \sim 2.5$ ms. Laser excitation on the $6p_{1/2} - 7s_{1/2}$ transition was used to spin-align the atoms, which were then subjected to parallel $\vec{\mathcal{E}}$- and $\vec{\mathcal{B}}$-fields. The applied electric field had magnitude $\mathcal{E} \approx 120$ kV/cm. Following a classic Ramsey measurement sequence that transformed the energy splitting between spin up and spin down into the relative population of one spin state, the final state was read out using laser-induced fluorescence. The small electrode spacing required to obtain a large $\mathcal{E}$-field limited the size of the atomic beam. So, fluorescence was collected from only a small area of ~ 0.025 cm^2 of the beam at the end of the interaction region. Over the course of ~ 35 hours of running, a total of $N \sim 10^{14}$ atoms were detected. This led to a final statistical uncertainty in the EDM-induced frequency shift of about $\delta\omega \approx 2\pi \times 10$ μHz.

Since the size of the effect being sought is so small, it is worth talking a little about systematic errors in EDM measurements of this type. One typical fear is errors due to leakage currents. That is: when you apply a large $\mathcal{E}$-field, currents will inevitably flow along the insulating spacers that support the electrodes. These currents create a magnetic field $\mathcal{B}_{\text{leak}}$ that can lead to Zeeman shifts (or equivalently, Larmor precession) that reverse along with the direction of the applied $\vec{\mathcal{E}}$-field. In turn, $\mathcal{B}_{\text{leak}}$ couples to the magnetic moment of the atom—which is $\sim 10^{17}$ times larger than the EDM! Remarkably, it turns out that with substantial effort and care, the leakage currents can be controlled at a sufficient level to keep this from becoming a dominant error.

Another common systematic error in EDM experiments can arise from what is known as a motional magnetic field. That is: if a particle such as an atom moves with nonrelativistic velocity $\vec{v}$ in an electric field $\vec{\mathcal{E}}$, in its rest frame it experiences a magnetic field $\vec{\mathcal{B}}_{\text{mot}} = \vec{v} \times \vec{\mathcal{E}}/c$. Again, this leads to a magnetic field that reverses along with $\vec{\mathcal{E}}$. It does not give rise to a shift linear in $\mathcal{E}$ if the applied fields $\vec{B}$ and $\vec{\mathcal{E}}$ are exactly parallel (since then $\vec{\mathcal{B}}_{\text{mot}} \perp \vec{B}$ exactly, and the total magnetic field magnitude does not change when $\vec{\mathcal{E}}$ reverses), but can lead to substantial errors if the applied are even slightly misaligned. For example, in the Tl experiment the $\vec{\mathcal{E}}$- and $\vec{\mathcal{B}}$-fields had to be parallel to within about a nanoradian. To account for this effect, the Tl experiment used counter-propagating atomic beams to measure the motional field effect (by looking at the difference in EDM-like signals for the two beams with opposite values of $\vec{v}$) and null it away (by tilting the $\vec{\mathcal{B}}$-field to make it as parallel to $\vec{\mathcal{E}}$ as possible). In the end, it was possible to rule out systematic errors at a level roughly comparable to the statistical uncertainty in the experiment.

The result of the experiment was consistent with a vanishing EDM, within the range of combined statistical and systematic uncertainties—roughly, it was possible to conclude that $\delta\omega \lesssim 2\pi \times 20\ \mu\mathrm{Hz}$. This is quite comparable to the size of the frequency shift we had estimated would arise from an additional Higgs boson, with mass near that of the Standard Model Higgs ($m_H \approx 125\ \mathrm{GeV/c^2}$) and interacting with a maximal CP-violating phase $\phi_{CP} \approx \pi/2$. This means that even fifteen years ago, long before the LHC began operation, this tabletop-scale AMO experiment had ruled out a perfectly viable prediction in many theoretical extensions to the Standard Model of particle physics. As we will discuss, more recent experiments have pushed considerably farther than this, and are increasingly sensitive to new particles with mass beyond the direct reach of the LHC.

2.7   Particle dipole moments as probes of particle physics

So far we have focused on new physics effects associated with the virtual exchange of "new" particles between the electrons and the nucleus in an atom. However, similarly important effects can arise due to virtual exchange of such "new" particles in a loop—for example, due entirely to the coupling of the new particles to electrons, even if these particles do not couple directly to nucleons. These loops of virtual particles can modify the magnetic dipole moments of elementary particles, or cause them to carry a permanent electric dipole moment. We explore these ideas next.

2.7.1   Non-symmetry violating dipole moments: magnetic moment

Let's start with the electron, which is arguably the simplest elementary particle. In particular, we will consider the effect of new particles on its magnetic moment, μ. We write this in the standard form $\vec{\mu} = g\mu_B \vec{s}$, where $\mu_B = \frac{1}{2}\frac{e}{m_e c} = \frac{\alpha}{2}ea_0$ is the Bohr magneton, and g is a dimensionless constant. Any measurement of a Cartesian component of $\vec{\mu}$ gives the answer $\pm g\mu_B/2$, so for brevity let's call this quantity the magnitude of the magnetic moment: that is, we define $\mu \equiv g\mu_B/2$.

For a point particle with $s = 1/2$ as described by the Dirac equation, $g = g_s = 2$ exactly, so $\mu = \mu_B$ exactly in this case. However, as you have surely heard, in quantum electrodynamics (QED) the electron's magnetic moment is modified due to radiative corrections. The simplest of these is described by a Feynman diagram where the electron emits and then reabsorbs a virtual photon, to form a loop (Fig. 2.5a). In the language of field theory that we sketched earlier, the interaction with the photon can be described by a coupling strength e at each vertex, and a propagator

$1/k^2$. Then, the change in the magnetic moment due to this interaction will be $\Delta\mu \propto e^2/k^2 \propto \alpha/k^2$. What is the correct proportionality constant? In the cgs units we are using here, magnetic and electric dipole moments have the same units, which can be expressed as charge*length. To get the dimensionality right, our expression for $\Delta\mu$ should have units of e*length $= e/$momentum. Since k is a momentum, that means we need to multiply by some momentum scale p and by e, i.e., $\Delta\mu \propto e\alpha p/k^2$. Now, the only physical momentum scale in the problem is the inverse Compton wavelength, $m_e c$. So, let's set both momenta, p and k, to that scale. Then we estimate $\Delta\mu \sim e\alpha(m_e c)/(m_e c)^2 \sim \alpha\frac{e}{m_e c} \sim \alpha\mu_B$. In quantum field theory calculations, the amplitudes associated with Feynman diagrams that contain a loop typically are suppressed by a numerical factor of $\sim 1/(2\pi)$, as compared to "tree-level" diagrams without the loop. Hence, we should really expect $\Delta\mu \sim \frac{\alpha}{2\pi}\mu_B$. When calculated with the full formalism of QED, the lowest-order correction described by this simple loop diagram is in fact precisely given by $\Delta\mu = \frac{\alpha}{2\pi}\mu_B$. This can be recast as a correction to the electron's g-factor, g_e, away from the Dirac value $g_s = 2$. The quantity $a_e \equiv \frac{g_e - 2}{2} \approx \frac{\alpha}{2\pi}$ is called the electron magnetic anomaly.

The Feynman diagram with one photon loop corresponds to the 1st order of perturbation theory. Additional loops correspond to higher orders of perturbations, and they also contribute. For example, there can be a loop within loop, or chained loops (Fig. 2.5b). By the same sort of arguments we made above, each can be expected to give a contribution suppressed by another factor of about $\frac{\alpha}{2\pi}$. The photon can also split into an e^+/e^- pair (Fig. 2.5c), which gives a similar contribution (though we haven't talked about fermion propagators, I will ask you to accept this). Overall, the sum of these effects gives a 2nd order correction to the magnetic anomaly, $\Delta a_e^{(2)}$, with magnitude $\Delta a_e^{(2)} \approx \left(\frac{\alpha}{2\pi}\right)^2$. Adding the effect of higher-order corrections, the magnetic anomaly can be written as a series expansion in powers of $\alpha/(2\pi)$: $a_e = C_1\left(\frac{\alpha}{2\pi}\right) + C_2\left(\frac{\alpha}{2\pi}\right)^2 + \cdots$, where $C_1 = 1$ [(Gabrielse, 2013)]. The higher-order coefficients in this power series expansion, C_i, can be determined from first-principles calculations in QED, and has been calculated precisely up to $i = 5$. This theoretical relationship between a_e and α

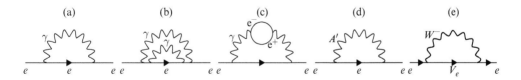

(a) (b) (c) (d) (e)

Fig. 2.5 Feynman loop diagrams contributing to the magnetic anomaly of the electron, a_e. (a) Simplest diagram in QED, describing virtual emission and reabsorption of a photon. (b) 2-loop diagram in QED, with two virtual photons. (c) A different 2-loop diagram in QED, where the virtual photon splits into a virtual electron-positron pair. (d) Potential contribution to a_e from virtual emission and reabsorption of a hidden photon, A'. The size of the contribution depends on the effective charge of the electron and on the hidden photon mass M. (e) Contribution to a_e from a known electroweak process, when the electron virtually splits into a W^- boson and an electron neutrino, ν_e. Although the coupling strength is nearly the same as for the electron-photon coupling shown in (a), the large mass of the W boson suppresses the effect of this diagram to below the present experimental accuracy.

is known with a remarkable accuracy of 0.03 ppb. The electron magnetic anomaly also has been measured with extraordinary accuracy: $\delta a_e^{\mathrm{exp}}/a_e \approx 0.3$ ppb. Comparison of the measured value and the calculated expression gives the most accurate value for α. Alternatively, by using independent measurements of α, it is possible to check for deviations from the predictions of QED. None are seen—they agree at the level of 1 part in 10^{-9}, i.e., 1 ppb.

We can use this information to set some limits on the properties of new particles that might in principle exist. For example, the kind of hidden photon that we discussed before can also appear in a loop that would modify the electron magnetic moment (Fig. 2.5d). If the hidden photon has mass M, the Feynman diagram will have a propagator $1/(k^2 + M^2 c^2)$. Now we can follow the same steps of logic we used to estimate the contribution to the magnetic moment from a photon loop. Again, here, the only momentum scale that defines the electron is $m_e c$, so we can expect $k \sim m_e c$. If the hidden photon couples to the electron with effective charge $e\chi$, where $\chi \ll 1$ is a dimensionless number, its one-loop diagram will give a change in magnetic moment of roughly $\Delta \mu_e \sim e \frac{\alpha}{2\pi} \chi^2 \frac{m_e c}{m_e^2 c^2 + M^2 c^2} \sim \frac{\alpha}{2\pi} \chi^2 \left(\frac{m_e^2}{m_e^2 + M^2} \right) \mu_B$. This is equivalent to a change in the magnetic anomaly of $\delta a_e = \delta \left(\frac{g_e - 2}{2} \right) \sim \frac{\alpha}{2\pi} \chi^2 \frac{m_e^2}{m_e^2 + M^2}$. Hence, for a hidden photon with light mass $M \ll m_e$, we can say $\frac{\alpha}{2\pi} \chi^2 \lesssim 10^{-9} \Rightarrow \chi^2 \lesssim 10^{-6}$. For $M > m_e$, the limits on χ get less strict as M increases. We can alternatively think of the electron a_e measurement as setting a limit on new particles with strong couplings such that $\chi^2 = 1$: their mass M must satisfy $M \gtrsim 150\,\mathrm{MeV}$. This sets limits on hidden photons that are comparable to information from other types of experiments—see the plots in fig. 1 of (Jaeckel and Roy, 2010) and fig. 6.3 of (Collar et al., 2012).

Heavier particles, including known particles such as $W^\pm$ or Z^0 bosons, can also appear in loops that affect the value of a_e (Fig. 2.5e). We can estimate the size of the effect simply from the same argument we gave above. All we need to know is the relative coupling strength, χ. The essence of electroweak unification is that the coupling strengths of electromagnetic and weak interactions are fundamentally the same, i.e. $\chi \sim 1$. The effect of exchanging "weak" bosons (Z^0 and $W^\pm$) appears weak only because their mass is so large, and this suppresses their effect at low energies. Hence, we can roughly estimate the change in the electron's magnetic anomaly due to one-loop diagrams with the exchange of a Z^0 or $W^\pm$ boson: $\delta a_e^{Z,W} \sim \frac{\alpha}{2\pi} \frac{m_e^2}{M_{Z,W}^2}$. With the known masses $M_{Z,W} \sim 100$ GeV, this implies that $\delta a_e^{Z,W} \sim 3 \times 10^{-14}$. This is roughly one order of magnitude below the current theoretical uncertainty in a_e, and about two orders of magnitude below the current experimental uncertainty (dominated by the uncertainty in α from other measurements).

I note in passing that this analysis explains why it is sensible to look for the effect of new particles by measuring the muon magnetic anomaly, a_μ, rather than the electron's [(Miller 2012)]. For particles with electromagnetic-strength couplings (such as Z^0 and $W^\pm$ bosons) and mass far above the muon's, $\delta a_\mu \sim \frac{\alpha}{2\pi} \frac{m_\mu^2}{M^2}$. Since the muon's mass is $\sim 200\times$ larger than the electron's, the relative sensitivity to changes in a, for new particles of the same high mass, is about 4×10^4 times larger for a_μ than for a_e. Of course, the experiment to measure a_μ is not as accurate as for a_e, but it is good enough to probe electroweak-scale physics. In fact, there is an interesting discrepancy, of just more than 3σ, between the predicted and observed values of a_μ. This discrepancy would naturally be explained either by a new particle with mass $M \sim M_{W,Z}$, or possibly one with lighter mass and weaker couplings $\chi < 1$ such as the hidden photon discussed earlier. A new experiment at Fermilab is seeking to improve the measurement accuracy of a_μ, and test whether this discrepancy persists [(Miller et al., 2012)].

2.7.2 Electron electric dipole moment: *CP*-violation

Now we return to a topic of high interest at present. We've seen that a T, P-violating interaction, such as psuedoscalar-scalar coupling from exchange of spin-0 bosons between electrons and the nucleus in an atom, can cause the atom to have a T, P-violating EDM, $\vec{d}$, along its angular momentum, $\vec{j}$. It turns out that this can happen not only for atoms, but also for fundamental particles such as the electron, or quarks, or the proton or neutron. In this case, the particle EDM $\vec{d}$ will be along the particle spin $\vec{s}$, such that $\vec{d} = d\vec{s}/s$. (This normalization ensures that if you measure any vector component of $\vec{d}$, you get the result $\pm d$.) Just as in the case of atoms, we need some new physics—new particles with *CP*-violating interactions—to make this happen. That is: in Dirac theory and even in QED, the electron EDM, d_e, is strictly zero. A nonzero value can be induced by a loop diagram, just like for the electron's anomalous magnetic moment (see Fig. 2.6). The typical magnitude of d_e, arising from a one-loop diagram containing a particle of mass M with coupling strength g to the electron, and associated with *CP*-violating phase $\phi_{\rm CP}$, is:

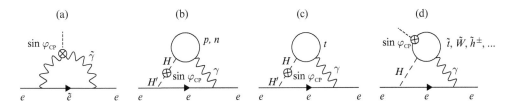

Fig. 2.6 Examples of loop diagrams giving rise to an electron EDM. (a) Typical diagram in a SUSY model. Here, the electron virtually splits into the SUSY partner of the electron, the "selectron" $\tilde{e}$, and the partner of the photon, the "photino" $\tilde{\gamma}$. The required *CP*-violation is associated with the mechanism that breaks supersymmetry, which acts as an effective *CP*-violating background field that couples to the SUSY partner particles (as denoted by the dashed line). (b) Electron EDM arising from *CP*-violating Higgs mixing. The same electron-nucleus interaction discussed earlier also gives rise to d_e via the two-loop diagram shown here. (c) Diagram analogous to (b), but with light u/d quarks in the proton and neutron replaced by a top quark, t. Because the t quark is so massive, its coupling to the Higgs is much larger than for u/d quarks, and this diagram gives a much larger contribution than (b) despite arising from the same underlying new physics. (d) Two-loop diagram analogous to combined effects from (a) and (c), but with new SUSY particles (partners of the t quark, W boson, charged Higgs bosons $h^\pm$, etc.) in the loop. Because the electron couples only to known Standard Model particles, this sort of diagram describes an extremely generic mechanism for generating d_e.

$$d_e \sim e \frac{g^2}{2\pi} \frac{m_e c}{m_e^2 c^2 + M^2 c^2} \sin\phi_{\mathrm{CP}} \sim \frac{g^2}{2\pi} \sin\phi_{\mathrm{CP}} \frac{m_e^2}{M^2} \frac{e}{m_e c}$$

$$\sim \frac{g^2}{2\pi} \sin\phi_{\mathrm{CP}} \frac{m_e^2}{M^2} \mu_B \sim \frac{g^2}{2\pi} \sin\phi_{\mathrm{CP}} \frac{m_e^2}{M^2} \left(\frac{\alpha}{2} e a_0\right). \tag{2.31}$$

Let's make a simple estimate for the size of d_e. Suppose the new particle has EM-strength couplings so that $g^2 \sim \alpha$, a maximal CP-violating phase so that $\sin\phi_{\mathrm{CP}} \sim 1$, and mass near that of the Z^0, $W^\pm$, and H bosons, $M \sim 100$ GeV. Then, $d_e/e \sim \frac{\alpha}{2\pi} \frac{(0.5 \text{ MeV})^2}{(100 \text{ GeV})^2} \left(\frac{\alpha}{2} a_0\right) \sim 10^{-16} a_0 \sim 5 \times 10^{-25}$ cm. As we will discuss, the current limit on the electron EDM is roughly $d_e \lesssim 10^{-28} e \cdot \text{cm}$. This means that these experiments are probing CP-violating physics up at a mass scale well above 100 GeV; in fact, the scale is approaching 10 TeV for particles in the sort of simple diagrams shown in Fig. 2.6a, if the CP-violating phase is maximal. This is already beyond limits placed on particle masses at the LHC, which are (at best) a few TeV [(Nakai and Reece, 2017; Feng, 2013; DeMille, Doyle, and Sushkov, 2017)]. Even in more generic scenarios such as those shown in Fig. 2.6c and 2.6d, the limits on the new particle masses have reached the few TeV scale (again, assuming maximal CP-violation). These AMO experiments are ongoing and in a period of rapid progress, which promises to allow them to far outstrip the mass reach of the LHC within a wide range of particle theory models. So, we will discuss these exeriments in considerable detail next.

By the way, the electron has a nonzero EDM in the Standard Model [(Pospelov and Ritz, 2014)]. This is in some sense inevitable, since CP-violation appears in the Standard Model. However, all Standard Model CP-violation is in the quark sector, so getting it to induce an electron EDM requires multiple loops in a Feynman diagram. In addition, the specific mechanism for CP-violation in the Standard Model causes the sum of all 2-loop and even the sum of all 3-loop diagrams to vanish. The 4-loop diagrams do not all sum to zero, but even here there is a strong cancellation in the sum over all loops (again due to the particular way that CP-violation enters the Standard Model). The resulting predicted value of d_e in the Standard Model is $d_e \sim 10^{-39} e \cdot \text{cm}$: not zero, but many orders of magnitude below the current sensitivity. It is, functionally, a null background. This means that any EDM observed in the lab must be from new physics, not from the Standard Model.

2.7.3 How to look for an electron EDM

How do you look for an electron EDM? You can't simply do what we described for an atom: if you apply an $\mathcal{E}$-field to an electron, it will accelerate away very quickly before its spin has a chance to precess. This leads to a natural idea: what if we embed the electron in a neutral atom (or molecule), and apply an $\mathcal{E}$-field to that system? It won't accelerate away, which was the goal—but then, what is the $\mathcal{E}$-field acting on the electron in this system? Here, when we speak of "the electron", we mean a valence electron, with unpaired spin, in an atom or molecule. Since the EDM $\vec{d}$ lies along the spin axis $\vec{s}$, the EDMs of all electrons in closed shells pair off and cancel, so we need not worry about them.

Now, back to the question at hand: what is the $\mathcal{E}$-field experienced by a valence electron in a neutral atom or molecule, when that system is exposed to an external $\mathcal{E}$-field? Suppose the electron experiences a purely electrostatic potential energy $V = V_{int} + V_{ext}$. Here, the subscript *int* refers to the potential from the internal field in the atom/molecule, i.e., the Coulomb field of nucleus and all the other electrons; *ext* refers to the potential from the applied $\mathcal{E}$-field. Then the total electric field that the electron experiences is $\vec{\mathcal{E}} = -\vec{\nabla}V$. But, we can write $H = \frac{p^2}{2m} + V$. So, written as an operator, $\vec{\mathcal{E}} = i\vec{p}V = i[H,\vec{p}]$. If the electron is in an energy eigenstate, i.e., an eigenstate of H, then the expectation value $\langle\vec{\mathcal{E}}\rangle$ of the electric field, i.e., the average value of the $\mathcal{E}$-field felt by the electron, is

$$\langle\vec{\mathcal{E}}\rangle \propto \langle[H,\vec{p}]\rangle = \langle\psi|H\vec{p} - \vec{p}H|\psi\rangle$$
$$= \langle\psi|H\vec{p}|\psi\rangle - \langle\psi|\vec{p}H|\psi\rangle = E\langle\psi|\vec{p}|\psi\rangle - E\langle\psi|\vec{p}|\psi\rangle = 0. \qquad (2.32)$$

This means that the average $\mathcal{E}$-field experienced by the valence electron is zero. Hence, there will be no energy shift due to the EDM: since $H_{d_e} = -\vec{d}_e \cdot \vec{\mathcal{E}}$, the average energy shift is $\langle H_{d_e}\rangle = -\vec{d}_e \cdot \langle\vec{\mathcal{E}}\rangle = 0$. This result, known as Schiff's theorem, really should not surprise us: after all, if there were an average $\mathcal{E}$-field on the electron, it would accelerate away! This obviously generalizes to the nucleus as well: that is, just like for the electron, no EDM of the nucleus will lead to an observable shift according to this argument. This makes our program—to embed some particle in a neutral atom, in order to observe its EDM—appear to be fruitless.

However, we've made some critical hidden assumptions that turn out to invalidate the argument made so far. One of these assumptions is that the electron moves nonrelativistically. Our argument employed the Schrödinger equation rather than the Dirac equation, and even in the Schrödinger formulation we did not include spin-orbit coupling or any other relativistic effect. This turns out to be critical for the electron. We will talk next about how relativistic motion leads to a nonzero energy shift due to the electron's EDM (eEDM).

An aside: this argument is also relevant to the question of detecting the EDM of a nucleus in an atom. Relativistic motion is not important for the nucleus (since its velocity, in an atom, is far smaller than that of the electron). However, we also implicitly assumed that the particle of interest is a point particle. This is very well true for the electron, but not for a nucleus. The finite size of the nucleus turns out to be the main reason why it is also possible to detect the EDM of a nucleus in a neutral atom or molecule—or rather, strictly speaking, something slightly different from a nuclear EDM but closely related to it, known as a Schiff moment. In these lectures we will only have time to talk about the case of the electron, but the nuclear Schiff moment is in fact equally interesting and important.

Now, back to the discussion of the eEDM. How does the electron's relativistic motion affect our argument about the energy shift, due to the eEDM, of an electron bound in an atom that is subject to an external $\mathcal{E}$-field? The effects of relativity in an atom, to leading order, can be described by adding new terms to the Schrödinger equation: a spin-orbit coupling term, a Darwin interaction term, and a relativistic kinetic energy term. It turns out that none of these change the result, however. The effect that *does* matter is subtle. It arises because the EDM is a length-like Lorentz vector (since an EDM has dimensions of charge×distance, and charge is a Lorentz scalar while distance is a Lorentz vector). Hence, when the electron moves the eEDM, $\vec{d}_e$, undergoes a relativistic length contraction [(Commins, Jackson, and DeMille, 2007)].

The physical interpretation for why this length contraction can lead to a nonvanishing interaction of the eEDM with an electric field is as follows. When we neglected this length contraction, we said that $\langle H_{d_e} \rangle = -\vec{d}_e \cdot \langle \vec{\mathcal{E}} \rangle$, which vanishes since $\langle \vec{\mathcal{E}} \rangle = 0$. However, due to the length contraction, the operator $\vec{d}_e$ is not a constant. Rather, it has a nontrivial spatial dependence since the velocity of the electron changes as it moves in the Coulomb potential of the nucleus plus core electrons. Likewise, of course, the part of the total electric field $\vec{\mathcal{E}}$ due to the Coulomb field also varies over position in the atom. It is generally true in quantum mechanics that for any operators A and B that are not both simple constants, the expectation value of the product of the operators is not equal to the product of the expectation values: $\langle AB \rangle \neq \langle A \rangle \langle B \rangle$. Similarly, here—due to the length contraction—we can no longer assume that $\langle \vec{d}_e \cdot \vec{\mathcal{E}} \rangle = \langle \vec{d}_e \rangle \cdot \langle \vec{\mathcal{E}} \rangle = \vec{d}_e \cdot \langle \vec{\mathcal{E}} \rangle$.

We can take into account the effect of the length contraction mathematically, in a fairly simple way. We use the standard notations $\vec{\beta} = \vec{v}/c = \vec{p}/(m_e c)$ for the dimensionless relativistic velocity, and $\gamma = 1/\sqrt{1-\beta^2}$ for the Lorentz factor. A moving object with length $\vec{L}$ along $\vec{v}$ is contracted to L/γ. It proves useful to rewrite this using the relation $\frac{1}{\gamma} = 1 - \frac{\gamma}{1+\gamma}\beta^2$. Then, the interaction of the length-contracted eEDM with a total electric field $\vec{\mathcal{E}}$ (the sum of an external, applied field and the intra-atomic field) takes the form

$$H_{\text{EDM}}^{\text{rel}} = -\vec{d}_e^0 \cdot \vec{\mathcal{E}} + \frac{\gamma}{1+\gamma}\vec{\beta}\cdot\vec{\mathcal{E}}\vec{\beta}\cdot\vec{d}_e^0, \tag{2.33}$$

where d_e^0 is the electron EDM in its rest frame. The first term on the right hand side in Eq. 2.33 is the nonrelativistic EDM interaction, whose expectation value vanishes, by Schiff's theorem. The second term, $\delta H_{\text{EDM}} = \frac{\gamma}{1+\gamma}\vec{\beta}\cdot\vec{\mathcal{E}}\vec{\beta}\cdot\vec{d}_e^0$, is a relativistic correction that results in a nonzero net interaction energy for the eEDM. If we expand this in powers of the small parameter β, the lowest-order term is $\delta H_{\text{EDM}} \approx \vec{\beta}\cdot\vec{\mathcal{E}}\vec{\beta}\cdot\vec{d}_e$.

The residual interaction due to the relativistic correction term can be expressed in terms of an "effective electric field", $\vec{\mathcal{E}}_{\text{eff}}$, defined such that $\langle \delta H_{\text{EDM}} \rangle = -\langle \vec{d}_e \rangle \cdot \vec{\mathcal{E}}_{\text{eff}}$. Detailed calculations using the Dirac equation and the proper relativistic version of the electron EDM show that this "effective electric field" within an atom or molecule can be significantly larger in magnitude than the applied external field. However, we can estimate the size of $\mathcal{E}_{\text{eff}}$ without explicit use of the Dirac equation.

First, consider the effect of the relativistic correction term, in the absence of any external field. The contribution to this term due to the internal $\mathcal{E}$-field only—call it $\delta H_{\text{EDM}}^{\text{int}}$—acts as a perturbation on the atomic structure. We can write this in the form

$$\delta H_{\text{EDM}}^{\text{int}} \approx \left(\vec{\beta}\cdot\vec{\mathcal{E}}_{\text{int}}\right)\left(\vec{\beta}\cdot\vec{d}_e\right) = d_e\left(\frac{\vec{p}\cdot\hat{r}}{m_e c}\mathcal{E}_{\text{int}}(r)\right)\left(\frac{\vec{p}}{m_e c}\cdot\vec{s}\right), \tag{2.34}$$

where $\mathcal{E}_{\text{int}}(r)\hat{r}$ is the intra-atomic Coulomb field. It is evident by inspection that $\delta H_{\text{EDM}}^{\text{int}}$ is a P-odd and T-odd scalar. This is much like the Hamiltonian $H_{H'}$ that described a CP-violating Higgs interaction. In just the same way, $\delta H_{\text{EDM}}^{\text{int}}$ mixes opposite-parity states with the same value of total angular momentum j, such as $s_{1/2}$- and $p_{1/2}$- states, but leads to no first-order energy shift.

Just as for the case of the CP-violating Higgs interaction, then, the eEDM results in a net atomic EDM (see Fig. 2.3). This means that an energy shift proportional to d_e will arise by applying an external electric field $\mathcal{E}_{\text{ext}}$ to the atom. As we did for the CP-violating Higgs interaction, you can think of this as creating an energy shift of the atomic EDM that was induced by the eEDM. Alternatively, you can think of $\mathcal{E}_{\text{ext}}$ as polarizing the atom, so that the electron feels a net, effective internal $\mathcal{E}$-field. This second way proves to be quite useful, so let's take that approach here.

Applying $\mathcal{E}_{\text{ext}}$ leads to mixing of the $s_{1/2}$- and $p_{1/2}$- states. Let's say that $\vec{\mathcal{E}}_{\text{ext}}$ is in the z-direction. The ground state $s_{1/2}$ is perturbed by $\vec{\mathcal{E}}_{\text{ext}}$, with a wavefunction (to first order in $\mathcal{E}_{\text{ext}}$) given by

$$|\widetilde{s_{1/2},m}\rangle = |s_{1/2},m\rangle + \frac{\langle p_{1/2},m\,|e\mathcal{E}_{\text{ext}}z|\,s_{1/2},m\rangle}{E_s - E_p}|p_{1/2},m\rangle. \tag{2.35}$$

Here, we continue to use a toy mmodel atom with only one $p_{1/2}$ excited state. Then, we can write

$$|\widetilde{s_{1/2},m}\rangle \approx |s_{1/2},m\rangle + \eta_{\text{St}}|p_{1/2},m\rangle. \tag{2.36}$$

Here we have defined the Stark mixing coefficient

$$\eta_{\text{St}} \sim \frac{e\mathcal{E}_{\text{ext}}\langle r\rangle_{sp}}{E_s - E_p}. \tag{2.37}$$

The perturbed atomic state $|\widetilde{s_{1/2},m}\rangle$ is polarized, with induced electric dipole moment $\langle D_z\rangle$ given by

$$
\begin{aligned}
\langle D_z\rangle &= \langle \widetilde{s_{1/2},m}\,|ez|\,\widetilde{s_{1/2},m}\rangle \\
&= \eta_{\text{St}}\langle s_{1/2},m\,|ez|\,p_{1/2},m\rangle + \eta_{\text{St}}\langle p_{1/2},m\,|ez|\,s_{1/2},m\rangle \\
&\sim 2\eta_{\text{St}}e\langle r\rangle_{sp} \sim \eta_{\text{St}}\langle D\rangle_{\text{max}},
\end{aligned} \tag{2.38}
$$

where $\langle D\rangle_{\text{max}} = e\langle r\rangle_{sp}$ is the maximum possible dipole moment of the atom (see Fig. 2.3). The quantity $\mathcal{P} \equiv \langle D_z\rangle/\langle D\rangle_{\text{max}}$ is a dimensionless measure of the degree to which the atom is polarized; hence we will refer to $\mathcal{P}$ simply as the polarization of the atom. In the weak-field limit, $\mathcal{P} = 2\eta_{\text{St}} \propto \mathcal{E}_{\text{ext}}$. For typical values $\langle r\rangle_{sp} \sim a_0$ and $E_s - E_p \sim e^2/a_0$, $\mathcal{P} \sim \mathcal{E}_{\text{ext}}/(e/a_0^2)$, where $e/a_0^2 \sim 1$ GV/cm is the atomic unit of electric field. Such a large field is not possible to apply in the lab, so we always expect $\mathcal{P} \ll 1$ and $\mathcal{P} \propto \mathcal{E}_{\text{ext}}$ for atoms. The largest fields that you can apply in the lab are $\mathcal{E}_{\text{ext}} \sim 100$ kV/cm, so that the largest attainable value of $\mathcal{P}$ in atoms is $\mathcal{P}_{\text{max}} \sim 10^{-4}$. However, if you *could* go far beyond the weak-field limit, full mixing (in this toy 2-level model) would yield the perturbed wavefunction

$$|\widetilde{s_{1/2},m}\rangle \approx \frac{1}{\sqrt{2}}|s_{1/2},m\rangle + \frac{1}{\sqrt{2}}|p_{1/2},m\rangle, \tag{2.39}$$

and this would lead to $\mathcal{P} \approx 1$.

Now, let's see what nonzero polarization of the atom means for the effective $\mathcal{E}$-field acting on the eEDM. Assume the spin is in the $\hat{z}$ direction. We need to calculate

$$\langle \delta \mathcal{H}_{\mathrm{EDM}}^{\mathrm{rel}}\rangle = \langle \widetilde{s_{1/2},m}| \, d_e \left(\frac{\vec{p}}{m_e c} \cdot \vec{\mathcal{E}} \right) \left(\frac{\vec{p}}{m_e c} \cdot \vec{s} \right) |\widetilde{s_{1/2},m}\rangle$$

$$\sim 2\eta_{\mathrm{St}} d_e \langle s_{1/2},m| \left(\frac{\vec{p}}{m_e c} \cdot \vec{\mathcal{E}} \right) \left(\frac{\vec{p}}{m_e c} \cdot \vec{s} \right) |p_{1/2},m\rangle. \tag{2.40}$$

It is useful to rewrite this in the suggestive form

$$\langle \delta \mathcal{H}_{\mathrm{EDM}}^{\mathrm{rel}}\rangle = -d_e \mathcal{P} \mathcal{E}_{\mathrm{int}}^{\mathrm{eff}} = -d_e \mathcal{E}^{\mathrm{eff}}. \tag{2.41}$$

Here, we replaced $2\eta_{\mathrm{St}}$ with its more general form $\mathcal{P}$, and we defined both the effective *internal* electric field that would act on the eEDM in a fully polarized atom,

$$\mathcal{E}_{\mathrm{int}}^{\mathrm{eff}} \equiv -\langle s_{1/2},m| \left(\frac{\vec{p}}{m_e c} \cdot \vec{\mathcal{E}} \right) \left(\frac{\vec{p}}{m_e c} \cdot \vec{s} \right) |p_{1/2},m\rangle, \tag{2.42}$$

and the overall effective electric field that acts on the eEDM in the general case,

$$\mathcal{E}^{\mathrm{eff}} \equiv \mathcal{P} \mathcal{E}_{\mathrm{int}}^{\mathrm{eff}}. \tag{2.43}$$

Let's now estimate the matrix element that determines the effective internal field $\mathcal{E}_{\mathrm{int}}^{\mathrm{eff}}$. Near the nucleus, both $\mathcal{E} = Ze/r^2$ and $p = \sqrt{2m(E-V)} \approx \sqrt{-2mV}$ are large in a semiclassical picture, so we only need to focus on what happens in that region. The first term is

$$\frac{\vec{p}}{m_e c} \cdot \vec{\mathcal{E}} \sim \frac{1}{m_e c} \frac{\partial}{\partial r} \frac{Ze}{r^2} \sim \frac{1}{m_e c} \frac{Ze}{r^3}. \tag{2.44}$$

The second term acts on the p state wavefunction:

$$\left(\frac{\vec{p}}{m_e c} \cdot \vec{s} \right) |p_{1/2},m\rangle \sim \frac{1}{m_e c} \frac{\partial}{\partial r} R_p(r). \tag{2.45}$$

Recall that we wrote expressions for the valence electron wavefunctions near the nucleus in a neutral atom with nuclear charge Z, in Eqs. 2.19 and 2.20; from those, we find that the matrix element in Eq. 2.42 is

$$\mathcal{E}_{\text{int}}^{\text{eff}} \sim \int \frac{Z^{1/2}}{a_0^{3/2}} \frac{1}{m_e c} \frac{Ze}{r^3} \left(\frac{1}{m_e c} \frac{Z^{3/2}}{a_0^{5/2}} \right) r^2 dr$$

$$\sim Z^3 \frac{e}{a_0^4} \frac{1}{m_e^2 c^2} \int \frac{1}{r} dr \sim Z^3 \frac{e}{a_0^2} \frac{1}{(m_e c/\alpha)^2} \frac{1}{m_e^2 c^2} \int \frac{1}{r} dr. \tag{2.46}$$

The integral in this expression diverges if we take the limits from $r = 0$ to $r = \infty$. However, it does not make sense to extend the lower limit to below the radius of the nucleus, R_N, since inside the nucleus the Coulomb field stops increasing and eventually drops to zero. So, we set the lower limit of the integral as $r_{\text{min}} \sim R_N$. Similarly, it does not make sense to extend the upper limit of the integral beyond roughly the radius of the innermost core electron, $R_{\text{core}} \sim a_0/Z$: beyond that radius, the Coulomb field of the nucleus is screened by the core electrons and our simple expressions for the valence electron's wavefunction near the (unscreened) nucleus are no longer valid. So, we set the upper limit of the integral as $r_{\text{max}} \sim R_{\text{core}}$. However, the value of the integral $\int_{r_{\text{min}}}^{r_{\text{max}}} \frac{1}{r} dr = \ln(r_{\text{max}}/r_{\text{min}})$ depends only logarithmically on these limits, so even large changes in their definition does not affect the answer very significantly. Of course, all of this can be taken care of thoroughly in a careful calculation, but this argument is sufficient for our purposes here.

Hence, we finally find that the effective internal field that would act on the eEDM in a fully polarized atom has magnitude given roughly by

$$\mathcal{E}_{\text{int}}^{\text{eff}} = \langle s_{1/2}, m | \left(\frac{\vec{p}}{m_e c} \cdot \vec{\mathcal{E}} \right) \left(\frac{\vec{p}}{m_e c} \cdot \vec{s} \right) | p_{1/2} \rangle \sim Z^3 \alpha^2 \frac{e}{a_0^2}. \tag{2.47}$$

Recall that the atomic unit of electric field is $e/a_0^2 \sim 1$ GV/cm. For the heaviest atoms, with $Z \sim 90$, the effective internal field is even larger: in this case, $\mathcal{E}_{\text{int}}^{\text{eff}} \sim Z^3 \alpha^2 e/a_0^2 \sim 100$ GV/cm. Of course, the effective field $\mathcal{E}_{\text{eff}}$ acting on the eEDM in a polarized atom is smaller than this, by a factor of $\mathcal{P}$, according to Eq. 2.43. This implies the important result:

$$\mathcal{E}_{\text{eff}} \sim \mathcal{P} Z^3 \alpha^2 \frac{e^2}{a_0}. \tag{2.48}$$

For atoms in an external electric field $\mathcal{E}_{\text{ext}}$, we found that $\mathcal{P} \sim \mathcal{E}_{\text{ext}}/(e^2/a_0) \ll 1$. So, we can conclude that the ratio of effective field acting on the eEDM in a partially polarized atom to the applied external field, referred to the "eEDM enhancement factor" R for the atom, is given by

$$R \equiv \mathcal{E}_{\text{eff}}/\mathcal{E}_{\text{ext}} \sim Z^3 \alpha^2. \tag{2.49}$$

For Tl atoms ($Z = 81$), detailed calculations show that $R \approx 600$; for Cs atoms ($Z = 55$), $R \approx 100$; and for Fr atoms ($Z = 87$), $R \approx 1000$. These numbers are systematically ~ 10 times larger than our handwaving estimate, mainly because the energy splittings in atoms like these are a few times smaller than our estimate

$E_p - E_s \sim e^2/a_0$ and because the dipole matrix elements are a few times larger than our estimate $\langle r \rangle_{sp} \sim a_0$.

We earlier remarked on the similarity between the observable energy shift due to the eEDM, and that due to the pseudoscalar-scalar e-N coupling we talked about before. Both give an energy shift between electron spin up and spin down relative to an applied $\vec{\mathcal{E}}$-field, and both are proportional to the polarization $\mathcal{P}$ of the atom. In fact, in any given experiment you cannot distinguish between one of these effects and the other—rather, the measurable quantity is a linear combination of the two effects. That's fine: both are fantastically small in the Standard Model of particle theory, so it is all the better to have a larger net to catch both possible new effects at once in a single measurement.

It turns out that in most (but not all) scenarios for new physics beyond the Standard Model, the eEDM is a somewhat better probe of physics than the e-N pseudoscalar-scalar interaction. Let's take as an example the CP-violating Higgs exchange scenario. Here, the exact same interaction gives rise to an eEDM via a two-loop Feynman diagram (Fig. 2.6b). In this case, however, there is also a similar diagram where the proton or neutron (composed of light u and d quarks) is replaced by a top quark, which has much stronger coupling to the Higgs boson due to its much larger mass (Fig. 2.6c). Considering this type of diagram, together with limits on the size of the eEDM, gives very strong limits on the CP-violating phase associated with exchange of mixed scalar-pseudoscalar Higgs bosons—already beyond the direct reach of the LHC for this type of scenario. Similar types of two-loop diagrams are very generic, and remove any requirement that new particles with CP-violating interactions couple directly to the electron (Fig. 2.6d). The mass reach of the eEDM to such indirectly-coupled particles is only $\sim \sqrt{1/(2\pi)}$ worse, in general, than for particles that couple directly to the electron via one-loop diagrams. This means that the eEDM is a very powerful probe of one currently popular idea for why the LHC is not seeing SUSY yet: the scenario known as "split SUSY". In this model, the SUSY partners of ordinary fermions can be very heavy, but the SUSY partners of Standard Model bosons (such as gluinos, the partners of gluons) must remain fairly light in order to keep the Higgs mass from getting too large. If the LHC sees gluino-like particles in the near future, but no other SUSY-like particles, the eEDM is the one of the few remaining ways to detect the presence of other, heavier SUSY particles.

2.8   Searching for parity and time-reversal violation with molecules

We've seen that for the eEDM and for T, P-violating e-N interactions, the observable energy shifts are proportional to the dimensionless degree of atomic polarization, $\mathcal{P}$. We've also seen that the $\mathcal{E}$-field needed to fully polarize an atom is prohibitively large. (There are a handful of exceptions where states are anomalously close together, but none of them has the other properties needed for an EDM measurement.) However, it has long been recognized that this restriction can be overcome by using polar molecules rather than atoms. In this section we discuss why this is so.

Since molecules are unfamiliar to many students in atomic physics, we begin with a few digressions on molecular structure. We restrict our discussion for now to simple, ionically bonded diatomic molecules of the form A^+P^-, where the atomic ion A^+ is alkali-like (e.g., Ba^+ or Yb^+) and heavy (so it has large Z), and the "partner" negative ion P^- has closed shells (e.g., F^-) so that we can treat it as structureless. Consider a situation in which the internuclear axis of such a polar molecule is fixed in space (not rotating). The "partner ion" P^- leads to a strong $\mathcal{E}$-field on the "atom of interest" A^+. This intramolecular $\mathcal{E}$-field, $\mathcal{E}_{\text{mol}}$, must be ~ 1 atomic unit: $\mathcal{E}_{\text{mol}} \sim e/a_0^2$. Hence, $\mathcal{E}_{\text{mol}}$ will lead to strong mixing of the s, p, and even d, f, etc. orbitals of the valence electron in the ion A^+. Hence, we can write the ground state wavefunction of the electron in this molecule-fixed frame—i.e., the molecular orbital $|\psi\rangle_{\text{mol}}$—as a linear combination of these atomic orbitals:

$$|\psi\rangle_{\text{mol}} = a_s|s\rangle + a_p|p, M = 0\rangle + a_d|d, M = 0\rangle + \cdots, \tag{2.50}$$

where the projection M is defined relative to the internuclear axis of the molecule, $\hat{Z}$. Due to the large value of $\mathcal{E}_{\text{mol}}$, we can expect the coefficients to have values $a_s \lesssim 1$, $a_p \lesssim a_s$, $a_d \lesssim a_p$, etc. That is, both a_s and a_p should be not much smaller than 1. This corresponds to the limit of strong polarization that we seek in order to have a large T, P-violating energy shift. The effective polarization of the s- and p- orbitals inside the molecule, which can lead to such an energy shift, is given here by $\mathcal{P}_{\text{int}}^{\text{eff}} \approx 2a_s a_p$. Realistic calculations of the weighting coefficients a_ℓ for YbF and BaF molecules each give $a_s \approx 0.9$ and $a_p \approx 0.4$, corresponding to $\mathcal{P}_{\text{int}}^{\text{eff}} \approx 0.7$. That is indeed strong polarization!

However, molecules do not naturally orient their axes in a particular direction in the lab. Rather, in the absence of any external fields, the molecule (and hence its axis $\hat{Z}$) can freely rotate. More precisely, the expectation value of the projection of the internuclear axis along any given direction—for example, the lab $\hat{z}$ axis—vanishes: $\langle \hat{Z} \cdot \hat{z} \rangle = 0$. This means that, for a rotating molecule, the intra-molecular electric field $\vec{\mathcal{E}}_{\text{mol}} = \mathcal{E}_{\text{mol}}\hat{Z}$ also averages to zero in the lab frame. So, to take advantage of the huge intra-molecular field for an EDM experiment, it is necessary to orient the molecule along the lab-frame $\hat{z}$ axis—that is, we must polarize the molecule in the lab. The reason that molecules are interesting for EDM experiments is because such polarization is much easier to achieve in molecules than it is in atoms. So, now we discuss how polarization of molecules works.

The quantum number M we used above is usually referred to in molecular physics as the quantum number $\Lambda \equiv \vec{L} \cdot \hat{Z}$, where $\vec{L}$ is the electron's orbital angular momentum. For a state derived primarily from an atomic s-orbital, as we have assumed here, the only possible value is $\Lambda = 0$. Since we are considering a molecule consisting of an alkali-like atom plus a closed-shell ion, the molecule has total electron spin $S = 1/2$. Analogous to the notation $^{2S+1}L_J$ in atoms, in molecules states are labeled as $^{2S+1}\Lambda$, and here we are discussing a $^2\Sigma$ state. We can hence label the molecular-frame state as $|\Lambda\rangle_{\text{mol}}$. Of course, there are states with the same value of Λ but different radial wavefunctions; however, we will only care about one set of radial wavefunctions (in this case, the ground state), so we won't bother to include radial quantum numbers in our notation. Also, since any contribution to the orbital angular momentum is constrained to precess rapidly about the Z-axis (so that $M = \Lambda = 0$), there is no way to couple the spin to the orbital angular momentum; hence the spin, in this case, can point in any direction relative to the internuclear axis $\hat{Z}$.

Next, we introduce even more molecular notation. For the case we are considering, where the spin is not coupled to the molecular frame, we define the rotational + orbital angular momentum, J, via the relation $\vec{J} = \vec{R} + \vec{\Lambda}$, where $\vec{R}$ is the mechanical (end-over-end) rotational angular momentum of the molecule, and $\vec{\Lambda} = \Lambda\hat{Z}$. For the freely rotating molecule, the energy eigenstates are states of definite J, with projection m_J along the laboratory $\hat{z}$ axis. Suppose the molecular $\hat{Z}$ axis is rotated to polar angle (θ, ϕ) in the laboratory coordinate system (x, y, z). Then the electron orbital in the lab frame can be obtained by applying a rotation operator to the orbital in the molecule-fixed frame, $|\psi\rangle_{\text{mol}} \equiv |\Lambda\rangle_{\text{mol}}$. In particular, standard textbooks [(Sakurai and Napolitano, 2011; Brown and Carrington, 2003)] define the rotation operator that acts to rotate a state with angular momentum projection M (here, Λ) along $\hat{Z}$ to a state with angular momentum projection m_J along z, while preserving the total angular momentum J. This rotation operator, usually written as $\mathcal{D}^J_{m_J,\Lambda}(\theta,\phi)$, describes the rotational + orbital angular momentum part of the molecular wavefunction. That is: if the state of the rotating molecule has quantum numbers J, m_J, and Λ, then the probability amplitude to find the internuclear axis $\hat{Z}$ pointing in the direction (θ, ϕ) is just $\mathcal{D}^J_{m_J,\Lambda}(\theta,\phi)$. In the particular case we are discussing here, where $\Lambda = 0$, the rotation operators/wavefunctions are nothing other than the usual spherical harmonic functions: $\mathcal{D}^J_{m_J,0}(\theta,\phi) = Y_J^{m_J}(\theta,\phi)$. This is exactly the angular wavefunction for a simple rigid rotor, with Hamiltonian $H_{\text{rot}} = J^2/(2I)$ (where I is the moment of intertia of the rotor), in the state $|J, m_J\rangle$. With this form of the wavefunction, it is easy to verify that, as anticipated, $\langle \hat{Z} \cdot \hat{z} \rangle = \langle \cos\theta \rangle = 0$, so that the expectation value $\langle \vec{\mathcal{E}}_{\text{mol}} \rangle$ vanishes as well.

Now, we can work out how to polarize the molecule in the lab. A polar molecule has (by definition) a static electric dipole moment $\vec{D} = D\hat{Z}$ along its axis. For molecules like the ones we are considering, $D \sim ea_0$. The rotational wavefunctions $|Jm_J\rangle = Y_J^{m_J}(\theta,\phi)$ are states with alternating parity, $(-1)^J$. The Stark interaction of this intramolecular dipole moment with an external electric field, $\vec{\mathcal{E}}_{\text{ext}} = \mathcal{E}_{\text{ext}}\hat{z}$, is described by the Hamiltonian $H_{\text{St}} = -\vec{D} \cdot \vec{\mathcal{E}}_{\text{ext}} = -D\mathcal{E}_{\text{ext}}\cos\theta$. This mixes only states with the same value of m_J, but opposite parity. The mixing of rotational states $|J, m_J\rangle$ and $|J', m_J\rangle$ is determined by the matrix element $\langle J', m_J | H_{\text{St}} | J, m_J \rangle = -D\mathcal{E}_{\text{ext}} \int Y_{J'}^{m_J} \cos\theta Y_J^{m_J} d\Omega \sim D\mathcal{E}_{\text{ext}}\delta_{J',J\pm1}$. The molecular polarization due to the external field, $\mathcal{P}^{\text{ext}}_{\text{mol}}$, is sensibly defined as $\mathcal{P}^{\text{ext}}_{\text{mol}} = \langle \hat{Z} \cdot \hat{z} \rangle$. If the wavefunction of the ground state, with the field applied, is written as

$$|\psi\rangle_{\text{rot}} = a_0|J = 0, m_J = 0\rangle + a_1|J = 1, m_J = 0\rangle + \cdots, \quad (2.51)$$

then—in the crude approximation of ignoring the contribution of rotational levels with $J \geq 2$—for this state, $\mathcal{P}^{\text{ext}}_{\text{mol}} \approx 2a_0a_1$.

This is all, so far, very similar to our discussion of electrical polarization of atoms. The main difference is that the energy of rotational states, E_{rot}, is much smaller than the energy of electronic states in atoms. To see this, consider a molecule with reduced mass $M = Am_p$, where m_p is the proton mass. For our prototypical molecule with a heavy positive ion and a light negative ion, $A \sim 20$ is roughly the atomic mass of the light ion. This type of molecule will have moment of inertia $I = MR^2$, where $R \sim a_0$ is the spacing between the atoms in the molecule. Then, for a molecule in an eigenstate with angular momentum $J \sim 1$, the rotational energy is

$$E_{\text{rot}} \sim \frac{J^2}{2I} \sim \frac{1}{Ma_0^2} = \frac{m_e}{M} \frac{1}{m_e a_0^2} \sim \frac{1}{A} \frac{m_e}{m_p} \frac{e^2}{a_0} \sim \frac{1}{20} \frac{1}{2000} \frac{e^2}{a_0} \lesssim 10^{-4} \frac{e^2}{a_0}. \qquad (2.52)$$

Hence, the energy denominator that enters in the perturbative description of the polarizability (see Eq. 2.37) is many orders of magnitude smaller for molecules than for atoms. Therefore, the $\mathcal{E}$-field needed to fully mix adjacent rotational states (for example, the ground state $J = 0$ and first excited state $J = 1$) is smaller than that needed to mix atomic s and p states by the same factor. For a typical polar molecule of the type we are considering, a field strength $\mathcal{E}_{\text{ext}} \sim 30$ kV/cm is sufficient to almost fully mix the $J = 0$ and $J = 1$ states, i.e., to achieve molecular polarization $\mathcal{P}_{\text{ext}}^{\text{mol}} \gtrsim 1/2$.

Now, finally, we can see how large T, P-odd energy shifts arise in molecules. Our expression for the T, P-odd energy shift due to the eEDM in Eq. 2.41 should be replaced by

$$\langle \delta \mathcal{H}_{\text{EDM}}^{\text{rel}} \rangle = -d_e \, \mathcal{P}_{\text{ext}}^{\text{mol}} \, \mathcal{P}_{\text{int}}^{\text{eff}} \mathcal{E}_{\text{int}}^{\text{eff}}. \qquad (2.53)$$

That is, the expression for the effective field acting on the eEDM, 2.43, should be replaced by

$$\mathcal{E}^{\text{eff}} = \mathcal{P}_{\text{ext}}^{\text{mol}} \, \mathcal{P}_{\text{int}}^{\text{eff}} \mathcal{E}_{\text{int}}^{\text{eff}} \sim \mathcal{P}_{\text{ext}}^{\text{mol}} \, \mathcal{P}_{\text{int}}^{\text{eff}} Z^3 \alpha^2 \frac{e}{a_0^2}. \qquad (2.54)$$

We have seen that both the "internal" polarization $\mathcal{P}_{\text{int}}^{\text{eff}}$ and the "external" polarization $\mathcal{P}_{\text{ext}}^{\text{mol}}$ can be very near their maximum values of 1. Thus, in molecules, the effective field $\mathcal{E}^{\text{eff}}$ can nearly reach the theoretical maximum value $\mathcal{E}_{\text{int}}^{\text{eff}} \sim Z^3 \alpha^2 e/a_0^2 \sim 100$ GV/cm. This is, again, ~ 4 orders of magnitude larger than the values of $\mathcal{E}^{\text{eff}}$ that can be achieved in atoms. Because of this, many of the most recent experiments to search for the eEDM use polar molecules instead of atoms.

Now we can outline a simple concept for how to perform an eEDM experiment with molecules. First, pick a heavy $^2\Sigma$ molecule of the type we have discussed, such as YbF or BaF (both of which are both in active use now). Then, "just" do the same type of experiment as was outlined for Tl atoms! Unfortunately, when you begin to think about "just" doing this, you will immediately see there are some difficulties. The first is that molecules with unpaired spins, as we have been discussing, are by definition chemical free radicals. This means they are thermodynamically likely to form other types of molecules; for example, YbF_3 is chemically stable while YbF is not. This means you need some way to break (or form) bonds—typically, a very high temperature and/or a nonequilibrium chemical environment. This requirement held back the field for many years: the yield of heavy free radicals in molecular beams was simply too small to be useful until fairly recently. In addition, the small rotational energies that lead to the desired high external polarizability also lead to a broad distribution of population among rotational levels, at any finite temperature T. The fraction, f_0, in the lowest rotational level, with $J = 0$, in the Boltzmann distribution is $f_0 \sim 1/(2Ik_BT)$, where k_B is the Boltzmann constant. For example, at $T = 300$ K typically $f_0 \sim 1/1000$. This means that only $\sim 10^{-3}$ of the already low available flux is available to contribute to the signal for an EDM measurement.

The group of Ed Hinds at Imperial College, London, managed to overcome many of these difficulties. They recently reported a measurement of the eEDM with sensitivity better than in the Tl atom experiment, using what was—aside from the use of molecules rather than atoms—a similar approach [(Hudson et al., 2011; Kara et al., 2012)]. The Imperial College experiment used the molecule YbF. Yb has $Z = 70$, and YbF provides a large effective internal field, $\mathcal{P}_{\text{int}}^{\text{eff}}\mathcal{E}_{\text{int}}^{\text{eff}} \approx 26$ GV/cm. To form a beam of YbF, Yb vapor (produced by laser ablation of Yb metal) was allowed to chemically react with SF_6 gas, which is a small admixture in a strong supersonic flow of argon carrier gas. Isenthalpic expansion of the flow cools the molecules via to temperature $T \sim 4$ K, where $f_0 \sim 1/10$ of the molecules are in the desired $J = 0$ rotational state. The beam must be pulsed, with small duty cycle, to reduce the gas load on the vacuum pumps to acceptable levels. The time-averaged brightness per state of this beam is many orders of magnitude smaller than the atomic Tl beam used in the earlier experiments at Berkeley. In addition, the velocity was $\sim 2\times$ larger than for the Tl atomic beam, due to the supersonic argon expansion; hence the spin coherence time, after traveling a similar distance, was $\sim 2\times$ smaller. Nevertheless, by using a larger solid angle of the beam (possible since the external $\mathcal{E}$-field is much smaller, only ~ 10 kV/cm rather than ~ 120 kV/cm) and taking advantage of the much larger value of $\mathcal{E}^{\text{eff}}$ ($\sim 250\times$ larger than in Tl, for the external $\mathcal{E}$-fields used, which yielded $\mathcal{P}_{\text{ext}}^{\text{mol}} \approx 1/2$), they achieved slightly better sensitivity than the Tl experiment.

2.8.1   New developments in molecule-based EDM experiments

This line of work motivated the development of new types of molecular beam sources. The ideal source for EDM experiments would be bright, cold, and slow. In the mid-2000s, John Doyle and his group at Harvard developed exactly such a source, which they refer to as the "hydrodynamically enhanced cryogenic buffer gas beam" source [(Hutzler, Lu, and Doyle, 2012)]. Here, hot molecules are injected (usually via laser ablation of a solid precursor) into a cell filled with inert buffer gas (usually He or Ne) held at cryogenic temperature (typically ~ 1–4 K for He or ~ 16 K for Ne). The molecules rapidly equilibrate to the temperature of the buffer gas, via collisions. The molecules and buffer gas exit the cell through an aperture and form a beam. With a correct choice of buffer gas density, cell size, and aperture diameter, the molecules are entrained in the flow of buffer gas through the cell and extracted into the beam with good efficiency ($\gtrsim 10\%$). In this regime of operation, the buffer gas flow outside the aperture is mildly supersonic, leading to further cooling of the molecules in the resulting expansion (down to $\lesssim 1$K for He and ~ 4K for Ne). This means that the fraction of molecules in the $J = 0$ state is $f_0 \gtrsim 10\%$. The kinematics of the expansion also lead to a low divergence of heavy molecules in the beam. In practice, the flux of heavy polar molecules in their ground state, as needed for EDM experiments, is $\sim 1000\times$ larger than from any previous method, and in addition the forward velocity of the beam is $\sim 3 - 4$ times slower than in the type of room-temperature supersonic sources used previously. Several experiments in the field now rely on beams produced with this technique.

In the meantime, new ideas also emerged for how to use molecules more effectively. One that has proved especially fruitful is to use molecules with two valence electrons, rather than just one. The second electron serves as a "co-traveler" that does not affect the EDM physics directly, but does modify the molecular structure in useful ways. One immediate consequence of this change is that it is no longer necessary to use nonequilibrium chemistry to make molecules with unpaired spins. Instead, you can start with closed-shell molecules (in an electronic state with $^1\Sigma$ symmetry) and use optical pumping to drive them to a triplet state, where both electrons are unpaired. Since we need heavy molecules anyway, the spin-orbit coupling needed to connect singlet and triplet states is large enough that, with a high-power laser, this type of pumping can be done efficiently. In addition, the lowest-lying triplet levels in such molecules can be quite long-lived, since their decays are suppressed by the usual singlet-triplet selection rule. In favorable cases, the lifetime can be at least as long as the time typically needed for molecules to fly through a molecular beam apparatus— in other words, in these cases there is little disadvantage to using a metastable state for the measurement.

Let's now discuss the structure of molecules with two unpaired electrons. We will see that there is a qualitatively new energy level structure, unlike any that occurs in atoms, that proves extremely useful for EDM measurements. Here, of course, the electronic angular momentum (denoted as $\vec{J}_e = \vec{L} + \vec{S}$) takes on only integer values. Consider specifically the case where $\vec{J}_e = 1$. (An example of this would be the molecular state associated with binding a 2-electron alkaline-earth atom, in its $nsnp^3 P_1$ metastable state, to a closed-shell O^{2-} ion.) In the molecular frame, the intra-molecular $\mathcal{E}$-field acting on the valence electrons will cause a tensor Stark shift of the sublevels of J_e. The projections $M_e = \vec{J}_e \cdot \hat{Z} = \pm 1$ along the molecular Z-axis must be degenerate if we ignore any symmetry-violating effects, since the states with $M_e = \pm 1$ are mirror images of each other. However, the $M_e = 0$ state can be far away in energy due to the large intra-molecular $\mathcal{E}$-field. In fact, typically this level will be about as far away from the degenerate $M_e = \pm 1$ pair as would any other electronic orbital state, since the intramolecular $\mathcal{E}$-field is about one atomic unit in size. The quantity M_e is a good quantum number, generally given the label Ω. Two-electron states are labeled via the notation $^{2S+1}\Lambda_\Omega$.

This picture is, of course, in the molecular frame. What happens in the lab frame, with a rotating molecule? We need to include the effect of the mechanical rotational angular momentum $\vec{R}$, which has Hamiltonian $H_{\text{rot}} = R(^2/(2I)) = BR^2$, where $B = 1/(2I)$ is called the rotational constant. But the total angular momentum is $\vec{J} = \vec{R} + \vec{J}_e$, so $H_{\text{rot}} = B(\vec{J} - \vec{J}_e)^2 = BJ^2 - 2B\vec{J} \cdot \vec{J}_e + BJ_e^2$, where the last term is a constant for all rotational states within a given electronic state where J_e is fixed. The first term is just what we wrote before. However, for each value of J, here there are two states: one with $\vec{J}_e$ parallel to $\hat{Z}$ (the state with $\Omega > 0$), and the other with $\vec{J}_e$ antiparallel to $\hat{Z}$ ($\Omega < 0$). However, these are *not* eigenstates of parity, so they cannot be the energy eigenstates (since are not yet considering the influence of any exotic new physics).

Eigenstates of the parity operator, P, can be constructed as superpositions of the eigenstates of Ω (which we write as $|\pm\Omega\rangle$). This is because, in the molecular frame,

$$P|\Omega\rangle = P|\vec{J}_e \cdot \hat{Z}\rangle = \left|\vec{J}_e \cdot (-\hat{Z})\right\rangle = |-\Omega\rangle, \tag{2.55}$$

so that

$$P\left[\frac{|\Omega\rangle \pm |-\Omega\rangle}{\sqrt{2}}\right] = \frac{|-\Omega\rangle \pm |\Omega\rangle}{\sqrt{2}} = \pm\left[\frac{|\Omega\rangle \pm |-\Omega\rangle}{\sqrt{2}}\right]. \tag{2.56}$$

The lab-frame wavefunctions associated with these parity eigenstates can be labeled by the quantum number $p = \pm 1$ (which describes the parity in the molecule-fixed frame), the magnitude of Ω, and the total angular momentum J and its lab-frame projection m. These are related to the molecule-fixed frame wavefunctions above via the relation

$$||\Omega|, p; J, m\rangle = \mathcal{D}^J_{\Omega,m}(\theta,\phi)\left[\frac{|\Omega\rangle + p|-\Omega\rangle}{\sqrt{2}}\right]. \tag{2.57}$$

The pair of states with the same value of J but with $p = \pm$, which have parity quantum number $P = p * (-1)^J$ in the lab frame, are referred to as "Ω-doublet" levels.

In the molecular frame, this pair of states appeared to be exactly degenerate. However, the rotation of the molecule breaks this degeneracy. The second term in the rotational Hamiltonian, $-2B\vec{J}\cdot\vec{J}_e$, is responsible for this effect: it is a Coriolis coupling, which arises because we are dragging the electronic angular momentum $\vec{J}_e$ around with the rotating molecular axis $\hat{Z}$, while holding the component $\vec{\Omega} = (\hat{Z}\cdot\vec{J}_e)\hat{Z}$ perpendicular to $\vec{R}$. Note that nothing analogous to this takes place in atoms, where angular momenta instead always couple such that each component freely precesses about the vector sum. Note also that in the molecular case, the total angular momentum J must satisfy $J \geq |\Omega|$, since the pure rotational angular momuntum $\vec{R}$ is perpendicular to $\vec{\Omega}$ and cannot act to cancel it.

Now, let's work out explicitly how this Coriolis term breaks the degeneracy of the two levels with different signs of p. The operator $\vec{J}\cdot\vec{J}_e = J_z J_{ez} + (J_+ J_{e-} + J_- J_{e+})/2$ can, via the final term, couple levels with $\Omega = \pm 1$ to an energetically distant level with $\Omega = 0$. Keeping careful track of signs while evaluating the matrix elements of this operator leads to the result that the coupling is present for one of the linear combinations of $\Omega = \pm 1$ that corresponds to a parity eigenstate, and vanishes for the orthogonal combination. The energy splitting Δ_Ω between the Ω-doublet parity eigenstates then arises in 2nd order of perturbation theory, and has magnitude

$$\Delta_\Omega \sim \frac{(2BJJ_e)^2}{E(J_e = 1) - E(J_e = 0)} \sim \frac{BJ_e^2}{(e^2/a_0)}BJ^2$$

$$\sim \frac{m_e}{Am_p}BJ^2 \sim 10^{-4}BJ^2 \sim 2\pi \times 1 \text{ MHz} * J(J+1). \tag{2.58}$$

There is an intuitive way to visualize what is going on here, if you imagine that $\vec{J}_e$ could be oriented at any angle Θ relative to the internuclear axis $\hat{Z}$. Then, the energy as a function of Θ will have the form of a double-well potential, with a maximum at $\Theta = \pi/2$ corresponding to the higher-energy $\Omega = 0$ state, and minima at $\Theta = 0$ and π corresponding to the degenerate $\Omega = \pm 1$ states (see Fig. 2.7). Tunneling through this barrier (here, due to the Coriolis interaction) leads to a pair of nearly degenerate ground-state wavefunctions that are symmetric/antisymmetric superpositions of states with $\Theta = 0$ and π, i.e., superpositions of states with $\Omega = \pm 1$.

The most interesting case for eEDM experiments, in a state with $|\Omega| = 1$, happens for $J = 1$ levels, when an external $\mathcal{E}$-field $\vec{\mathcal{E}}_{\text{ext}} = \mathcal{E}_{\text{ext}} \hat{z}$ is applied. It is certainly plausible that $\vec{\mathcal{E}}_{\text{ext}}$ can mix the close-lying pair of Ω-doublet states, since they have opposite parity. However, this case is unfamiliar enough that you might not take that for granted; hence, we work out the details here. As usual, the Hamiltonian for the Stark interaction is given by $H_{\text{St}} = -\vec{D} \cdot \vec{\mathcal{E}}_{\text{ext}} = -D\mathcal{E}_{\text{ext}} \hat{Z} \cdot \hat{z}$. We need to evaluate matrix elements of the form $-D\mathcal{E}_{\text{ext}} \langle Jm\Omega' | \hat{Z} \cdot \hat{z} | Jm\Omega \rangle$, since the Ω-doublet levels with the same value of J have by far the smallest energy splitting in the system. Happily, these matrix elements can be worked out, given the wavefunctions we wrote above and using standard results from angular momentum algebra. (This includes some relations having to do with the rotation operators that are perhaps less than familiar, but which are still just something you can look up [(Brown and Carrington, 2003)]). In the end, the relevant matrix elements have the form

$$\left\langle Jm \pm \Omega | \hat{Z} \cdot \hat{z} | Jm\Omega \right\rangle$$

$$= (-1)^{J-m}(-1)^{J\mp\Omega} \begin{pmatrix} J & 1 & J \\ -m & 0 & m \end{pmatrix} (2J+1) \begin{pmatrix} J & 1 & J \\ \mp\Omega & 0 & \Omega \end{pmatrix} \langle \pm\Omega | \hat{Z} | \Omega \rangle_{\text{mol}}$$

$$= (-1)^{-2J}(2J+1) \frac{1}{\sqrt{J(1+J)(1+2J)}} \, m \frac{1}{\sqrt{J(1+J)(1+2J)}} \, \Omega \, \delta_{\pm\Omega,\Omega}$$

$$= J(J+1)m\Omega \, \delta_{\pm\Omega,\Omega}. \tag{2.59}$$

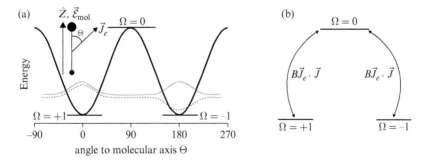

Fig. 2.7 Intuitive picture of the mechanism leading to Ω-doublet substructure. (a) Effective double-well potential, leading to symmetric and antisymmetric combinations of $\Omega = \pm 1$ states. See the text for details. (b) Proper quantum-mechanical description, showing the coupling of nominally degenerate $\Omega = \pm 1$ states to a distant state with $\Omega = 0$ via the Coriolis coupling.

Because this is proportional to the sign of Ω, off-diagonal matrix elements between the parity eigenstates (which are, again, even/odd superpositions of these eigenstates of Ω) will survive, while diagonal matrix elements will vanish (as they must, due to the parity selection rule). So, finally we can construct the matrix of the Stark Hamiltonian, within this subspace of Ω-doublet levels (expressed in the basis of energy and parity eigenstates):

$$H_{\text{St}} = \begin{pmatrix} -\Delta_\Omega/2 & -\frac{D\mathcal{E}_{\text{ext}}}{J(J+1)}m \\ -\frac{D\mathcal{E}_{\text{ext}}}{J(J+1)}m & \Delta_\Omega/2 \end{pmatrix}. \tag{2.60}$$

For the case of $J = 1$, this has energy eigenvalues, $E_\pm$, given by

$$E_\pm = \pm\sqrt{\left(\frac{\Delta_\Omega}{2}\right)^2 + \left(\frac{D\mathcal{E}_{\text{ext}}}{2}\right)^2 m^2}. \tag{2.61}$$

Note that in the high-field limit, such that $D\mathcal{E}_{\text{ext}} \gg \Delta_\Omega$, the levels are fully mixed, and the Stark shift becomes linear in $|\mathcal{E}_{\text{ext}}|$. Because of the tiny splitting Δ_Ω, it does not take a very large value of $\mathcal{E}_{\text{ext}}$ to enter this regime—typically, even $\mathcal{E}_{\text{ext}} \sim$ 10 V/cm is sufficient. Such a small field is dramatically easier to apply in the lab than even the field $\mathcal{E}_{\text{ext}} \sim 30$ kV/cm needed to polarize molecules *without* this Ω-doublet substructure.

In the high-field limit, the energy eigenstates are the fully parity-mixed states, $|Jm \pm\Omega\rangle$. That is: these states really are fully polarized, in the sense that they have $\langle \hat{Z} \cdot \hat{z} \rangle = 1$. The pair of states with $m = \pm 1$ that shifts down in energy when $\mathcal{E}_{\text{ext}}$ is applied must have $\langle \vec{D} \rangle$ parallel to $\vec{\mathcal{E}}_{\text{ext}}$ so that the energy $-\vec{D} \cdot \vec{\mathcal{E}}_{\text{ext}}$ is negative; this pair has molecular polarization $\mathcal{P}_{\text{ext}}^{\text{mol}} = 1$. The pair of $m = \pm 1$ states that goes up in energy must have $\langle \vec{D} \rangle$ antiparallel to $\vec{\mathcal{E}}_{\text{ext}}$, and $\mathcal{P}_{\text{ext}}^{\text{mol}} = -1$. However, the $m = 0$ states don't mix or shift at all (neglecting the much smaller effect from mixing with other, much more distant levels with different values of J); instead, these states retain their character as parity eigenstates. The associated energy level structure is shown in Fig. 2.8.

This structure has some remarkable features in addition to the ability to get full polarization with a very modest size of $\mathcal{E}_{\text{ext}}$. Most striking is that now there are (nominally) identical states with equal but opposite molecular polarization relative to $\vec{\mathcal{E}}_{\text{ext}}$. This makes it possible to do an EDM measurement (i.e., to measure the energy shift between $m = \pm 1$ sublevels) either in one polarized state or the other, by spectroscopically resolving the levels and doing the necessary state preparation and readout in only one or the other of the pairs of $m = \pm 1$ states. The direction of the internuclear axis $\hat{Z} \cdot \hat{z}$, and hence the effective electric field $\vec{\mathcal{E}}_{\text{int}}^{\text{eff}}$ acting on the eEDM, has opposite sign in these two states. Hence, switching between measurements in these two pairs of states has the same effect on the EDM signal as reversing the direction of external field $\vec{\mathcal{E}}_{\text{ext}}$ [(DeMille, Bay, Bickman, Kawall, Hunter, Krause, Maxwell, and Ulmer, 2001)]. Having this "extra" way to reverse the sign of the T, P-odd signal of

(a)

(b)

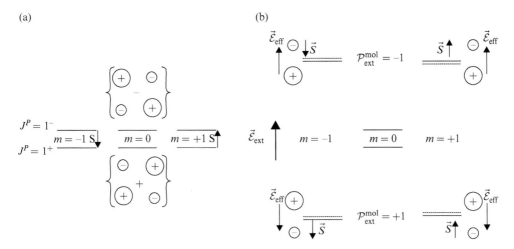

Fig. 2.8 Ω-doublet structure for a $J = 1$ state with $|\Omega| = 1$. (a) Sublevels in the absence of an electric field. (b) Sublevels in the presence of a field $\mathcal{E}_{\text{ext}}$ strong enough to reach $|\mathcal{P}_{\text{ext}}^{\text{mol}}| = 1$. The dashed lines show the additional shifts of the levels that would be present for an eEDM $\vec{d}_e$ in the direction of the spin $\vec{S}$.

interest suppresses many potentially dangerous systematic errors, often by very large factors.

2.8.2   State of the art for electron EDM: the ACME experiment

These new features were used in a recent experiment, known as ACME, that significantly improved the state-of-the art sensitivity to the eEDM and T, P-violating e-N interactions [(Baron et al., 2017)]. ACME uses ThO molecules; the atomic number of Th is $Z = 90$. ThO has two valence electrons, and the ground state has $^1\Sigma$ symmetry. The EDM measurement is done in the first excited state, a metastable triplet state labeled $H^3\Delta_1$ that has lifetime $\tau \approx 2$ ms. The effective field acting on the eEDM in this state is $\mathcal{E}_{\text{eff}} \sim 80$ GV/cm [(Denis and Fleig, 2016; Skripnikov, 2016)]. This state has the Ω-doublet structure described above. It also has an additional, very nice feature first suggested by a team from JILA (E. Meyer, J. Bohn, and E. Cornell). That is: to first order, the magnetic moment of this state vanishes, which suppresses systematic errors from magnetic effects as well as noise due to fluctuating magnetic fields [(Meyer, Bohn, and Deskevich, 2006; Leanhardt et al., 2011)]. This suppression can be seen as follows. From the state label $^3\Delta_1$, you can see that the orbital angular momentum projection along the internuclear axis is $\Lambda = 2$, and the projection of the total electronic angular momentum $\vec{J}_e$ along this axis is $\Omega = 1$. The only way this can occur is if the projection of the spin $S = 1$ along the the internuclear axis, $\Sigma = \vec{S} \cdot \hat{Z}$, is in the direction opposite to $\vec{\Lambda} = \Lambda\hat{Z}$. Since the orbital angular momentum has magnetic g-factor $g_L = 1$ while the spin has $g_S \cong 2$, the magnetic moments due to the antiparallel orbital and spin angular momenta almost exactly cancel. (Note that since we know the energy eigenstates are linear combinations of states with $+\Omega$

and $-\Omega$, we can choose whichever sign is convenient for our discussion; the only important fact here is the *relative* signs of the constituent angular momenta.)

It is also useful to consider what is happening with the individual electrons in this type of molecular state. The ThO molecule can be modeled crudely as $Th^{2+}O^{2-}$. The Th^{2+} ion has two valence electrons in a $7s6d$ orbital configuration, and O^{2-} has closed shells. The $7s$ valence electron has the properties we described for single-electron molecular orbitals before: it has orbital angular momentum projection $\lambda_1 = 0$ and total angular momentum projection $\omega_1 = 1/2$. Hence its spin projection is $\sigma_1 = 1/2$. To end up with a $^3\Delta_1$ state, the second valence electron must have $\lambda_2 = -2$, $\omega_2 = -3/2$, and $\sigma_2 = +1/2$, so that $\Lambda = \lambda_1 + \lambda_2 = -2$ and $\Omega = \omega_1 + \omega_2 = -1$. The first electron has magnetic moment $\mu_1 = g_s\omega_1\mu_B = \mu_B$, while the second electron has $\mu_2 = (g_s\sigma_2 + g_\ell\lambda_2)\mu_B = -\mu_B$, cancelling that due to the first electron. Of course, it is also the presence of the second electron, with $\lambda_2 = 2$, that is responsible for the Ω-doublet structure.

As an aside: you may wonder why we take the first electron's spin to be pinned to the molecular axis here, while before we did not. This is due to the following logic. First, due to the same arguments as in atoms about the exchange interaction, the molecular energy eigenstate has a well-defined total spin $S = 1$. This means that the spins of the first and second electrons are linked. The second electron, which arises from an atomic $6d$ orbital, has a strong spin-orbit interaction. This pins both electron spins to the atomic orbital angular momentum $\vec{L}$. Finally, through the strong intramolecular $\mathcal{E}$-field, $\vec{L}$ is pinned to the internuclear axis as described before. This means, finally, that the spin of the first electron is also pinned to this axis.

Now, back to the specific features of the ACME experiment. It uses Ne buffer gas at ~ 16 K, which cools the molecules to ~ 4 K as it expands to form a beam. A further $\sim 2\times$ improvement in the flux of a particular rotational level of the ground state (here, the $J=1$ level) is achieved by optical pumping of population from the $J=0$, 2, and 3 states into $J=1$, using several lasers. The beam has forward velocity $v \approx 200$ m/s, about $2\times$ slower than the atomic beam in the Tl experiment. The resulting molecular beam brightness in a single sublevel of the $J=1$ state was $B_{mol} \sim 5 \times 10^{13}$/sr/s—some 4 orders of magnitude smaller than in the atomic Tl experiment, yet nearly 1000 times greater than had been available in the YbF experiment. The ground state population was transferred to the metastable $H^3\Delta_1$ state by optical pumping via a short-lived intermediate state with $^3\Pi_0$ symmetry and $J = 0$, in the presence of the polarizing electric field $\mathcal{E}_{ext} \sim 40 - 150$ V/cm. In this weak field, both the ground state and excited state (which do *not* have Ω-doublet substructure) still have eigenstates of good parity, but the $H^3\Delta_1$ state is fully polarized and its eigenstates have fully mixed parity. Hence, decay of the $^3\Pi_0$ $J = 0$ state populates both the upper and lower states of the $H^3\Delta_1$, $J = 1$ Ω-doublet pair, i.e., the states with $\mathcal{P}_{ext}^{mol} = \pm 1$.

Preparation of a spin-aligned state, and readout of the final spin alignment direction, are both done optically. For both these processes, a laser is tuned to resonance with the transition from one pair of $m = \pm 1$ states with a given molecular polarization $\mathcal{P}_{ext}^{mol} = \pm 1$, to an isolated excited state with $J' = 1, m' = 0$ and a definite parity P'. Through a straightforward (but tedious) application of angular momentum algebra, it is possible to show that light with linear polarization

$\epsilon = \cos(\phi)\hat{x} + \sin(\phi)\hat{y}$ couples to a particular "bright" superposition of the spin states, $|\psi_b(\phi)\rangle \propto e^{-i\phi}|m = +1\rangle + qe^{-i\phi}|m = -1\rangle$, with $q = \pm 1$ determined by the value of $P' \cdot \mathcal{P}_{\text{ext}}^{\text{mol}}$. The orthogonal "dark" state, $|\psi_d(\phi)\rangle \propto e^{-i\phi}|m = +1\rangle - qe^{-i\phi}|m = -1\rangle$, is unaffected by the light. Hence, excitation by laser light with polarization $\epsilon = \hat{x}$ acts to deplete the "bright" state, and leave behind population in the orthogonal "dark" state, $|\psi_d(\phi = 0)\rangle \propto |m = +1\rangle - q|m = -1\rangle$. This is the desired initial spin-aligned state. The state then evolves for time $T \sim 1$ ms as it flies ~ 20 cm through the apparatus, till it reaches a probe laser beam. During this time, the states acquire a differential phase $\Phi_{EDM} = -d_e\mathcal{E}_{\text{eff}}T$ due to interaction of the eEDM with the effective electric field, so that the final state before probing is $|\psi_f\rangle \propto e^{-i\Phi_{EDM}}|m = +1\rangle - qe^{-i\Phi_{EDM}}|m = -1\rangle$. The probe laser excites this state with polarization that alternates rapidly between $\phi = \pm\pi/4$, and the induced fluorescence from each polarization state is collected via a system of lenses, optical fiber bundles, and photomultiplier tubes. The difference between the signals for the two probe polarizations is proportional to $\sin(\Phi_{EDM})$.

In the end, ACME detected $\sim 10^3$ photoelectrons/pulse from ~ 1 cm^2 of the molecular beam, at a repetition rate of 50 Hz, over ~ 100 hours, for a total of $N \sim 2 \times 10^{10}$ photoelectrons. It did not see any evidence of a T, P-violating signal, but set the best limit to date on the eEDM: $d_e < 9 \times 10^{-29}e$ ·cm [(Baron et al., 2017)]. As is standard practice in this field, this limit assumes that any T, P-violating signal arises entirely from the eEDM rather than from a possible pseudoscalar-scalar e-N interaction. More formally, the ACME result can be taken as a limit on a particular linear combination of the eEDM d_e and the dimensionless pseudoscalar-scalar interaction strength, C_S.

This result was published in 2014, but in the meantime ACME is improving many aspects of the experiment for better sensitivity in future measurements [(Panda et al., 2016)]. We now use a probe transition where the emitted fluorescence wavelength can be detected with higher quantum efficiency, and we have improved the efficiency of our collection optics, for a gain of ~ 4 in signal. We opened our detection region to ~ 4 cm^2 and shortened the parts of the beam prior to the interaction region, increasing the solid angle of the molecular beam we intercept overall by a factor of ~ 8. We also now use Stimulated Raman Adiabatic Passage (STIRAP) rather than optical pumping to populate the desired superposition of sublevels in the $H^3\Delta_1$ state; this yields a gain of ~ 10 in signal. Altogether, this should allow us to increase our sensitivity to d_e by a factor of $\sim 10 - 20$, by sometime in 2018. This would correspond to probing for the existence of new particles with mass well beyond the range accessible at the LHC, as long as these particles have CP-violating interactions that are not deeply suppressed.

2.8.3 Future prospects for electron EDM experiments

There is at present a very steep curve of improvement in this well-motivated class of searches for new physics. Basically, anything that can improve the signal size and/or the coherence time in experiments with heavy polar molecules is a potential avenue for improvement. Some improvements of this type seem very likely to be implemented. For example, ACME is planning to add an electric or magnetic lens to focus the molecular beam, which can give another factor of $\sim 5 - 20$ increase in signal. In parallel, the Imperial College YbF experiment is now using a cryogenic buffer gas beam source and rotational cooling, which they project will give them enough sensitivity to surpass the original ACME limit [(Rabey et al., 2016)].

However, there are also more revolutionary changes on the near horizon. For example, it has recently become possible to implement optical cycling transitions in molecules [(Shuman et al., 2009)]. This was long considered not to be feasible, because of the branching to different vibrational and rotational levels when a molecular electronic orbital decays after being excited by a laser. However, it has been understood that for certain molecules, on certain transitions, this type of cycling can be achieved [(Di Rosa, 2004)]. In both the YbF and ACME results described above, each molecule scattered only ~ 1 photon, and these photons were detected with modest overall efficiency of $\lesssim 10\%$. Hence, even scattering a few photons can help detect molecules more efficiently. Both ACME and YbF are planning to use cycling schemes for their laser-induced fluorescence detection in coming rounds of the experiments, with the goal to increase signal sizes by a factor of 10–100 (up till 100% detection efficiency per molecule is achieved).

Of course, even more cycling, if possible, enables laser cooling [(Shuman, Barry, and DeMille, 2010)] and even trapping [(Barry et al., 2014)] of molecules, with a potentially huge associated increase in coherence time. Only a few molecular species can be coaxed to scatter the many thousands of photons needed for effective cooling, but, happily, both YbF and BaF (which will be used in a new experiment at Uni. Groningen) are among these species. Both these experiments plan to take advantage of transverse cooling to brighten their molecular beams, and have the potential to push their sensitivity well beyond the current ACME result. The Imperial College team also has long-term plans to use a laser-cooled fountain of YbF [(Tarbutt, et al., 2013)] for very long coherence times and even better sensitivity to the eEDM.

A closely related program is also underway at JILA, using trapped molecular ions (HfF^+) in a metastable $^3\Delta_1$ state [(Cairncross et al., 2017)]. Here, they take advantage of the very small external $\mathcal{E}$-field needed to polarize molecules in Ω-doublet states, and apply this field so that it rotates quickly in time. This keeps the molecular ions polarized, while also keeping them trapped. The JILA group has demonstrated spin coherence times in excess of 700 ms. In this experiment, you *cannot* reverse the external $\mathcal{E}$-field: it must always point towards the center of the trap. However, the effective internal $\mathcal{E}$-field can still be reversed spectroscopically. Using ions intrinsically limits the density and hence also the number of molecules in a given trap volume, due to Coulomb interactions between the ions. Nevertheless, the first generation of this experiment recently reported results with sensitivity very near that of the original ACME limit [(Cairncross et al., 2017)]. Going forward, there are clear ways for the JILA group to improve their state preparation and detection efficiencies, as well as the volume of the trap. They may also be able to use a different species, ThF^+, which has slightly larger $\mathcal{E}_{\text{eff}}$. In ThF^+, the $^3\Delta_1$ state is the ground state, so that the spin coherence time cannot be limited by the lifetime of this state. It is not clear what the ultimate limiting sensitivity will be for this approach.

In the even longer term, it is not hard to imagine an ultimate eEDM experiment. Such an experiment would employ a large, dense sample of ultracold molecules, held in an optical trap, with very long spin coherence times. Much as is being pursued for atomic clocks, here techniques of spin squeezing [(Hosten et al., 2016)] might allow sensitivity that scales with $1/N$ (where N is the number of molecules) rather than $1/\sqrt{N}$ as in the standard quantum limit described earlier. It does not seem crazy to imagine obtaining sensitivity to d_e that is $\gtrsim 4$ orders of magnitude better than the current ACME limit, using such an approach. This would correspond to probing for the existence of particles with masses above 1000 TeV! This is one of the very few plausible ways to probe particle physics at such a high energy scale, and as such seems very ripe with the possibility for a new, fundamental discovery [DeMille, Doyle, and Sushkov, 2017].

2.9 Final remarks

We have discussed some current topics of interest at the intersection of particle physics and atomic/molecular/optical physics. However, there are many things we did not have time to dicuss, but which are nevertheless important and interesting. A comprehensive review of this field, broadly defined, was recently published in (Safronova et al., 2017). However, I highlight here some types of experiments that I find particularly exciting for their ability to probe for new physics at very high energy scales.

Near the top of this list is the search for hadronic CP-violation, via the EDMs of neutrons or protons or via CP-violating interactions between neutrons and protons in a nucleus. These underlying effects give rise to T, P-violating energy shifts much like those described for the eEDM, but now linked to nuclear spin rather than electron spin. The precise mechanism that leads to observable energy shifts in atoms or molecules, in a way that evades Schiff's theorem, is different from that described for the eEDM, and would take similar effort to explain. However, the bottom line is that the best experiments of this type are, like the eEDM searches we discussed, also sensitive to new CP-violating physics at the TeV scale. However, here the sensitivity is related to hadronic particles, which can be affected by new interactions that couple to the "color charge" of the strong interaction. Hence, such experiments are complementary to eEDM searches, which are most sensitive to effects in the leptonic sector.

The most sensitive experiment of this type, by far, is a long-running project at the University of Washington searching for an induced EDM of the ^{199}Hg atom, along its nuclear spin $\vec{I}$ [(Graner, et al., 2016)]. For this type of T, P-violating effect, much as for the eEDM, the much larger polarizability of molecules enhances the observable signal by several orders of magnitude relative to the case of atoms like Hg [(Cho, Sangster, and Hinds, 1991; Khriplovich and Lamoureaux, 1997)]. A new project to take advantage of this amplification, called CeNTREX, is now under construction in my own lab. A competing experimental approach uses ^{225}Ra atoms, where the atomic nucleus itself has a molecule-like structure that enhances the effect of CP-violating interactions inside the nucleus [(Parker, et al., 2015; Bishof, et al., 2016)].

In the recent past, experiments to measure parity (but not time-reversal) violating effects in atoms were very much at the front line of particle physics. Here, the interaction is due to the virtual exchange of Z^0 bosons between a valence electron and the nucleus. The experimental methods are rather different from those we have discussed, and we will not delve into them here. The most accurate of these experiments was a measurement of P-violation in atomic Cs, performed by Carl Wieman and his group at JILA, with an accuracy of $\sim 0.3\%$ [(Wood et al., 1997)]. When combined with very accurate theoretical calculations [(Porsev, Beloy, and Derevianko, 2010; Dzuba, Berengut, Flambaum, and Roberts, 2012)], this experiment was interpreted as a precise measurement of the coupling strength of the Z^0 boson to electrons and nucleons. By comparing the measured value of this coupling to the value predicted by the Standard Model, this result also was sensitive to certain classes of new particles with mass of ~ 1 TeV [(Porsev, Beloy, and Derevianko, 2010)]. Improving the physics reach of this type of measurement would require both a more accurate measurement, and a more accurate theoretical calculation of the size of the effect.

The JILA Cs experiment focused mainly on the so-called "weak charge", Q_W, which encodes the strength of the P-violating coupling between the axial current of the electron and the vector current of nucleons. However, there is also a P-violating coupling between the vector current of the electron and the axial current of nucleons, which is numerically small in the Standard Model and so far measured with rather poor accuracy ($\sim 70\%$ of the predicted value) [(The Jefferson Lab PVDIS Collaboration, et al., 2014)]. An experiment is underway to measure this coupling, by taking advantage of an enhanced signal due to this effect in molecules (somewhat analogous to the enhancement of EDM signals) [(DeMille et al., 2008)]. Such a measurement also could be, in principle, sensitive to different classes of new particles near the TeV scale.

In summary: precision measurements in AMO physics are capable of providing information about new physics appearing at very high energy scales, well beyond the direct reach of any current or conceived high-energy collider. This is an exciting time for this part of the field, with rapid improvements from new ideas and new techologies happening now and for the foreseeable future. Perhaps, with some luck, one of these AMO experiments will be among the first to detect whatever new phenomena nature has in store for us at very high energies.

References

Baron, J. et al. (2014). Order of magnitude smaller limit on the electric dipole moment of the electron. *Science*, **343**(6168), 269–72.

Baron, J. et al. (2017). Methods, analysis, and the treatment of systematic errors for the electron electric dipole moment search in thorium monoxide. *New J. Phys.*, **19**(7), 073029.

Barr, S. M. (1993). A review of CP Violation in atoms. *Int. J. Mod. Phys. A*, **8**(2), 209–36.

Barr, S. M., and Zee, A. (1990). Electric dipole moment of the electron and of the neutron. *Phys. Rev. Lett.*, **65**(1), 21–4.

Barr, S. M. (1992). T-and P-odd electron-nucleon interactions and the electric dipole moments of large atoms. *Phys. Rev. D*, **45**(11), 4148.

Barry, J. F. et al. (2014). Magneto-optical trapping of a diatomic molecule. *Nature*, **512**, 286–9.

Bishof, M. et al. (2016). Improved limit on the ^{225}Ra electric dipole moment. *Phys. Rev. C*, **94**(2), 025501.

Brown, J. M., and Carrington, A. (2003). *Rotational Spectroscopy of Diatomic Molecules*. Cambridge University Press.

Budker, D., Kimball, D. F., and DeMille, D. P. (2004). *Atomic Physics: An Exploration Through Problems and Solutions*. Oxford University Press, Oxford.

Cairncross, W. B. et al. (2017). Precision measurement of the electron's electric dipole moment using trapped molecular ions. *Phys. Rev. Lett.*, **119**, 153001.

Cho, D., Sangster, K., and Hinds, E. A. (1991). Search for time-reversal-symmetry violation in thallium fluoride using a jet source. *Phys. Rev. A*, **44**(5), 2783–99.

Collar, J. I. et al. (2012). New light, weakly coupled particles. In *Fundamental Physics at the Intensity Frontier*, 131–58.

Commins, E. D., Jackson, J. D., and DeMille, D. P. (2007). The electric dipole moment of the electron: an intuitive explanation for the evasion of Schiff's theorem. *Am. J. Phys.*, **75**(6), 532.

Delaunay, et al. (2017). Probing atomic Higgs-like forces at the precision frontier. *Phys. Rev. D*, **96**, 093001.

DeMille, D. et al. (2001). Search for the electric dipole moment of the electron using metastable PbO. In *Art and Symmetry in Experimental Physics*, Volume 596, 72–83. AIP.

DeMille, D. et al. (2008). Using molecules to measure nuclear spin-dependent parity violation. *Phys. Rev. Lett.*, **100**, 023003.

DeMille, D., Doyle, J. M., and Sushkov, A. O. (2017). Probing the frontiers of particle physics with tabletop-scale experiments. *Science*, **357**, 990–4.

Denis, M., and Fleig, T. (2016). In search of discrete symmetry violations beyond the standard model: thorium monoxide reloaded. *J. Chem. Phys.*, **145**(21), 214307.

Di Rosa, M. D. (2004). Laser-cooling molecules. *Eur. Phys. J. D*, **31**, 395–402.

Dzuba, V. A. et al. (2012). Revisiting parity nonconservation in Cesium. *Phys. Rev. Lett.*, **109**(20), 203003.

Engel, J., Ramsey-Musolf, M. J., and van Kolck, U. (2013). Electric dipole moments of nucleons, nuclei, and atoms: the standard model and beyond. *Prog. Part. Nucl. Phys.*, **71**(0), 21–74.

Feng, J. L. (2013). Naturalness and the status of supersymmetry. *Annu. Rev. Nucl. Part. Sci.*, **63**(1), 351–82.

Gabrielse, G. (2013). The standard model's greatest triumph. *Phys. Today*, **66**(12), 64.

Goldhaber, A. S., and Nieto, M. M. (2010). Photon and graviton mass limits. *Rev. Mod. Phys.*, **82**, 939–79.

Graham, P. W. et al. (2015). Experimental searches for the axion and axion-like particles. *Annu. Rev. Nucl. Part. Sci.*, **65**, 485–514.

Graner, B. et al. (2016). Reduced limit on the permanent electric dipole moment of Hg 199. *Phys. Rev. Lett.*, **116**(16), 1–5.

Griffiths, D. J. (2017). *Introduction to Electrodynamics*. Cambridge University Press.

Hosten, O. et al. (2016). Measurement noise 100 times lower than the quantum-projection limit using entangled atoms. *Nature*, **529**, 505–8.

Hudson, J. J. et al. (2011). Improved measurement of the shape of the electron. *Nature*, **473**(7348), 493–6.

Hutzler, N. R., Lu, H-I., and Doyle, J. M. (2012). The buffer gas beam: an intense, cold, and slow source for atoms and molecules. *Chem. Rev.*, **112**(9), 4803–27.

Ibrahim, T., and Nath, P. (2008). *CP* violation from standard model to strings. *Rev. Mod. Phys.*, **80**, 577–631.

Jackson, J. D. (1998). *Classical Electrodynamics* (3rd edn). Wiley, New York.

Jaeckel, J., and Roy, S. (2010). Spectroscopy as a test of Coulomb's law: a probe of the hidden sector. *Phys. Rev. D*, **82**, 125020.

Jung, M., and Pich, A. (2014). Electric dipole moments in two-Higgs-doublet models. *J. High Energy Phys.*, **2014**(4), 76.

Kara, D. M. et al. (2012). Measurement of the electron's electric dipole moment using YbF molecules: methods and data analysis. *New J. Phys.*, **14**(10), 103051.

Khriplovich, I. B., and Lamoureaux, S. K. (1997). *CP Violation without Strangeness*. Springer Verlag, Berlin.

King, W. H. (1984). *Isotope Shifts in Atomic Spectra*. Plenum Press, New York.

Landau, L. D., and Lifshitz, E. M. (1977). *Quantum Mechanics*. Butterworth-Heinemann, Oxford.

Leanhardt, A. E. et al. (2011). High-resolution spectroscopy on trapped molecular ions in rotating electric fields: a new approach for measuring the electron electric dipole moment. *J. Mol. Spectrosc.*, **270**(1), 1–25.

Meyer, E., Bohn, J., and Deskevich, M. (2006). Candidate molecular ions for an electron electric dipole moment experiment. *Phys. Rev. A*, **73**(6), 062108.

Miller, J. P. et al. (2012). Muon ($g - 2$): experiment and theory. *Annu. Rev. Nucl. Part. Sci.*, **62**(1), 237–64.

Mohr, P. J., Newell, D. B., and Taylor, B. N. (2016). CODATA recommended values of the fundamental physical constants: 2014. *Rev. Mod. Phys.*, **88**, 035009.

Nakai, Y., and Reece, M. (2017). Electric dipole moments in natural supersymmetry. *J. High Energy Phys.*, **2017**(8), 31.

Panda, C. D. et al. (2016). Stimulated raman adiabatic passage preparation of a coherent superposition of ThO $H^3\Delta_1$ states for an improved electron electric-dipole-moment measurement. *Phys. Rev. A*, **93**(5), 52110.

Parker, R. H. et al. (2015). First measurement of the atomic electric dipole moment of Ra 225. *Phys. Rev. Lett.*, **114**(23), 233002.

Patrignani, C. et al. (2016). Review of particle physics. *Chin. Phys.*, **C40**(10), 100001.

Porsev, S. G., Beloy, K., and Derevianko, A. (2010). Precision determination of weak charge of ^{133}Cs from atomic parity violation. *Phys. Rev. D*, **82**(3), 36008.

Pospelov, M., and Ritz, A. (2005). Electric dipole moments as probes of new physics. *Ann. Phys. (N. Y).*, **318**(1), 119–69.

Pospelov, M., and Ritz, A. (2014). CKM benchmarks for electron electric dipole moment experiments. *Phys. Rev. D*, **89**(5), 056006.

Rabey, I. et al. (2016). Sensitivity improvements to the YbF electron electric dipole moment experiment. *Bull. Am. Phys. Soc.*

Regan, B. et al. (2002). New limit on the electron electric dipole moment. *Phys. Rev. Lett.*, **88**(7), 18–21.

Safronova, M. et al. (2018). Search for new physics with atoms and molecules. *Rev. Mod. Phys.*, **90**, 025008.

Sakurai, J. J., and Napolitano, J. (2011). *Modern Quantum Mechanics*. Addison-Wesley, Boston.

Shuman, E. S., Barry, J. F., and DeMille, D. (2010). Laser cooling of a diatomic molecule. *Nature*, **467**, 820–3.

Shuman, E. S. et al. (2009). Radiative force from optical cycling on a diatomic molecule. *Phys. Rev. Lett.*, **103**(22), 223001.

Skripnikov, L. V. (2016). Combined 4-component and relativistic pseudopotential study of ThO for the electron electric dipole moment search. *J. Chem. Phys.*, **145**(21), 214301.

Streater, R. F., and Wightman, A. S. (2000). *PCT, Spin and Statistics, and All That*. Princeton Univ. Press, Princeton.

Tarbutt, M. R. et al. (2013). Design for a fountain of YbF molecules to measure the electron's electric dipole moment. *New J. Phys.*, **15**, 053034.

The Jefferson Lab PVDIS Collaboration, Wang, D. et al. (2014). Measurement of parity violation in electron-quark scattering. *Nature*, **506**, 67–70.

Trodden, M. (1999). Electroweak baryogenesis. *Rev. Mod. Phys.*, **71**(5), 1463–99.

Wood, C. S. et al. (1997). Measurement of parity nonconservation and an anapole moment in cesium. *Science*, **275**(5307), 1759–63.

3

Molecular-physics aspects of cold chemistry

FRÉDÉRIC MERKT

Physical Chemistry Laboratory, ETH Zurich
CH-8093 Zurich, Switzerland

Merkt, F. "Molecular-physics aspects of cold chemistry." In *Current Trends in Atomic Physics*. Edited by Antoine Browaeys, Thierry Lahaye, Trey Porto, Charles S. Adams, Matthias Weidemüller, and Leticia F. Cugliandolo. Oxford University Press (2019).
© Oxford University Press. DOI: 10.1093/oso/9780198837190.003.0003

Chapter Contents

3.1   Introduction

These notes cover the content of several lectures given at the 2016 Physics Summer School in Les Houches. Their purpose was to give an introduction to selected aspects of molecular physics relevant to cold chemistry for an audience of young researchers with research interests centered in quantum optics and atomic physics. To avoid a repetition of the content of lectures on molecular structure and molecular spectroscopy, which is covered by numerous textbooks, e.g., (Herzberg, 1989, 1991a,; Brown and Carrington, 2003; Lefebvre-Brion and Field, 2004; Quack and Merkt, 2011), I chose to place the emphasis of the lectures on three aspects of cold molecules: (1) cold few-electron molecules as ideal systems for the comparison of precise measurements and (almost) exact calculations, (2) experimental methods to generate cold samples of few-electron molecules, and (3) studies of reactions involving cold few-electron molecules. This choice allowed me to bypass the main difficulty one faces when trying to provide, in a few hours, a comprehensive description of molecular physics and cold chemistry and which lies in the very large diversity of molecular properties and chemical behavior. At the same time, it enabled me to present simple principles that might be applicable, or be extended, to a wider range of problems.

3.1.1   Chemistry with ultracold atoms and with cold molecules: simple considerations

Cold and ultracold chemistry, one of the topics of the summer school, can be broadly defined as the field devoted to the study of chemical reactions involving low-temperature atomic and molecular samples. The convention has established itself to call samples ultracold and cold if their temperatures lie below or above 1 mK, respectively (Krems et al., 2009; Bell and Softley, 2009; Friedrich and Doyle, 2009). The cold regime extends to temperatures of a few K. Whereas laser cooling is the king's way to generate ultracold samples, cold samples can be produced by a broad range of methods.

The periodic table consists of 118 elements and the nuclide chart lists more than 2900 nuclide, 254 of which are known to be stable and about 340 of which occur naturally. 13924 different diatomic molecules can be produced from these 118 elements and the number of their possible isotopic forms exceeds $4.2 \cdot 10^6$. The number of molecules that can be made out of N atoms literally explodes with N. In addition to the rapid increase in the number of molecules that result from different combinations of atoms, one must also consider that molecules consisting of more than two atoms can exist in several isomeric forms. For instance, HCN and HNC both consist of the same atoms but have different chemical properties. C_3 exists as a linear triatomic molecule but also as a cyclic molecule. Chiral molecules, i.e., molecules that do not possess a plane of symmetry and thus cannot be superimposed on their mirror image, consist of enantiomers having the same isolated-molecule Hamiltonian (neglecting contributions from the electroweak interaction (Quack et al., 2008)) but distinct optical properties. Finally, molecules with indistinguishable atoms may exist in several nuclear-spin isomers. For instance, hydrogen (H_2) exists as para (nuclear-spin singlet, $I = 0$) or ortho (nuclear-spin triplet, $I = 1$) nuclear-spin isomers.

One may ask which molecules are relevant in the field of cold and ultracold chemistry: few-electron diatomic molecules? Large molecules of relevance in biology, such as proteins, DNA strands, and prebiotic molecules? The molecules that have been detected in molecular clouds in the interstellar medium? Molecules with an energy level structure susceptible to provide information relevant to fundamental physics, e.g., by permitting the measurement of a permanent dipole moment of the electron, by enabling a more accurate determination of fundamental constants, or by revealing energy differences in chiral molecules caused by the parity-violating electroweak force, ..., (see, e.g., Quack et al., (2008), DeMille, (2015), Ubachs et al. (2016))? This question can only be answered in very general terms. There must be prospects of generating new insights or obtaining new information by exploiting the low temperatures of the samples, such as obtaining more accurate spectroscopic data by exploiting the reduced Doppler effect and the longer measurement times, of uncovering and explaining unexpected reactivity patterns, or of exploiting specific quantum-mechanical effects to control the outcome of reactions.

Laser-cooling schemes have been developed for only a small fraction of atoms, primarily alkali-metal atoms, alkaline-earth metal atoms, and transition-metal atoms, and for very few diatomic molecules, such as SrF (Norrgard et al., 2016). Consequently, ultracold chemistry is so far a discipline restricted to only very specific classes of compounds, i.e., the atoms and molecules that can be laser cooled and the molecules that can be made out of ultracold atoms through association processes (see, e.g., (Fioretti et al., 1998)). Consequently, only a very restricted space of chemical diversity and behavior can be explored at ultracold temperatures. Experiments in ultracold chemistry are, however, well suited to highlight general patterns of behavior at low temperature, such as the role of individual collision partial waves, of particle-exchange symmetry, and of quantum effects in heavy-particle collisions. They also offer the possibility of controlling chemical reactions through long-range interactions and external electric fields (see, e.g., Ni et al. (2010)).

A broader range of molecules can be prepared in the "cold" regime, and several methods different from laser cooling are available for this purpose: Cold samples of polar molecules and of paramagnetic molecules can be produced by multistage Stark (Bethlem et al., 1999) and Zeeman (Vanhaecke et al., 2007; Narevicius et al., 2008) deceleration of supersonic beams, respectively. Cold-molecule samples can also be generated by sympathetic cooling (Doyle et al., 1995), by billiard-like collisions (Elioff et al., 2003), by deceleration in optical fields (Fulton et al., 2004), by Rydberg-Stark deceleration (Procter et al., 2003), by velocity filtering in thermal samples (Rangwala et al., 2003), in molecular centrifuges (Chervenkov et al., 2014), or by optoelectrical cooling, with which samples of ultracold polar molecules have been produced recently (Prehn et al., 2016). The first scientific applications of these samples are becoming possible and research on cold and ultracold chemistry has gained momentum in the past years, despite the difficulties that are still associated with the cooling of their translational and internal degrees of freedom.

3.1.2 Cold chemistry with simple cold molecules

The generation of cold-molecule samples and the experimental control of their translational and internal degrees of freedom have opened up the possibility to study chemical reactions at temperatures below 1 K. In this temperature range, the thermal collision energy, $k_B T$, is of similar magnitude as the characteristic energy shifts and splittings resulting from the weakest intramolecular interactions and internal-energy contributions, such as the rotational and electron-spin fine structures, the hyperfine structure, weak nonadiabatic interactions, relativistic effects, and even radiative corrections, also known as quantum-electrodynamics (QED) corrections or Lamb shifts. These weak interactions, which are typically neglected in the theoretical treatments of reaction kinetics and dynamics at higher temperatures, may thus play a significant role in chemical reactions at very low temperatures.

The calculation of nonadiabatic, relativistic, and radiative contributions to the energy-level structure of atoms and molecules is a nontrivial task. Accurate calculations of these corrections have only been carried out for a handful of few-electron atoms and molecules. For the one-, two-, and three-electron molecules H_2^+, H_2, He_2^+, the magnitudes of these corrections to the dissociation energies of the lowest bound levels are listed in Table 3.1, where the relativistic and radiative corrections are listed in order of increasing power of their dependence on the fine structure constant α.

Characteristic hyperfine-structure, fine-structure, rotational and vibrational intervals in the ground electronic state of these molecules are presented in Table 3.2.[1] For more "chemical" polyatomic molecules, such as water, ammonia, or benzene, the

Table 3.1 Contributions to the dissociation energies D_0 of H_2^+, H_2, and He_2^+ determined *ab initio*. All values are given in cm^{-1}.

Contribution	H_2^+	H_2	He_2^+
Born-Oppenheimer	21375.930356^a	$36112.59273158\ ^f$	$19954.583^{j,k}$
Adiabatic correction	3.262071^b	$5.7709817(3)^f$	$2.120^{j,l}$
Nonadiabatic correction	0.099912^c	$0.4340331(1)^g$	—
Nonrelativistic energy	21379.29234^d	$36118.7977463(2)^h$	$19116.116^{j,m}$
α^2	0.13859^e	$-0.5318(3)^i$	—
α^3	-0.08017^e	$-0.1948(2)^i$	—
α^4	$-5.5 \cdot 10^{-4\,e}$	$-0.002065(6)^i$	—
Higher terms	$3 \cdot 10^{-5\,e}$	$8.7(59) \cdot 10^{-5\,i}$	—
QED corr. (total)	0.05790^e	$-0.72858(60)^i$	—
Total	$21379.35024(6)^e$	$36118.0691(6)^i$	—

[a]: From a Born-Oppenheimer calculation of $E_I^{BO}(H_2^+)$, as described by Beyer and Merkt (2016b), after subtraction of R_∞; [b]: From the adiabatic corrections of the ground states of H_2^+ and H; [c]: Difference between the nonrelativistic energy and the sum of the Born-Oppenheimer energy and the adiabatic correction; [d]: From Korobov (2006b) and references therein; [e]: Calculated from the expressions reported in Korobov (2008), as described in Sprecher et al. (2011).; [f]: From Pachucki and Komasa (2014); [g]: From Pachucki and Komasa (2015) and (2016a); [h]: From Pachucki and Komasa (2016); [i]: From Puchalski et al. (2016); [j]: From Tung et al. (2012); [k]: D_e; [l]: Calculated at R_e; [m]: Includes the Born-Oppenheimer energy, the adiabatic correction, and an estimate of the nonadiabatic correction through the introduction of an effective R-independent reduced mass.

[1] The values in Tables 3.1 and 3.2 are given as wave numbers $\tilde{\nu} = \Delta E/(hc)$ in the unit cm^{-1} and can be converted into frequency equivalents using $\nu = c\tilde{\nu} = \Delta E/h$ ($1\,cm^{-1} \approx 30\,GHz$) and in temperature equivalents using $T = hc\tilde{\nu}/k_B = \Delta E/k_B$ ($1\,cm^{-1} \approx 1.44\,K$).

Table 3.2 Characteristic hyperfine (hf), fine-structure, rotational (N), and vibrational (v) intervals in the ground electronic state of H_2^+, H_2, and $^4He_2^+$. The rovibrational levels are labeled (v, N). All values are given in cm^{-1}.

Interval	H_2^+	H_2	He_2^+
(1,0)-(0,0)	$2191.15021898(7)^a$	4161.1662^e	$1628.380(4)^f$
(0-2)-(0,0)	$174.23671(7)^b$	354.3732^e	$70.9379(8)^g$
(0-3)-(0,1)	$288.85900(8)^c$	587.0320^e	—
Main hf splitting	0.0462^d	—	—
(Fermi-Contact)			
Spin-rotation splittings	$\approx 2 \cdot 10^{-3\,d}$	—	$\approx 10^{-4\,h}$

[a]: From Korobov et al. (2014); [b]: From Ch. Haase et al. (2015); [c]: From Arcuni et al. (1990); [d]: From Osterwalder et al. (2004); [e]: From Komasa et al. (2011); [f]: From Tung et al. (2012); [g]: From Semeria et al. (2016); [h]: Estimated from the value calculated for $^3He\,^4He^+$ Yu et al. (1989).

relative positions of rotational and vibrational energy levels is accurately known from microwave- and/or infrared-spectroscopic investigations, but reliable calculations of nonadiabatic, relativistic and QED corrections to the energy-level structure of such molecules are currently out of reach.

The ability to carry out close-to-exact calculations in few-electron molecules makes them attractive as test systems for cold-molecule experiments, not so much because of a particular chemical relevance (although low-temperature reactions involving the few-electron atoms H, He$^+$, He, ... and molecules H_2^+, H_2, HeH$^+$, He_2^+, H_3^+ ... are of key importance in astrophysics and astrochemistry (Oka, 2013)), but primarily because of the prospect of reaching a fundamental understanding of low-temperature chemical processes by comparing precise experiments and exact calculations. This aspect has been the guiding principle in preparing these lecture notes.

The structure of these notes is as follows: after an introduction to the quantum mechanical treatment of few-electron molecules in Sec. 3.2, basic aspects of cold chemistry are summarized in Sec. 3.3, with emphasis on ion-molecule reactions. Several such reactions are exothermic and without potential-energy barrier separating reactants and products so that they are usually fast, even at very low temperatures. Sec. 3.4 briefly reviews the methods we use to generate cold samples of H_2^+, H_2, and metastable He$_2$. Sec. 3.5 finally presents two examples of low-temperature reactions involving few-electron reactants, the $H + H^+ \leftrightarrow H_2^+$ reaction system and the $H_2 + H_2^+ \rightarrow H_3^+ + H$ reaction.

3.2 Introduction to the quantum-mechanical treatment of few-electron molecules

The accurate determination of energy levels in few-electron molecules is based either on the direct solution of the Schrödinger equation by treating the electrons and nuclei simultaneously using variational methods (Bishop, 1974; Taylor et al., 1999; Karr and Hilico, 2006; Korobov, 2006b; Korobov et al., 2014; Pachucki and Komasa, 2016) and artificial-channel-scattering methods (Moss, 1993), or on high-level electronic-structure calculations starting with the Born-Oppenheimer approximation (Kołos and Wolniewicz, 1963; Peek, 1965a; Hunter and Pritchard, 1967; Kolos, 1969; Bishop and Wetmore, 1973; Wolniewicz and Poll, 1978, 1986; Wolniewicz and Orlikowski, 1991; Piszczatowski et al., 2009; Pachucki and Komasa, 2009, 2010, 2015). The latter method requires the evaluation of the effects of the coupling between electronic and nuclear motion by variational or perturbative methods. Relativistic effects can be evaluated in separate fully relativistic calculations (Reiher and Wolf, 2015), or starting from nonrelativistic calculations using either variational or perturbative methods (Howells and Kennedy, 1990; Moss, 1993; Korobov, 2008). The QED corrections are typically evaluated using the second route (Bukowski et al., 1992; Korobov, 2004, 2006a; Piszczatowski et al., 2009; Puchalski et al., 2016). The results of these calculations are usually presented as tables of dissociation energies for the bound vibrational and rotational levels of the ground state. They are obtained as sums of energies calculated either directly or on the basis of the Born-Oppenheimer approximation, and a series of corrections that take into account the effects of the coupled nature of the electron and nuclear motions (adiabatic and nonadiabatic corrections), relativistic effects (relativistic corrections) and the interaction with the zero-point electromagnetic field (radiative corrections). The terminology used to designate these corrections needs getting used to. For comparison with experimental results, transition frequencies can then be derived by taking differences between calculated level energies.

Calculations for the one-electron molecules H_2^+, HD^+, and D_2^+ are essentially exact (Karr and Hilico, 2006; Korobov, 2006b; Korobov et al., 2014), i.e., their accuracy approaches the fundamental limit imposed by the uncertainty of the proton-to-electron mass ratio. Comparison of experimental with theoretical results in these molecules thus has the potential to lead to improved values of this ratio.

For the two-electron molecules H_2, HD, and D_2, the accuracy of the best calculations (Piszczatowski et al., 2009; Pachucki and Komasa, 2009, 2010, 2015, 2016) is comparable with that of the most recent experiments (Liu et al., 2009, 2010; Sprecher et al., 2010, 2011). For instance, calculated and experimental values of the dissociation energies agree within the reported uncertainties of about 30 MHz for all three molecules. The reduced accuracy of the calculations, compared to the case of H_2^+, HD^+, and D_2^+, is a consequence of the necessity to treat electron-correlation effects. Currently, rapid progress is being made towards obtaining essentially exact theoretical results also for these two-electron molecules (Pachucki and Komasa, 2016; Puchalski et al., 2016).

For the three-electron molecule He_2^+, the most recent calculations of the energy level structure have only partially evaluated the nonadiabatic corrections and estimated the magnitude of the relativistic and QED corrections. Rovibrational levels of the X $^2\Sigma_u^+$ ground electronic state were calculated with an estimated uncertainty of about 200 MHz (Tung et al., 2012). Experimental results suggest that this estimate might be too optimistic (Sprecher et al., 2014; Jansen et al., 2016).

The next section provides a qualitative introduction to the main steps of the calculations (see also Carrington and Kennedy (1984) and Leach and Moss (1995)) and explains the terminology used in the literature.

3.2.1 The Born-Oppenheimer approximation with the example of H_2^+

The Born-Oppenheimer approximation represents the basis for most *ab initio* calculations of molecular structure. The approximation, which relies on the fact that electrons move much faster than nuclei, separates the electronic and nuclear motions by first solving the Schrödinger equation of the electronic motion for fixed nuclear configurations, leading to Born-Oppenheimer electronic potential-energy curves for diatomic molecules, and to Born-Oppenheimer electronic potential-energy surfaces for polyatomic molecules. The Schrödinger equation of the nuclear motion on a given Born-Oppenheimer potential-energy surface is then solved, neglecting interactions with other electronic states.

Effects beyond the Born-Oppenheimer approximation are usually only included when the potential-energy surfaces associated with different electronic states lie energetically very close. In this case, the characteristic period of the electronic motion, which is given by the inverse spacing between the Born-Oppenheimer potential-energy surfaces, can become as long as, or even longer than, the characteristic periods of the nuclear motion. A breakdown of the approximation and a strong coupling of nuclear and electronic degrees of freedom, called vibronic coupling, result. Prominent examples of vibronic coupling in polyatomic molecules are those arising in the vicinity of conical intersections between two or more potential energy surfaces (Domcke et al., 2004). In molecules with symmetrical nuclear configurations, such conical intersections result in the effects known as Jahn-Teller and Renner-Teller effects (Bersuker, 2006). In other molecules, such intersections are accidental but result in fast and complex dynamics.

Deviations from the Born-Oppenheimer solutions are typically small when the electronic states are energetically well separated. However, small is meant here in comparison to the distance between different electronic states. The deviations may be large compared to molecular rotational energies and, in cold molecules, in comparison to $k_B T$. They are also typically larger than the precision of modern spectroscopic measurements, even when the ground state is separated from all electronically excited states by 10 eV or more, as is the case for H_2 and H_2^+. Calculations in these systems with the accuracy of about $k_B T$ required for the full interpretation of cold-chemistry experiments thus necessitate the inclusion of effects beyond the Born-Oppenheimer approximation.

The general procedure to determine the adiabatic, nonadiabatic, relativistic, and radiative corrections to energies calculated within the Born-Oppenheimer approximation is now briefly outlined with the example of H_2^+.[2]

The starting point is the nonrelativistic Hamiltonian

$$\hat{H} = -\frac{\hbar^2}{2}\left(\frac{\nabla_1^2}{m_1} + \frac{\nabla_2^2}{m_2}\right) - \frac{\hbar^2}{2m_e}\nabla_e^2 + \frac{e^2}{4\pi\epsilon_0}\left(\frac{1}{R} - \frac{1}{r_{1e}} - \frac{1}{r_{2e}}\right), \tag{3.1}$$

---

[2] In this case, which involves only three particles, it is possible and also advantageous to solve the Schrödinger equation without making the Born-Oppenheimer approximation (Korobov 2006b, Korobov, 2008). Its simplicity, however, makes H_2^+ well suited to illustrate the treatment of molecular structure based on the Born-Oppenheimer approximation.

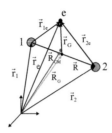

Fig. 3.1 Definition of the coordinates used in the treatment of the H_2^+, HD^+ and D_2^+.

where the indices 1, 2, and e designate the two nuclei and the electron, respectively, and the symbols have their usual meaning. The relevant coordinates are defined in Fig. 3.1, where $\vec{R}_G$ and $\vec{R}_{CM}$ give the positions of the geometric center of the nuclei and the molecular center of mass, respectively. The coordinate system adequate for the Born-Oppenheimer treatment of molecular hydrogen has its origin at the geometric center halfway between the two nuclei

$$\vec{R}_G = \vec{r}_1 + \frac{\vec{r}_2 - \vec{r}_1}{2}. \tag{3.2}$$

The separation of the translational motion of the molecule from the internal degrees of freedom and the separation of electronic and nuclear motions is achieved with the substitutions (see Fig. 3.1):

$$\vec{r}_G = \vec{r}_e - \vec{R}_G, \tag{3.3}$$

$$\vec{R} = \vec{r}_2 - \vec{r}_1, \tag{3.4}$$

and

$$\vec{R}_{CM} = \frac{m_1}{M}\vec{r}_1 + \frac{m_2}{M}\vec{r}_2 + \frac{m_e}{M}\vec{r}_e, \tag{3.5}$$

with $M = m_1 + m_2 + m_e$. The relation between $(\vec{r}_G, \vec{R}, \vec{R}_{CM})$ and $(\vec{r}_1, \vec{r}_2, \vec{r}_e)$ is thus

$$\begin{pmatrix} \vec{r}_G \\ \vec{R} \\ \vec{R}_{CM} \end{pmatrix} = \begin{pmatrix} -1/2 & -1/2 & 1 \\ -1 & 1 & 0 \\ \frac{m_1}{M} & \frac{m_2}{M} & \frac{m_e}{M} \end{pmatrix} \begin{pmatrix} \vec{r}_1 \\ \vec{r}_2 \\ \vec{r}_e \end{pmatrix}. \tag{3.6}$$

Eqs. 3.3–3.5 can be used to express the ∇_i ($i = 1, 2, e$) operators in Eq. 3.1 in terms of the new coordinates $\vec{r}_G$, $\vec{R}$ and $\vec{R}_{CM}$, i.e.,

$$\nabla_1 = \left(\frac{\partial r_G}{\partial r_1}\right)\nabla_{r_G} + \left(\frac{\partial R}{\partial r_1}\right)\nabla_R + \left(\frac{\partial R_{CM}}{\partial r_1}\right)\nabla_{R_{CM}}, \tag{3.7}$$

with analogous expressions for ∇_2 and ∇_e. Evaluating the partial derivatives using Eqs. 3.2–3.5, one obtains the Jacobi matrix relating the nabla opertors

$$\begin{pmatrix} \nabla_1 \\ \nabla_2 \\ \nabla_e \end{pmatrix} = \begin{pmatrix} -1/2 & -1 & \frac{m_1}{M} \\ -1/2 & 1 & \frac{m_2}{M} \\ 1 & 0 & \frac{m_e}{M} \end{pmatrix} \begin{pmatrix} \nabla_{r_G} \\ \nabla_R \\ \nabla_{R_{CM}} \end{pmatrix}, \tag{3.8}$$

from which the Laplace operators in Eq. 3.1 can be determined to be

$$\nabla_1^2 = \frac{\nabla_{r_G}^2}{4} + \nabla_R^2 + \left(\frac{m_1}{M}\right)^2 \nabla_{R_{CM}}^2 + \nabla_{r_G}\nabla_R - \frac{m_1}{M}\nabla_{r_G}\nabla_{R_{CM}} - \frac{2m_1}{M}\nabla_R\nabla_{R_{CM}}, \tag{3.9}$$

$$\nabla_2^2 = \frac{\nabla_{r_G}^2}{4} + \nabla_R^2 + \left(\frac{m_2}{M}\right)^2 \nabla_{R_{CM}}^2 - \nabla_{r_G}\nabla_R - \frac{m_2}{M}\nabla_{r_G}\nabla_{R_{CM}} + \frac{2m_2}{M}\nabla_R\nabla_{R_{CM}}, \tag{3.10}$$

$$\nabla_e^2 = \nabla_{r_G}^2 + \left(\frac{m_e}{M}\right)^2 \nabla_{R_{CM}}^2 + \frac{2m_e}{M}\nabla_{r_G}\nabla_{R_{CM}}. \tag{3.11}$$

The kinetic-energy part of the Hamiltonian (first three terms in Eq. 3.1) can therefore be expressed as

$$\frac{\hat{T}}{\hbar^2} = -\frac{\nabla_{R_{CM}}^2}{2M} - \frac{\nabla_{r_G}^2}{2m_e} - \frac{\nabla_R^2}{2\mu} - \frac{\nabla_{r_G}^2}{8\mu} - \frac{\nabla_{r_G}\nabla_R}{2\mu_\alpha}, \tag{3.12}$$

where $\mu = \frac{m_1 m_2}{m_1+m_2}$ and $\mu_\alpha = \frac{m_1 m_2}{m_2-m_1}$. The first term in Eq. 3.12 describes the kinetic energy of the translation of the molecule in free space, the second term the electron kinetic energy, and the third term the kinetic energy associated with the rovibrational motion of the nuclear framework. The last two terms represent kinetic-energy contributions resulting from the coupled nuclear and electronic degrees of freedom. The last term vanishes for the homonuclear molecules H_2^+ and D_2^+ but is significant in HD^+. This term is sometimes referred to as the g/u-mixing term because it couples states of *gerade* and *ungerade* electronic symmetry.

After separation of the center-of-mass motion, the Schrödinger equation describing the internal motion can be written as

$$\left(\left[-\frac{\hbar^2\nabla_{r_G}^2}{2m_e} - \frac{e^2}{4\pi\epsilon_0}\left(\frac{1}{r_{1e}} + \frac{1}{r_{2e}} - \frac{1}{R}\right)\right] - \hbar^2\left[\frac{\nabla_R^2}{2\mu} + \frac{\nabla_{r_G}^2}{8\mu} + \frac{\nabla_{r_G}\nabla_R}{2\mu_\alpha}\right]\right)\Psi_{int}(\vec{R},\vec{r}_G)$$
$$= E_{int}\Psi_{int}(\vec{R},\vec{r}_G). \tag{3.13}$$

3.2.2 The Born-Oppenheimer solution

In the Born-Oppenheimer approximation, only the terms in the first square bracket are retained and the equation describing the electronic motion is solved for fixed values of the internuclear distance R. The resulting equation

$$\left(-\hbar^2\frac{\nabla^2_{r_G}}{2m_e}+V\right)\psi_i(R;\xi,\eta,\chi)=E_i(R)\psi_i(R;\xi,\eta,\chi),\qquad(3.14)$$

where

$$V=-\frac{e^2}{4\pi\epsilon_0}\left(\frac{1}{r_{1e}}+\frac{1}{r_{2e}}-\frac{1}{R}\right),\qquad(3.15)$$

describes the motion of the electron in the presence of two stationary protons and is separable in the prolate spheroidal coordinates ξ, η and χ, where $\xi=(r_{1e}+r_{2e})/R$, $\eta=(r_{1e}-r_{2e})/R$, and χ is the angle of $\vec{r}_G$ around the internuclear axis. The procedure to determine the eigenvalues $E_i(R)$ and eigenfunctions

$$\psi_i(R,\Lambda;\xi,\eta,\chi)=g_i(R,\Lambda;\xi)f_i(R,\Lambda;\eta)n(\Lambda;\chi)\qquad(3.16)$$

is described, e.g., by Peek (1965b), Hunter and Pritchard (1967), and Carrington and Kennedy (1984). The solutions (see Beyer and Merkt (2016b)) can be expressed as

$$g_i(R,\Lambda;\xi)=(\xi^2-1)^{\Lambda/2}(\xi+1)^{(R/p)-\Lambda-1}\exp\left(-p\xi\right)\sum_{n-0}^{\infty}g_{n,i}(R)\left(\frac{\xi-1}{\xi+1}\right)^n,\qquad(3.17)$$

$$f_i(R,\Lambda;\eta)=\sum_{s=0}^{\infty}f_{s,i}(R)P^{\Lambda}_{\Lambda+s}(\eta),\text{ and}\qquad(3.18)$$

$$n(\Lambda;\chi)=\frac{1}{\sqrt{2\pi}}\exp(i\Lambda\chi),\qquad(3.19)$$

where $P^{\Lambda}_{\Lambda+s}(\eta)$ are associated Legendre polynomials and $p=R\sqrt{-(U^{\mathrm{BO}}(R)-1/R)/2}$. The expansion coefficients $g_{n,i}$ in Eq. 3.17 and $f_{s,i}$ in Eq. 3.18 of solution i are determined by requiring that the eigenvalue A simultaneously satisfies

$$\mathbf{G}\cdot\mathbf{g}=-A\mathbf{g}$$
$$\mathbf{F}\cdot\mathbf{f}=A\mathbf{f},\qquad(3.20)$$

where the tridiagonal matrices $\mathbf{G}$ and $\mathbf{F}$ are given in terms of R, p, and Λ (Hunter, 1967).

The solutions depend parametrically on R and are labeled by (i) the quantum number Λ associated with the projection of the electron orbital angular momentum along the internuclear axis and which is a consequence of the cylindrical symmetry of the electrostatic potential, and (ii) the symmetry (g or u) of the electronic wavefunction under inversion of the electronic coordinates through the geometric center. The dependence of the lowest two eigenvalues on R in the region near the dissociation into H(1s) and H$^+$ fragments is depicted in Fig. 3.2a. The two functions represent the Born-Oppenheimer potentials of the ground X$^+$ $^2\Sigma^+_g$ (solid gray curve)

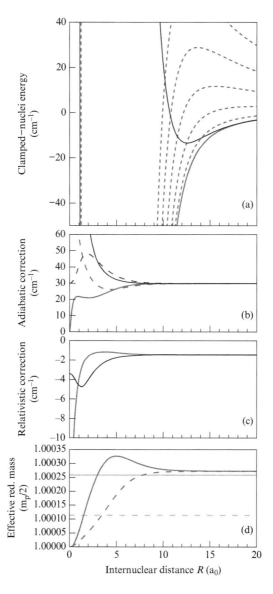

Fig. 3.2 (a) Born-Oppenheimer potential energy function of the ground $X^+ \ ^2\Sigma_g^+$ (solid gray) and first excited $A^+ \ ^2\Sigma_u^+$ (solid black) electronic states of H_2^+, HD^+, and D_2^+ in the vicinity of the $H(1s) + H^+$ dissociation threshold. The dashed curves represent the sum of the Born-Oppenheimer potential and the centrifugal potential for $N = 2$, 4, 6, 8, and 10 in the $X^+ \ ^2\Sigma_g^+$ ground state of para H_2^+. (b) Adiabatic corrections for H_2^+. The full and dashed curves represent the $\langle \hat{H}_1' \rangle$ and $\langle \hat{H}_2' \rangle$ terms (see text). (c) Radiative corrections for H_2^+ from Howells and Kennedy (1990) (d) Reduced masses of the vibrational and rotational motion in the X^+ state of H_2^+ for the effective consideration of nonadiabatic interactions with distant states while still retaining the concept of a potential energy function. Light gray: from Moss (1996); dark gray: from Jaquet and Kutzelnigg (2008). From Beyer and Merkt (2016b).

and first excited A+ $^2\Sigma_u^+$ (solid black curve)[3] electronic states of H_2^+. Because the nuclei are clamped in the Born-Oppenheimer approximation, the potentials and the symmetry labels are independent of isotopic substitution.

3.2.3 Adiabatic and nonadiabatic corrections

The effects of the terms in the second square bracket in Eq. 3.13 are evaluated in the basis of the electronic Born-Oppenheimer solutions. Using the Born Ansatz, the general solution can be written as

$$\Psi_{\text{int}}(\vec{R};\vec{r}_G) = \sum_t F_t(\vec{R})\psi_t(R,\vec{r}_G) = \frac{1}{R}\sum_t \phi_t(\vec{R})\psi_t(R;\vec{r}_G) \qquad (3.21)$$

Inserting Eq. 3.21 in Eq. 3.13, multiplying both sides from the left with ψ_s^*, and integrating over the electronic coordinates gives

$$\int \psi_s^*(R;\vec{r}_G)\left[-\hbar^2\frac{\nabla_{r_G}^2}{2m_e}+V\right]\sum_t F_t(\vec{R})\psi_t(R;\vec{r}_G)d\vec{r}_G$$

$$+\int \psi_s^*(R;\vec{r}_G)\left[-\hbar^2\left(\frac{\nabla_R^2}{2\mu}+\frac{\nabla_{r_G}^2}{8\mu}+\frac{\nabla_{r_G}\nabla_R}{2\mu_\alpha}\right)\right]\sum_t F_t(\vec{R})\psi_t(R;\vec{r}_G)d\vec{r}_G$$

$$= F_s(\vec{R})E_s(R)+\int \psi_s^*(R;\vec{r}_G)\left[-\hbar^2\left(\frac{\nabla_R^2}{2\mu}+\frac{\nabla_{r_G}^2}{8\mu}+\frac{\nabla_{r_G}\nabla_R}{2\mu_\alpha}\right)\right]F_s(\vec{R})\psi_s(R;\vec{r}_G)d\vec{r}_G$$

$$+\int \psi_s^*(R;\vec{r}_G)\left[-\hbar^2\left(\frac{\nabla_R^2}{2\mu}+\frac{\nabla_{r_G}^2}{8\mu}+\frac{\nabla_{r_G}\nabla_R}{2\mu_\alpha}\right)\right]\sum_{t\neq s} F_t(\vec{R})\psi_t(R;\vec{r}_G)d\vec{r}_G$$

$$= \int \psi_s^*(R;\vec{r}_G)E_{\text{int}}\sum_t F_t(\vec{R})\psi_t(R;\vec{r}_G)d\vec{r}_G = E_{\text{int}}F_s(\vec{R}). \qquad (3.22)$$

In this expression, which represents a set of coupled equations, the terms coupling the electronic and nuclear motions have been separated in diagonal terms (second term of the third line of Eq. 3.22), i.e., terms acting only on the electronic state "s", and off-diagonal terms (fourth line of Eq. 3.22), i.e., terms that couple different electronic states.

In the Born-Oppenheimer approximation, all terms coupling the electronic and nuclear motions are neglected. The total wavefunction is thus

$$\Psi_{\text{int},s}(\vec{R},\vec{r}_G) = F_s^{\text{BO}}(\vec{R})\psi_s(R;\vec{r}_G) = \frac{1}{R}\phi_s(\vec{R})\psi_s(R;\vec{r}_G), \qquad (3.23)$$

---

[3] The left superscript of the molecular term symbol stands for the spin multiplicity ($2S + 1$) and the right superscript gives the parity of the electronic function, which is equivalent to the symmetry under reflection through a plane containing the internuclear axis (Lefebvre-Brion and Field, 2004)

and the nuclear motion in the potential $E_s(R) = U_s^{\mathrm{BO}}(R)$ is described by the equation

$$\left[-\hbar^2 \left(\frac{\nabla_R^2}{2\mu}\right) + U_s^{\mathrm{BO}}(R)\right] F_s^{\mathrm{BO}}(\vec{R}) = E_{\mathrm{int},s}^{\mathrm{BO}} F_s^{\mathrm{BO}}(\vec{R}). \tag{3.24}$$

Expressing ∇_R^2 in spherical coordinates (R, θ, Φ) and using the Born Ansatz (see Eq. 3.21) enables one to separate the radial $(R,$ vibrational$)$ motion from the angular $(\theta, \Phi,$ rotational$)$ motion

$$\left[-\frac{\hbar^2}{2\mu}\frac{\partial^2}{\partial R^2} + \frac{\hbar^2 N(N+1)}{2\mu R^2} + U_s^{\mathrm{BO}}(R)\right] \phi_{i,N}(R,\theta,\Phi) = E_{\mathrm{int}}^{\mathrm{BO}} \phi_{i,N}(R,\theta,\Phi), \tag{3.25}$$

where N is the rotational-angular-momentum quantum number. The second term in the square bracket can be regarded as a centrifugal contribution to the potential and modifies it as indicated for the X^+ state by the dashed lines in Fig. 3.2a, which correspond, in order of increasing energies, to $N = 2, 4, 6, 8,$ and 10. The modified potentials now have a potential barrier, called centrifugal barrier, separating the short-range and long-range regions. These barriers and possible metastable states (or shape resonances) located energetically between the dissociation asymptote and the barrier maxima and having large probability densities on the left of the barrier maximum play an important role in low-temperature processes, as will be discussed in more detail in Sec. 3.3 and 3.5. The total molecular wavefunction is a product of electronic (label s) and nuclear rovibrational (labels i, N) functions

$$\Psi_{\mathrm{int},s,i,N}(\vec{R},\vec{r}_{\mathrm{G}}) = \frac{1}{R}\psi_s(R;\vec{r}_{\mathrm{G}})\phi_{i,N}(R,\theta,\Phi)/R. \tag{3.26}$$

In the adiabatic approximation, the off-diagonal terms in Eq. 3.22 are neglected, so that each electronic state can be treated separately. The adiabatic correction is given by the second term of the third line of Eq. 3.22. Because the Born-Oppenheimer solutions have either g or u symmetry, the g/u-mixing term $-\nabla_{r_{\mathrm{G}}}\nabla_R/(2\mu_\alpha)$ of the operator does not contribute to the integral. The eigenvalues and eigenfunctions are determined as in the case of the Born-Oppenheimer approximation, but now using

$$\left[-\hbar^2 \left(\frac{\nabla_R^2}{2\mu}\right) + U_s^{\mathrm{BO}}(R) + \int \psi_s^*(R;\vec{r}_{\mathrm{G}})\left[-\hbar^2 \left(\frac{\nabla_R^2}{2\mu} + \frac{\nabla_{r_{\mathrm{G}}}^2}{8\mu}\right)\right]\psi_s(R;\vec{r}_{\mathrm{G}})d\vec{r}_{\mathrm{G}}\right] F_s^{\mathrm{Ad}}(\vec{R})$$
$$= \left[-\hbar^2 \left(\frac{\nabla_R^2}{2\mu}\right) + U_s^{\mathrm{Ad}}(R)\right] F_s^{\mathrm{Ad}}(\vec{R}) = E_{\mathrm{int},s}^{\mathrm{Ad}} F_s^{\mathrm{Ad}}(\vec{R}). \tag{3.27}$$

instead of Eq. 3.24. This equation can be brought in the same form as Eq. 3.25, with the only difference that the adiabatic potential $U_s^{\mathrm{Ad}}(R) = U_s^{\mathrm{BO}}(R) + \langle \hat{H}_1'\rangle(R) + \langle \hat{H}_2'\rangle(R)$ is used instead of the Born-Oppenheimer potential $U_s^{\mathrm{BO}}(R)$. The two contributions of the adiabatic correction to $U_s^{\mathrm{Ad}}(R)$ can be evaluated separately as

$$\langle \hat{H}_1'\rangle(R) = \int \psi_s^*(r;R)\left[-\hbar^2 \frac{\nabla_R^2}{2\mu}\right]\psi_s(r;R)d\tau \tag{3.28}$$

and

$$\langle \hat{H}'_2 \rangle (R) = \int \psi_s^*(r;R) \left[-\hbar^2 \frac{\nabla_r^2}{8\mu} \right] \psi_s(r;R) \mathrm{d}\tau, \tag{3.29}$$

and are depicted as full and dashed lines in Fig 3.2b for the X^+ (gray) and A^+ (black) states. The adiabatic corrections are typically a few tens of cm^{-1} (or about $2 \cdot 10^6$ MHz) in the ground state of H_2^+, which is very large compared to the precision and accuracy of 1 MHz that can almost routinely be reached experimentally, but very small (i.e., about 10^{-3} times less) compared to the dissociation energy of H_2^+ $(D_0(H_2^+) = 21\,379.350232(49)\,\mathrm{cm}^{-1}$ (see (Moss, 1993) and (Liu et al., 2009)). The adiabatic corrections partially cancel out in the determination of transition frequencies and energy intervals. For instance, the adiabatic correction to the dissociation energy of H_2^+ is only 3.262 cm^{-1} (see Table 3.1) because it is the difference between the adiabatic correction of the H(1s) atom at the H(1s) + H$^+$ dissociation limit (59.765 cm^{-1}) and the adiabatic correction of the X^+ $^2\Sigma_g^+(v^+ = 0, N^+ = 0)$ ground state of H_2^+ (56.503 cm^{-1}).

The adiabatic approximation enables one to partially include the effects of the terms coupling nuclear and electronic degrees of freedom and at the same time to retain the concept of a potential energy function. The approximation is good when the electronic states under consideration are isolated, as is the case for the ground electronic states of H_2^+ and D_2^+. It is less accurate in the case of the ground state of HD$^+$ because the g/u-mixing operator that is neglected in the adiabatic approximation effectively couples the X^+ $^2\Sigma_g^+$ and A^+ $^2\Sigma_u^+$ states. In HD$^+$, the adiabatic approximation even breaks down completely at large internuclear distances where the Born-Oppenheimer potentials of these two states become almost degenerate (See Fig. 3.2a).

The determination of nonadiabatic corrections requires the solution of the system of coupled equations (Eq. 3.22), or the use of a method that includes the nuclear motion in the basis function. The largest nonadiabatic contributions are those from energetically close-lying electronic states and, in some cases, the corrections can be reliably evaluated using a subset of electronic states. Because several electronic states contribute to the solution, the concept of a potential-energy curve (or surface) to describe the structure and dynamics of molecules breaks down. Nonadiabatic effects are frequent in electronically excited states of diatomic molecules and are ubiquitous in polyatomic molecules. They are at the origin of the processes of predissociation, autoionization, and internal conversion. In diatomic molecules, they couple electronic states with the selection rule $\Delta\Lambda = 0, \pm 1$ in Hund's angular-momentum coupling cases (a) and (b) and $\Delta\Omega = 0, \pm 1$ in Hund's angular momentum coupling case (c). In homonuclear diatomic molecules such as H_2^+ and D_2^+, for which the g/u-mixing term vanishes, nonadiabatic couplings preserve the g/u symmetry. In these molecules, g/u mixing can be caused by the hyperfine interaction. Numerous methods and computer codes have been developed to evaluate nonadiabatic couplings and corrections. Their effects on the structure and dynamics of few-electron molecules can be calculated with extreme precision (Piszczatowski et al., 2009; Pachucki and Komasa, 2009, 2015, 2016). Nevertheless, the treatment of nonadiabatic couplings remains involved and cumbersome. An exception is the case where all nonadiabatic couplings are to energetically distant states. In this case, which is encountered in the electronic ground states of H_2^+ and H_2, nonadiabatic corrections can be treated approximately while retaining the concept of a potential energy function. This is achieved by introducing effective, and if necessary R-dependent, reduced masses for the vibrational $[-\frac{\hbar^2}{2\mu_{vib}}\frac{\partial^2}{\partial R^2}$ term in Eq. 3.25] and the rotational $[\frac{\hbar^2 N(N+1)}{2\mu_{rot}R^2}$ term in Eq. 3.25] motion. Possible choices for the reduced masses reported in the literature for H_2^+ (Moss, 1996; Jaquet and Kutzelnigg, 2008) are presented in Fig. 3.2d (see also Sec. 3.5.2).

3.2.4   Relativistic and radiative corrections

The dominant contributions to the relativistic corrections come from the electronic motion and are proportional to the square of the fine-structure constant α. The starting point for their determination is the Breit equation in the Pauli approximation, which enables one to calculate the relativistic corrections using the solutions of the nonrelativistic Hamiltonian (Bethe and Salpeter, 1957). Five terms $\hat{H}_i^{rel}$, $i = 1$–5 contribute, which take into account the variation of mass with velocity ($i = 1$), the retarded interaction involving the electrons ($i = 2$), the spin-orbit coupling, which is zero for Σ states ($i = 3$), the spin-spin interaction, which vanishes in one-electron molecules ($i = 4$), and the Dirac term ($i = 5$).

The most important contributions being diagonal in the adiabatic states, they can be directly added to the adiabatic potentials $U_s^{Ad}(R)$ as corrections

$$\Delta U_s^{rel}(R) = \int \psi_s(R;\vec{r}_G)^* \left(\sum_i \hat{H}_i^{rel}\right) \psi_s(R;\vec{r}_G)d\vec{r}_G. \tag{3.30}$$

The corrected rovibrational eigenvalues are obtained directly from the modified potential. Further efforts are required to include the relativistic corrections in nonadiabatic calculations. The relativistic corrections evaluated by Howells and Kennedy (1990) are depicted in Fig. 3.2c.

The calculation of quantum electrodynamics corrections becomes progressively more challenging with increasing number of electrons. The leading terms of these corrections are proportional to α^3. For their evaluation in one- and two-electron molecules, I refer to the work of Bukowski et al. (1992), Korobov (2008), Piszczatowski et al. (2009), and Puchalski et al. (2016).

3.3   Basic aspects of cold ion-molecule chemistry

Chemical reactions involve breaking and/or forming the chemical bonds that hold atoms together in molecules. The rate v_i of elementary reactions, i.e., reactions leading from reactants to products in a single step, such as unimolecular reactions (isomerization reactions)

$$A \rightarrow B, \tag{3.31}$$

or bimolecular reactions

$$C + D \to E + F, \tag{3.32}$$

are described by rate equations, such as

$$v_1 = k_1 N_A = -dN_A/dt = dN_B/dt, \tag{3.33}$$

for unimolecular reactions, and

$$v_2 = k_1 N_C N_D = -dN_C/dt = -dN_D/dt = dN_E/dt = dN_F/dt \tag{3.34}$$

for bimolecular reactions, where N_i ($i = A - F$) is the number density of atom or molecule i, and k_j ($j = 1, 2$) is the rate coefficient of reaction j. The SI units of k_1 and k_2 are s^{-1} and m^{3}s^{-1}. For many reactions and over broad ranges of temperature, the rate coefficients are well described by Arrhenius' empirical equation

$$k_i = A_i \exp(-E_{\text{act},i}/(k_B T)), \tag{3.35}$$

where A_i is a preexponential factor and $E_{\text{act},\,i}$ is the activation energy of reaction i (in J). The exponential term is also predicted by transition-state theory, which leads to the expressions

$$k_1 = \frac{k_B T}{h} \frac{q^{\ddagger}}{q_A} \exp(-E_{0_1}/(k_B T)), \tag{3.36}$$

and

$$k_2 = \frac{RT}{h} \frac{q^{\ddagger}}{q_C q_D} \exp(-E_{0_2}/(k_B T)), \tag{3.37}$$

for uni- and bi-molecular reactions, respectively. In Eqs. 3.36 and 3.37, $q^{\ddagger}$, q_A, q_C, and q_D represent the molecular partition functions per unit volume of the transition state and the reactants A, C, and D, respectively, and E_{0_i} is the energy difference between the lowest energy level of the transition state and the reactants (see Fig. 3.3a). If $E_{\text{act},\,i}$ and E_{0_i} are positive, the exponential terms in Eqs. 3.35–3.37 go to zero as $T \to 0$.

A chemical reaction may nevertheless proceed at 0 K if there is no barrier along the reaction coordinate (see Fig. 3.3b) or if the reactants are converted into products by quantum-mechanical tunneling through the potential barrier. These two situations are therefore of central importance in cold chemistry. Barrier-free reactions are typically encountered when the reactants are highly reactive compounds, such as free radicals or ions, and capture models are often used to estimate the reaction rates in this case (Clary, 1985; Troe, 1987; Clary, 1990; Troe, 1996). Several bimolecular reactions involving ions have long been known as examples of barrier-free exothermic reactions. A prototypical example of such a reaction is the $H_2^+ + H_2 \to H_3^+ + H$ reaction (Oka, 2013).

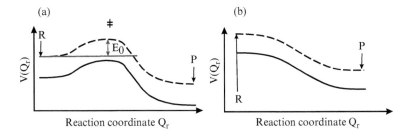

Fig. 3.3 Cut of the potential energy hypersurface along the reaction coordinate Q_r without (full lines) and with (dashed lines) inclusion of the zero-point energy contribution for (a) a reaction with a potential barrier separating the reactants (R) and the products (P) and (b) a barrier-free exothermic reaction.

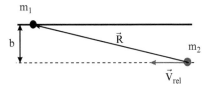

Fig. 3.4 Impact parameter (b) and asymptotic relative velocity $\vec{v}_{rel}$ of an ion (mass m_1) colliding with a neutral molecule (mass m_2).

At low temperatures or low collision energies, quantum-mechanical tunneling through the centrifugal barriers in the interaction potential of the reactants becomes important and can significantly affect the reaction rates. The simplest examples are the recombination of H^+ and H to form H_2^+ and the reverse reaction, the dissociation of H_2^+ near the H $+H^+$ dissociation threshold. These two reaction systems, which have only recently been studied at low temperatures, will be used as illustrations in Sec. 3.5, after introducing basic aspects of capture models in the remainder of this section and presenting the methods of generating the relevant cold samples in Sec. 3.4.

3.3.1 Ion-neutral reactions at low temperatures: Langevin capture models

Numerous reactions between ions (e.g., H_2^+) and polarizable molecules (e.g., H_2) are exothermic barrier-free reactions. At long range, the reactants interact through the ion–induced-dipole term of the electric multipole expansion series. Langevin's classical capture model is found to provide a good description of the reaction rates if the reactions take place with 100 % probability upon close encounter of the reactants. A close encounter is possible only if the collision energy is larger than the centrifugal barrier in the intermolecular potential. An analytical expression of the rate coefficient of such reactions can be determined by considering

- the angular momentum $\vec{L}$ associated with the collision (see Fig. 3.4)

$$\vec{L} = \mu \vec{R} \times \frac{\mathrm{d}\vec{R}}{\mathrm{d}t} \quad \text{and} \quad L = \mu v_{rel} b, \tag{3.38}$$

where μ is the reduced mass $m_1 m_2/(m_1 + m_2)$ of the collision system, $\vec{R}$ the relative position vector of the colliding molecules, $\vec{v}_{\mathrm{rel}}$ their asymptotic relative velocity vector, and b the impact parameter as defined in Fig. 3.4,
- the asymptotic kinetic energy of the collision $E_{\mathrm{coll}} = \mu v_{\mathrm{rel}}^2/2$,
- the charge q of the ion ($q = +e$ in the cases treated here), and
- the polarizability α of the neutral molecule.

Disregarding the internal energy of the collision partners, the total energy of the collision is

$$E_{\mathrm{coll}} = \frac{1}{2}\mu \left(\frac{\mathrm{d}R}{\mathrm{d}t}\right)^2 + V(R) + \frac{L^2}{2\mu R^2} = \frac{1}{2}\mu \left(\frac{\mathrm{d}R}{\mathrm{d}t}\right)^2 + V_{\mathrm{eff},L}(R), \qquad (3.39)$$

where the potential $V_{\mathrm{eff},L}(R)$ includes the centrifugal-potential term $\frac{L^2}{2\mu R^2}$, which is analogous to the second term in the square bracket of Eq. 3.25. In the Langevin capture model, one considers the interaction between the charge of the ion (e) and the dipole $\vec{d}_{\mathrm{ind}}$ of the neutral molecule induced by the field $\vec{F}_{\mathrm{ion}}$ of the ion at the position of the neutral molecule

$$\vec{d}_{\mathrm{ind}} = \alpha \vec{F}_{\mathrm{ion}} = 4\pi\epsilon_0 \alpha' \vec{F}_{\mathrm{ion}}, \qquad (3.40)$$

where α is the polarizability (SI unit C·m^2/V) and $\alpha' = \alpha/(4\pi\epsilon_0)$ the polarizability volume (SI unit m^3) of the neutral species. The interaction potential $V(R)$ is

$$V(R) = -\frac{1}{2}\alpha F_{\mathrm{ion}}^2 = -\frac{1}{2}4\pi\epsilon_0 \alpha' \frac{e^2}{(4\pi\epsilon_0 R^2)^2} = -\frac{\alpha' e^2}{8\pi\epsilon_0 R^4}, \qquad (3.41)$$

where F_{ion} is the electric field strength at the position of the neutral molecule originating from the ion.

Fig. 3.5 shows a model long-range potential in the region of the $-1/R^4$ attractive branch of the ion–induced-dipole interaction (thick gray line) and the effects of the centrifugal potential for several values of the angular momentum L (thin dashed gray lines). The black line in Fig. 3.5 indicates a possible value of the collision energy E_{coll}. Classically, i.e., without considering quantum mechanical tunneling, a close encounter of the reactants (capture) is only possible for angular-momentum values such that the barrier maximum is located energetically below E_{coll}, i.e., in the case of Fig. 3.5 only for angular momenta that are smaller than or equal to the value corresponding to the fourth dashed line. E_{coll} thus determines the maximal value $L_{\max}$ for which the reaction can take place, and via the relation

$$L_{\max} = \mu v_{\mathrm{rel}} b_{\max}, \qquad (3.42)$$

the maximal value $b_{\max}$ of the impact parameter leading to a reaction. The reaction cross section σ is therefore (see Fig. 3.4)

$$\sigma = \pi b_{\max}^2, \qquad (3.43)$$

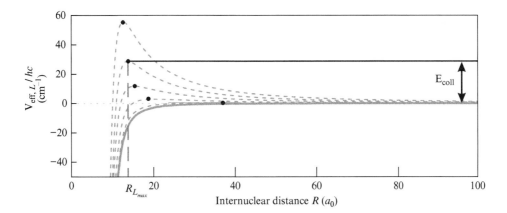

Fig. 3.5 Model long-range potentials based on parameters for the $H + H^+$ collision illustrating the $-1/R^4$ attractive branch of the ion–induced-dipole interaction (thick gray line) and the effects of the centrifugal potential for selected values of the angular momentum L. The black line indicates a possible value of the collision energy.

and can be related to the collision energy by equating E_{coll} to the maximum value of the effective potential, i.e., to $V_{\text{eff},L_{\max}}(R_{L_{\max}})$ (see dashed vertical gray line in Fig. 3.5). This is achieved in a straightforward manner using Eqs. 3.39 and 3.42, which can be used to obtain

$$V_{\text{eff},L_{\max}}(R) = -\frac{\alpha' e^2}{8\pi\epsilon_0 R^4} + \frac{L_{\max}^2}{2\mu R^2} = -\frac{\alpha' e^2}{8\pi\epsilon_0 R^4} + \frac{\mu v_{\text{rel}}^2 b_{\max}^2}{2R^2}, \tag{3.44}$$

and imposing the condition

$$\frac{\mathrm{d}V_{\text{eff},L_{\max}}(R)}{\mathrm{d}R} = 0, \tag{3.45}$$

which yields

$$R_{L_{\max}} = \sqrt{\frac{2\alpha' e^2}{4\pi\epsilon_0 \mu v_{\text{rel}}^2 b_{\max}^2}} \tag{3.46}$$

and

$$V_{\text{eff},L_{\max}}(R_{L_{\max}}) = \frac{\pi\epsilon_0 \mu^2 v_{\text{rel}}^4 b_{\max}^4}{2\alpha' e^2}. \tag{3.47}$$

From $V_{\text{eff},L_{\max}}(R_{L_{\max}}) = E_{\text{coll}} = \mu v_{\text{rel}}^2/2$, the cross section is found to be

$$\sigma = \pi b_{\max}^2 = \sqrt{\frac{\pi\alpha' e^2}{\epsilon_0 \mu v_{\text{rel}}^2}} = \sqrt{\frac{\pi\alpha' e^2}{2\epsilon_0 E_{\text{coll}}}}. \tag{3.48}$$

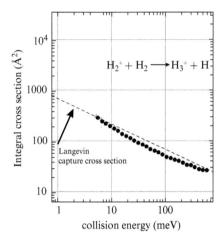

Fig. 3.6 Cross section of the $H_2^+ + H_2 \to H_3^+ + H$ reaction, as measured by Glenewinkel-Meyer and Gerlich (1997) (black dots). The dashed line represents the cross section predicted with the classical Langevin capture model.

The rate coefficient of a bimolecular reaction is the product of the cross section and the relative asymptotic velocity. For a reaction obeying Langevin capture, the rate coefficient k_L is therefore given by

$$k_L = \sigma v_{\text{rel}} = \sqrt{\frac{\pi \alpha' e^2}{\epsilon_0 \mu}}, \tag{3.49}$$

and is independent of the collision energy. For reactions involving reactants in thermal equilibrium, the thermal Langevin-capture rate coefficient $k_L(T)$ is thus independent of the temperature. The $H_2^+ + H_2 \to H_3^+ + H$ reaction represents a classical example of a reaction that closely follows Langevin-capture behavior (Oka, 2013). As illustration, Fig. 3.6 compares the cross section of this reaction measured by (Glenewinkel-Meyer and Gerlich, 1997) (black dots) with the cross section calculated with Eq. 3.48 (dashed line). Until recently, experimental data were not available below 5 meV, or about $k_B \cdot 60$ K.

3.3.2   Breakdown of classical Langevin-capture models at low temperature

Although Fig. 3.5 has been drawn for a quantized angular momentum, the Langevin-capture model is a purely classical model. The classical limit corresponds to large collision energies or high temperatures, for which a very large number of collisional partial waves contribute to the reaction rate. As the temperature or the collision energy decreases, fewer and fewer partial waves contribute until none of the partial waves satisfies the condition $E_{\text{coll}} \geq V_{\text{eff}, L_{\max}}(R_{L_{\max}})$, except $L = 0$. This limit is known as the s-wave scattering limit, or the Bethe-Wigner limit. In this limit, one knows from the Bethe-Wigner law (see, e.g., Landau and Lifshitz (1977)) that the capture rate coefficients of such reactions must deviate from the classical Langevin behavior and reach their quantum (q) s-wave-scattering values. These values can be expressed as "universal" ratios $\frac{k_{q,j}}{k_L}$ to the Langevin-capture rate coefficients (the index j designates the rotational level of the neutral molecule). "Universal" refers here to the fact that the s-wave scattering cross section is entirely determined by the long-range potential and scales, for a $1/R^4$ Langevin-capture potential, with the system parameters μ and α, as in Eq. 3.48.

At very low collision energies, for which the classical Langevin model clearly breaks down, the de Broglie wavelength associated to the relative velocity of the collision partners is larger than the size of the reacting atoms or molecules. For collisions of homonuclear diatomic molecules in their ground rotational state ($j = 0$) with ions, for instance, $\frac{k_{q,j=0}}{k_L} = 2$ (Vogt and Wannier, 1954; Fabrikant and Hotop, 2001; Dashevskaya et al. (2005); Gao, 2011). For $j \geq 1$ rotational levels of the neutral molecule, the quadrupole of the rotating molecule must also be considered. The charge-quadrupole interaction and Coriolis interactions between levels of different values of the projection quantum number associated with the diatomic-molecule rotational angular momentum onto the collision axis lead to $\frac{k_{q,j}}{k_L}$ ratios that differ from 2 (Dashevskaya et al., 2005). Strong deviations from Eq. 3.49 are also expected at low collision energies when the neutral molecule possesses a permanent dipole moment (Clary, 1990; Troe, 1996; Auzinsh et al., 2013a,) or has an open-shell electronic structure (Wickham and Clary, 1993; Auzinsh et al., 2008; Klippenstein et al., 2010).

As the temperature increases, the number of partial waves contributing to the scattering rapidly increases until the relative motion of the reactants becomes classical. (Gao, 2010) and (2011) has used generalized multichannel quantum-defect theory to predict the universal behavior of the capture rate coefficients of ion-molecule collisions on a $- 1/R^4$ potential in this range. His calculations clearly reveal the transition from quantum ($k/k_L = 2$) to classical ($k/k_L = 1$) capture as the number of contributing partial waves gradually increases. His calculated energy-dependent rate coefficient shows weak oscillations as a function of the collision energy, with maxima corresponding to the maxima of the contributions of individual partial waves (see Figs. 1 and 2 of (Gao, 2011)). Similar oscillations have also been predicted by (Dashevskaya et al., 2016) (see also Sec. 3.5.1 below), who had previously also predicted significant additional deviations at very low collision energies in systems where the polarizable neutral species is a homonuclear diatomic molecule in a rotationally excited state (see Dashevskaya et al. (2005) and discussion above).

The transition from the quantum to the classical regime of relative motion of the reactants takes place in the millikelvin range for ion-molecule reactions involving the lightest species and at even lower temperatures for heavier species and has not been observed experimentally so far. At such low temperatures, the rotational motion of the neutral reactant is frozen or locked by the anisotropic intermolecular potential. As the temperature increases and reaches the range 1-20 K, the rotational degrees of freedom gradually unlock themselves but remain strongly perturbed by the intermolecular potential (Dashevskaya et al., 2005; Maergoiz et al., 2009). This range is particularly interesting: The rate coefficients are predicted to be strongly influenced by the intermolecular potential and by subtle effects, such as Coriolis interactions arising from the coupling between the relative motion and the rotation of the neutral molecule. In this range, which also remains unexplored experimentally, the hindered rotation of the neutral molecule often leads to strongly enhanced rate coefficients (Dashevskaya et al., 2005; Maergoiz et al., 2009).

At higher temperatures, beyond about 20 K for reactions involving light molecules, the rotation of the neutral molecule gradually becomes classical. The expressions derived by Su and Chesnavich from trajectory simulations (Su and Chesnavich, 1982) become adequate to describe the rate coefficients in this range, as has now been verified experimentally in measurements using buffer-gas-cooled ion traps (Barlow et al., 1986); (Gerlich, 2008); (Wester, 2009) and uniform supersonic flows [CRESU method, for Cinétique de Réaction en Ecoulement Supersonique Uniforme (Rowe et al., 1985; Marquette et al., 1985; Rowe et al., 1995; Smith and Rowe, 2000)], (see discussion in Maergoiz et al. (2009)). The vibrational motion of the neutral molecule only becomes classical at temperatures or collision energies where the effects of the short-range part of the intermolecular potential become dominant and Langevin-capture models, which only consider long-range interactions, break down.

3.4 Cold samples by supersonic-beam deceleration methods

The methods we use to generate cold molecules rely on the use of pulsed supersonic beams. Such beams are formed when a gas is allowed to expand from a high-pressure reservoir into vacuum through the orifice of a pulsed valve (Scoles, 1992). The adiabatic expansion leads to a rapid cooling of the internal (vibrational and rotational) degrees of freedom and to gas pulses propagating at supersonic velocities in the expansion direction. In the experiments, one usually uses a skimmer to select the central part of the beam, which is the densest and also the coldest. In this way, one also strongly reduces the velocity distribution perpendicular to the beam propagation axis. The velocity distribution along the propagation axis is also narrow and usually corresponds to temperatures around 1 K. The beam velocity, however, is typically high, although it can be controlled to some extent by adjusting the temperature of the gas reservoir and by entraining the molecules of interest in a specific carrier gas, usually a rare gas. The final velocity of the beam is then determined by the mass M of the carrier gas, with which it scales as $1/\sqrt{M}$, and the temperature T_{res} of the reservoir, with which it scales as $\sqrt{T_{\text{res}}}$. Their high directionality and the almost collision-free environment make supersonic beams ideal for use in spectroscopic and reaction-dynamics experiments, and indeed such beams have been and still are an essential experimental tool of gas-phase physical chemistry and chemical physics (see, e.g., Anderson (1974), Levy (1980), Rettner et al. (1984), and Scoles (1992)).

A significant advance in the control of the final velocity of supersonic beams has been made with the introduction of supersonic-beam-deceleration methods relying on the use of time-dependent inhomogeneous electric and magnetic fields. Such methods exploit the Stark and Zeeman effects in atoms and molecules. They were pioneered for polar molecules by (Bethlem et al., 1999) and were extended to paramagnetic atoms and molecules (Vanhaecke et al., 2007; Hogan et al., 2007; Narevicius et al.,

2008) and to Rydberg atoms and molecules (Procter et al., 2003; Vliegen et al., 2004). The principles and applications of beam-deceleration methods have been reviewed extensively in the past years (Hogan et al., 2011; van de Meerakker et al., 2012; Narevicius and Raizen, 2012).

In the presence of electric and magnetic fields, the energy levels of an atom or a molecule are shifted according to the Stark and Zeeman effects and the shifts are called Stark shifts and Zeeman shifts, respectively. The Stark shifts display a linear or a quadratic dependence on the electric field strength F at low fields, depending on whether the quantum states coupled by the field are degenerate or not. The Zeeman shifts tend to be linear at low magnetic field strength B, the magnetic field having the effect of lifting the degeneracy of the magnetic sublevels in free space. Quantum systems undergoing a linear Stark or Zeeman effect can be thought of as having an electric ($\vec{\mu}_{el}$) or magnetic ($\vec{\mu}_{mag}$) dipole moment which directly couples to the applied external field. The linear shifts can be expressed as

$$E_{Stark} = -\vec{\mu}_{el} \cdot \vec{F}, \quad E_{Zeeman} = -\vec{\mu}_{mag} \cdot \vec{B}. \tag{3.50}$$

A quantum state the energy of which is lowered (raised) in the presence of a field is often referred to as a "red-shifted" (blue-shifted) state, or a "high-field-seeking" (low-field-seeking) state. The latter designation can be understood from the fact that particles undergoing a Stark or a Zeeman effect are subject to a force $\vec{f}$ in inhomogeneous fields

$$\vec{f}_{Stark} = -\nabla E_{Stark}, \quad \vec{f}_{Zeeman} = -\nabla E_{Zeeman} \tag{3.51}$$

which is proportional to the field gradient. In inhomogeneous fields, particles in low-(high-)field-seeking states are accelerated in the direction of decreasing (increasing) field strength. If the force is directed perpendicularly to the beam propagation axis, it can be exploited to deflect an atomic beam or split it into different components, as in the well-known experiments of (Gerlach and Stern, 1922). If the acceleration vector points in the direction parallel or antiparallel to the beam propagation axis, the particles in the beam are accelerated or decelerated in the longitudinal direction. The magnetic moments associated with nuclear spins are much smaller than those arising from electron spins so that the forces originating from the nuclear Zeeman effect are to a good approximation negligible compared to those arising from the electron Zeeman effect.

The Stark shifts of typical polar molecules in their ground or low-lying metastable states are of the order of 1 cm^{-1} for the electric fields of a several kV/cm used in the experiments, and so are the Zeeman shifts of paramagnetic molecules in magnetic fields of a few T. Such shifts only represent a small fraction of the kinetic energy of the atoms and molecules in supersonic beams. Consequently, full deceleration requires the repetition of deceleration cycles in successive stages of a decelerator as the beam propagates. Care has to be taken that the multistage deceleration remains stable both in longitudinal and transverse directions, as discussed by van de Meerakker et al. (2006), Wiederkehr et al. (2010), Hogan et al. (2011), and van de Meerakker et al. (2012), to which I refer for details.

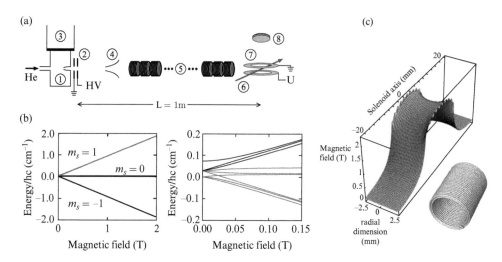

Fig. 3.7 (a) Schematic representation of the supersonic beam source of metastable He_2 molecules and the multistage Zeeman decelerator used to slow down the beam (not to scale). 1: Pulsed valve body and orifice, 2: Discharge electrodes, 3: Cryogenic-liquid container, 4: skimmer, 5: multistage Zeeman decelerator, 6: UV laser beam, 7: extraction electrodes, 8: MCP. The supersonic beam propagation direction, the UV laser beam, and the ion extraction direction are mutually orthogonal. From Motsch et al., (2014). (b) Zeeman effect in the $N'' = 1$ rotational ground state of $He_2(a\,^3\Sigma_u'')$. The low-field-seeking states, for which the deceleration pulse sequence is optimized, are indicated in gray. The low-field regime of the Zeeman effect is depicted on the right panel. From Motsch et al., (2014). (c) Distribution of magnetic field strength generated by the depicted coil operated at a current of 300 A. Adapted from Hogan et al., (2007).

The following sections briefly describe the instruments we use to generate cold samples of few-electron atoms and molecules. H atoms are paramagnetic in their (1s) $^2S_{1/2}$ ground state. He and He_2 molecules are paramagnetic in their metastable triplet state [(1s)(2s) $^3S_1$ state for He, and $(1\sigma_g)^2(1\sigma_u)(2s\sigma_g)$ a $^3\Sigma_u^+$ state for He_2]. All three species can be decelerated in a multistage Zeeman decelerator (Vanhaecke et al., 2007; Motsch et al., 2014). H_2 molecules are nonpolar and diamagnetic in their $(1\sigma_g)^2X\,^1\Sigma_g^+$ ground state and the properties (in particular the lifetimes) of known triplet states of H_2 are not suitable for multistage Zeeman deceleration. To decelerate H_2 molecules, we use Rydberg-Stark deceleration, which exploits the very large electric dipole moments of Rydberg-Stark states (Procter et al., 2003; Hogan et al., 2009).

3.4.1 Multistage Zeeman deceleration and trapping

The multistage Zeeman decelerator we employ to produce cold metastable He_2 molecules is depicted in Fig. 3.7 and described in detail by Motsch et al. (2014), upon which the following text is based.

The pulsed supersonic beam is created by expanding helium gas into vacuum from a reservoir held at high stagnation pressure (typically 2–6 bar) through the 250 μm-diameter ceramic orifice of a modified Even-Lavie valve (Even et al., 2000). To efficiently cool the valve, its entire housing, originally made of stainless steel, was replaced by copper components. The valve assembly is mounted on a reservoir which can be filled with cryogenic liquids, either liquid nitrogen for operation at 77 K, or liquid helium for operation down to 4 K. The temperature of the valve assembly is monitored with a silicon diode and can be regulated.

The metastable helium molecules are generated in a plasma formed by striking a discharge between two cylindrical electrodes separated by 2 mm and directly attached to the front plate of the valve. To obtain a clean pulse of metastable He_2, the discharge is ignited by applying a typically 5 μs long, $-$ 600 V pulse to the front electrode as soon as the density of the helium beam is sufficiently high. In the discharge, the metastable $(1s)(2s)^1S_0$ and $^3S_1$ states of atomic helium are populated, and the metastable helium dimer molecules are formed. To facilitate ignition of the discharge during the short duration of the gas pulse (typically a few tens of microseconds), the discharge is seeded with electrons emitted from a hot tungsten filament (Halfmann et al., 2000; Wiederkehr et al., 2011). The discharge and nozzle operation parameters (electric potentials, timing, stagnation pressure) are optimized by measuring the yield of He_2^* which we monitor by photoionization followed by mass-selective detection of He_2^+ in a linear time-of-flight mass spectrometer depicted schematically on the right side of Fig. 3.7a.

After passing through a skimmer, the beam of metastable He_2 molecules enters the Zeeman decelerator, which consists of an array of solenoids through which currents of up to 300 A are pulsed on and off, as described in detail by (Wiederkehr et al., 2011).

The deceleration relies on the conversion of the kinetic energy of the metastable He_2 molecules into potential (Zeeman) energy when the molecules enter the solenoids and experience a growing magnetic field. The magnetic field distribution generated by a coil operated at a current of 300 A is depicted in Fig. 3.7c and the level shifts resulting from the Zeeman effect of the $N = 1$ rotational level of the a $^3\Sigma_u^+(v = 0)$ state of He_2 are displayed in Fig. 3.7b. The states affected by the deceleration are those marked in gray and correspond to the $J = 0$ ($M_J = 0$) and $J = 2$ ($M_J = 2, 1$) spin-rotational components, which correlate at high fields to the $m_S = 1$ Zeeman component of the electron-spin triplet. Molecules in these low-field-seeking states slow down as they move toward the center point of the solenoid, where the magnetic field along the decelerator axis is maximal. The current through the coil is switched off before this point is reached so that the packet of decelerated molecules is bunched in the propagation direction. In this way, the faster molecules in the beam are located closer to the solenoid center than the slower molecules when the field is turned off. Consequently they are decelerated more, which keeps the packet of decelerated molecules together as it propagates from one solenoid to the next, where the process is repeated. This bunching principle is the same as that used in multistage Stark deceleration (Bethlem et al., 1999). The number of coils, the currents that are pulsed through the coils, and their switch-off times depend on the desired final velocity and are determined in particle-trajectory simulations.

An important feature of the magnetic field distribution is the fact that the field grows as one moves away from the decelerator axis. The low-field-seeking molecules, for which the deceleration is optimized, are therefore also radially confined close to the decelerator axis, and it is possible to operate the decelerator in a phase-stable manner,

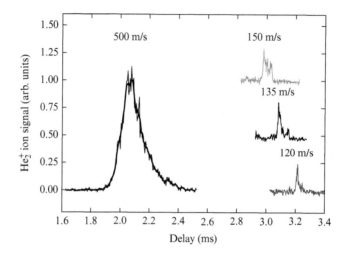

Fig. 3.8 Time-of-flight profiles of He_2^* detected 60 mm beyond the last stage of the decelerator obtained by detection through UV photoionization. The black trace with a broad TOF distribution centered at 500 m/s corresponds to the undecelerated beam. The three distributions presented at times of flight between 2.9 ms and 3.3 ms were obtained using deceleration pulse sequences designed to decelerate molecules with initial velocities of 505 m/s, 500 m/s, and 495 m/s, respectively. From Motsch et al., (2014).

in transverse and longitudinal directions, as explained in detail by Wiederkehr et al. (2010).

Fig. 3.8 presents distributions of flight times of the metastable He_2 molecules from the entrance of the decelerator to the point where they are detected by photoionization. The black trace with a broad TOF distribution centered at 500 m/s corresponds to the undecelerated beam. The three distributions presented at times of flight between 2.9 ms and 3.3 ms were obtained using deceleration pulse sequences designed to decelerate molecules with initial velocities of 505 m/s, 500 m/s, and 495 m/s, respectively. The final velocities of the decelerated molecules, as determined by comparing the measured time-of-flight profiles with the results of numerical particle-trajectory simulations, are indicated above the traces and are 150, 135, and 120 m/s, respectively.

The multistage decelerator also acts as a quantum-state filter that selects only low-field-seeking molecules. In the case of the a $^3\Sigma_u^+$ state of He_2, each rotational level (rotational quantum number N) is split into three spin-rotational components by the spin-spin and the spin-rotational interactions, with total angular momentum quantum number $J = N$, $N \pm 1$ (Lichten et al., 1974). The spin-rotational components with $J = N$ are high-field-seeking and are rejected by the decelerator, those with $J = N - 1$ are low-field-seeking and are transmitted by the decelerator and those with $J = N + 1$ have both low and high-field-seeking magnetic components and are only partially transmitted (see Fig. 3.7b). This property turns out to be very useful when assigning spectra. Indeed, the comparison of spectra obtained with and without operating the decelerator enables the straightforward assignment of the spin-rotational components. This is illustrated in Fig. 3.9, which compares a segment of the Rydberg spectrum of

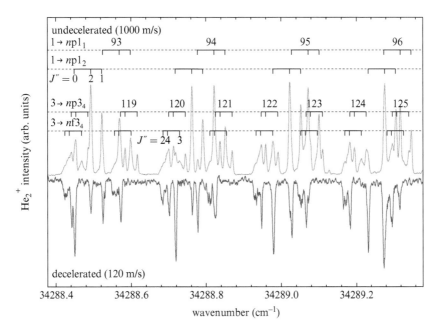

Fig. 3.9 Comparison between Rydberg spectra of He$_2$ obtained from nondecelerated (1000 m/s) and decelerated (120 m/s) metastable-He$_2$ samples. J'' designates the total angular-momentum quantum number of the metastable a $^3\Sigma_u^+(v''=0,N'')$ rotational levels of He$_2$ and corresponds to the vector addition $\vec{J}'' = \vec{N}'' + \vec{S}$. The transitions are labeled $N'' \to$ nppN_N^+, where N^+ is the rotational quantum number of the He$_2^+$X$^{+2}\Sigma_u^+(v^+=0)$ ion-core level to which the Rydberg series converge, N is the quantum number resulting from the vector addition $\vec{N} = \vec{N}^+ + \vec{\ell}$, and n is the principal quantum number. The values of n are given above the triplet assignment bars reflecting the relative positions of the $J'' = N'', N'' \pm 1$ triplet structure of the metastable levels. From Jansen et al., (2015).

metastable He$_2$ obtained with (lower, downward-pointing trace) and without (upper, upward-pointing trace) deceleration. The lines of the latter spectrum that are missing in the former spectrum are those originating from the spin-rotational components with $J = N$, as discussed in more detail by Jansen et al. (2015).

Other advantages of the deceleration, which become crucial in precision spectroscopic measurements, are the reduction of the Doppler width and the longer measurement times that are made possible by the reduced beam velocity. Both effects contribute to reduce the widths of the measured transitions and to achieve a higher precision in the determination of transition frequencies. Systematic errors in the transition frequencies originating from possible deviations of the supersonic-beam propagation direction from the assumed propagation direction are also reduced by the deceleration, and so are systematic uncertainties from the second-order (relativistic) Doppler effect.

By extrapolating the Rydberg series of metastable He$_2$ measured in spectra such as that depicted in Fig. 3.9, one can determine the relative positions of the rotational and vibrational levels of the X$^+$ $^2\Sigma_u^+$ ground state of He$_2^+$ (Ginter and Ginter, 1980;

Table 3.3 Rotational term values of the X^+ $^2\Sigma_u^+(\nu^+=0)$ state of $^4\mathrm{He}_2^+$ obtained by Rydberg-series extrapolation from the Rydberg spectrum of He$_2$ and comparison with the term values calculated *ab initio* by Tung et al. (2012). All values are given in cm^{-1} relative to the $N^+=1$ ground state.

N^+	Tung et al. (2012) *ab initio*	Semeria et al. (2016) Rydberg series extrapolation
1	0	0
3	70.936	70.9379 ± 0.0008
5	198.359	198.3647 ± 0.0008
7	381.822	381.8346 ± 0.0008
9	620.683	620.7021 ± 0.0009
11	914.112	914.1367 ± 0.0008
13	1261.089	1261.1242 ± 0.0008
15	1660.420	1660.4627 ± 0.0009
17	2110.736	2110.7932 ± 0.0009
19	2610.505	2610.5744 ± 0.0009

Raunhardt et al., 2008; Sprecher et al., 2014). These positions are very difficult to determine in spectroscopic measurements carried out on the He$_2^+$ ion because $^4\mathrm{He}_2^+$ does not have electric-dipoled allowed rotational and vibrational transitions. The only direct spectroscopic data obtained on He$_2^+$ available in the literature were obtained in studies of $^3\mathrm{He}^4\mathrm{He}^+$ by infrared spectroscopy by Yu and Wing (1987). Unlike $^4\mathrm{He}_2^+$, $^3\mathrm{He}^4\mathrm{He}^+$ has a permanent electric dipole moment resulting from the displacement of the center of charge from the center of mass, which makes rotational and vibrational transitions observable. To illustrate the possibility of determining the level structure of $^4\mathrm{He}_2^+$ from measurements such as those presented in Fig. 3.9, Table 3.3 presents the rotational term values of $^4\mathrm{He}_2^+$ extracted from the Rydberg spectrum of He$_2$ (Semeria et al., 2016) and compares them with the results of the latest *ab initio* quantum-chemical calculations (Tung et al., 2012), which include adiabatic corrections but only partially consider nonadiabatic corrections and neglect relativistic and radiative corrections. The comparison enables one to quantify the magnitude of the correction terms not evaluated in the theoretical treatment, which increase rapidly with the rotational energy and are about 2.3 GHz at $N^+=19$.

3.4.2 Rydberg-Stark deceleration, deflection, and trapping

Before discussing the principles of Rydberg-Stark deceleration, this section starts with a short introduction to Rydberg states taken with only few modifications from Wörner and Merkt (2011). Rydberg states are electronically excited states, in which one of the electrons (the Rydberg electron) has been excited to a hydrogen-like orbital having a principal quantum number n larger than the quantum number of the valence shell. Rydberg states can thus be regarded in good approximation as consisting of a positively charged ion core around which the Rydberg electron orbits at large distances. Many properties of these states can therefore be understood from the properties of the electronic states of the hydrogen atom. The expectation value of the distance between the electron and the core increases as n^2, and the amplitude of the Rydberg electron wavefunction in the immediate vicinity of the core decreases as $n^{-3/2}$ so that the electron density in the region of the positively charged core decreases as n^{-3}. The electron density in the core region also decreases rapidly with the orbital angular momentum quantum number ℓ because of the centrifugal barrier in the electron-ion-core interaction potential, which is proportional to $\ell(\ell + 1)$.

Rydberg states with a given value of ℓ but different values of n $(n > \ell)$ form infinite series of electronic states called Rydberg series. The energetic positions of the different members of a given Rydberg series can be described in good approximation by Rydberg's formula

$$E_{n\ell m} = E_{\mathrm{I}}(\alpha^+) - \frac{hcR_{\mathrm{M}}}{(n - \delta_\ell)^2}, \tag{3.52}$$

where $E_{\mathrm{I}}(\alpha^+)$ represents the energy of a given quantum state α^+ of the ionized atom (or molecule), R_{M} is the mass-corrected Rydberg constant, and δ_ℓ is the quantum defect, which is to a good approximation constant in a given series. δ_ℓ only appreciably differs from zero in s, p, and d Rydberg states and rapidly decreases with increasing ℓ value.

Fig. 3.10 depicts the energy level structure characteristic of Rydberg states of the hydrogen atom (panel a), of atoms having more than one electron (panel b), and of molecules (panel c) at high n values. In atoms with more than one electron, the energy level structure resembles closely that of the hydrogen atom, with the difference that low-ℓ states are displaced to lower energies because the Rydberg electron is exposed to an increasing nuclear charge when it penetrates through the inner electron shells (their quantum defect is positive). In molecules, the situation is additionally complicated by the fact that series of core-penetrating and nonpenetrating Rydberg states converge on every rotational (denoted by N^+ in Fig. 3.10), vibrational (denoted by v^+), and electronic state of the molecular cation. Because the potential that binds the Rydberg electron to the positively charged ion core can be roughly approximated by a Coulomb potential, the Rydberg-electron wavefunctions in atoms and molecules are labeled by the same quantum numbers.

Most physical properties of Rydberg states scale as integer powers of n. For instance, their polarizability scales as n^7, the classical radius of the electron orbit as n^2, the threshold field for field ionization as n^{-4}, the spacing between adjacent states of a given Rydberg series as n^{-3}, the absorption cross section from the ground state as n^{-3}, the transition moment between neighboring Rydberg states of the same series as n^2, ... (Gallagher, 1994).

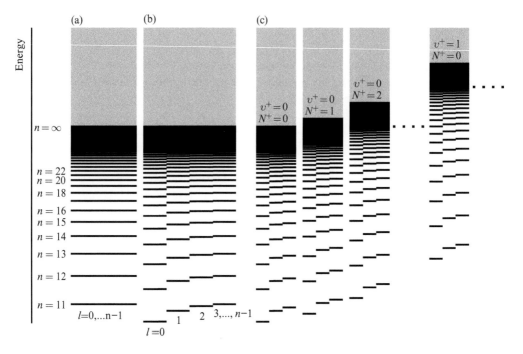

Fig. 3.10 Energy level structure of the Rydberg states of (a) the hydrogen atom, (b) polyelectronic atoms, and (c) molecules.

Because of the degenerate nature of all $|n\ell m >$ states having the same n value in the hydrogen atom at zero field, the Stark effect leads to a linear splitting of the degenerate energy levels (the linear Stark effect) represented in Fig. 3.11a for $n = 5$. Because the electric field breaks the central symmetry, the orbital angular-momentum quantum number ℓ ceases to be a good quantum number. Instead, states of different angular-momentum quantum number ℓ but the same value of m are mixed by the electric field. The atoms are polarized by the electric field, as illustrated in Fig. 3.11b, in which the electron density in a plane containing the z axis, which is the axis along which the electric-field vector points, is displayed for the five $n = 5$, $m = 0$ levels. The energy levels are more conveniently labeled by the parabolic quantum numbers n_1 and n_2 that arise in the solution of the Schrödinger equation of the H atom in parabolic coordinates (Gallagher, 1994) than by ℓ, because the $|nn_1 n_2 m >$ basis functions are adapted to the cylindrical symmetry of the problem. The parabolic coordinates (ξ, η, and ϕ) of the Rydberg electron are related to the spherical and cartesian coordinates through $\xi = r + z$, $\eta = r - z$ and $\phi = \tan^{-1}(x/y)$.[4] The quantum numbers n_1 and n_2 can each take the values between 0 and $n - 1$. The energy levels are given to first order of perturbation theory by

---

[4] Note the analogy with H_2^+ discussed in Sec. 3.2.1. Both problems have in common that the electron moves in a cylindrically symmetric potential.

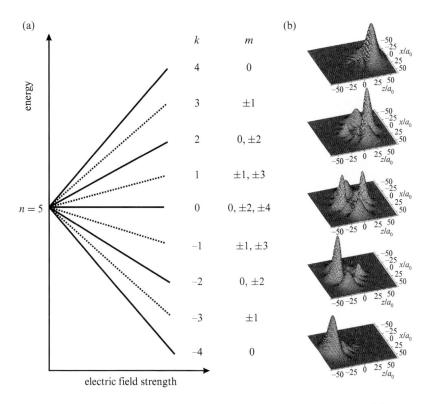

Fig. 3.11 The linear Stark effect in the $n = 5$ state of atomic hydrogen. (a) Energy level diagram. (b) Electron density in the xz plane of the $m = 0$ states. The position of the nucleus is at the origin of the coordinate system. k represents the difference $n_1 - n_2$. From Wörner and Merkt (2011).

$$E_{n n_1 n_2 m} = -\frac{hcR_{\mathrm{H}}}{n^2} + \frac{3}{2}ea_0(n_1 - n_2)nF. \qquad (3.53)$$

and depend linearly on the electric field strength F. To label the states, it is useful to use the difference k between n_1 and n_2. For given values of n and $|m|$, k takes values ranging from $- (n - |m| - 1)$ to $(n - |m| - 1)$ in steps of 2 (see Fig. 3.11a). The plots of the electron density represented in Fig. 3.11b enable one to see that all states except the $k = 0$ states have electric dipole moments and to understand why the states with a positive value of k are shifted to higher energies by the field whereas those with a negative value of k are shifted to lower energies. The states with largest Stark shifts have an electic dipole moment μ_{el} of $1.5ea_0n^2$ which, at $n = 30$, exceeds 3000 Debye.

Rydberg-Stark deceleration exploits the very large dipole moments of Rydberg-Stark states to exert forces on Rydberg atoms and molecules (see Eq. 3.51). These forces can be so large that supersonic beams containing light atoms and molecules can be decelerated to zero velocity and trapped using a single deceleration stage (Vliegen et al., 2007; Hogan and Merkt, 2008).

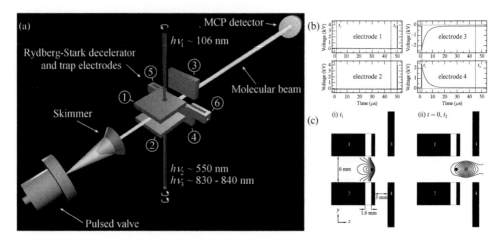

Fig. 3.12 (a) Schematic diagram of the experimental setup used to decelerate and trap cold H_2 Rydberg molecules. (b) Time dependence of the potentials applied to electrodes 1, 2, 3, and 4 for deceleration and trapping experiments. Excitation takes place at $t = 0$ μs and detection at $t = 50$ μs. The exponential decay of the potentials applied to electrodes 3 and 4 have a time constant $\tau_{1/e} = 3.65$ μs and an amplitude of 2.3 kV. (c) Section of the trap in the plane defined by the laser and supersonic beams, with lines of constant electric field strength during deceleration (time t_1) (i) from 100 V/cm to 1000 V/cm in steps of 100 V/cm and before and after deceleration (times $t = 0$ and t_2) (ii) from 30 V/cm to 300 V/cm in steps of 30 V/cm indicated in (b). The VUV, VIS, and IR laser beams propagate along the y direction through two holes in electrodes 1 and 2. Adapted from Seiler et al. (2011).

The single-stage decelerator we use to decelerate and trap Rydberg atoms and molecules is depicted schematically in Fig. 3.12a. With this device, cold samples of H Rydberg atoms and H_2 Rydberg molecules have been trapped (Hogan and Merkt 2008; Hogan et al. 2009; Seiler et al. 2011a). The device consists of six electrodes, four of which (electrodes 1–4) are used for deceleration and trapping in a quadrupole-trap configuration. The remaining two electrodes (electrodes 5 and 6) serve the purpose of closing the electric trap in the third dimension.

The experiments begin by photoexcitation of the atoms or molecules in the beam to Rydberg-Stark states in the homogeneous-field region between electrodes 1 and 2. When molecules are decelerated, it is essential to excite Rydberg-Stark states of core-nonpenetrating character to avoid the decay of the Rydberg states by predissociation or autoionization. In the case of H_2, we use the three-photon excitation sequence

$$nkm[\text{X}^{+\,2}\Sigma_g^+\,(0,0)] \leftarrow \text{I}\,^1\Pi_g(0,2) \leftarrow \text{B}\,^1\Sigma_u^+\,(3,1) \leftarrow \text{X}\,^1\Sigma_g^+\,(0,0). \tag{3.54}$$

to access the Rydberg states through the intermediate B and I states. These states are convenient because they have predominantly p and d character, respectively, and enable one to access the nonpenetrating f states of H_2 under field-free conditions and the nonpenetrating $m = 3$ Stark states through strong electronic transitions analogous to the excitation $nf \leftarrow 3d \leftarrow 2p \leftarrow 1s$ in the H atom. To restrict excitation to the

$m = 3$ Stark states, which do not have any s, p, or d character, the laser radiation used for all three steps of the excitation sequence is chosen to be circularly polarized with the same helicity. The Rydberg-Stark states produced in this way have lifetimes of more than 100 μs, which is amply sufficient for deceleration and trapping (Seiler et al., 2011b, 2016).

Deceleration and trapping is achieved by applying time-dependent potentials to the electrodes 1–4 with the time dependence depicted in Fig. 3.12b. At the beginning of the deceleration process, these potentials generate a very strong field gradient at the position of the atoms (given as black dot in Fig. 3.12c) opposing the motion of the Rydberg molecules (or atoms) prepared optically in low-field-seeking Rydberg-Stark states (see left part of Fig. 3.12c). As the atoms are decelerated and move forward, the potential decreases so that the deceleration proceeds at approximately constant field and adiabatically. The final potential values are chosen so as to form a quadrupolar electric trap at the location (gray dot) of the Rydberg molecules at the end of the deceleration (see right part of Fig. 3.12c). The Rydberg molecules then remain in the electric trap until they decay by radiative or nonradiative processes, or by collisions with ground-state or Rydberg molecules, as described in detail by Seiler et al., (2012, 2016).

The trapping process can be detected by recording spectra of the trapped molecules, as illustrated in Fig. 3.13. This figure compares the spectrum of the transition to the $n = 22$, $l = 3$ Rydberg state of H_2 recorded with the excitation sequence (3.54) under field-free conditions (panel (a)), in the presence of an electric field of 278 V/cm but without trapping (panel (b)), and after deceleration and trapping for 50 μs with time-dependent potential of maximal values of 1.7 and 2.3 kV (panels (c) and (d), respectively). The field-free spectrum consists of only one line because the circularly polarized laser radiation restricts the excitation to f Rydberg states only. All optically accessible $n = 22$, $m = 3$ Stark states are observed in panel (b). Panels (c) and (d) show that only low-field-seeking Stark states with a sufficiently large electric dipole moment are trapped (see Eq. 3.53). The field gradients achieved with 2.3 kV are larger than with 1.7 kV and, consequently, H_2 molecules in $k = 10$ Stark states are more efficiently decelerated and trapped, as can be seen by comparing panels (c) and (d).

A weakness of the deceleration and trapping methods illustrated by Figs. 3.12 and 3.13 is that the deceleration is achieved along the propagation axis of the supersonic beam. The loss of Rydberg atoms or molecules by collisions with other atoms and molecules in the beam, and particularly with those in the tail of the gas pulse, is significant. Moreover, the nondecelerated atoms and molecules in the gas beam render the investigation of cold collisions involving the decelerated atoms or molecules extremely difficult. The nondecelerated particles are also directed at the collision target, making it difficult to distinguish processes involving the decelerated molecules from those involving nondecelerated molecules. This problem can be overcome by deflecting the atoms or molecules during deceleration and by loading them into off-axis electric traps, as demonstrated for H Rydberg atoms and H_2 Rydberg molecules by Seiler et al., (2011b) and Seiler et al. (2012), respectively.

The Rydberg-Stark decelerators and traps described in this section rely on the motion of electric traps and differ in this respect from the multistage Stark and

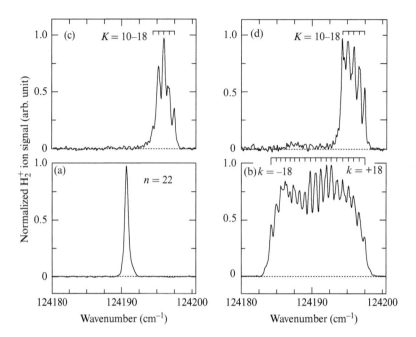

Fig. 3.13 (a) Field-free spectrum of $n = 22$, $\ell = 3$, $m = 3$ Rydberg state of H_2 detected by pulsed field ionization $t_{ion} = 3$ μs after photoexcitation. (b) Spectrum of $n = 22$, $m = 3$ Stark manifold recorded in an electric field of 278 V/cm with detection at $t_{ion} = 3$ μs. Spectrum of $n = 22$, $m = 3$ Stark states of H_2 detected at $t_{ion} = 50$ μs following deceleration and trapping with pulsed potentials of (c) 1.7 kV, and (d) 2.3 kV applied for deceleration and showing only low-field-seeking $k = 10 - 18$ Rydberg-Stark states. Adapted from Hogan et al. (2009).

Zeeman decelerators described in Sec. 3.4.1. They were the first realization of moving-trap decelerators, which have now been developed both for polar molecules (Oster-walder et al., 2010) and paramagnetic atoms and molecules (Trimeche et al., 2011; Lavert-Ofir et al., 2011). Chip-based moving-trap decelerators have also been developed and rely on the application of time-dependent potentials on parallel electrodes located at the surface of chips (Meek et al., 2008).

3.4.3   Rydberg-Stark deceleration, deflection, and trapping using surface-electrode decelerators

This section describes the operational principle of surface-electrode Rydberg-Stark decelerators. Such decelerators were first described by (Hogan et al., 2012; All-mendinger et al., 2013), from which the following text is inspired.

The surface-electrode decelerator for Rydberg atoms and molecules (Hogan et al., 2012; Allmendinger et al., 2013) depicted in Fig. 3.14 consists of 44 parallel surface electrodes printed on a circuit board and was used to decelerate and trap He Rydberg atoms. The width of 0.5 mm and center-to-center spacing $d_z = 1.0$ mm between the electrodes in z direction was chosen to match the decelerator acceptance to the

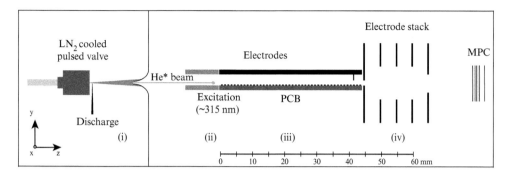

Fig. 3.14 Schematic diagram of the 44-electrode Rydberg-Stark decelerator used to decelerate beams of helium atoms moving initially at a velocity of 1200 m/s. (i) Source chamber; (ii) photoexcitation region; (iii) surface-electrode decelerator; (iv) detection region. From Allmendinger et al. (2013).

phase-space characteristics of the Rydberg-atom beam. The ends of the electrodes along the x dimension are widened to reduce the oscillatory motion of the potential minima in this dimension during deceleration and so avoid heating of the decelerated sample.

The Rydberg-Stark states are prepared in a homogeneous-field region preceding the decelerator in the same manner as described in Sec. 3.4.2. When the Rydberg atoms or molecules approach the decelerator, they are loaded into electric traps moving at the speed of the supersonic beam. The surface-electrode decelerator is operated by applying six different oscillating electrical potentials V_i to the first six surface electrodes (with index $i = 1 - 6$) in Fig. 3.14 and repeating the sequence from the seventh electrode on. These potentials (called waveforms hereafter) have the general form $V_i = (-1)^i V_0 [1 + \cos(\omega t + \phi_i)]$, where $2V_0$ is the peak-to-peak amplitude, ω is the oscillation frequency and $\phi_i = (i - 1)2\pi/3$. Using this configuration, a set of moving electric traps, separated by a distance of $3d_z = 3$ mm, is generated above the decelerator surface. The depth of the electric traps can be adjusted by changing the value of V_0 or by changing the dipole moment of the Rydberg-Stark states, i.e., by exciting Rydberg-Stark states of specific n and k values with the lasers. The switch-on time of the waveforms is selected so as to ensure that all Rydberg atoms prepared optically are loaded into a single trap. As the potentials oscillate, this trap moves in the positive z direction with a velocity $v_z = 3d_z\omega/(2\pi)$. Acceleration/deceleration, a_z, is achieved by applying a linear frequency chirp $\Delta\omega(t)$ to the waveforms such that

$$\omega(t) = \omega_0 + \Delta\omega(t) = \omega_0 + (2\pi/3d_z)a_z t, \tag{3.55}$$

where ω_0 corresponds to the initial velocity v_0. The acceleration and the linear chirp are therefore connected by the linear relation

$$a_z = 3d_z\Delta\omega(t)/(2\pi t). \tag{3.56}$$

The experiments are carried out with a pulsed supersonic beam of pure helium. Metastable He atoms are produced near the orifice of the pulsed valve, using the method described in Sec. 3.4.1 for the production of metastable He_2 molecules. They are excited to Rydberg-Stark states in an homogeneous-field region between two parallel plates located just before the surface-electrode decelerator. The Rydberg atoms then approach the decelerator at the speed of the supersonic beam, i.e., 1200 m/s when the valve is cooled to 77 K with liquid N_2. The waveforms are switched on only when the atoms are above the chip to avoid undesirable heating by the gradients of the fringe fields at the edges of the chip. During the switch-on and -off processes, ω remains constant and V_0 is ramped linearly.

The operation of the decelerator is illustrated in Fig. 3.15, which shows the distributions of flight times needed by the atoms to move from the position of photoexcitation (gray dot in Fig. 3.14) to the middle of the electrode stack located at the end of the decelerator, where the Rydberg atoms are field ionized by applying a large potential difference across the electrode stack. The figure compares a set of

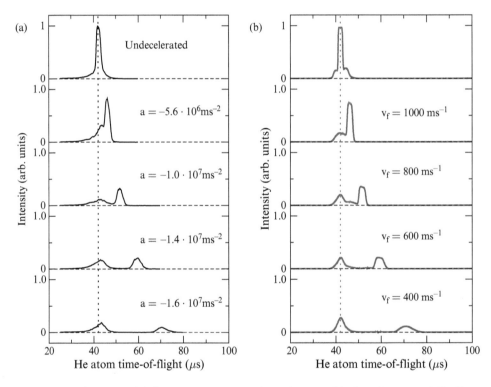

Fig. 3.15 Deceleration of helium atoms prepared in the $n = 30$, $k = 23$, $m = 0$ Rydberg-Stark state. The deceleration was achieved using waveforms designed for accelerations of $-5.6 \cdot 10^6$, $-1.0 \cdot 10^7$, $-1.4 \cdot 10^7$, and $-1.6 \cdot 10^7$ ms^{-2}, and final velocities of 1000, 800, 600, and 400 m/s, respectively, as indicated above the corresponding traces. The traces were obtained by monitoring the pulsed-field ionization signal at the end of the decelerator as a function of the delay time between laser excitation and the application of the field-ionization pulse. (a) Experimental results. (b) Results of particle-trajectory simulations. The position corresponding to the undecelerated beam (velocity of 1200 m/s) is marked by a dashed vertical line. From Allmendinger et al. (2013).

flight-time distributions corresponding to decelerations of Rydberg He atoms from $v_i = 1200$ m/s to final velocities between 1000 m/s and 400 m/s with distributions generated by numerical particle-trajectory simulations (panel (b)). The different measurements only differ in the chirp used to achieve the desired final velocity. The time-of-flight distribution shown in the top panel corresponds to the undecelerated beam of Rydberg atoms, which is detected at a time of ~ 42 μs (vertical dotted line). The final velocities of the atoms can be inferred from the simulations because these do not involve any adjustable parameters and accurately reproduce the experimental results.

To deflect Rydberg-atom (or Rydberg-molecule) beams during deceleration, we use a curved surface decelerator depicted schematically in the left panel of Fig. 3.16. It was designed to deflect the initial beam by $10°$ (Allmendinger et al., 2014). The principle of this device is identical to that described above. Because of the moving-trap nature of the deceleration, the Rydberg atoms (or molecules) follow the decelerator surface.

Typical results obtained by operating this device as deflector are presented in the images labeled (a)–(c) in the right panel of Fig. 3.16. The experiments were carried out with H_2 Rydberg molecules which were prepared using the three-photon excitation sequence (3.54).

The three images displayed in Fig. 3.16 were obtained using an microchannel-plate (MCP) detector coupled to a phosphor screen and a CCD camera. The signal was generated by the impact of the long-lived H_2 Rydberg molecules on the MCP surface. The deflected beam corresponds to the sharp signals at the bottom of the

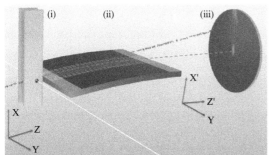 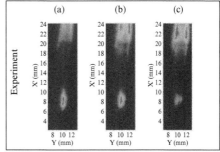

Fig. 3.16 Left panel: Schematic representation of the surface-electrode deflector used to deviate H_2 Rydberg molecules in a supersonic beam by $10°$ from the initial beam-propagation axis. (i) Photoexcitation region where the Rydberg molecules are prepared in the homogeneous electric field generated by applying dc electric potentials to two parallel planar electrodes separated by 4 mm. (ii) 50-mm-long printed circuit board with 50 equidistant copper electrodes. The first 4-mm-long and the last 3.5-mm-long sections along the beam-propagation axis are planar whereas the middle 42.5-mm-long section forms an arc of a circle with a radius of 244 mm. (iii) MCP detector with anode connected to a phosphor screen and a CCD camera for imaging. Right panel: Measured spatial distributions of the H_2 Rydberg molecules ($n = 31$, $k = 7 - 11$) detected at the imaging MCP detector following deflection at a constant velocity of 1290 m/s using waveform amplitudes of (a) 80 V, (b) 60 V, and (c) 40 V. From Allmendinger et al. (2014).

images whereas the more diffuse signals at the top of the images originate from Rydberg molecules that could not be loaded into the moving traps. The depth of the moving traps and thus the trap-loading efficiency can be adjusted by changing the amplitude V_0 of the waveforms. When the deflector is operated with $V_0 = 80$ V, almost all Rydberg molecules are deflected, but the number of deflected molecules rapidly decreases with decreasing V_0 value. Although low values of V_0 lead to less dense beams of deflected molecules, the temperature of the deflected sample, which we found to be directly proportional to the depth of the moving traps, is lower.

Deceleration or acceleration can be achieved during deflection, which makes this device ideally suited to the study of low-energy collisions in merged-beam experiments, as will be demonstrated in Sec. 3.5.1.

3.4.4   Ion-neutral reactions within a Rydberg-electron orbit

Studies of ion-molecule reactions in buffer-gas-cooled ion traps (Barlow, Luine, and Dunn, 1986; Gerlich, 2008; Wester, 2009) and in uniform supersonic flows (Rowe et al., 1985; Marquette et al., 1985; Rowe et al., 1995; Smith and Rowe, 2000), represent the state of the art in low-temperature ion-molecule chemistry (Smith, 2011). These methods can be used to obtain rate coefficients down to around 20 K. Estimates of rate coefficients at such low temperatures can also be obtained by extrapolation from measurements carried out at higher temperatures using a suitable capture model (Snow and Bierbaum, 2008). The study of ion-molecule reactions at very low temperatures or very low collision energies is complicated by stray electric fields, which accelerate ions. In the presence of a stray field of 1 mV/cm, which corresponds to a potential difference of only 1 mV across an experimental volume with a diameter of 1 cm, ions distributed over this volume are accelerated to kinetic energies of up to 1 meV (or 8 cm^{-1}, or 12 K), and this effect prevents studies of collisions at low collision energies or low temperatures.

In their pioneering investigation of the $H_2^+ + H_2 \rightarrow H_3^+ + H$ reaction, Glenewinkel-Meyer and Gerlich (1997) therefore only measured the cross section down to collision energies $E_{coll}/k_B = 60$ K. Their data, presented as black dots in Fig. 3.6, nicely illustrate that the classical Langevin capture model provides a good approximation for this reaction (see Sec. 3.3.1). To reach lower temperatures or collision energies, the effects of stray fields have to be eliminated. Laser cooling of alkaline-earth-metal atoms in ion traps can be used to cool ions to the ground motional state in the trap. The ions are then confined in space and the effect of stray fields is reduced.

Cold molecular-ion samples, down to the millikelvin range, can be produced by sympathetic cooling with laser-cooled ions in ion crystals (Drewsen, 2015; Bell and Softley, 2009; Willitsch, 2012; Heazelwood and Softley, 2015; Rugango et al., 2015). Collisions of such ions with neutral molecules in slow, velocity-selected beams are currently used to study the kinetics of ion-molecule reactions at collision energies corresponding to a few Kelvin,[5] primarily limited by the mean velocity of the neutral

---

[5] Temperature-dependent rate coefficients can be obtained from collision-energy-dependent rate coefficients by averaging over the distribution of thermally occupied levels; see, e.g., (Bowers, 1979; Clary, 1985).

beam (Bell and Softley, 2009; Willitsch, 2012; Heazelwood and Softley, 2015). Still lower temperatures can be reached in experiments in which laser-cooled atoms such as Na or Rb are co-trapped with laser-cooled ions such as Ca^+ or Yb^+ (Hall and Willitsch, 2012; Grier et al., 2009; Rellergert et al., 2011; Ratschbacher et al., 2012; Huber et al., 2014), but at the price of a limited chemical diversity. The potential of these methods for observing new low-temperature phenomena in ion-molecule chemistry remains to be demonstrated.

The strategy we pursue to study low-temperature ion-molecule chemistry consists of investigating the ion-molecule reactions within the orbit of a highly excited Rydberg electron. In a Rydberg state of sufficiently high principal quantum number n, the Rydberg electron effectively shields the ion core from stray electric fields without significantly affecting the properties (in particular the reactivity) of the ion core. The equivalence of an ion-molecule reaction

$$A^+ + B \rightarrow C^+ + D \tag{3.57}$$

with the corresponding reaction within the orbit of a highly excited Rydberg electron

$$A^* + B \rightarrow C^* + D, \tag{3.58}$$

where the asterisk designates a Rydberg state of high principal quantum number n, has been examined in previous studies (Pratt et al., 1994; Dehmer and Chupka, 1995; Wrede et al., 2005; Dai et al., 2005; Matsuzawa, 2010; Matsusawa, 2013; Allmendinger et al., 2016a). While discrepancies may play a role in specific cases, e.g., for differential cross sections using Rydberg states having a nonnegligible penetrating character (Hayes and Skodje, 2007), there is agreement that the equivalence between reactions (3.57) and (3.58) is approached more and more closely, the higher the principal quantum number n and the orbital angular momentum l become (Matsusawa, 2013). Rydberg-Stark states such as those we use in our deceleration and trapping experiments (see Sec. 3.4.2 and 3.4.3) hardly have any core-penetrating character and fulfill this condition well.

The approach we are currently developing to study ion-neutral reactions at low collison energies exploits this equivalence in combination with our ability to control the translational motion of Rydberg-atom or Rydberg-molecule beams. This approach is described in detail in Allmendinger etal. (2016a), on which the following paragraphs are based. The apparatus we use is depicted schematically in Fig. 3.17 (Allmendinger et al., 2016a). It consists of differentially-pumped beam-source chambers and a reaction chamber containing a time-of-flight mass spectrometer. The beam-source chambers are separated from the reaction zone by a 6-cm-long region containing a Rydberg-Stark surface-electrode deflector. Two pulsed collimated supersonic beams containing the species of interest are formed in the beam-source chambers using liquid-nitrogen-cooled pulsed valves and initially propagate at an angle of $10°$. The adiabatic expansion of the gases from the high-pressure zone behind the nozzle orifice into vacuum cools the rotational and vibrational degrees of freedom of the molecules so that typically only the vibrational ground state and very few rotational states are occupied. After passing a skimmer, the two beams enter a high-vacuum chamber and

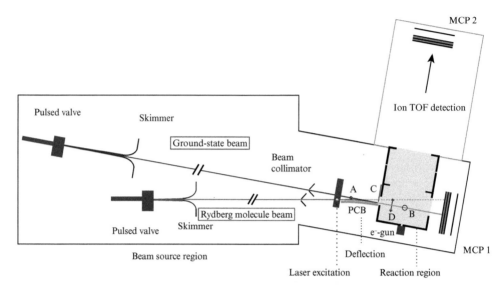

Fig. 3.17 Schematic representation of the merged-beam apparatus used in our studies of the $H_2^+ + H_2$ reaction at low collision energies, with the two skimmed supersonic beams initially propagating at an angle of $10°$, the Rydberg-Stark deflector made of a curved printed circuit board (PCB) and used to merge the beams after laser excitation, the reaction region, and the linear time-of-flight spectrometer used to detect reactants and products separately. (A) Laser excitation spot used in measurements of the velocity distribution of the undeflected beam of ground-state H_2 molecules. (B) Region in which the density profile of the ground-state molecules in the undeflected beam is measured by electron-impact ionization induced by an electron gun. (C) Aperture used to prevent the undeflected Rydberg molecules from entering the reaction region. (D) Movable microchannel-plate detector used for measurements of the velocity distribution and the lifetimes of the Rydberg molecules. (MCP1) and (MCP2) Microchannel-plate detectors to monitor the flight times of Rydberg H_2 molecules and the ion-time-of-flight spectra, respectively. From Allmendinger et al. (2016a).

are further collimated, so that the transverse velocity distribution of the particles in the beam is centered around 0 with a very narrow distribution. Just before the two beams cross, the molecules in one of these two beams are photoexcited from their ground state to long-lived nkm Rydberg-Stark states with principal quantum number n in the range 20–40 and the ion core in a selected rotational-vibrational level. For instance, to study reactions of H_2^+ ions, we use the resonant three-photon excitation sequence (3.54). This excitation sequence and the three lasers used for the successive excitation steps have been described by Hogan et al. (2009) and Seiler et al. (2011a). Utilizing circularly polarized laser radiation of the same helicity for all three steps of the excitation sequence enables one to access nonpenetrating Rydberg-Stark states with $m = 3$ and thus $\ell \geq 3$, as explained in Sec. 3.4.2.

The surface-electrode deflector, consisting of a set of electrodes on a curved printed circuit board presented in Sec. 3.4.3 (see also Allmendinger et al. (2014)), is used to deflect the Rydberg molecules and to merge the two supersonic beams. A shutter, labeled C in Fig. 3.17, prevents undeflected Rydberg molecules from entering the reaction zone. The central relative velocity of the two beams is set either by adjusting the temperature of the pulsed valve generating the ground-state H_2 beam while keeping the velocity of the Rydberg H_2 beam constant, or by adjusting the waveforms of the electric potentials applied to the electrodes of the deflector, as explained in Sec. 3.4.3.

Because a precise knowledge of the velocity and density distributions of the two beams is crucial for the determination of the energy dependence of the cross sections, they are measured independently by laser photoionization and electron-impact ionization with an electron gun. The velocity distribution of the ground-state H_2 beam is determined by photoexciting the H_2 molecules to long-lived Rydberg states at the spot labeled A in Fig. 3.17 and measuring their time-of-flight distribution to the microchannel-plate detector labeled "MCP1" in Fig. 3.17. The three-dimensional velocity distribution of the beam is reconstructed from measurements in which the delay between the valve-opening time and the laser-photoexcitation pulse is systematically varied over the entire duration of the gas pulse. We use this distribution together with the well-characterized velocity distribution of the Rydberg molecules (see Allmendinger et al. (2014)) as input for the simulation of the trajectories of the ground-state H_2 and the Rydberg H_2 molecules in the vacuum chambers.

The merged beams enter a reaction zone located within a cylindrical stack made of three electrodes, with which the reaction products are extracted towards an imaging detector ("MCP2" in Fig. 3.17) in a direction perpendicular to the merged-beam-propagation axis and detected mass selectively. The relative cross sections are determined by monitoring the yield of product ions as a function of the relative velocity of the two beams.

The experiment exploits the following experimental features:

- The production of a short, velocity-selected and localized packet of Rydberg molecules in the reaction zone following laser excitation and Rydberg-Stark deflection and deceleration.
- The short (13μ s long) gas pulses containing the ground-state neutral molecules generated by the pulsed valve and the long propagation distance (about 1 m) from the nozzle orifice to the reaction zone ensure that the faster molecules in the beam are well separated from the slower ones when they reach the reaction zone. The velocity dispersion that naturally takes place during propagation enables a considerable improvement of the collision-energy resolution of the measurements. The same advantage was elegantly exploited in earlier studies of neutral-neutral reactions (Henson et al., 2012; Jankunas et al., 2014).
- The merged-beam geometry enables one to tune the relative velocity of the two beams, and thus the collision energy, down to zero by either varying the temperature of the valve producing the H_2-ground-state beam or by varying the velocity of the H_2 Rydberg molecules during deflection with optimized potential waveforms.

First results on the $H_2^+ + H_2 \rightarrow H_3^+ + H$ reaction obtained with this experimental approach are presented in Sec. 3.5.1.

3.5 Examples

3.5.1 The $H_2 + H_2^+ \rightarrow H_3^+ + H$ reaction

This section summarizes our recent studies of the of the $H_2 + H_2^+ \rightarrow H_3^+ + H$ reaction. These studies were described in Allmendinger et al. (2016a), from which the following paragraphs are inspired and to which I refer for additional information. The experiments were carried out with the methods described in Sec. 3.4.3 and 3.4.4.

In a first set of measurements (Allmendinger et al., 2016a), we have (1) verified the spectator role of the Rydberg electron by measuring the reaction rate for different values of the principal quantum number of the Rydberg-Stark states prepared initially, and (2) measured the relative reaction cross section in the range of collision energies from $E_{coll} = k_B \cdot 60$ K, which corresponds to the lowest values studied by Glenewinkel-Meyer and Gerlich (1997), down to $k_B \cdot 5$ K. The experimental results are depicted as dark and light gray dots in Fig. 3.18, where they are compared with the earlier results of Glenewinkel-Meyer and Gerlich (1997) mentioned above, and the results of classical-molecular-dynamics-with-quantum-transitions calculations of Sanz-Sanz et al. (2015) (dashed gray line). The dashed black line, with a slope of -1/2 characteristic of classical Langevin cross sections, is drawn to guide the eye. In this range of collision energies, the experimental data still closely follows classical Langevin-capture behavior. Quantum mechanical effects in ion-molecule reactions of the Langevin-capture type thus appear

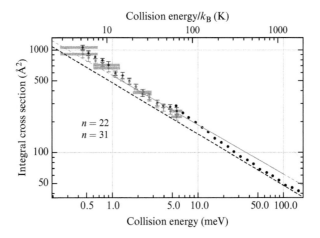

Fig. 3.18 Double-logarithmic plot of the absolute reaction cross section of the $H_2^+ + H_2 \rightarrow H_3^+ + H$ reaction. The black data points are the absolute cross-section measurements of Glenewinkel-Meyer and Gerlich (1997). The dashed black line and the gray line represent a pure Langevin-capture behavior and the classical-molecular-dynamics-with-quantum-transitions calculations of Sanz-Sanz et al. (2015), respectively. The light gray and dark gray dots represent our measurements using $n = 31$ and $n = 22$ H_2 Rydberg states, respectively, and scaled so as to match the cross section calculated by Sanz-Sanz et al. (2015) (their equation (27)) at the collision energy $E_{col}/k_B = 30$ K. The vertical error bars indicate the standard deviation of the measured cross section and the horizontal bars represent the energy resolution of our measurements for representative points. From Allmedinger et al. (2016a).

negligible, even at collision energies as low as $k_B \cdot 5$ K. Given that H_2^+ and H_2 are very light molecules, this observation indicates just how good classical Langevin-capture describes fast barrierless exothermic reactions.

In work carried out since the Les Houches summer school, we have extended this measurement to collision energies below 1 K and have observed first systematic deviations from classical Langevin behavior at collision energies below $k_B \cdot 1$ K. The experimental data, presented as ratios of the observed rate coefficients k to the Langevin capture rate coefficient k_L, are depicted in Fig. 3.19.

The data reveal a gradual increase of the rate coefficient at collision energies below 0.1 meV ($k_B \cdot 1.2$ K). At the lowest collision energies we could reach, i.e., about $k_B \cdot 0.3$ K, we observe a rate coefficient larger than k_L by about 15%, limited by the collision-energy resolution of our experiment, which is indicated by the horizontal gray bars in Fig. 3.19.

In parallel to this experimental work, Dashevskaya et al. (2016) have carried out a detailed theoretical investigation of the reaction in the low-temperature range. They predicted the energy-dependent rate coefficients of the reaction $H_2^+ + H_2 \rightarrow H_3^+ + H$ for a ground-state H_2 sample consisting of a mixture of 25% para-H_2 molecules in the $j = 0$ rotational level and 75% ortho-H_2 molecules in the $j = 1$ rotational level, corresponding to the gas sample used in our experiments (see the green curve labeled $\bar{\chi}$ in their fig. 2, which is reproduced as full black line in Fig. 3.19). These

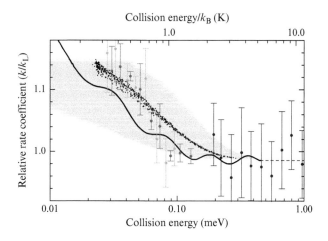

Fig. 3.19 Comparison of the energy-dependence of the measured relative rate coefficients $k(E)/k_L$ (gray dots) to the calculation by Dashevskaya et al. (2016) for normal H_2 (75% H_2 in $j = 1$ and 25% H_2 in $j = 0$) at fixed collision energies (solid line) and for collision energies averaged over the simulated experimental energy distributions (black dots, gray bars indicate one standard deviation). Light gray dots: two-pulse sequence ($\Delta t = 7$ μs) and H_2^* Rydberg beam central velocity $v(H_2^*) = 1800$ m/s. Mid-gray dots: single-pulse sequence for $v(H_2^*) = 1700$ m/s. Dark gray dots: single-pulse sequence, $v(H_2^*) = 1540$ m/s. The absolute scaling of each experimental data set was chosen to minimize the deviation from the simulation. The simulation (black dots) is based on the experimental parameters of the two-pulse measurement (light gray dots), but the result is very similar for the other measurements. From Allmendinger et al. (2016b).

calculated rate coefficients were used to simulate the experimental results on the basis of the velocity and density distributions measured in our experiments with a particle-trajectory simulation program (Allmendinger et al., 2016b). The H_3^+ ion yield predicted in this way is depicted as small black dots in Fig. 3.19.

The simulation indicates that the observed increase of the reaction rate at low collision energy is consistent with the energy dependence of the rate coefficient for the reaction $H_2^+ + H_2 \rightarrow H_3^+ + H$ calculated by Dashevskaya et al. (2016) when averaged over the experimental distribution of collision energies (black dots and gray bars in Fig. 3.19). Our experimental results thus reveal for the first time a pronounced and rather sudden departure of the rate of the $H_2^+ + H_2 \rightarrow H_3^+ + H$ reaction from the behavior predicted on the basis of the classical Langevin-capture model at low collision energies.

The excellent agreement with the calculations of Dashevskaya et al. (2016) validates their predictions in the range of collision energies probed by our experiment and leads to the conclusion that the mechanism responsible for the observed enhancement of the rate coefficient at low collision energies has its origin in the interaction between the charge of H_2^+ and the rotational quadrupole moment of the ground state of ortho-H_2 ($j = 1$). This interaction, which scales with the intermolecular separation R as $1/R^3$, leads to an anisotropic modification of the long-range scattering potential, which is dominated by the isotropic charge–induced-dipole coupling (falling off as $1/R^4$). A priori all orientations of the rotating H_2 molecule (or, equivalently, all values of the projection quantum number $\omega = 0, \pm 1$ of j onto the collision axis) have equal probabilities. At large collision energies, no (re)locking of the ground-state H_2 intrinsic angular momentum takes place, and the anisotropic contributions average out. At low collision energies, however, the collision complex can follow the minimum energy trajectory in the anisotropic potential adiabatically, leading to a "locking" of the intrinsic rotation of the H_2 molecule to the collision axis and thus to an enhanced rate coefficient (see Fig. 3.19). In the limit of zero collision energy, which is not probed yet with sufficient precision in our experiments, the rate constant is predicted to approach a 0 K (Bethe-Wigner) limit of about 3.6 k_L, as given by $(\frac{1}{4} \cdot 2 + \frac{3}{4} \cdot 4.18)k_L$ for a $\frac{1}{4} - \frac{3}{4}$ para-ortho H_2 mixture at low temperature (Dashevskaya et al., 2016).

3.5.2   The $H + H^+ \leftrightarrow H_2^+$ reaction system

In a reaction of the Langevin-capture type, only the long-range part of the potential needs to be considered when predicting the reaction rate coefficient, and the structure of the molecular potential at short range is disregarded. This part of the potential has not even been drawn in Fig. 3.5. Consequently, shape resonances (i.e., resonances originating from continuum states having a large probability density at internuclear distances shorter than the positions of the centrifugal barriers in the potentials (see Eq. 3.25 and gray dots in Fig. 3.5) do not play any role. However, because the potentials typically have steep repulsive walls at short range, one could expect that quasi-bound molecular states exist above the dissociation limit, with wavefunctions predominantly contained between the repulsive wall and the potential barriers. These quasi-bound molecular states, called shape resonances, could be populated at the appropriate collision energies and affect the association rates.

Classically, these resonances are not accessible in collisions because tunneling through the barriers is excluded. In a quantum-mechanical system, they cannot be excluded a priory, and, indeed, such resonances are known to affect the rate coefficients of many reactions at low temperatures (see, e.g., Henson et al. (2012) and Jankunas et al. (2014) for recent examples involving reactions between neutral species). One way of rationalizing the complete absence of resonances in reactions of the Langevin-capture type is by imagining that (i) the decay of the capture complex is very fast, so that the resonances are broadened, and (ii) the resonance density at the collision energy is very large, so that resonances are not distinguishable from a continuum and hardly affect the cross sections and rate coefficients. One can also imagine that once the barrier is passed (by tunneling), the reaction takes place before the incoming wave can be reflected by the repulsive wall, thus preventing the formation of a resonance.

Whereas the $H_2^+ + H_2 \rightarrow H_3^+ + H$ reaction follows Langevin-capture behavior closely (see Sec. 3.5.1), the $H + H^+ \leftrightarrow H_2^+$ reaction system is an example of a reaction system for which Langevin-capture behavior cannot be expected: even if the reaction is barrier-free and strongly exothermic (the dissociation energy of H_2^+ is $D_0 = 21379.35024(6)$ cm^{-1} (Mohr et al., 2008; Korobov, 2008), the formation of H_2^+ requires energy to be taken away from the $H + H^+$ collision system, either from a third body (three-body recombination) or through emission of a photon (radiative association). At low gas density, the former process is negligible. Radiative association is an inefficient process because continuum→bound radiative transitions have very small cross sections. In the absence of g/u mixing effects by the hyperfine interaction, the only electric-dipole-allowed continuum→ bound transitions in H_2^+ are electronic transitions from the continuum of the X^+ state to one of the four bound states of the A^+ state (i.e., $v^+ = 0$, $N^+ = 0 - 2$ and $v^+ = 1$, $N^+ = 0$ (Moss, 1993; Carbonell et al., 2003)) and from the continuum of the A^+ state to one of the 481 bound levels of the X^+ state (Moss, 1993). These transitions are very unlikely to take place during a collision. Consequently, the condition of 100% reaction efficiency required for the Langevin-capture model to be valid is not met. The density of resonances is not expected to be particularly high and those located far below the barrier maxima should be sharp, their width being given by the tunneling rate through the barrier. The radiative association is thus expected to depend on the positions of the bound levels and the shape resonances of H_2^+ and thus on the entire molecular potentials.

The situation for para H_2^+ is illustrated in Fig. 3.20. Only five bound levels of para H_2^+ [the $(v^+, N^+) = (17, 6)$, (18,0), (18,2) and (19,0) of the X^+ state and the (0,1) level of the A^+ state] and one shape resonance [the (18,4) resonance] are located in the region within the interval $[-25$ cm^{-1}, 5 cm$^{-1}]$ of the dissociation limit.

The remainder of this section summarizes our recent experimental and theoretical studies of the $H + H^+ \leftrightarrow H_2^+$ reaction system. These studies were described in Beyer and Merkt (2016a) and (2016b), which the following paragraphs follow closely and to which I refer for additional information.

Experimentally, we have studied the dissociation reaction $H_2^+ \rightarrow H + H^+$ using the technique of pulsed-field-ionization zero-kinetic-energy photoelectron spectroscopy (Reiser et al., 1988; Hollenstein et al., 2001). Specifically, we have measured the ionization yield resulting from the field ionization of very high ($n \approx 250$) Rydberg states of H_2 as the laser-excitation frequency was scanned in the vicinity of the

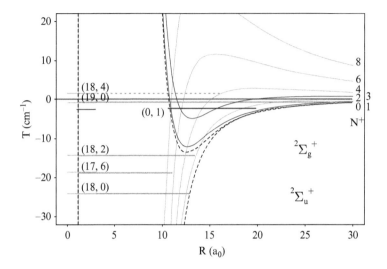

Fig. 3.20 Effective potential energy functions of the X^+ $^2\Sigma_g^+$ and A^+ $^2\Sigma_u^+$ states of H_2^+ determined as explained in Sec. 3.2.1 and used to calculate the bound states and shape resonances of H_2^+ (see also Beyer and Merkt (2016a, 2016b)). The dashed lines are the corrected Born-Oppenheimer potential, the thin gray and black lines include the centrifugal potential terms for the X^+ and A^+ states, respectively. The horizontal full and dashed lines give the positions of the bound levels and the shape resonances, respectively, using the labels (v^+, N^+).

dissociative-ionization limit of H_2, leading to the fragments $H(1s) + H^+ + e^-$. In these states, the Rydberg electron can be regarded as being decoupled from the H_2^+ ion core and as being a spectator of the $H_2^+ \rightarrow H + H^+$ dissociation (see discussion accompanying Eqs. 3.57 and 3.58). Because the Rydberg electron in these high Rydberg states is bound by less than 6 GHz and the binding energies can be accurately included in the analysis, the position of the lines observed in the spectra can be reliably interpreted as energy differences between the levels of H_2^+ and the ground state of H_2, i.e., the $(v = 0, N = 0)$ and the $(0,1)$ rovibrational levels of the $X\ ^1\Sigma_g^+$ state for para and ortho H_2, respectively.

To access the weakly-bound states of H_2^+ from the ground state of H_2, we used a resonant three-photon excitation sequence via the intermediate $B\ ^1\Sigma_u^+$ $(v = 9)$ and $H\bar{H}\ ^1\Sigma_g^+(v = 11)$ states in order to gradually increase the internuclear distance and thus achieve larger Franck-Condon overlap factors than would have been possible by direct single-photon excitation from the X ground state with the excitation sequence in Eq. 3.54.

The spectra we recorded for para and ortho H_2 are presented in the upper and lower panels of Fig. 3.21, respectively. The spectra provide an almost complete map of the energy levels of H_2^+ in the vicinity of the $H^+ + H$ dissociation threshold. Next to numerous bound rotational levels associated with vibrational levels of the X^+ state with quantum number v^+ in the range 14–19, as given by the assignment bars, the spectra also contain transitions to the $v^+ = 0$, $N^+ = 0$-2 bound levels of the A^+ state. The onset of the $H + H^+$ dissociation continuum is clearly recognizable in both

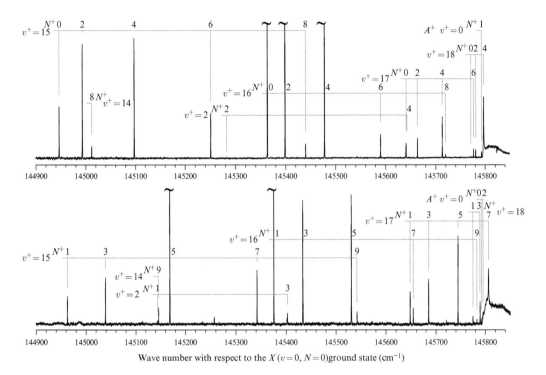

Fig. 3.21 PFI-ZEKE photoelectron spectra of H_2 recorded near the dissociative-ionization threshold from the $\bar{H}$ $(v = 11, N = 2)$ level (upper panel) and the $\bar{H}$ $(v = 11, N = 3)$ level (lower panel). The wave numbers are given with respect to the X $(v = 0, N = 0)$ ground state of H_2. The vertical scale is linear and in arbitrary units. The background signal below the dissociative-ionization threshold corresponds to zero signal intensity. Taken from Beyer and Merkt (2016b).

spectra and so are two shape resonances, corresponding to the (17,7) and (18,4) levels of the X^+ state.

Theoretically, we have calculated the level positions, and in the case of the shape resonances also the resonance widths, using the effective one-dimensional potentials including the effects of the adiabatic, nonadiabatic, relativistic, and radiative corrections to the Born-Oppenheimer approximation discussed in Sec. 3.2.1 (see Fig. 3.2).

To calculate the positions and widths of the resonances, we integrated Eq. 3.25 (after replacing the Born-Oppenheimer potential by the corrected potential) from low to high R values at successive energies in the continuum and counted the number of nodes in the wavefunctions. At the energies at which a new node appeared, which indicates a resonance, we integrated the wavefunction in the vicinity of these energies to large R and matched the nuclear wavefunction with its asymptotic form

$$\lim_{R\to\infty} \phi(R;k) = A_k kR\left[j_{N^+}(kR)\cos\delta_{N^+} - n_{N^+}(kR)\sin\delta_{N^+}\right], \tag{3.59}$$

where j_{N+} and n_{N+} are the spherical Bessel functions and $k = \sqrt{2\mu(E - U(R = \infty))}/\hbar$. Energy-normalized continuum wavefunctions were obtained by scaling the amplitude to (Fano and Rau, 1986)

$$A_k = \frac{1}{\hbar}\sqrt{\frac{2\mu}{\pi k}}. \tag{3.60}$$

The phase shift for a given energy $\delta_{N+}(E)$ was obtained from the values of the wavefunction at the two outermost points R_a and $R_b = R_{\max}$ of the grid of internuclear distances used in the integration with

$$\tan \delta_{N+} = \frac{K j_{N+}(k R_a) - j_{N+}(k R_b)}{K n_{N+}(k R_a) - n_{N+}(k R_b)}; \quad K = \frac{R_a \phi_{N+}(R_b)}{R_b \phi_{N+}(R_a)}. \tag{3.61}$$

The energy grid in the vicinity of a resonance was adapted by requiring a certain number of points per phase jump π.

The position E_{res} and the full widths at half maximum $\Gamma^{(\text{BW})}$ of the resonances and the parameters $a^{(j)}$ $(j = 0 - 2)$ describing an energy-dependent background phase shift were determined in a nonlinear least-squares fit of a Breit-Wigner(BW)-like formula (Smith 1971)

$$\delta_{N+}(E) = \sum_{j=0}^{2} a_{N+}^{(j)} E^j + \arctan\left[\frac{\Gamma^{(\text{BW})}/2}{E_{\text{res}} - E}\right]. \tag{3.62}$$

The comparison of dissociation energies (the values are negative for shape resonances) and widths calculated at various levels of approximation (see Table 3.4) indicates that the effects of the relativistic and radiative corrections are minor (less than 30 MHz), whereas the adiabatic (about 15 GHz for the (17,7) resonance) and nonadiabatic (about 4 GHz) corrections are significant. The experimental and calculated dissociation energies agree within the experimental uncertainties. The calculated and measured widths of the (18,4) resonance of para H_2^+ are in agreement but, surprisingly, the experimental value of the width of the (17,7) resonance of ortho H_2^+ is almost three times larger than the calculated value.

Table 3.4 Dissociation energies and widths of the two shape resonances of H_2^+ observed experimentally calculated at different levels of approximation: Born-Oppenheimer (BO), adiabatic (AD), nonadiabatic (NA), nonadiabatic with relativistic corrections (NArel), nonadiabatic with relativistic and radiative corrections (NArelrad). All values are in cm^{-1}. See Beyer and Merkt (2016b) for details.

v^+	N^+		BO	AD	NA	NArel	NArelrad	Exp.
18	4	E_{res}	−1.991	−1.879	−1.876	−1.879	−1.878	−1.84(4)
		Γ	0.259	0.206	0.193	0.194		0.21(7)
17	7	E_{res}	−11.699	−11.210	−11.094	−11.104	−11.104	−11.08(6)
		Γ	0.232	0.177	0.162	0.163		0.56(8)

Although we do not have an explanation for this discrepancy, we plan to investigate whether the nuclear spin ($I = 1$ in ortho H_2^+) and the resulting hyperfine interaction might cause an additional broadening compared to para H_2^+, which has $I = 0$.

The bound rotational levels of the A^+ state of H_2^+ have $N^+ \leq 2$ whereas the X^+ shape resonance with the lowest N^+ value is the (17,4) resonance, with $N^+ = 4$. We conclude from this investigation that, in the absence of g/u mixing by the hyperfine interaction, radiative association through the shape resonances of H_2^+ is forbidden in the electric-dipole approximation, which implies the selection rule $\Delta N^+ = \pm 1$ for transitions between to Σ states. The formation of $H_2{}^+$ by radiative association is thus an extraordinarily inefficient process.

3.6 Conclusions

This series of lectures has reviewed basic aspects of molecular physics of relevance in the context of experiments with cold molecules. Cold samples offer the possibility of carrying out experiments with a high degree of control over the internal degrees of freedom of molecules and over the relative motion of reaction partners. Few-electron molecules have the advantage that their properties can be calculated with high precision by the methods of *ab initio* quantum chemistry. Experiments with cold few-electron molecules enable one (i) to obtain precise and accurate experimental data, which can then be used to quantify the effects of the different levels of approximation made in the theoretical treatment, and (ii) to potentially detect discrepancies with the results of "exact" calculations that might reveal unknown phenomena.

After an introduction to some of the techniques used in precise calculations of the properties of few-electron molecules, the methods with which cold samples of one-, two-, three-, and four-electron molecules can be generated in the laboratory have been presented and two examples of recent cold-chemistry experiments with such samples have been discussed. Both examples involve reactions between ionic and neutral species containing hydrogen as sole element: The $H + H^+ \rightarrow H_2^+$ and the $H_2 + H_2^+ \rightarrow H_3^+ + H$ reaction systems; both reactions have in common that they involve singly-charged cations (H^+ and H_2^+) and nonpolar neutral species (H and H_2). Consequently, they have similar long-range attractive potentials scaling with the internuclear distance as $1/R^4$. Both reactions are strongly exothermic and barrier-free; both reactions are relevant to molecular astrophysics, the former being a possible route to form H_2^+ in the early universe, the latter being one of the key steps in reaction cycles in interstellar clouds (Oka, 2013).

The results were discussed in the context of the classical Langevin capture model for fast exothermic, barrier-free ion-molecule reactions, which was briefly recapitulated. This model predicts that the rate coefficient k_L is independent of the collision energy or the temperature. Despite the analogies listed above, the two reactions reveal opposite behavior: the $H_2 + H_2^+ \rightarrow H_3^+ + H$ reaction is an extremely rapid reaction and closely follows Langevin-capture kinetics down to very low temperature (Allmendinger et al., 2016a). It is only below 1 K that deviations from classical Langevin-capture behavior become noticeable. At such low temperatures, the reaction rates begin to rise above the Langevin-capture value as a result of the interaction between the quadrupole of the $N = 1$ rotational level of ortho H_2 with the charge of H_2^+. This rise, which could clearly be observed experimentally

(Allmendinger et al., 2016b), confirmed theoretical work (Dashevskaya et al., 2016), which, in addition, predicted the evolution of the rate coefficient down to its 0 K s-wave scattering limit, i.e., $2k_L$ and $4.2k_L$ for reactions with para H_2 ($N = 0$) and ortho H_2 ($N = 1$), respectively. Quantum-mechanical effects of the relative motion of the reactants thus significantly enhance the reaction rates at very low temperatures. Langevin capture excludes the possibility of shape resonances affecting the reaction rates. For this reason, the exact computation of the potential energy surface may not be as crucial as for other types of reactions. The knowledge of the polarizability and the quadrupole moment of the neutral molecule is, however, essential for accurate predictions.

The $H + H^+ \rightarrow H_2^+$ reaction is an extremely slow reaction, which, at low gas densities, takes place following a radiative association mechanism. Langevin capture models are completely inadequate to describe the reaction rate in this case. Instead, shape resonances are expected to play a role and affect the cross sections at low temperatures. We have observed the shape resonances of H_2^+ for the first time by studying the reverse reaction, the dissociation of H_2^+ very near the dissociation threshold, and measured their positions and widths (Beyer and Merkt, 2016a, 2016b). We also determined their positions and widths in calculations that included adiabatic, nonadiabatic, relativistic, and radiative corrections to the Born-Oppenheimer approximation and investigated the effects of various approximations. To reliably predict the positions and widths of the resonances, it is crucial to carry out calculations beyond the Born-Oppenheimer approximation, and include at least adiabatic and nonadiabatic corrections. The role of relativistic and radiative corrections was found to be minor. From our experiments and calculations, we proved that the radiative association of H^+ and H to form H_2^+ is forbidden in the electric-dipole approximation and that, therefore, this reaction is extraordinarily slow (Beyer and Merkt, 2016a, 2016b). The radiative association reaction forming HD^+ is expected to be much faster because nonadiabatic interactions effectively break the g/u symmetry at long range and make continuum$\rightarrow$bound transitions much more probable.

The study of ion-neutral reactions at very low temperature or very small collision energies is very challenging because even very small stray electric fields inevitably heat the ions. To avoid this problem, we have developed an experimental approach in which the reactions take place within the orbit of a highly excited Rydberg electron, which shields the reaction from stray electric fields without affecting its outcome. We expect that this approach will greatly facilitate studies of low-temperature ion-molecule chemistry in the future.

Acknowledgments

The material presented in these lecture notes stems from work carried out in collaboration with many other scientists. The method of Rydberg-Stark deceleration was developed initially in collaboration with Prof. T. P. Softley (Oxford and Birmingham), and then in the realm of the PhD theses of Edward Vliegen, Christian Seiler, Pitt Allmendinger, Matija Zesko, and Katharina Höveler, of the postdoctoral research of Johannes Deiglmayr and Ondrej Tkac, and of the habilitation of Stephen D. Hogan. The method of Zeeman deceleration was initially developed in collaboration with Prof. Beat H. Meier (ETH Zurich) and Dr. Nicolas Vanhaecke (CNRS, Orsay, then Fritz Haber Intitute, Berlin), and then in the realm of the dissertations of Alex Wiederkehr and Luca Semeria, of the postdoctoral research of Michael Motsch and Paul Jansen, and the habilitation of Stephen D. Hogan. The spectroscopic studies of H_2 and of H_2^+ have been carried out in collaboration with Dr. Christian Jungen (CNRS, Orsay) and Prof. Wim Ubachs (VU Amsterdam) in the realm of the dissertations of Andreas Osterwalder, M. Sommavilla, Hans Jakob Wörner, Daniel Sprecher, Christa Haase, and Maximilian Beyer, and of the postdoctoral research of Gregory Greetham, Jinjun Liu, and Helen Cruse. The work on He_2 and He_2^+ was carried out in the realm of the dissertations of Matthias Raunhardt, Daniel Sprecher, and Luca Semeria, and the postdoctoral research of Nicolas Vanhaecke, Jinjun Liu, Martin Schäfer, Michael Motsch, and Paul Jansen. I thank them all for their important contributions. I particularly thank Maximilian Beyer, whose work on the shape resonances of H_2^+ is reviewed in Sec. 3.5.2, and Pitt Allmendinger, Johannes Deiglmayr, Otto Schullian, and Katharina Höveler, whose work on the $H_2^+ + H_2$ reaction is reviewed in Sec. 3.5.1. I thank Josef A. Agner, Hansjürg Schmutz, Markus Andrist, Bruno Lambillotte, and René Gunzinger for their work in the development of the instrumentation used in our studies. Our work was generously funded over many years by the Swiss National Science Foundation and by the ETH Zurich, and also by an ERC advanced grant. Finally, I thank Maximilian Beyer and Johannes Deiglmayr for the careful reading of the manuscript and many discussions concerning its content.

References

Allmendinger, P., Agner, J. A., Schmutz, H., and Merkt, F. (2013). Deceleration and trapping of a fast supersonic beam of metastable helium atoms with a 44-electrode chip decelerator. *Phys. Rev. A*, **88**, 043433:1–8.

Allmendinger, P., Deiglmayr, J., Agner, J. A., Schmutz, H., and Merkt, F. (2014). Surface-electrode decelerator and deflector for Rydberg atoms and molecules. *Phys. Rev. A*, **90**, 043403.

Allmendinger, P., Deiglmayr, J., Höveler, K., Schullian, O., and Merkt, F. (2016a). New method to study ion-molecule reactions at low temperatures and applications to the $H_2^+ + H_2 \rightarrow H_3^+ + H$ reaction. *ChemPhysChem*, **17**, 3596–608.

Allmendinger, P., Deiglmayr, J., Schullian, O., Höveler, K., Agner, J. A., Schmutz, H., and Merkt, F. (2016b). Observation of enhanced rate coefficients in the $H_2^+ + H_2 \rightarrow H_3^+ + H$ reaction at low collision energies. *J. Chem. Phys.*, **145**, 244316.

Anderson, J. B. (1974). Molecular beams from nozzle sources. In *Molecular Beams and Low Density Gas Dynamics* (ed. P. P. Wegener), 1–91. Marcel Dekker.

Arcuni, P. W., Fu, Z. W., and Lundeen, S. R. (1990). Energy difference between the $(\nu = 0, R = 1)$ and the $(\nu = 0, R = 3)$ states of H_2^+, measured with interseries microwave spectroscopy of H_2 Rydberg states. *Phys. Rev. A*, **42**(11), 6950–3.

Auzinsh, M., Dashevskaya, E. I., Litvin, I., Nikitin, E. E., and Troe, J. (2008). Nonadiabatic transitions between lambda-doubling states in the capture of a diatomic molecule by an ion. *J. Chem. Phys.*, **128**, 184304.

Auzinsh, M., Dashevskaya, E. I., Litvin, I., Nikitin, E. E., and Troe, J. (2013a). Quantum effects in the capture of charged particles by dipolar polarizable symmetric top molecules. I. General axially nonadiabatic channel treatment. *J. Chem. Phys.*, **139**, 084311.

Auzinsh, M., Dashevskaya, E. I., Litvin, I., Nikitin, E. E., and Troe, J. (2013*b*). Quantum effects in the capture of charged particles by dipolar polarizable symmetric top molecules. II. Interplay between electrostatic and gyroscopic interactions. *J. Chem. Phys.*, **139**, 144315.

Barlow, S. E., Luine, J. A., and Dunn, G. H. (1986). Measurement of ion/molecule reactions between 10 and 20 K. *Int. J. Mass Spectrom. Ion Proc.*, **74**, 97–128.

Bell, M. T., and Softley, T. P. (2009). Ultracold molecules and ultracold chemistry. *Mol. Phys.*, **107**, 99.

Bersuker, I. B. (2006). *The Jahn-Teller effect*. Cambridge University Press.

Bethlem, H. L., Berden, G., and Meijer, G. (1999). Decelerating neutral dipolar molecules. *Phys. Rev. Lett.*, **83**, 1558–61.

Beyer, M., and Merkt, F. (2016*a*). Observation and calculation of the quasibound rovibrational levels of the electronic ground state of H_2^+. *Phys. Rev. Lett.*, **116**, 093001.

Beyer, M., and Merkt, F. (2016*b*). Structure and dynamics of H_2^+ near the dissociation threshold: a combined experimental and computational investigation. *J. Mol. Spectrosc.*, **330**, 147–57.

Bishop, D. M. (1974). Non-adiabatic calculations for H_2^+, HD^+ and D_2^+. *Mol. Phys.*, **28**, 1397–1408.

Bishop, D. M., and Wetmore, R. W. (1973). Vibrational spacings for H_2^+, D_2^+ and H_2. *Mol. Phys.*, **26**, 145–57. See also Mol. Phys. 27, 279 (1974).

Bowers, M. T. (ed.) (1979). *Gas Phase Ion Chemistry: Vol. 1 and 2*. Academic Press.

Brown, John M. and Carrington, Alan (2003). *Rotational Spectroscopy of Diatomic Molecules*. Cambridge University Press.

Bukowski, R., Jeziorski, B., Moszyński, R., and Kolos, W. (1992). Bethe logarithm and Lamb shift for the hydrogen molecular ion. *Int. J. Quant. Chem.*, **42**, 287–319.

Carbonell, J., Lazauskas, R., Delande, D., Hilico, L., and Kiliç, S. (2003). A new vibrational level of the H_2^+ molecular ion. *Europhys. Lett.*, **64**, 316–22.

Carrington, A., and Kennedy, R. A. (1984). Chapter 26 - Spectroscopy and structure of the hydrogen molecular ion. In *Ions and Light* (ed. M. T. Bowers), 393–442. Academic Press, Cambridge, MA.

Chervenkov, S., Wu, X., Bayerl, J., Rohlfes, A., Gantner, T., Zappelfeld, M., and Rempe, G. (2014). Continuous centrifuge decelerator for polar molecules. *Phys. Rev. Lett.*, **112**, 013001.

Clary, D. C. (1985). Calculations of rate constants for ion-molecule reactions using a combined capture and centrifugal sudden approximation. *Mol. Phys.*, **54**, 605–18.

Clary, D. C. (1990). Fast chemical reactions: theory challenges experiment. *Annu. Rev. Phys. Chem.*, **41**, 61–90.

Dai, D., Wang, C. C., Wu, G., Harich, S. A., Song, H., Hayes, M., Skodje, R. T., Wang, X., Gerlich, D., and Yang, X. (2005). State-to-state dynamics of high-n Rydberg H-atom scattering with D_2. *Phys. Rev. Lett.*, **95**, 013201.

Dashevskaya, E. I., Litvin, I., Nikitin, E. E., and Troe, J. (2005). Rates of complex formation in collisions of rotationally excited homonuclear diatoms with ions at very low temperatures: application to hydrogen isotopes and hydrogen-containing ions. *J. Chem. Phys.*, **122**, 184311.

Dashevskaya, E. I., Litvin, I., Nikitin, E. E., and Troe, J. (2016). Relocking of intrinsic angular momenta in collisions of diatoms with ions: capture of $H_2(j = 0, 1)$ by H_2^+. *J. Phys. Chem.*, **145**, 244315.

Dehmer, P. M., and Chupka, W. A. (1995). Rydberg state reactions of atomic and molecular hydrogen. *J. Phys. Chem.*, **99**, 1686–99.

DeMille, D. (2015). Diatomic molecules, a window onto fundamental physics. *Phys. Today*, **68**, 34–40.

Domcke, W., Yarkony, D. R., and Köppel, H. (eds.) (2004). *Conical intersections: electronic structure, dynamics and spectroscopy.* Adv. Ser. in Phys. Chem., Vol. 15. World Scientific Pub Co Inc.

Doyle, J. M., Friedrich, B., Kim, J., and Patterson, D. (1995). Buffer-gas loading of atoms and molecules into a magnetic trap. *Phys. Rev. A*, **52**, R2515–R2518.

Drewsen, M. (2015). Ion Coulomb crystals. *Physica B*, **460**, 105.

Elioff, M. S., Valentini, F. F., and Chandler, D. W. (2003). Subkelvin cooling NO molecules via "billiard-like" collisions with argon. *Science*, **302**, 1940.

Even, U., Jortner, J., Noy, D., Lavie, N., and Cossart-Magos, C. (2000). Cooling of large molecules below 1 K and He clusters formation. *J. Chem. Phys.*, **112**, 8068–71.

Fabrikant, I. I., and Hotop, H. (2001). Low-energy behavior of exothermic dissociative electron attachment. *Phys. Rev. A*, **63**, 022706.

Fano, U., and Rau, A. R. P. (1986). *Atomic Collisions and Spectra.* Academic Press, Cambridge, MA.

Fioretti, A., Comparat, D., Crubellier, A., Dulieu, O., Masnou-Seeuws, F., and Pillet, P. (1998). Formation of cold Cs_2 molecules through photoassociation. *Phys. Rev. Lett.*, **80**, 4402–5.

Friedrich, B., and Doyle, J. M. (2009). Why are cold molecules so hot? *Chem. Phys. Chem.*, **10**, 604–23.

Fulton, R., Bishop, A. I., and Barker, P. F. (2004). Optical Stark decelerator for molecules. *Phys. Rev. Lett.*, **93**(24), 243004.

Gallagher, T. F. (1994). *Rydberg Atoms.* Cambridge University Press.

Gao, B. (2010). Universal model for exoergic bimolecular reactions and inelastic processes. *Phys. Rev. Lett.*, **105**, 263203.

Gao, B. (2011). Quantum Langevin model for exoergic ion-molecule reactions and inelastic processes. *Phys. Rev. A*, **83**, 062712.

Gerlach, W., and Stern, O. (1922). Der experimentelle Nachweis der Richtungsquantelung im Magnetfeld. *Z. Phys.*, **9**, 349–52.

Gerlich, D. (2008). The study of cold collisions using ion guides and traps. In *Low Temperatures and Cold Molecules* (ed. I. M. W. Smith), 121–74. Imperial College Press, London.

Ginter, D. S., and Ginter, M. L. (1980). The spectrum and structure of the He_2 molecule. Multichannel quantum defect analyses of the triplet levels associated with $(1\sigma_g)^2(1\sigma_u)np\lambda$. *J. Mol. Spectrosc.*, **82**, 152–75.

Glenewinkel-Meyer, T., and Gerlich, D. (1997). Single and merged beam studies of the reaction $H_2^+(v=0,1; j=0,4) + H_2 \rightarrow H_3^+ + H$. *Israel J. Chem.*, **37**, 343.

Grier, A. T., Cetina, M., Oručević, F., and Vuletić, V. (2009). Observation of cold collisions between trapped ions and trapped atoms. *Phys. Rev. Lett.*, **102**, 223201.

Haase, Ch., Beyer, M., Jungen, Ch., and Merkt, F. (2015). The fundamental rotational interval of para-H_2^+ by MQDT-assisted Rydberg spectroscopy of H_2. *J. Chem. Phys.*, **142**, 064310.

Halfmann, T., Koensgen, J., and Bergmann, K. (2000). A source for a high-intensity pulsed beam of metastable helium atoms. *Meas. Sci. Technol.*, **11**, 1510–14.

Hall, F. H. J., and Willitsch, S. (2012). Millikelvin reactive collisions between sympathetically cooled molecular ions and laser-cooled atoms in an ion-atom hybrid trap. *Phys. Rev. Lett.*, **109**, 233202.

Hayes, M. Y., and Skodje, R. T. (2007). Dynamics of the Rydberg electron in $H^* + D_2 \rightarrow D^* + HD$ reactive collisions. *J. Chem. Phys.*, **126**, 104306.

Heazelwood, B. R., and Softley, T. P. (2015). Low-temperature kinetics and dynamics with Coulomb crystals. *Ann. Rev. Phys. Chem.*, **66**, 475–95.

Henson, A. B., Gersten, S., Shagam, Y., Narevicius, J., and Narevicius, E. (2012). Observation of resonances in Penning ionization reactions at sub-Kelvin temperatures in merged beams. *Science*, **338**, 234–38.

Herzberg, G. (1989). *Molecular Spectra and Molecular Structure, Volume I, Spectra of Diatomic Molecules* (2nd edn). Krieger Publishing Company, Malabar, FL.

Herzberg, G. (1991*a*). *Molecular Spectra and Molecular Structure, Volume II, Infrared and Raman Spectra of Polyatomic Molecules*. Krieger Publishing Company, Malabar, FL.

Herzberg, G. (1991*b*). *Molecular Spectra and Molecular Structure, Volume III, Electronic Spectra and Electronic Structure of Polyatomic Molecules* (2nd edn). Krieger Publishing Company, Malabar, FL.

Hogan, S. D., Allmendinger, P., Sassmannshausen, H., Schmutz, H., and Merkt, F. (2012). A surface-electrode Rydberg-Stark decelerator. *Phys. Rev. Lett.*, **108**, 063008.

Hogan, S. D. and Merkt, F. (2008). Demonstration of three-dimensional electrostatic trapping of state-selected Rydberg atoms. *Phys. Rev. Lett.*, **100**, 043001.

Hogan, S. D., Motsch, M., and Merkt, F. (2011). Deceleration of supersonic beams using inhomogeneous electric and magnetic fields. *Phys. Chem. Chem. Phys.*, **13**, 18705–23.

Hogan, S. D., Seiler, Ch., and Merkt, F. (2009). Rydberg-state-enabled deceleration and trapping of cold molecules. *Phys. Rev. Lett.*, **103**, 123001.

Hogan, S. D., Sprecher, D., Andrist, M., Vanhaecke, N., and Merkt, F. (2007). Zeeman deceleration of H and D. *Phys. Rev. A*, **76**, 023412.

Hollenstein, U., Seiler, R., Schmutz, H., Andrist, M., and Merkt, F. (2001). Selective field ionization of high Rydberg states: application to zero-kinetic-energy photoelectron spectroscopy. *J. Chem. Phys.*, **115**, 5461–9.

Howells, M. H., and Kennedy, R. A. (1990). Relativistic corrections for the ground and first excited states of H_2^+, HD^+ and D_2^+. *J. Chem. Soc., Faraday Trans.*, **86**, 3495.

Huber, T., Lambrecht, A., Schmidt, J., Karpa, L., and Schaetz, T. (2014). A far-off-resonance optical trap for a Ba^+ ion. *Nature Comm.*, **5**, 5587.

Hunter, G., and Pritchard, H. O. (1967). Born-Oppenheimer separation for three–particle systems. II. Two-center wavefunctions. *J. Chem. Phys.*, **46**, 2146–52.

Jankunas, J., Bertsche, B., Jachymski, K., Hapka, M., and Osterwalder, A. (2014). Dynamics of gas phase $Ne^* + NH_3$ and $Ne^* + ND_3$ Penning ionisation at low temperatures. *J. Chem. Phys.*, **140**, 244302.

Jansen, P., Semeria, L., Esteban Hofer, L., Scheidegger, S., Agner, J. A., Schmutz, H., and Merkt, F. (2015). Precision spectroscopy in cold molecules: the first rotational interval of He_2^+ and metastable He_2. *Phys. Rev. Lett.*, **115**, 133202.

Jansen, P., Semeria, L., and Merkt, F. (2016). High-resolution spectroscopy of He_2^+ using Rydberg-series extrapolation and Zeeman-decelerated supersonic beams of metastable He_2. *J. Mol. Spectrosc.*, **322**, 9–17.

Jaquet, R., and Kutzelnigg, W. (2008). Non-adiabatic theory in terms of a single potential energy surface. The vibration–rotation levels of H_2^+ and D_2^+. *Chem. Phys.*, **346**, 69–76.

Karr, J. Ph., and Hilico, L. (2006). High accuracy results for the energy levels of the molecular ions H_2^+, D_2^+ and HD^+, up to $J = 2$. *J. Phys. B: At. Mol. Opt. Phys.*, **39**, 2095–105.

Klippenstein, S. J., Georgievskii, Y., and McCall, B. J. (2010). Temperature dependence of two key interstellar reactions of H_3^+: $O(^3P) + H_3^+$ and $CO + H_3^+$. *J. Phys. Chem. A*, **114**, 278.

Kolos, W. (1969). Some accurate results for three-particle systems. *Acta Phys. Acad. Sci. Hung.*, **27**, 241–52.

Kołos, W., and Wolniewicz, L. (1963). Nonadiabatic theory for diatomic molecules and its application to the hydrogen molecule. *Rev. Mod. Phys.*, **35**, 473–83.

Komasa, J., Piszczatowski, K., Lach, G., Przybytek, M., Jeziorski, B., and Pachucki, K. (2011). Quantum electrodynamics effects in rovibrational spectra of molecular hydrogen. *J. Chem. Theory Comput.*, **7**, 3005–15.

Korobov, V. I. (2004). Bethe logarithm for the hydrogen molecular ion HD^+. *Phys. Rev. A*, **70**, 012505.

Korobov, V. I. (2006a). Bethe logarithm for the hydrogen molecular ion $H_2{}^+$. *Phys. Rev. A*, **73**, 024502.

Korobov, V. I. (2006b). Leading-order relativistic and radiative corrections to the rovibrational spectrum of H_2^+ and HD^+ molecular ions. *Phys. Rev. A*, **74**, 052506.

Korobov, V. I. (2008). Relativistic corrections of $m\alpha^6$ order to the rovibrational spectrum of H_2^+ and HD^+ molecular ions. *Phys. Rev. A*, **77**, 022509.

Korobov, V. I., Hilico, L., and Karr, J.-P. (2014). Theoretical transition frequencies beyond 0.1 ppb accuracy in H_2^+, HD^+, and antiprotonic helium. *Phys. Rev. A*, **89**, 032511.

Krems, R. V., Stwalley, W. C., and Friedrich, B. (ed.) (2009). *Cold Molecules: Theory, Experiment, Applications*. CRC Press, Taylor & Francis Group.

Landau, L. D., and Lifshitz, E. M. (1977). *Quantum Mechanis*. Pergamon, Oxford.

Lavert-Ofir, E., Gersten, S., Henson, A. B., Shani, I., David, L., Nafevicius, J., and Narevicius, E. (2011). A moving magnetic trap decelerator: a new source of cold atoms and molecules. *New J. Phys.*, **13**, 103030.

Leach, C. A., and Moss, R. E. (1995). Spectroscopy and quantum mechanics of the hydrogen molecular cation: a test of molecular quantum mechanics. *Ann. Rev. Phys. Chem.*, **46**, 55–82.

Lefebvre-Brion, H., and Field, R. W. (2004). *The Spectra and Dynamics of Diatomic Molecules*. Elsevier, Cambridge, MA.

Levy, D. H. (1980). Laser spectroscoy of cold gas-phase molecules. *Ann. Rev. Phys. Chem.*, **31**, 197.

Lichten, W., McCusker, M. V., and Vierima, T. L. (1974). Fine structure of the metastable $a^3\Sigma_u^+$ state of the helium molecule. *J. Chem. Phys.*, **61**, 2200–12.

Liu, Jinjun, Salumbides, Edcel J., Hollenstein, Urs, Koelemeij, Jeroen C. J., Eikema, Kjeld S. E., Ubachs, Wim, and Merkt, Frédéric (2009). Determination of the ionization and dissociation energies of the hydrogen molecule. *J. Chem. Phys.*, **130**(17), 174306.

Liu, J., Sprecher, D., Jungen, Ch., Ubachs, W., and Merkt, F. (2010). Determination of the ionization and dissociation energies of the deuterium molecule (D_2). *J. Chem. Phys.*, **132**, 154301.

Maergoiz, A. I., Nikitin, E. E., and Troe, J. (2009). Capture of asymmetric top dipolar molecules by ions: rate constants for capture of H_2O, HDO, and D_2O by arbitrary ions. *Int. J. Mass Spectr.*, **280**, 42–9.

Marquette, J. B., Rowe, B. R., Dupeyrat, G., Poissant, G., and Rebrion, C. (1985). Ion polar-molecule reactions - a CRESU study of He^+, C^+, N^+ +H_2O at 27 K, 68 K and 163 K. *Chem. Phys. Lett.*, **122**, 431.

Matsusawa, M. (2013). Effective elimination of space charge effects in ion-beam experiments. *Chin. J. Phys.*, **51**, 1184–91.

Matsuzawa, M. (2010). Highly excited Rydberg electron as a spectator to an ion-molecule reaction. *Phys. Rev. A*, **82**, 054701.

Meek, S. A., Bethlem, H. L., Conrad, H., and Meijer, G. (2008). Trapping molecules on a chip in traveling potential wells. *Phys. Rev. Lett.*, **100**, 153003.

Mohr, P. J., Taylor, B. N., and Newell, D. B. (2008). CODATA recommended values of the fundamental physical constants: 2006. *Rev. Mod. Phys.*, **80**, 633–730.

Moss, R. E. (1993). Calculations for the vibration-rotation levels of H_2^+ in its ground and first excited electronic states. *Mol. Phys.*, **80**, 1541–54.

Moss, R. E. (1996). On the adiabatic and non-adiabatic corrections in the ground electronic state of the hydrogen molecular action. *Mol. Phys.*, **89**, 195–210.

Motsch, M., Jansen, P., Agner, J. A., Schmutz, H., and Merkt, F. (2014). Slow and velocity-tunable beams of metastable He_2 by multistage Zeeman deceleration. *Phys. Rev. A*, **89**, 043420.

Narevicius, E., Libson, A., Parthey, C. G., Chavez, I., Narevicius, J., Even, U., and Raizen, M. G. (2008). Stopping supersonic oxygen with a series of pulsed electromagnetic coils: a molecular coilgun. *Phys. Rev. Lett.*, **100**, 093003.

Narevicius, E., and Raizen, M. G. (2012). Toward cold chemistry with magnetically decelerated supersonic beams. *Chem. Rev.*, **112**, 4879–89.

Ni, K.-K., Ospelkaus, S., Wang, D., Quemener, G., Neyenhuis, B., de Miranda, M. H. G., Bohn, J. L., Ye, J., and Jin, D. S. (2010). Dipolar collisions of polar molecules in the quantum regime. *Nature*, **464**, 1324.

Norrgard, E. B., McCarron, D. J., Steinecker, M. H., Tarbutt, M. R., and DeMille, D. (2016). Submillikelvin dipolar molecules in a radio-frequency magneto-optical trap. *Phys. Rev. Lett.*, **116**, 063004.

Oka, T. (2013). Interstellar H_3^+. *Chem. Rev.*, **113**, 7388.

Osterwalder, A., Meek, S. A., Hammer, G., Haak, H., and Meijer, G. (2010). Deceleration of neutral molecules in macroscopic traveling traps. *Phys. Rev. A*, **81**, 051401.

Osterwalder, A., Wüest, A., Merkt, F., and Jungen, Ch. (2004). High-resolution millimeter wave spectroscopy and multichannel quantum defect theory of the hyperfine structure in high Rydberg states of molecular hydrogen H_2. *J. Chem. Phys.*, **121**(23), 11810–38.

Pachucki, K., and Komasa, J. (2009). Nonadiabatic corrections to rovibrational levels of H_2. *J. Chem. Phys.*, **130**, 164113.

Pachucki, K., and Komasa, J. (2010). Rovibrational levels of HD. *Phys. Chem. Chem. Phys.*, **12**, 9188–96.

Pachucki, K., and Komasa, J. (2014). Accurate adiabatic correction in the hydrogen molecule. *J. Chem. Phys.*, **141**, 224103.

Pachucki, K., and Komasa, J. (2015). Leading order nonadiabatic corrections to rovibrational levels of H_2, D_2, and T_2. *J. Chem. Phys.*, **143**, 034111.

Pachucki, K., and Komasa, J. (2016). Schrödinger equation solved for the hydrogen molecule with unprecedented accuracy. *J. Chem. Phys.*, **144**, 164306.

Peek, J. M. (1965*a*). Eigenparameters for the $1s\sigma_g$ and $2p\sigma_u$ orbitals of H_2^+. *J. Chem. Phys.*, **43**, 3004–6.

Peek, J. M. (1965*b*). Use of approximate functions in evaluating the Born matrix element for H_2^+. *Phys. Rev.*, **139**, A1429–A1431.

Piszczatowski, K., Lach, G., Przybytek, M., Komasa, J., Pachucki, K., and Jeziorski, B. (2009). Theoretical determination of the dissociation energy of molecular hydrogen. *J. Chem. Theory Comput.*, **5**, 3039–48.

Pratt, S. T., Dehmer, J. L., Dehmer, P. M., and Chupka, W. A. (1994). Reactions of Rydberg states of molecular hydrogen. *J. Chem. Phys.*, **101**, 882–90.

Prehn, A., Ibrugger, M., Glöckner, R., Rempe, G., and Zeppenfeld, M. (2016). Optoelectrical cooling of polar molecules to submillikelvin temperatures. *Phys. Rev. Lett.*, **116**, 063005.

Procter, S. R., Yamakita, Y., Merkt, F., and Softley, T. P. (2003). Controlling the motion of hydrogen molecules. *Chem. Phys. Lett.*, **374**, 667–75.

Puchalski, M., Komasa, J., Czachorowski, P., and Pachucki, K. (2016). Complete $\alpha^6 m$ corrections to the ground state of H_2. *J. Chem. Phys.*, **143**, 034111.

Quack, M., and Merkt, F. (ed.) (2011). *Handbook of high-resolution spectroscopy*. Volume 1–3. John Wiley & Sons, New York.

Quack, M., Stohner, J., and Willeke, M. (2008). High-resolution spectroscopic studies and theory of parity violation in chiral molecules. *Ann. Rev. Phys. Chem.*, **59**, 741–69.

Rangwala, S. A., Junglen, T., Rieger, T., Pinkse, P. W. H., and Rempe, G. (2003). Continuous source of translationally cold dipolar molecules. *Phys. Rev. A*, **67**, 043406.

Ratschbacher, L., Zipkes, C., Sias, C., and Köhl, M. (2012). Controlling chemical reactions of a single particle. *Nat. Phys.*, **8**, 649–52.

Raunhardt, M., Schäfer, M., Vanhaecke, N., and Merkt, F. (2008). Pulsed-field-ionization zero-kinetic-energy photoelectron spectroscopy of metastable He_2: ionization potential and rovibrational structure of He_2^+. *J. Chem. Phys.*, **128**, 164310.

Reiher, M., and Wolf, A. (2015). *Relativistic Quantum Chemistry, 2nd Edition*. WILEY-VCH Verlag GmbH, Berlin.

Reiser, G., Habenicht, W., Müller-Dethlefs, K., and Schlag, E. W. (1988). The ionization energy of nitric oxide. *Chem. Phys. Lett.*, **152**, 119–23.

Rellergert, W. G., Sullivan, S. T., Kotochigova, S., Petrov, A., Chen, K., Schowalter, S. J., and Hudson, E. R. (2011). Measurement of a large chemical reaction rate between ultracold closed-shell ^{40}Ca atoms and open-shell $^{174}Yb^+$ ions held in a hybrid atom-ion trap. *Phys. Rev. Lett.*, **107**, 243201.

Rettner, C. T., Marinero, E. E., Zare, R. N., and Kung, A. H. (1984). Pulsed free jets: novel nonlinear media for generation of vacuum ultraviolet and extreme ultraviolet radiation. *J. Phys. Chem.*, **88**, 4459–65.

Rowe, B., Canosa, A., and Le Page, V. (1995). FALP and CRESU studies of ionic reactions. *Int. J. Mass. Spec.*, **149/150**, 573 96.

Rowe, B. R., Marquette, J. B., Dupeyrat, G., and Ferguson, E. E. (1985). Reactions of He^+ and N^+ ions with several molecules at 8 K. *Chem. Phys. Lett.*, **113**, 403.

Rugango, R., Goeders, J. E., Dixon, T. H., Gray, J. M., Khanyile, N. B., Shu, G., Clark, R. J., and Brown, K. R. (2015). Sympathetic cooling of molecular ion motion to the ground state. *New J. Phys.*, **17**, 035009.

Sanz-Sanz, C., Aguado, A., Roncero, O., and Naumkin, F. (2015). Non-adiabatic couplings and dynamics in proton transfer reactions of H_n^+ systems: application to $H_2 + H_2^+ \rightarrow H + H_3^+$ collisions. *J. Chem. Phys.*, **143**, 234303.

Scoles, Giacinto (ed.) (1988–92). *Atomic and Molecular Beam Methods*. Oxford University Press. 2 vols.

Seiler, Ch., Agner, J. A., Pillet, P., and Merkt, F. (2016). Radiative and collisional processes in translationally cold samples of hydrogen Rydberg atoms studied in an electrostatic trap. *J. Phys. B: At. Mol. Phys.*, **49**, 094006.

Seiler, Ch., Hogan, S. D., and Merkt, F. (2012). Dynamical processes in Rydberg-Stark deceleration and trapping of atoms and molecules. *Chimia*, **66**, 208–11.

Seiler, Ch., Hogan, S. D., and Merkt, F. (2011*a*). Trapping cold molecular hydrogen. *Phys. Chem. Chem. Phys.*, **13**, 19000–12.

Seiler, Ch., Hogan, S. D., Schmutz, H., Agner, J. A., and Merkt, F. (2011*b*). Collisional and radiative processes in adiabatic deceleration, deflection and off-axis trapping of a Rydberg atom beam. *Phys. Rev. Lett.*, **106**, 073003.

Semeria, L., Jansen, P., and Merkt, F. (2016). Precision measurement of the rotational energy-level structure of the three-electron molecule He_2^+. *J. Chem. Phys.*, **145**, 204301.

Smith, I. W. M. (2011). Laboratory astrochemistry: gas-phase processes. *Annu. Rev. Astron. Astrophys.*, **49**, 29–66.

Smith, I. W. M., and Rowe, B. (2000). Reaction kinetics at very low temperatures: laboratory studies and interstellar chemistry. *Acc. Chem. Res.*, **33**, 261–8.

Smith, K. (1971). *The Calculation of Atomic Collision Processes*, Chapter 1. Wiley.

Snow, T. P., and Bierbaum, V. M. (2008). Ion chemistry in the interstellar medium. *Annu. Rev. Anal. Chem.*, **1**, 229–59.

Sprecher, D., Jungen, Ch., Ubachs, W., and Merkt, F. (2011). Towards measuring the ionisation and dissociation energies of molecular hydrogen with sub-MHz accuracy. *Faraday Discuss.*, **150**, 51–70.

Sprecher, D., Liu, J., Jungen, Ch., Ubachs, W., and Merkt, F. (2010). The ionization and dissociation energies of HD. *J. Chem. Phys.*, **133**, 111102.

Sprecher, D., Liu, J., Krähenmann, T., Schäfer, M., and Merkt, F. (2014). High-resolution spectroscopy and quantum-defect model for the gerade triplet np and nf Rydberg states of He_2. *J. Chem. Phys.*, **140**, 064304.

Su, T., and Chesnavich, W. J. (1982). Parametrization of the ion-pair molecule collision rate-constant by trajectory calculations. *J. Chem. Phys.*, **76**, 5183.

Taylor, J. M., Yan, Zong-Chao, Dalgarno, A., and Babb, J. F. (1999). Variational calculations on the hydrogen molecular ion. *Mol. Phys.*, **97**, 25–33.

Trimeche, A., Bera, M. N., Cromiéres, J.-P., Robert, J., and Vanhaecke, N. (2011). Trapping of a supersonic beam in a traveling magnetic wave. *Eur. Phys. J. D*, **65**, 263–71.

Troe, J. (1987). Statistical adiabatic channel model for ion molecule capture processes. *J. Chem. Phys.*, **87**, 2773.

Troe, J. (1996). Statistical adiabatic channel model for ion-molecule capture processes. II. Analytical treatment of ion-dipole capture. *J. Chem. Phys.*, **105**, 6249.

Tung, W.-C., Pavanello, M., and Adamowicz, L. (2012). Very accurate potential energy curve of the He_2^+ ion. *J. Chem. Phys.*, **136**, 104309.

Ubachs, W., Bagdonaite, J., Salumbides, E. J., Murphy, M. T., and Kaper, L. (2016). Colloquium: search for a drifting proton-electron mass ratio from H_2. *Rev. Mod. Phys.*, **88**, 021003.

van de Meerakker, S. Y. T., Bethlem, H. L., Vanhaecke, N., and Meijer, G. (2012). Manipulation and control of molecular beams. *Chem. Rev.*, **112**, 4828–78.

van de Meerakker, S. Y. T., Vanhaecke, N., Bethlem, H. L., and Meijer, G. (2006). Transverse stability in a Stark decelerator. *Phys. Rev. A*, **73**, 023401.

Vanhaecke, N., Meier, U., Andrist, M., Meier, B. H., and Merkt, F. (2007). Multistage Zeeman deceleration of hydrogen atoms. *Phys. Rev. A*, **75**, 031402(R).

Vliegen, E., Hogan, S. D., Schmutz, H., and Merkt, F. (2007). Stark deceleration and trapping of hydrogen Rydberg atoms. *Phys. Rev. A*, **76**, 023405.

Vliegen, E., Wörner, H. J., Softley, T. P., and Merkt, F. (2004). Nonhydrogenic effects in the deceleration of Rydberg atoms in inhomogeneous electric fields. *Phys. Rev. Lett.*, **92**(3), 033005.

Vogt, E., and Wannier, G. H. (1954). Scattering of ions by polarization forces. *Phys. Rev.*, **95**, 1190–8.

Wester, R. (2009). Radiofrequency multipole traps: tools for spectroscopy and dynamics of cold molecular ions. *J. Phys. B: At. Mol. Phys.*, **42**, 154001.

Wickham, A. G., and Clary, D. C. (1993). Rate coefficient expressions for reactions of molecules in $^2\pi$ electronic states at low temperatures. *J. Chem. Phys.*, **98**, 420.

Wiederkehr, A. W., Hogan, S. D., and Merkt, F. (2010). Phase stability in a multistage Zeeman decelerator. *Phys. Rev. A*, **82**, 043428.

Wiederkehr, A. W., Motsch, M., Hogan, S. D., Andrist, M., Schmutz, H., Lambillotte, B., Agner, J. A., and Merkt, F. (2011). Mutistage Zeeman deceleration of metastable neon. *J. Chem. Phys.*, **135**, 214202.

Willitsch, S. (2012). Coulomb-crystallised molecular ions in traps: methods, applications, prospects. *Int. Rev. Phys. Chem.*, **31**, 175–99.

Wolniewicz, L., and Orlikowski, T. (1991). The $1s\sigma_g$ and $2p\sigma_u$ states of the H_2^+, D_2^+ and HD^+ ions. *Mol. Phys.*, **74**, 103–11.

Wolniewicz, L., and Poll, J. D. (1978). Ab initio nonadiabatic vibrational energies of the hydrogen molecule ion. *J. Mol. Spectrosc.*, **72**, 264–74.

Wolniewicz, L., and Poll, J. D. (1986). On the higher vibration-rotational levels of HD^+ and H_2^+. *Mol. Phys.*, **59**, 953–64.

Wörner, H. J., and Merkt, F. (2011). Fundamentals of electronic spectroscopy. In *Handbook of High-Resolution Spectroscopy* (ed. M. Quack and F. Merkt), Volume 1, 175–262. John Wiley & Sons, New York.

Wrede, E., Schnieder, L., Seekamp-Schnieder, K., Niederjohanna, B., and Welge, K. H. (2005). Reactive scattering of Rydberg atoms: $H^* + D_2 \rightarrow HD + D^*$. *Phys. Chem. Chem. Phys.*, **7**, 1577–82.

Yu, N., and Wing, W. H. (1987). Observation of the infrared spectrum of the helium molecular ion ($^3He^4He$)$^+$. *Phys. Rev. Lett.*, **59**, 2055–8. See erratum in Phys. Rev. Lett., **60**, 2445 (1988).

Yu, N., Wing, W. H., and Adamowicz, L. (1989). Hyperfine structure in the infrared spectrum of $^3Hc^4Hc^+$. *Phys. Rev. Lett.*, **62**, 253–6.

4
Frequency combs and precision spectroscopy of atomic hydrogen

Thomas Udem

Max-Planck-Institute of Quantum Optics
Garching, Germany

Udem, T. "Frequency combs and precision spectroscopy of atomic hydrogen." In *Current Trends in Atomic Physics*. Edited by Antoine Browaeys, Thierry Lahaye, Trey Porto, Charles S. Adams, Matthias Weidemüller, and Leticia F. Cugliandolo. Oxford University Press (2019).
© Oxford University Press. DOI: 10.1093/oso/9780198837190.003.0004

Chapter Contents

A laser frequency comb allows the phase coherent conversion of the very rapid oscillations of visible light of some 100's of THz down to frequencies that can be handled with conventional electronics. This capability has enabled the most precise laser spectroscopy experiments yet, making it possible to test quantum electrodynamics, to determine fundamental constants, and to construct an optical atomic clock. We will review the development of the frequency comb, derive its properties, and discuss its application for high resolution spectroscopy of atomic hydrogen.

4.1 The frequency comb

4.1.1 Introduction

Out of all measurable quantities the utmost precision is achieved with time and frequency. The two are actually equivalent as the former is usually measured by counting the cycles of a periodic process, while the latter is determined by the time between the cycles. To use this accuracy for other measurable quantities, it is desirable to map them to time or frequency. The best example of such a mapping is the 1983 redefinition of the meter by fixing the value of the speed of light c. Together with the definition of the SI second, the measurement of the frequency of light ν can determine a length in terms of the wavelength $\lambda = c/\nu$. A length standard on this basis is realized with the aid of an interferometer. The redefinition of the meter took place when it became technically possible to count the very high frequencies of light (several hundred THz).

Not only allow these ultrafast counters the realization of the meter, they also form the clockwork mechanism for an optical clock. Comparing the various types of clocks, such as sundials, pendulum clocks, quartz clocks, and cesium clocks, one finds that the faster the periodic process they use, the more accurate the clocks get. This is not by chance, but because faster oscillators slice time into finer intervals, just like a finer ruler allows to measure distances more accurately (this consideration leaves aside possible systematic uncertainties). The SI second is derived from the cesium ground state hyperfine splitting and defined to be exactly 9 192 631 770 Hz. Hence the second hand of a cesium clock is advanced every 9 192 631 770 cycles, again expressing the tight relationship between frequency and time. The operation of the clock requires a sufficiently fast counter as the clockwork mechanism.

With the invention of the laser in 1960, clean optical waves at several 100 THz could be generated. This triggered a discussion on how the pace of the clocks "pendulum" could be further increased into the optical regime. In 1982 Hans Dehmelt suggested the use of a narrow transition in a stored ion for this purpose (Dehmelt, 1982)[1], a method that has been realized almost twenty years later (Diddams et al., 2001) and is still competing for the best clock performance. However, this proposal was somewhat ahead of its time because the optical counters at those days only existed in form of

---

[1] We are giving only the most appropriate references in this introduction. More citations may be given in the main text.

large and complex frequency chains (see Sec. 4.1.2). One could merely use them during the brief occasional operation to calibrate suitable optical reference lines.

Highly accurate time scales are maintained by the national standard institutes, by continuously comparing their clocks using satellites and in recent times also dedicated optical fiber links. The radio frequency cesium clocks may win or loose a couple of nanoseconds within a months. The optical clocks can do much better, but require a reliable optical counter. The optical frequency combs have taken over this task (Udem et al., 2004). Spectrally they consists of a large number of continuous wave modes with identical frequency separations. The name derives from that spectrum. Modern frequency combs can operate autonomously for months while being as compact as a desktop computer. Unlike the frequency chains, that have been specifically designed for one particular optical frequency, the frequency comb can measure many different frequencies simultaneously. This includes the possibility of determining ratios of optical frequencies (Diddams et al., 2001), a feature that is important for comparing optical frequency standards of different types, i.e. of vastly different frequencies.

Initially the frequency combs have been used inversely to the optical clock. With a radio frequency clock at the low frequency end, it allows to measure the frequency of a laser that is used for spectroscopy (Reichert et al., 2000). By measuring transition frequencies in atomic hydrogen very accurately and comparing with likewise precise theoretical predictions, quantum electrodynamics can be tested and values for fundamental constants are obtained. By observing optical frequency ratios over an extended period of time one can search for possible slow variations of fundamental constants, such as the fine structure constant (Rosenband et al., 2008). Advances in optical frequency standards fueled by the optical clock development allowed improved of tests of relativity (Reinhardt et al., 2007). With an accuracy at the 17th decimal place, the gravitational red-shift at an elevation difference of only 33 cm could be measured (Chou et al., 2010). Rather than testing general relativity, one may believe in it and measure elevation differences by comparing remote frequency standards. While approaching the 18th decimal place, such a relativistic geodesy (Takano et al., 2016) is accurate to within 1 cm without moving any hardware as with traditional leveling.

The frequency combs have more to offer besides being a frequency converter between the optical and the radio frequency domain. Although it is not the subject of this review, a few of these applications should be mentioned here. The most practical source of a frequency comb is a mode-locked laser that emits a pulse train with a repeating envelope. When generating single peaked pulses with a carrier frequency that is constant across the pulse (no chirp), one can define the carrier–envelope phase. As it turns out, the pulse to pulse evolution (Xu et al., 1996) of this phase can be controlled using the frequency comb techniques discussed in Sec. 4.1.3. For very short pulses, the position of the carrier relative to the pulse envelope plays a role in nonlinear interactions and becomes decisive for high order harmonics (Baltuška et al., 2003). Aligning the peak of the pulse with one of the maximum excursions of the carrier wave ("cos–pulse"), produces the largest field excursions and hence the highest order harmonics. This extreme field excursion lasts only for a very brief time, significantly shorter than a half cycle of the optical carrier. Under these conditions isolated attosecond pulses are generated in the extreme ultraviolet (Goulielmakis et al., 2008).

While frequency combs have been used to measure the frequency of a narrow band continuous wave spectroscopy laser, it is also possible to perform direct frequency comb spectroscopy (Stowe et al., 2008). This has several advantages if one wants to record a large spectral region at once, for example to investigate molecular spectra. Unlike other broadband light sources, a frequency comb may be resonantly coupled into a cavity for increased sensitivity (Bernhardt et al., 2010). On the other hand it is also possible to use an individual mode to excite a spectrally isolated dipole allowed transition in an atom (Gerginov et al., 2005). While it is interesting to verify in this way that the individual modes of the comb indeed represent continuous wave laser lines, this method is probably of limited practical use. One reason is the large number of non-resonant spectral modes, that cause AC Stark shifts without contributing to the signal. The situation is quite different when exciting a two-photon transition with a frequency comb. In that case the energy of photons from pairs of modes may add up, such that the full power of all modes are used, while maintaining the linewidth of a single mode. The total excitation rate can be the same as for a continuous wave laser of the same average power (Baklanov and Chebotayev, 1977). However, there does not appear to be a real advantage over a continuous wave laser unless no such laser exists. The large peak intensity of a pulse train that generates the frequency comb may be used to efficiently drive high harmonic generation to obtain a frequency comb in the extreme ultraviolet (Pupeza et al., 2013). This would offer to perform high resolution laser spectroscopy in this region for the first time (Herrmann et al., 2009).

Another application that is mentioned here but otherwise not treated in this contribution is in astronomy. Observing the spectra of astronomical objects reveals a great deal of information about them like the chemical composition of the photosphere of stars. Extrasolar planets can be detected by periodic variation of the Doppler shift of the associated lines, because the planet and the host star orbit their common center of mass. Typically the host star is too bright and the exoplanets too faint to be seen directly. For these applications astronomical spectrographs need very good calibration that needs to be reproducible over the periods of interest. To discover an Earth-Sun like system one would need to resolve a periodic Doppler shift that corresponds to a velocity modulation of 9 cm/sec. This extremely precise calibration needs to be reproducible over a period of one year. Frequency comb calibration has recently gotten close to these requirements (Wilken et al., 2012).

4.1.2   The history of the frequency comb

4.1.2.1   Optical beat notes

Before the first lasers became a reality in 1960 (Maiman, 1960; Javan et al., 1961), it was not clear as to whether these new devices would emit a classical wave with a defined amplitude and phase. Assuming stimulated emission to be a quantum amplifier, it would have seemed natural that the resulting radiation to be in a Fock state, i.e. with a well defined amplitude but undefined phase. No stable beat note would be expected when two laser beams were brought into interference. This is indeed *not* the case as already suggested in A. Javan's first laser paper (Javan et al., 1961). In 1963 Roy Glauber introduced the coherent state into quantum optics (sometimes also

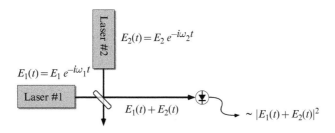

Fig. 4.1 Superimposing two laser beams with a beam splitter. A beat note at the frequency difference is obtained with a photo detector at either exit ports of the beam splitter.

called the Glauber states) (Glauber, 1963). These states are the most classical states in the sense that there is a fair split of uncertainties between energy (photon number distribution) and phase. Indeed, two carefully overlapped laser beams generate a clean beat note signal (Javan et al., 1962) that is best observed with a radio frequency spectrum analyzer. This beat note occurs at the difference of the two laser frequencies ω_1 and ω_2 when their superimposed fields fall on a photo detector (see Fig. 4.1).

With the two fields of the lasers that are proportional to $E_1(t)$ and $E_2(t)$ respectively, the photo detector delivers the following signal proportional to the intensity

$$\left| \vec{E}_1 e^{-i\omega_1 t} + \vec{E}_2 e^{-i\omega_2 t + i\varphi} \right|^2 = E_1^2 + E_2^2 + 2\vec{E}_1 \cdot \vec{E}_2 \cos\left((\omega_2 - \omega_1)t - \varphi\right), \qquad (4.1)$$

where the vector properties and an additional arbitrary phase φ, that reflects the choice $t=0$ and optical path lengths, have been taken into account in addition to Fig. 4.1.

The beat note signal is given by the third term in Eq. 4.1 which needs to be investigated in more detail. The sketch in Fig. 4.1 neglects the transverse extension and the wavefront curvature of the laser beam, i.e. its dependence of the amplitudes E_1 and E_2 on the position $\vec{r}$. To take this into account, the optical powers of the three terms in Eq. 4.1 are rewritten as

$$P_1 = c\varepsilon_o \int\int \vec{E}_1^2(\vec{r}) dA \qquad P_2 = c\varepsilon_o \int\int \vec{E}_1^2(\vec{r}) dA \qquad (4.2)$$

$$P_{12} = 2c\varepsilon_o \cos\left((\omega_2 - \omega_1)t - \varphi\right) \int\int \vec{E}_1(\vec{r}) \cdot \vec{E}_2(\vec{r}) dA. \qquad (4.3)$$

The integrals extend over the photo sensitive surface of the detector. The reduction of the beating term in form of the overlap integral is called the antenna theorem (Siegman, 1966). It is clear that this integral vanishes for orthogonal transverse modes or orthogonal polarizations. In practice, this integral becomes very small for only a small angle tilt, of say two otherwise identical TEM$_{00}$ modes. It is easy to see that the spatial phase perpendicular to the beams varies by more than π and mostly cancels the overall beat note if the tilt angle θ is more than λ/w_0, where w_0 is the beam radius. Similar arguments can be made if the two beams are axially matched but have different beam waists or positions of the these waists. One possibility to achieve close to perfect spatial mode matching is to maximize the coupling of the overlapped beams through a single mode fiber.

Since signal strength can always be boosted with an electronic amplifier, an important aspect of the beat note detection is the signal-to-noise ratio (Kingston, 1978; Protopopov, 2009). The latter is unaffected by a perfect amplifier. The current supplied by the photo detector results from the number of photons per second at the optical power P, multiplied by the elementary charge and the quantum efficiency of the detector η

$$I = \frac{e\eta P}{h\nu}, \tag{4.4}$$

where $\omega_1 \approx \omega_2 = 2\pi\nu$. The current is converted into a voltage with a load resistor R. Assuming perfect spatial mode matching, the square average electrical powers of the three terms in Eq. 4.1 are given by:

$$R\langle I_1^2 \rangle = R\left(\frac{e\eta}{h\nu}\right)^2 P_1^2, \quad R\langle I_2^2 \rangle = R\left(\frac{e\eta}{h\nu}\right)^2 P_2^2, \quad R\langle I_{12}^2 \rangle = 2R\left(\frac{e\eta}{h\nu}\right)^2 P_1 P_2. \tag{4.5}$$

It is important to note that the electrical power of the signal is proportional to the *square* of the optical power, as this is often confused. The last term comes with a factor 2 instead of 2^2 as might be expected from Eq. 4.1 because we need to take the root mean square.

An important noise contribution is the (white) shot noise caused by the quantum nature of light. The number of photons $n = P\tau/(h\nu)$ that arrive in time intervals of length τ has a Poissonian distribution if the light field is in a coherent state. Hence the optical noise power due to the variation between time intervals is given by the square root of the mean photon number $\langle n \rangle$. The electrical noise power is obtained as

$$R\langle I_{sn}^2 \rangle = R\left(\frac{e}{\tau}\sqrt{\eta\langle n \rangle}\right)^2 = 2R\frac{e^2\eta B_w}{h\nu}P, \tag{4.6}$$

where the inverse of the detection times is now called the bandwidth of the system $B_w = 1/2\tau$ (see Nyquist theorem). The so-called "standard quantum limit"[2] for the signal-to-noise ratio is obtained by using Eq. 4.4, Eq. 4.5, and Eq. 4.6 together with the signal and noise generating terms in Eq. 4.1. The latter is due to the averaged full power that reaches the detector, i.e. the first two terms of Eq. 4.1. The last term generates the signal in case of a beat note detection. Hence the signal-to-noise ratio in terms of electrical power is given by:

---

[2] This limit can be surpassed with squeezed light. This does *not* violate the quantum (Heisenberg) limit.

$$\frac{S}{N} = \frac{\eta}{h\nu B_w} \frac{P_1 P_2}{P_1 + P_2}. \tag{4.7}$$

The noise appears in the denominator and is proportional to the bandwidth B_w. This means we are dealing with white noise as expected from the infinitely short photon clicks (Fourier transform of the δ-function). In general the denominator contains the full power that reaches the detector, because all of which contributes to the noise. This is important when generating beat notes with a frequency comb that typically comes with a large number of modes that do not contribute to the signal but to the noise (Reichert et al., 1999). Other white noise sources may be added to the denominator such as Johnson noise of the load resistor, the noise due to the dark current of the photo detector and noise contribution by the amplifiers that may be used to boost the signal. This is described in more detailed treatments of the heterodyne detection, i.e. beat note generation, in the textbooks by R. H. Kingston (Kingston, 1978) and V. V. Protopopov (Protopopov, 2009). It should be noted that the latter book is focused on the 10.6 μm radiation from a CO_2 laser where no detectors with low dark current exist.

The optical powers that appear in Eq. 4.7 are the ones that reach the detector. One can re-express this in terms of the optical powers before the beam splitter by replacing $P_1 \to TP_1$ and $P_2 \to (1-T)P_2$ in Eq. 4.7 with the beam splitter transmission T and reflection $(1-T)$. The optimum beam splitter is found by maximizing the signal-to-noise ratio:

$$T_{opt} = \frac{\sqrt{P_2}}{\sqrt{P_1} + \sqrt{P_2}}. \tag{4.8}$$

An adjustable beam splitter can be constructed with the help of two polarizers and a rotatable half wave plate (left hand side of Fig. 4.2). Such a design can be improved to form an optical balanced detector that is sketched at the right hand side of Fig. 4.2. In the case of equal laser powers one wins a factor of 2 (3 dB) for the signal-to-noise ratio with the balanced detector. This is because the signal doubles

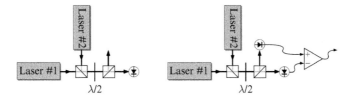

Fig. 4.2 Left: Adjustable beam splitter to experimentally optimize the signal-to-noise ratio according to Eq. 4.8. In contrast to Fig. 4.1 the lasers are assumed to emit orthogonal polarized light, such that the full power of both beams exit through one port of the first polarizing beam splitter. No interference will take place until the field vectors are projected on a common axis defined by the second polarizing beam splitter. **Right:** Optical balanced detector that uses the full power. It employs a differential amplifier and makes use of the 180° phase shift of the beat notes leaving the two ports of the second beam splitter (must be like that to obey energy conservation).

and gives four times the electrical power while the noise electrical power only doubles. The latter can be understood by noting that eventually all photons are detected, and the photon noise is anti-correlated in the two detectors (correlated at the amplifier output) because each photon that appears at one port is missing at the other. Balanced detectors are commonly used to detect squeezing of a very weak beam by mixing it with a strong laser, called the "local oscillator" in this context. In that case the signal-to-noise ratio of the balanced receiver with exactly $T = 50\%$ is identical to the one port detector with $T \to 1$ so that it seems to have no advantage (Schumaker, 1987). However, often one deals with other noise sources in addition to quantum noise. For example, power fluctuations of the lasers show up as correlated noise in the two detectors and hence cancel out in the output of the amplifier. The effective cancellation of this technical noise could be an important advantage. A 30 dB suppression of spurious radio frequency pick-up noise is readily achievable in practice.

4.1.2.2   *Optical phase–locked loops*

A phase-locked loop forces the phase of one oscillator to stay in phase with another by electronic feedback. In practice this feedback loop is not infinitely fast, so it cannot keep the relative phase at exactly zero at all times. However, the phase-locked loop can keep this phase at zero on average. Very often this property is good enough to make the frequency of the two oscillators identical to a very high degree. As an example, assume that the relative phase is locked to a vanishing average but with an rms fluctuation of $\Delta\varphi$, where this fluctuation possesses a white spectrum.[3] The frequency uncertainty after a measurement time τ is then given by $\Delta\nu = \Delta\varphi/\tau$. Even with a rather large rms fluctuation of say $\Delta\varphi = 2\pi$ and a phase–locked optical frequency of say 10^{14} Hz, the relative agreement of the frequencies of the two oscillators is 10^{-14} after only one second of measurement time. It further improves as the inverse of the measurement time τ. After only 100 sec it reaches the relative uncertainty of the current best cesium atomic clock. Therefore, the frequency uncertainties associated with non perfect phase-locked loops can almost always be neglected.[4] A frequency lock does not have this property.

Phase-locked loops are a standard technique in the radio frequency domain (Gardner, 2005). Therefore lots of literature exists that can be adapted to the optical domain, i.e. for phase locking lasers to each other. The general scheme is sketched in Fig. 4.3. A phase detector receives as inputs the signals from two oscillators and generates a voltage that is proportional to the phase difference between them. This voltage is then fed back with the proper sign to one of the oscillators that needs to have a control voltage that changes its frequency. In radio frequency technologies such device is generally referred to as a voltage controlled oscillator. By the feedback a

---

[3] A random process is best described in the Fourier domain. Even when the exact time dependence of a signal like $\varphi(t)$ cannot be given, the power spectral density, i.e. the square modulus of its Fourier transform $|\tilde{\varphi}(\omega)|^2$, may still be a well defined quantity. After all the Fourier transform is some sort of time average.

[4] This is very different when one needs to stabilize the phase of an optical carrier, say relative to the pulse envelope as described in Sec. 4.1.3.

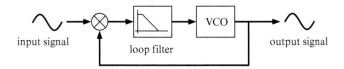

Fig. 4.3 Phase-locked loop: A phase detector ($\otimes$) generates a voltage that is proportional to the relative phase between the input signal and the voltage controlled oscillator (VCO). This voltage is used to control the frequency of the VCO such that it stays in phase with the input signal. A loop filter, most of the time a lowpass filter, is used to make the loop stable and to average over high frequency noise of the input signal.

lagging (leading) phase difference is compensated for by speeding up (slowing down) the controlled oscillator until the phase difference is back to zero. The important feature here is that this phase compensation can take place *after* a phase perturbation has occurred. Hence the feedback system is not required to act instantaneously (which is impossible in practice) in order to zero the phase difference *on average*.

There are several devices that can act as a phase detector. A radio frequency mixer is probably the most common one in the radio frequency domain. It generates a voltage that is proportional to the product of two input voltages (hence the symbol: $\otimes$). If these input voltages are given by $\sin(\omega_1 t + \varphi)$ and $\cos(\omega_2 t)$, the output will be $\sin((\omega_1 - \omega_2)t + \varphi) + \sin((\omega_1 + \omega_2)t + \varphi)$. The way this is turned into a phase detector is by using a lowpass filter (in addition to the loop filter) to cancel the second term. In the phase-locked condition we will have $\omega_1 = \omega_2$ on average, so that the signal will be $\propto \sin(\varphi)$ which can be approximated by $\sin(\varphi) \approx \varphi$.

The phase-locked loop will generate an exact copy of the input signal at the output, if the feedback speed (bandwidth) is infinite and the input signal possesses only phase noise (i.e., no amplitude noise). Obviously this situation is not very useful. The finite bandwidth provided by the low pass filter in Fig. 4.3 leads to a tracking of all phase perturbations by the voltage controlled oscillator within this bandwidth and averaging of faster phase noise. In the radio frequency domain, such a phase-locked loop may be used as a bandpass filter that could be extremely narrow even at very large signal frequencies. Once in lock, this filter even tracks the signal which is handy for low orbit satellite communication with largely varying Doppler shifts.

The mixer based phase detector is very simple but comes with a rather limited range. The phase-locked loop needs to be fast enough to keep the phase fluctuations $|\Delta\varphi| < \pi/2$, otherwise the feedback sign flips and the phase-locked loop will slip one or several cycles before it relocks to another zero crossing of $\sin(\varphi)$. For precision measurements these cycle slip events have to be avoided or, if this is not possible, detected (see below) so that the affected data can be removed. When phase locking two lasers to each other, the beat note takes over the role of the voltage controlled oscillator.[5] A voltage input to one of the lasers is required to change its frequency and hence the frequency of the beat note. There are many possibilities to turn a laser

---

[5] One may think that it is a formidable task to control the extremely large optical frequency. However, what is indeed controlled is the frequency difference between two laser. If these lasers are stable, one does not need a large bandwidth for locking.

into a voltage controlled oscillator. A piezo mounted end mirror for example allows to adjusts the laser frequency via its cavity length. In the same way an intra cavity etalon might be used. The frequency of a laser diode can be directly controlled via the injection current and so on.

In particular for piezo driven frequency control of lasers that lack a very high intrinsic stability, one frequently encounters the problem that the control bandwidth is insufficient to keep the in-lock phase fluctuations within $\pm\pi/2$. In this case one may resort to a different type of phase detector with a larger phase range. Of course any type of phase detector must have a limited phase range, because there is no principle limit on the measured phase difference but the possible output voltage is certainly limited. A common large range phase detector is based on a forward/backward counter (Prevedelli et al., 1995). A cycle at one of its input increases the number that is stored in an internal digital memory, while a cycle at the other input decreases this number. With the two inputs at the same frequency, the stored number stays constant. Subtracting an offset from this number and conversion to an analog signal produces a voltage proportional to the phase difference of the two input signals. The range of this digital phase detector is determined by the number of bits of the internal digital memory. Usually only a few (say 4) bits are required but in principle this number could be larger to increase the phase range. Another commonly used phase detector is an integrated circuit, MC12040 by Motorola with a phase range of $\pm2\pi$. This device turns into a frequency detector when it reaches its phase limit in order to provide at least the proper sign for relocking.

A large range phase detector in connection with a low servo bandwidth allows for large phase fluctuation $\Delta\varphi$ without actually loosing cycles. The consideration above explains that such a sloppy lock may be tolerated, even for precision frequency applications.[6] The in-lock phase fluctuations merely increase the required measurement time to reach a certain accuracy limit, but does not affect that limit. Very often it is far easier to increasing the range of the phase detector than the bandwidth of the laser frequency control. This is why the large range phase detectors are wide spread in optical phase-locked loops.

The optical phase-locked loop described here actually does not directly lock the phase between two lasers to each other. Instead it locks the phase of the beat note between two lasers to a radio frequency, the local oscillator. Of course the radio frequency offset has to be known and taken into account (with the proper sign!) for the data analysis. The requirement for its precision is rather moderate because it only *adds* to the optical frequencies. The reason why one usually employs this heterodyne method rather than locking the lasers to zero offset frequency (homodyne) is the larger noise at lower frequencies. Typically, the low frequency noise of radio frequency electronics is much smaller than that of lasers. Therefore, it is advantageous to generate an AC beat note and then mix it to DC with a radio frequency source for locking. Besides working above the low frequency noise of the laser, the radio frequency offset has another advantage when it comes to verifying the proper functioning of the phase-

---

[6] In some cases that is even desired, for example if one wants to lock a narrow linewidth laser to an accurate but unstable laser.

locked loop. This is done by counting the in-lock beat note with a radio frequency counter, that ideally uses a different bandpass filter than the phase detector. If no cycle slips occur, this counter should display the predefined radio frequency offset. Setting a reasonable threshold, one can use this criterion to reject data points that are possibly affected by cycle slipping events (Udem et al., 1998). It may of course also happen that this counter produces a wrong result, in particular if the signal-to-noise ratio of the beat note is limited. In this false alarm event, some good data points might get deleted. However, what is important here, is that cycle slip events can only enter the data if the phase-locked loop and the counter make the *same* mistake *at the same time*. One can estimate that rate from the cycle slip rate and, if higher immunity is required, a second counter may be employed (Zimmermann et al., 2004). Fig. 4.4 shows an example of a data set with a rather large cycle slip rate (for illustration) due to a limited signal-to-noise ratio. Insufficient control bandwidth may also lead to a large cycle slip rate, but can be coped with by a larger range phase detector. High quality optical phase-locked loops may operate without cycle slips for hours or even days. Even such a low cycle slip rate may be limiting, depending on the accuracy goal. Counting the frequency of a modern optical lattice clock for example, say with a relative accuracy of 10^{-18} at a frequency of 10^{15} Hz, one may not loose a single cycle out of 10^{18} optical cycles. This means one undetected cycle slip per hour already spoils

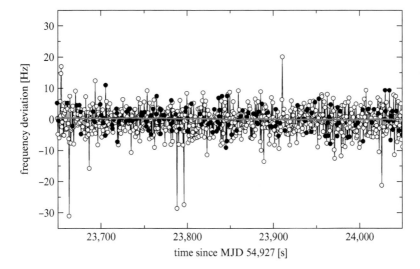

Fig. 4.4 Cycle slip detection: two independent frequency combs are used to simultaneously measure the frequency of a cavity stabilized laser. The data shows a record of the difference, that should be zero for perfectly operating frequency combs. The data fluctuates because the phase-locked loops operate with a finite bandwidth and because cycle slipping events take place. To detect the latter redundancy counters are used and readings where these counters disagree by more than 0.5 Hz are marked with hollow circles. It can be seen that all outliers are tagged as cycle slips in this way along with other data points that appear to be unbiased. These are probably not cycle slips but counter errors due to the limited signal-to-noise ratio. However, it is advisable to remove these data points if one needs to average (solid curve) significantly below the variance of this. MJD = Modified Julian date.

the clocks accuracy. Of course one also wants to have a safety margin here. Therefore, cycle slip detection is essential in this case, even with rock solid optical phase-locked loops.

In a typical laser spectroscopy setting one uses a narrow linewidth laser that is stabilized to a high finesse cavity. Phase locking the spectroscopy laser to a frequency comb is usually not advisable, unless the comb itself is locked to a high finesse cavity. With the standard locking scheme (see Sec. 4.1.3), the frequency comb provides an accurate (at long time average) but unstable reference. Therefore, counting the beat note of the spectroscopy laser with the frequency comb instead of phase locking it is the better choice to obtain more precise spectroscopic data. If the counted beat frequency is not known a priori, it cannot be used to detect counter errors. Nevertheless, redundancy can be employed in pretty much the same way. To do this, two counters are used in parallel, ideally operating with two different bandpass filters to clean up the beat notes at their inputs. To prevent counter errors entering the data, one rejects all readings where the two redundant counters deviate by more than some threshold. An additional complication emerges here when the beat note frequency is fluctuating a lot. In that case the counters must be synchronized very carefully. Some radio frequency counters allow to measure a frequency ratio of two inputs. These may be used as a redundancy check by feeding the same beat note to both inputs (again with different bandpass filters) and then require this ratio to be close enough to 1 for a valid data point. The synchronization problem can be solved with the gate of the ratio counter opening before the beat note counter and ending thereafter. It should be mentioned though that some counters operate with a reduced resolution when measuring a frequency ratio. Which of these methods works, has to be tested under the actual conditions. In particular, it should be tested if the final result of a measurement does not change within the statistical uncertainty when the cycle slip thresholds is reduced.

4.1.2.3   Interval divider

The largest frequency difference that can be measured with the beat note method is limited by the bandwidth of the fastest photo detectors, amplifiers, and counters. This is not a hard limit, but exceeding more than a few 100 GHz becomes extremely tedious. An optical frequency interval divider permits to increase this maximum (Hänsch, 1989; Telle et al., 1990). This device has played an important role in early phase coherent optical frequency measurements. As illustrated in Fig. 4.5, an optical frequency interval divider receives two input laser frequencies f_1 and f_2. The sum frequency $f_1 + f_2$ and the second harmonic of a third laser $2f_3$ are created in nonlinear crystals. The radio frequency beat signal between them at $2f_3 - (f_1 + f_2)$ is used to phase lock the third laser[7] at the midpoint $f_3 = (f_1 + f_2)/2$. Here and in the following we ignore the radio frequency offset added by the heterodyne phase-locked loop for simplicity.

---

[7] Likewise the frequency of any of the other lasers can be used for this phase lock.

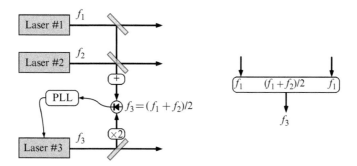

Fig. 4.5 Left: Principle of an optical frequency interval divider: The frequencies f_1 and f_2 of two input lasers are added in a nonlinear crystal. The second harmonic of a third laser $2f_3$ is phase-locked (PLL) to the sum frequency $f_1 + f_2$. The frequency f_3 thus divides the interval $f_1 - f_2$ into two equal sections. **Right:** symbol that is used in Fig. 4.11.

There is no principle limit of the size of the optical frequency interval that can be divided by 2 with this method. The optical frequency interval dividers can be arranged in series to successively divide a large interval. With n stages in place, the frequency interval is coherently divided by factor 2^n. While the first stage needs 3 lasers, any of the following stages can use two lasers of the previous stage, so that only one more laser has to be added. This principle was used for a coherent comparison of the hydrogen 1S-2S transition frequency at 121 nm (see next section) with a very accurate laser frequency standard based on a methane stabilized helium–neon laser at 3.39 μm. The 28-th harmonic of this laser would agree with the hydrogen transition frequency if there wasn't a gap of 8 THz. At the time of this measurement the frequency combs were not available yet, so that a cascade of 4 optical frequency interval stages was used to convert this gap into a measurable frequency beating (Udem et al., 1997).

4.1.2.4 *Frequency chains*

Historically the first optical phase-locked loops were used to operate harmonic frequency chains. A harmonic frequency chain consists of concatenated oscillators with successively higher frequencies and have been used to phase coherently multiply a radio frequency, provided by a cesium clock, all the way up to the optical domain. Intermediate nonlinear elements are used for frequency multiplication. Because nonlinear elements are usually weak, signal amplification is required in each stage to provide enough power to drive the next nonlinear element. Optical phase-locked loops are used for that purpose by locking a powerful (transfer) oscillator to the weak harmonic signal from the previous stage. To transverse the wide spectrum from the radio frequency domain all the way up to optical frequencies, vastly different oscillators, nonlinear elements, and detectors had to be used. The first harmonic frequency chains of that type were constructed at the end of the 1960s and reached up to sub-millimeter wavelength (Hocker et al., 1967). In 1973 the highest frequency that was obtained, was the one of the above mentioned methane stabilized helium–neon laser at 3.39 μm (88.376 THz) (Evenson et al., 1973). The possibility to measure "absolute"[8] optical frequencies was a very important step and eventually gave rise to assign a defined value to the speed of light within the SI system of units (1983). Thereafter the meter is realized with an interferometer using a laser frequency in combination with the relation $\lambda = c/\nu$. Proceeding even further up the frequency scale was very tedious and it took until 1996 to reach the frequency of visible light for the first time in this way (Schnatz et al., 1996). Shortly after an extension to the vacuum UV was realized in our lab (Udem et al., 1997) to access the 121 nm 1S-2S transition in atomic hydrogen. Fig. 4.6 shows these two frequency chains next to each other. For reasons given in Sec. 4.1.2, the phase coherence of the chain means that the accuracy of the frequency conversion is unlimited for infinite averaging times, provided that there are no cycle slips. The two parts of the chain were indeed coherent. This is true even for the beat notes that are counted rather than phase locked. After all a counter can be seen as measuring the phase between the input signal and its reference that derives from the same cesium clock which drives the frequency chain. However, since the two frequency chains were located at different labs, the only way to connect them at that time was to shuttle the methane stabilized helium–neon laser between them. Even though this laser had the best reproducibility of all stabilized lasers at that time, it was the weakest link that limited the accuracy of the cesium–hydrogen comparison. The reason is that only frequency, not phase could be transferred by transporting this laser.

Even though the harmonic frequency chains were a great improvement in measurement technology, they were extremely tedious, complex and most of them operated only occasionally. Another disadvantage was that only one optical frequency could be accessed with a dedicated frequency chain. It was not practical to build an optical clock on this basis. Instead they were used for precise frequency calibration of stabilized lasers, so that these lasers could then be reproduced elsewhere and used to realize the meter or to measure other optical frequencies of interest. A nice overview of forty years of developments in optical frequency metrology touching the beginning of frequency combs is given by J. L. Hall (Hall, 2000).

A new idea how to perform "absolute" optical frequency measurements was given by T. W. Hänsch in 1989 (Hänsch, 1989). Rather than brute force multiplication of a radio frequency all the way up to the optical domain, he suggested to measure the frequency gap between an optical frequency f and its second harmonic $2f$. Since $f = 2f - f$ this would give the "absolute" optical frequency. A chain of frequency interval dividers was proposed for this purpose. To measure the hydrogen Balmer-β line at 486 nm ($2f = 617$ THz) as suggested, one would have required 14 interval dividers, i.e. 16 phase-locked lasers to reduce the frequency gap to 38 GHz. At first glance this appears no more simpler than the harmonic frequency chains. However, the advantage here is that essentially all stages would use the same technology, except maybe for the first two, and the convergence wavelength could be chosen. In this way it seemed possible to use only compact and easy to handle diode lasers throughout the divider chain. In addition, with improving capabilities to measure the large remaining frequency gap,

---

[8] Measurements are always relative by comparing a quantity with an agreed upon common reference (the unit). Nevertheless the term "absolute frequency measurement" is wide spread, but it means *relative* to a cesium clock.

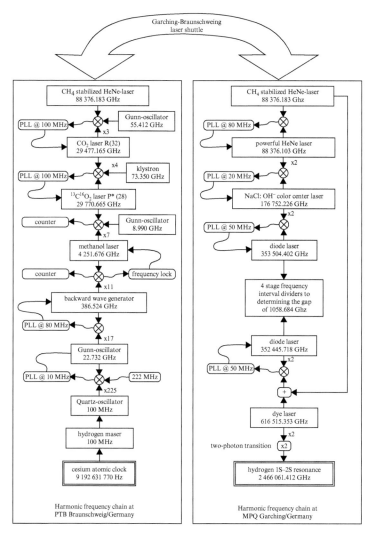

Fig. 4.6 Two harmonic frequency chains used to connect a methane stabilized He–Ne laser at 88.376 THz (3.39 μm) with the hydrogen 1S–2S transition at 2 466 THz (121 nm). One of them was operated by the Physikalisch–Technische Bundesanstalt at Braunschweig, Germany (left) and the other at the Max–Planck–Institut für Quantenoptik at Garching, Germany (right). The former connected a cesium atomic clock with the He–Ne laser which was then transported to Garching where it was used at the low frequency end and multiplied by 28. In this way it missed the frequency of the 1S–2S transition by 8 THz. This gap measured 1 THz at 353 THz (849 nm) and was further reduced by dividing it by 16 with the help of four optical frequency interval dividers. The final beat note at 66.2 GHz could be counted after mixing it with a radio frequency generated by a klystron. PLL: phase-locked loop; "$\times 2$": second harmonic generation; "+": sum frequency generation. The symbol $\otimes$ represents a mixer, just as in Sec. 4.1.2, however more than two signals (optical/radio frequencies) may be mixed. Metal-to-metal point contact diodes are used in the far infrared as extremely fast photo diodes, that simultaneously mix with harmonics of a radio frequency (Hocker et al., 1968).

the number of required stages would reduce. Therefore at least two groups, ours and the one of T. Ikegami (Ikegami et al., 1996), have started to construct such a divider chain. The latter experiment was actually based on a different scheme, that uses a parametric oscillator for interval division (Wong, 1992).

The main motivation for starting this effort was the development of modulator based optical frequency combs by M. Kourogi and coworkers (Kourogi et al., 1993). Frequency differences that exceed the bandwidth of the fastest photo detectors could be determined by generating two beat notes with the nearest modes of the comb and add to their sum frequency an integer multiple of the mode spacing. Soon these modulator based comb generators could measure optical intervals of a few THz (Saitoh et al., 1995) so that the number of necessary interval dividers reduced to about eight.[9] At that point an interval divider chain would have been a very good alternative to the clunky frequency chains. Then, towards the end of the 1990s, the comb generators reached a span of something[10] like 30 THz (Imai et al., 1998), reducing the number of necessary divider stages even further. This frequency comb was externally broadened using a dispersion-flattened optical fiber (see Sec. 4.1.3). Eventually this development yielded into an octave-spanning frequency comb with simultaneous active modes at frequencies f and $2f$. No more interval dividers were required (see Sec. 4.1.3).

4.1.3   Mode-locked lasers as comb generators

The spectral width of the modulator based comb generators is limited by group velocity dispersion which destroys the regular spacing of modes (see next section). Even more bandwidth can be generated with a device that already included dispersion compensation, gain and self-phase modulation: the Kerr lens mode-locked laser. Such a laser stabilizes the relative phase of many longitudinal cavity modes such that a solitary short pulse is formed. In the time domain this pulse propagates with its group velocity v_g back and forth between the end mirrors of the resonator. After each round trip a copy of the pulse is obtained at one of these mirrors that is partially transparent, like in any other conventional laser. Because of its periodicity, the pulse train generated this way produces a spectrum that consists of discrete modes that can be identified with the longitudinal cavity modes. The process of Kerr lens mode locking (KLM) introduced in the early 1990s (Spence et al., 1991), allows to generate pulses of 10 femtoseconds (fs) duration and below in a rather simple way. In contrast to KLM, active mode locking employs an intracavity modulator to generate sidebands at adjacent cavity modes. This method is subject to noise multiplication (see (Walls and DeMarchi, 1975) and Sec. 4.1.3) when it comes to very broad frequency combs and is therefore not further discussed here.

---

[9] We replaced the interval dividers in Fig. 4.6 with a modulator-based comb generator which then allowed fast alternation between isotopes and a precision measurement of the hydrogen–deuterium isotope shift of the 1S-2S transition (Huber et al., 1998)

[10] There are several ways to quantify the width of the comb. In this context the most practical measure would be to determine the maximum frequency separation that two lasers can be safely locked to the comb. This value is usually larger than the FWHM but not always given in the papers.

For Fourier limited pulses the bandwidth is given by the inverse of the pulse duration, with a correction factor of order unity that depends on the pulse shape and the way spectral and temporal widths are quantified (Diels and Rudolph, 2006). A 10 fs pulse for example will have a bandwidth of about $1/10$ fs $= 100$ THz, which is close to the optical carrier frequency, typically at 375 THz (800 nm) for a laser operated near the gain maximum of titanium:sapphire, a commonly used gain medium for this purpose. By virtue of the repeating pulses, this broad spectrum forms the envelope of the frequency comb. Depending on the length of the laser cavity, that determines the pulse round trip time, the pulse repetition rate ω_r is typically in the range of 100 MHz but 4 MHz through 10 GHz repetition rates have been realized. In any case ω_r is a radio frequency that is readily measured and stabilized. The usefulness of this comb critically depends on how constant the mode spacing is across the spectrum and to what precision it agrees with the readily measurable repetition rate. These questions will be addressed in the following sections.

For frequency comb generation mostly commercially available titanium:sapphire and fiber lasers are used. Both are sketched in Fig. 4.7 and are also described in textbooks like (Diels and Rudolph, 2006). Titanium:sapphire has a peak gain at 795 nm while erbium and ytterbium doped fibers have their maximum gain at 1550 nm and around 1030 nm respectively. All of these lasers support fs pulses. The titanium:sapphire laser, like most solid-state lasers, operates with moderate gain and low output coupling, i.e. with a high Q laser cavity. This leads to smaller background emission such as amplified spontaneous emission. Acoustic noise might be

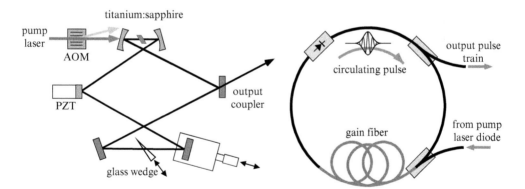

Fig. 4.7 Left: Compact ring titanium:sapphire laser. The ring design makes it immune to back reflections. Mode locking is achieved by a Kerr lens that increases the overlap with the pump beam and hence enables larger gain for the pulsed operation. A drawback is that this mechanism relies on precise alignment. This laser has four "knobs" to adjust the comb parameters: the acousto-optic modulator (AOM) that drains part of the pump beam, a glass wedge on a translation stage and the cavity length adjustment, manual and with a piezo transducer (PZT) (see also Sec. 4.1.3). **Right:** Fiber ring laser. The mode locking mechanism could be based on nonlinear polarization rotation in combination with a polarizer (not shown) or another usage of the Kerr effect. An optical diode enforces unidirectional operation. The control "knobs" can be the same as for titanium:sapphire with fiber stretchers adjusting the cavity length without the need of a free-space section. Electro-optic modulators for fast cavity length control can be implemented in both lasers.

dominating, depending on the mechanical construction of the laser cavity. In addition, most titanium:sapphire lasers require frequent re-alignment and hence do not operate autonomously for an extended time (with exemptions, see (Matos et al., 2006)). On the other hand the modes of the fiber lasers typically sit on a high frequency noise pedestal that does not significantly impact its usefulness as a comb generator for optical frequency metrology. The erbium doped fiber lasers emit in the telecom band, so that cheap and advanced optical components are available to build them. Ytterbium lasers on the other hand are readily scaled to high power thanks to the small quantum defect.

The mode-locking mechanism for fiber lasers is similar to the Kerr lens method, except that often nonlinear polarization rotation is used to favor the pulsed, high peak intensity operation. Up to a short free space section that can be built very stable, these lasers have no adjustable parts. This makes these lasers more reliable so that they can operate unattended for a very long time. As also discussed in Sec. 4.1.3, fiber lasers can be stably spliced to external fibers for spectral broadening, whereas titanium:sapphire lasers often suffer from mechanical and/or temperature drifts of the free space coupling to external fibers that require frequent re-alignment. Meanwhile, fiber laser based frequency combs are sufficiently robust to operate them remotely in space (Lezius et al., 2016).

Another method for frequency comb generation that is not a laser in itself, is based on four-wave mixing in a microresonator. Under suitable conditions, a continuous wave laser that is coupled into such a resonator gives rise to pairwise generation of sidebands that are resonant with the neighboring modes of the microresonator. These modes can again generate sidebands so that a very broad frequency comb is formed. This method is interesting for some applications, like in astronomy, because the microresonator generated combs naturally come with a very large mode spacing. However, so far they are not used routinely for optical frequency metrology. For a review and more references on this subject see (Kippenberg et al., 2011).

4.1.3.1 *Derivation of the comb from the cavity boundary conditions*

As in any laser the modes with wave number $k(\omega_n)$ and frequencies ω_n must obey the following boundary conditions of the resonator with length L:

$$2Lk(\omega_n) = n2\pi \qquad (4.9)$$

Here n is a large integer number identical to the number of half wavelengths within the resonator. Besides this propagation phase shift, there might be additional phase shifts, caused for example by diffraction (Gouy phase) or by the mirror coatings that are thought of being included in the above dispersion relation. This phase shift may depend on the wavelength of the mode, i.e. on the mode number n, and is not simple to determine accurately in practice. So we are seeking a description that lumps all the low accuracy quantities into readily measurable radio frequencies. For this purpose, dispersion is best be included by the following expansion of the wave vector about some mean frequency ω_m (not necessarily a cavity mode) according to Eq. 4.9:

$$2L \left[k(\omega_m) + k'(\omega_m)(\omega_n - \omega_m) + \frac{k''(\omega_m)}{2}(\omega_n - \omega_m)^2 + \ldots \right] = 2\pi n. \qquad (4.10)$$

The mode separation $\Delta\omega \equiv \omega_{n+1} - \omega_n$ is obtained by subtracting this formula from itself with n being replaced by $n + 1$:

$$2L \left[k'(\omega_m)\Delta\omega + \frac{k''(\omega_m)}{2} \left((\omega_{n+1} - \omega_m)^2 - (\omega_n - \omega_m)^2 \right) + \ldots \right] = 2\pi \qquad (4.11)$$

To obtain a constant mode spacing, as the defining requirement for a frequency comb, $\Delta\omega$ must be independent of n. This is the case if and only if all contributions of the expansion of $k(\omega)$ beyond the group velocity term $k'(\omega_m) = 1/\overline{v}_g(\omega_m)$ exactly vanish. The unwanted perturbing terms, or "higher-order dispersion"[11] terms, that contradict a constant mode spacing are those that deform the pulse as it travels inside the laser resonator. Therefore, the mere observation of a stable non-deformed pulse envelope stored in the laser cavity leads to a frequency comb with constant mode spacing given by

$$\Delta\omega = 2\pi \frac{\overline{v}_g}{2L} \qquad \text{with} \qquad \overline{v}_g = \frac{1}{k'(\omega_m)} \qquad (4.12)$$

with the round trip averaged group velocity[12] given by $\overline{v}_g$. Having all derivatives beyond $k'(\omega_m)$ vanishing, means that $k'(\omega_m)$ and $\overline{v}_g$ must be independent of frequency. Therefore, this derivation is independent of the particular choice of ω_m, provided it resides within the laser spectrum. The group velocity determines the cavity round trip time T of the pulse and therefore the pulse repetition rate $\omega_r = \Delta\omega$:

$$T^{-1} = \frac{\overline{v}_g}{2L} = \frac{\omega_r}{2\pi}. \qquad (4.13)$$

The frequencies of the modes ω_n of any frequency comb with a constant mode spacing ω_r can be expressed by (Eckstein, 1978; Wineland et al., 1989; Reichert et al., 1999; Udem et al., 2004):

$$\omega_n = n\omega_r + \omega_{CE} \qquad (4.14)$$

with a yet unknown frequency offset ω_{CE} common to all modes. As a convention we will now number the modes such that $0 \leq \omega_{CE} \leq \omega_r$. This means that ω_{CE}, like ω_r, resides in the radio frequency domain. Using Eq. 4.14 to measure the optical frequencies ω_n requires the measurement of ω_r, ω_{CE}, and n. The pulse repetition rate can be measured anywhere in the beam of the mode-locked laser. To determine the

---

[11] In the literature one often finds these higher-order terms simply referred to as "dispersion".
[12] The usual textbook approach ignores dispersion either as a whole ($\Delta\omega = 2\pi c/2L$) or just includes a constant refractive index: $\Delta\omega = 2\pi v_{ph}/2L$ with v_{ph} and c being the phase velocities with and without dispersive material respectively.

comb offset requires some more effort as will be detailed in Sec. 4.1.3. In practice the beam of a continuous wave laser, whose frequency is to be determined, is superimposed with the beam containing the frequency comb on a photo detector to record a beat note with the nearest comb mode. Knowing ω_r and ω_{CE}, the only thing missing is the mode number n. This may be determined by a coarse and simple wavelength measurement or by repeating the same measurement with slightly different repetition rates (Holzwarth et al., 2001a) (see Sec. 4.1.3).

Some insight on the nature of the frequency offset ω_{CE} is obtained by resolving Eq. 4.14 for ω_{CE} and using the cavity averaged phase velocity of the n-th mode $\bar{v}_p(\omega_n) = \omega_n/k(\omega_n)$. With the expansion of the wave vector in Eq. 4.10 and the fact that it ends with the group velocity term, one derives:

$$\omega_{CE} = \omega_n - n\omega_r = \omega_m \left(1 - \frac{\bar{v}_g}{\bar{v}_p(\omega_m)} \right). \qquad (4.15)$$

For this the frequency offset is independent of n, i.e. common to all modes, and vanishes if the cavity averaged group and phase velocities are identical. In such a case the pulse train possesses a strictly periodic field that produces a frequency comb containing only integer multiples of ω_r. In general though, this condition is not fulfilled and the comb offset frequency is related to the difference of the group and phase round trip time. For an unchirped pulse, i.e. a pulse that has a carrier frequency ω_c that is constant across the pulse, it makes sense to expand the wave vector around $\omega_m = \omega_c$ so that Eq. 4.15 becomes:

$$\omega_{CE}T = \Delta\varphi \qquad \text{with} \qquad \Delta\varphi = \omega_c \left(\frac{2L}{\bar{v}_g} - \frac{2L}{\bar{v}_p} \right). \qquad (4.16)$$

As discussed in more detail in the next section, it follows that the pulse envelope continuously shifts relative to the carrier wave in one direction. The shift per pulse round trip $\Delta\varphi$ is given by the advance of the carrier phase during a phase round trip time $2L/\bar{v}_p$ in a frame that travels with the pulse. The shift $\Delta\varphi$ per pulse round trip time $T = 2L/\bar{v}_g$ fixes the frequency comb offset. Hence it has been dubbed carrier-envelope (CE) offset phase (Udem, Th. (1997)).

How precise the condition of vanishing higher-order dispersion is fulfilled in practice, could be estimated by knowing that irregular phase variations between the modes on the order of 2π are sufficient, to completely destroy the stored pulse in the time domain. The phase between adjacent modes of a proper frequency comb advances as $\omega_r t$, i.e. typically by some 10^8 times 2π per second. An extra random cycle per second would offset the modes by only 1 Hz from the perfectly regular grid, but destroy the pulse in the same time. Compared to the optical carrier frequency this corresponds to a relative uncertainty of 3 parts in 10^{15} at most, which is already close to the best cesium atomic clocks. Experimentally, one observes the pulse to be intact for a much longer time than just one second (in some lasers it stays for months). In fact no deviations have been detected yet at a sensitivity of a few parts in 10^{16} (Udem et al., 1999a; Holzwarth et al., 2000; Diddams et al., 2002; Ma et al., 2004).

It should be noted that a more appropriate derivation of the frequency comb must include nonlinear shifts of the group and phase velocities, because this is used to lock the modes in the first place. In fact it has been shown that initiating the mode locking mechanism significantly shifts the cavity modes of a laser (Stenger and Telle, 2000). Of course the question remains how this mode locking process forces the cancellation of the pulse deforming dispersive contributions with this precision. Even though this is beyond the scope of this article and details may be found elsewhere (Krausz et al., 1992; Diels and Rudolph, 2006), the simplest explanation is given in the time domain: Mode locking, i.e. synchronizing the longitudinal cavity modes to sum up for a short pulse, is most commonly achieved with the help of the Kerr effect which is expressed in the time domain through

$$n(t) = n_0 + n_2 I(t). \tag{4.17}$$

Here n_2 denotes a small intensity $I(t)$ dependence of the refractive index $n(t)$. Generally a Kerr coefficient with a fs response time is very small, so that it becomes noticeable only for the high peak intensity of short pulses. Kerr lens mode locking uses this effect in two ways. The radially intensity variation of a Gaussian mode produces a lens that becomes part of the laser resonator for a short pulse. This resonator is designed such that it has larger losses without such a lens, i.e. for a superposition of randomly phased modes. In addition self-phase modulation through the intensity dependent refractive index can compensate the de-phasing action of higher-order dispersion. The result is a wave packet that does not deform as it travels. In general such a wave packet is called a soliton. The corresponding optical wave equation is called the nonlinear Schrödinger equation and includes the Kerr effect and the group velocity dispersion (that is canceled by the Kerr effect). Its analytic solution has a cosh-shape (Hasegawa and Tappert, 1973; Agrawal, 2013). The nice feature about this cancellation is that it is self-adjusting: Excess peak intensity leads to a larger frequency chirp[13] that does extend the spectral width and reduces the peak intensity through larger dispersion and vice versa. So that neither the pulse intensity nor the dispersion has to be matched exactly (which would not be possible) to generate a soliton. Of course it helps to pre-compensate higher-order dispersion as good as possible before initiating mode locking. It appears that for the shortest pulses, the emission bandwidth is basically given by the bandwidth this pre-compensation achieves (Sutter et al., 1999a,). Once mode locking is initiated, mostly by a mechanical disturbance of the laser cavity, the modes are pulled on the regular grid described by Eq. 4.14. Another way of thinking about this process is that self-phase modulation, that occurs with the repetition frequency, produces side bands on each mode that will injection lock the neighboring modes by mode pulling. This pulling produces the regular grid of modes and is limited by the injection locking range (Siegman, 1986), i.e. how well the pre-compensation of higher-order dispersion is done for the cold cavity.

---

[13] Which is that the carrier frequency changes across the pulse.

4.1.3.2   Derivation of the comb from the pulse train

Rather than considering intracavity dispersion, the cavity length and so on, one may simply investigate the emitted pulse train with little consideration how it is generated. For this one assumes that the electric field $E(t)$, measured for example at the output coupling mirror, can be written as the product of a periodic envelope function $A(t)$ and a carrier wave $C(t)$:

$$E(t) = A(t)C(t) + c.c. \qquad (4.18)$$

The envelope function defines the pulse repetition time $T = 2\pi/\omega_r$ by demanding $A(t) = A(t - T)$. The only thing about dispersion that should be added for this description, is that there might be a difference between the group velocity and the phase velocity inside the laser cavity. This will shift the carrier with respect to the envelope by a certain amount after each round trip. The electric field is therefore in general not periodic with T. To obtain the spectrum of $E(t)$ the Fourier integral has to be calculated:

$$\tilde{E}(\omega) = \int_{-\infty}^{+\infty} E(t) e^{i\omega t} dt. \qquad (4.19)$$

Separate Fourier transforms of $A(t)$ and $C(t)$ are given by:

$$\tilde{A}(\omega) = \sum_{n=-\infty}^{+\infty} \delta\left(\omega - n\omega_r\right) \tilde{A}_n \qquad \text{and} \qquad \tilde{C}(\omega) = \int_{-\infty}^{+\infty} C(t) e^{i\omega t} dt. \qquad (4.20)$$

A periodic frequency chirp imposed on the pulses is accounted for by allowing a complex envelope function $A(t)$. Thus the "carrier" $C(t)$ is defined to be whatever part of the electric field that is non-periodic with T. The convolution theorem allows us to calculate the Fourier transform of $E(t)$ from $\tilde{A}(\omega)$ and $\tilde{C}(\omega)$:

$$\tilde{E}(\omega) = \frac{1}{2\pi} \int_{-\infty}^{+\infty} \tilde{A}(\omega') \tilde{C}(\omega - \omega') d\omega' + c.c. = \frac{1}{2\pi} \sum_{n=-\infty}^{+\infty} \tilde{A}_n \tilde{C}\left(\omega - n\omega_r\right) + c.c. \qquad (4.21)$$

The sum represents a periodic spectrum in frequency space. If the spectral width of the carrier wave $\Delta\omega_c$ is much smaller than the mode separation ω_r, it represents a regularly spaced comb of laser modes just like Eq. 4.14, with identical spectral line shapes, namely the line shape of $\tilde{C}(\omega)$ (see Fig. 4.8). If $\tilde{C}(\omega)$ is centered at say ω_c, than the comb is shifted from containing only exact harmonics of ω_r by ω_c. The center frequencies of the mode members are calculated from the mode number n (Eckstein, 1978; Udem, 1997; Wineland et al., 1989; Reichert et al., 1999; Udem et al., 2004):

$$\omega_m = m\omega_r + \omega_c. \qquad (4.22)$$

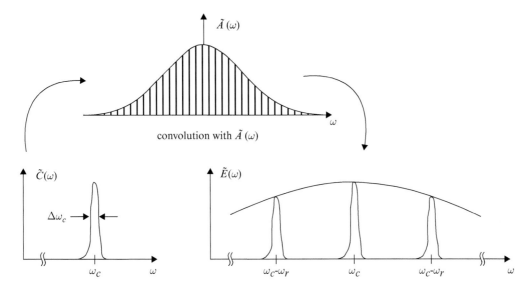

Fig. 4.8 The spectral shape of the carrier function (left), assumed to be narrower than the pulse repetition frequency ($\Delta\omega_c \ll \omega_r$), and the resulting spectrum according to Eq. 4.21 after modulation by the envelope function (right).

The measurement of the frequency offset ω_c (Udem, 1997; Hänsch et al., 1999; Reichert et al., 1999, 2000; Diddams et al., 2000; Jones et al., 2000; Holzwarth et al., 2000; Udem et al., 2004) as described below usually yields a value modulo ω_r, so that renumbering the modes will restrict the offset frequency to smaller values than the repetition frequency and again yields Eq. 4.14.

The individual modes can be separated with a suitable spectrometer if the spectral width of the carrier function is narrower than the mode separation: $\Delta\omega_c \ll \omega_r$. This condition is easy to satisfy, even with a free running laser. If a single mode is selected from the frequency comb one obtains a continuous wave. However, it is easy to show that a grating with sufficient resolution would be at least as large as the laser cavity, which appears unrealistic for a typical mode-locked laser. Fortunately, for experiments performed so far, it has never been necessary to resolve a single mode in the optical domain in this way (see Sec. 4.1.3).

Now let us consider two instructive examples of possible carrier functions. If the carrier wave is monochromatic $C(t) = e^{-i\omega_c t - i\varphi}$, its spectrum will be δ-shaped and centered at the carrier frequency ω_c. The individual modes are also δ-functions $\tilde{C}(\omega) = \delta(\omega - \omega_c)e^{-i\varphi}$. The frequency offset Eq. 4.22 is identified with the carrier frequency. According to Eq. 4.18 each round trip will shift the carrier wave with respect to the envelope by $\Delta\varphi = \arg(C(t-T)) - \arg(C(t)) = \omega_c T$ so that the frequency offset is given by $\omega_{CE} = \Delta\varphi/T$. In a typical laser cavity this pulse-to-pulse carrier-envelope phase shift is much larger than 2π, but measurements usually yield a value modulo 2π. The restriction $0 \leq \Delta\varphi \leq 2\pi$ is synonymous with the restriction $0 \leq \omega_{CE} \leq \omega_r$ introduced earlier. Fig. 4.9 sketches this situation in the time domain for a chirp free pulse train.

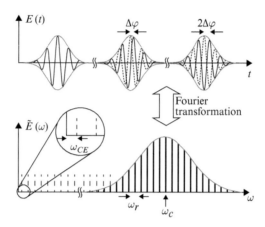

Fig. 4.9 Consecutive unchirped pulses ($A(t)$ real) with carrier frequency ω_c and the corresponding spectrum (not to scale). Because the carrier propagates with a different velocity within the laser cavity than the envelope (phase- and group velocity), the electric field does not repeat itself after one round trip. A pulse-to-pulse phase shift $\Delta\varphi$ results in an offset frequency of $\omega_{CE} = \Delta\varphi/T$ (Udem, 1997). The mode spacing is given by the repetition rate ω_r. The width of the spectral envelope is given by the inverse pulse duration up to a factor of order unity that depends on the pulse shape (the time-bandwidth-product of a Gaussian pulse for example is 0.441) (Diels and Rudolph, 2006).

As a second example consider a train of half-cycle pulses like:

$$E(t) = E_0 \sum_k e^{-\left(\frac{t-kT}{\tau}\right)^2}. \tag{4.23}$$

In this case the electric field would be repetitive with the round trip time. Therefore $C(t)$ is a constant and its Fourier transform is a δ-function centered as $\omega_c = 0$. If it becomes possible to build a laser able to produce a stable pulse train of that kind, all the comb frequencies would become exact harmonics of the pulse repetition rate. Obviously, this would be an ideal situation for optical frequency metrology.[14]

As these examples are instructive, it is important to realize that one neither relies on assuming a strictly periodic electric field nor that the pulses are unchirped. The strict periodicity of the spectrum, as stated in Eq. 4.21, and the possibility to generate beat notes between continuous lasers and single modes (Bramwell et al., 1985), are the only requirement that enables precise optical to radio frequency conversions.

In a real laser the carrier wave will not be a clean sine wave as in the above example. The mere periodicity of the field, allowing a pulse-to-pulse carrier envelope phase shift, already guarantees the comb-like spectrum. Very few effects can disturb that property. In particular, for an operational frequency comb, both ω_r and ω_{CE}

---

[14] It should be noted though that this is a rather academic example because such a pulse would be deformed quickly upon propagation since it contains vastly distinct frequency components with different diffraction. For example, the DC component that this carrier certainly has, would not propagate at all.

will be phase-locked so that slow drifts are compensated. The property that the comb method really relies on, is the mode spacing being constant across the spectrum. As explained above, even a small deviation from this condition will have very quick and devastating effects on the temporal pulse envelope. Not even an indefinitely increasing chirp could disturb the mode spacing constancy, as this can be seen as a constantly drifting carrier frequency, that does not perturb the spectral periodicity but shifts the comb as a whole. However, the phase of individual modes can fluctuate about an average value required for staying in lock with the rest of the comb. This will cause noise that can broaden individual modes as discussed in the next section.

4.1.3.3 Linewidth of a single mode

The modes of a frequency comb have to be understood as continuous laser modes. As such they posses a linewidth which is of interest here. Of course as usual in such a case, several limiting factors are effective at the same time. Nevertheless, it is instructive to derive the Fourier limited linewidth that is due to observing the pulse train for a limited number of pulses only. Following a derivation by A. E. Siegman (Siegman, 1986)[15] the linewidth of a train of N pulses can be derived. In accordance with the previous section, we assume that the pulse train consists of identical pulses $\mathcal{E}(t)$ separated in time by T and subjected to a pulse-to-pulse phase shift of $e^{i\Delta\varphi}$:

$$E(t) = \frac{E_0}{\sqrt{N}} \sum_{m=0}^{N-1} e^{im\Delta\varphi} \mathcal{E}(t - mT).$$
(4.24)

For decent pulse shapes, that fall off at least as $\propto 1/t$ from the maximum, this series converges even as N goes to infinity. Using the shift theorem

$$\mathcal{FT}\{\mathcal{E}(t-\tau)\} = e^{-i\omega\tau}\mathcal{FT}\{\mathcal{E}(t)\},$$
(4.25)

one can relate the Fourier transform of the pulse train $E(t)$ to the Fourier transform of a single pulse $\tilde{\mathcal{E}}(\omega)$. With the sum formula for the geometric series this becomes:

$$\tilde{E}(\omega) = \frac{E_0\tilde{\mathcal{E}}(\omega)}{\sqrt{N}} \sum_{m=0}^{N-1} e^{-im(\omega T - \Delta\varphi)} = \frac{E_0\tilde{\mathcal{E}}(\omega)}{\sqrt{N}} \frac{1 - e^{-iN(\omega T - \Delta\varphi)}}{1 - e^{-i(\omega T - \Delta\varphi)}}.$$
(4.26)

Now the intensity spectrum for N pulses $I_N(\omega)$ may be calculated from the spectrum of a single pulse $I(\omega) \propto |\tilde{\mathcal{E}}(\omega)|^2$:

$$I_N(\omega) = \frac{1 - \cos(N(\omega T - \Delta\varphi))}{N(1 - \cos(\omega T - \Delta\varphi))} I(\omega).$$
(4.27)

---

[15] The carrier envelope phase shift is ignored there but can easily be accounted for.

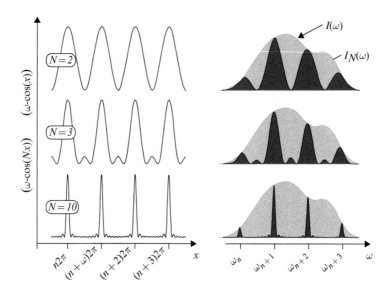

Fig. 4.10 Left: The function $(1 - \cos(Nx))/(1 - \cos(x))$ normalized to the peak for $N = 2$, 3 and 10 pulses. The maxima are at $x = \omega_n T - \Delta\varphi = 2\pi n$ with integer n. From that, and redefining $\omega_r = 2\pi/T$ and $\omega_{CE} = \Delta\varphi/T$ we derive again the frequency comb equation (Eq. 4.14). **Right:** The resulting frequency comb spectrum is given by the spectrum of a single pulse multiplied with the comb function at the left hand side.

Fig. 4.10 sketches this result for several N. As the spectrum of a single pulse is truly a continuum which is sinusoidally modulated for double pulses, sharper modes emerge as more pulses are added. This is similar to the diffraction from a grating that becomes sharper as more grating lines are illuminated. The spectral width of a single mode can now be calculated from Eq. 4.27 and may be approximated by $\Delta\omega \approx \sqrt{24}/TN$ for the pulse train observation time NT. Hence the Fourier limited linewidth is reduced after one second of observation time to $\sqrt{24}/2\pi$ Hz ≈ 0.78 Hz and further decreases as the inverse observation time. In the limit of an infinite number of pulses, the spectral shape of the modes approximate δ-functions peaked at $x = \omega T - \Delta\varphi \approx 2\pi n$

$$\frac{1}{2\pi} \lim_{N\to\infty} \frac{1 - \cos(Nx)}{N(1 - \cos(x))} \approx \frac{1}{\pi} \lim_{N\to\infty} \frac{1 - \cos(Nx)}{Nx^2} = \delta(x). \qquad (4.28)$$

The whole frequency comb becomes an equidistant array of δ-functions (a so-called Dirac comb):

$$I_N(\omega) \to I(\omega) \sum_n \delta(\omega T - \Delta\varphi - 2\pi n). \qquad (4.29)$$

The Fourier limit to the linewidth derived here is important when only a few pulses can be used for example for direct comb spectroscopy as will be discussed in Sec. 4.2.4.

On the other hand, when a large number of pulses contribute to the signal, for example when measuring the carrier envelope beat note, other limits enter. In most of these cases acoustic vibrations seem to set the limit, as can be seen by observing that the noise of the repetition rate of a non-stabilized laser dies off very steeply for frequencies above typical ambient acoustic vibrations around 1 kHz (Hollberg et al., 2001). Even if these fluctuations are controlled, as they can be with the best continuous wave lasers, a limit set by quantum mechanics in terms of the power dependent Schawlow-Townes formula applies. Remarkably the total power of *all* modes enters this formula to determine the linewidth of a *single* mode (Rush et al., 1984). In fact sub-hertz linewidths have been measured across the entire frequency comb when it is stabilized appropriately (Bartels et al., 2004; Swann et al., 2006).

4.1.3.4   *Testing the comb*

The arguments of the previous sections that lead to a perfectly regular frequency comb have to be experimentally verified when it comes to the utmost precision. The optical frequency divider played a decisive role for the first tests (Udem et al., 1999a). As sketched in Fig. 4.11 two laser diodes at 822.8 nm and 870.9 nm were phase-locked to modes of a frequency separated by more than 20 THz. The exact number of modes between the two lasers was not controlled, but obviously would be either an even or odd number. In the latter case[16] one expects another mode of the frequency comb right in the center, i.e. at the frequency that is generated by the optical frequency interval divider. In practice there is a well known offset due to the local oscillators used for phase-locking. This leads to a countable and known beating frequency (40 MHz in this case) between the center mode of the comb and the center frequency of the optical frequency divider.

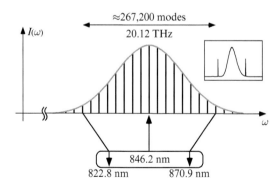

Fig. 4.11 The uniform distribution of the modes of a mode-locked laser is verified by comparison with an optical frequency interval divider (see Sec. 4.1.2). The inset shows the measured spectrum of the frequency comb together with the 822.8 nm and the 870.9 nm laser diodes drawn into it.

---

[16] If one finds an even number one flips the phase-locked loop on and off until it locks to an odd mode number difference.

To test frequency combs regularity this beat note has been recorded with gate times of 1, 10, and 100 seconds, which yields a resolution of 1 mHz, 0.1 mHz, and $10\,\mu$Hz respectively for one gate time. To ensure that lost cycles were detected, redundant counters were in place as described in Sec. 4.1.2. The weighted average of the results obtained using the various gate times gave -0.59 ± 0.48 mHz deviation from the expected frequency of 40 MHz. This verified the uniform distribution of the modes within a relative accuracy of 3.0×10^{-17} (see (Udem et al., 1999a) for more details). To rule out some asymmetric comb irregularities, that would evade a symmetric testing of Fig. 4.11, the measurement was repeated with a ten-times smaller frequency separation and all lasers that make up the optical interval divider positioned on either side of the peak of the combs spectral envelope. Again the comb regularity was confirmed with the same absolute accuracy but with a ten-times larger relative uncertainty, as expected due to the smaller frequency gap. No systematic effect was detected. This means the measured beat frequency average was always within one standard deviation at the expected frequency (for a perfectly regular mode spacing).[17]

Of course there are loopholes in this test that may lead to an excellent agreement, even though the frequency comb cannot convert between optical and the radio frequency domain with the same accuracy. Nevertheless, the mere possibility to isolate individual modes of the frequency comb by phase-locking another laser to it, ruled out a major concern about noise propagation (see next section and (Telle, 1996)).

A much better way of testing the combs capability to convert between the radio frequency and the optical domain is to operate two stabilized frequency combs (see Sec. 4.1.3) in parallel. Such a test is for example performed by using a common radio frequency source for both and comparing the generated optical frequencies. The first experiment of that type gave an agreement within a statistical uncertainty of 5.1×10^{-16} (Holzwarth et al., 2000). Other tests of this type followed with improved limits. Currently the best comb-comb comparison performed by L. A. M. Johnson and co-workers finds agreement within 5×10^{-18} (Johnson et al., 2015). Reaching this level of accuracy is not trivial and probably every individual comb system has to be tested in this way before one can be sure that it will support 18 digits. Unfortunately, too little attention is paid to testing the frequency combs that are used in optical clocks claiming uncertainties at the 10^{-18} level.[18] Probably the best way of operating a high level optical clock is to use two frequency combs for continuous verification of the clockwork mechanism.

Besides checking the regularity of the modes and a comb-comb comparison, there are other properties of a frequency comb that can be verified. One may test how well sum frequency generation and difference frequency generation agree for self-referencing (see Sec. 4.1.3) and finds agreement within 6.6×10^{-21} (Zimmermann et al., 2004). The capability of measuring optical frequency ratios with a frequency comb has been verified within 3×10^{-21} (Johnson et al., 2015) as well as other comparisons between

---

[17] A similar test has been performed with a modulator based optical frequency comb (Udem et al., 1998).

[18] Optical frequency standards have been compared at this level without counting. However, without accurate frequency counting such a device is not an optical clock. This is often confused.

optical frequencies at the 10^{-19} level (Ma et al., 2004). None of these tests verifies the comb's capability to convert between optical and radio frequencies.

4.1.3.5   *Generating an octave-spanning comb*

The width of the comb emitted by a fs laser can be significantly broadened in a single mode fiber by self-phase modulation (Agrawal, 2013). According to Eq. 4.17 assuming a single carrier wave ω_c, a pulse that has propagated the length l acquires a self induced phase shift of

$$\Phi_{NL}(t) = -n_2 I(t)\omega_c l/c \qquad \text{with } I(t) = \frac{1}{2}c\varepsilon_0 |A(t)|^2. \qquad (4.30)$$

For fused silica the nonlinear coefficient is comparatively small but almost instantaneous, even on the time scale of fs pulses. The rapidly varying phase delay generates new frequencies without affecting the pulse duration. The pulses gets chirped and are no longer at the Fourier limit so that the spectrum is broader than the inverse pulse duration. The extra frequencies are determined by the time derivative of the self induced phase shift $\dot{\Phi}_{NL}(t)$. With the envelope function in Eq. 4.18 we get:

$$A(t) \longrightarrow A(t)e^{i\Phi_{NL}(t)}. \qquad (4.31)$$

Because $\Phi_{NL}(t)$ has the same periodicity as $A(t)$, the comb structure of the spectrum is maintained and the derivations of Sec. 4.1.3 remain valid because periodicity of $A(t)$ was the only assumption made. An optical fiber is most appropriate for this process, because it can maintain the necessary small focus area over a virtually unlimited length. In practice, however, other pulse reshaping mechanisms, both linear and nonlinear, are present so that the above explanation is too simple.

Higher-order dispersion is usually limiting the effectiveness of self-phase modulation as it increases the pulse duration and therefore lowers the peak intensity after a propagation length of a few mm or cm for fs pulses. One can get a better picture if pulse broadening due to group velocity dispersion $k''(\omega_c)$ is included. To measure the relative importance of the two processes, the dispersion length L_D (the length that broadens the pulse by a factor $\sqrt{2}$) and the nonlinear length L_{NL} (the length that corresponds to the peak phase shift $\Phi_{NL}(t=0) = 1$) are used (Agrawal, 2013):

$$L_D = \frac{4\ln(2)\tau^2}{|k''(\omega_c)|} \qquad L_{NL} = \frac{cA_f}{n_2\omega_c P_0} \qquad (4.32)$$

where τ_0, A_f and $P_0 = \frac{1}{2}A_f c\varepsilon_0 |A(t=0)|^2$ are the initial pulse duration, the effective fiber core area and the pulse peak power. In the dispersion dominant regime $L_D \ll L_{NL}$, the pulses will disperse before any significant nonlinear interaction can take place. For $L_D > L_{NL}$ spectral broadening could be thought of as effectively taking place for a length L_D, even though the details are more involved. The total nonlinear phase shift can therefore be approximated by the number of nonlinear lengths within one dispersion length. As this phase shift occurs roughly within one pulse duration τ, the spectral broadening is estimated to be $\Delta\omega_{NL} = L_{NL}/(L_D\tau)$. As an example consider a silica single mode fiber (Newport F-SF) with $A_f = 26 \ \mu m^2$, $k''(\omega_c) = 281 \ fs/cm^2$, and $n_2 = 3.2 \times 10^{-16} \ cm^2/W$ that is seeded with $\tau = 73$ fs Gaussian pulses (FWHM intensity) at 905 nm with 225 mW average power and a repetition rate of 76 MHz (Reichert et al., 1999, 2000). In this case the dispersion length becomes 6.1 cm and the nonlinear length 35 mm. The expected spectral broadening of $L_{NL}/L_D\tau = 2\pi \times 44$ THz is indeed very close to the observed value (Reichert et al., 1999).

It turns out that within this model the spectral broadening is independent of the pulse duration τ because $P_0 \propto 1/\tau$. Therefore, using shorter pulses may not be effective for extending the spectral bandwidth beyond an optical octave as required for simple self-referencing (see Sec. 4.1.3). However, very efficient spectral broadening can be obtained in microstructured fiber[19] that can be manufactured with $k''(\omega_c) \approx 0$ around a design wavelength (Knight et al., 1996; Ranka et al., 2000; Russell, 2003). In this case the pulses are temporally broadened by other processes (linear and nonlinear) than group velocity dispersion as they propagate along the fiber. Eventually this will also terminate self-phase modulation and the dispersive length has to be replaced appropriately in the above analysis. At this point a whole set of effects enter such as Raman and Brillouin scattering, optical wave breaking and modulation instability. A review of so-called supercontinuum[20] generation in microstructured fibers can be found in (Dudley et al., 2006).

A microstructured fiber uses an array of submicron-sized air holes that surround the fiber core and run the length of a silica fiber to obtain a desired effective dispersion. This can be used to maintain the high peak power over an extended propagation length and to significantly increase the spectral broadening. With these fibers it became possible to broaden low peak power, high repetition rate lasers to beyond one optical octave as Fig. 4.12 shows.

A variant of the microstructured fibers are regular single mode fibers (or microstructured fibers) that have been pulled in a flame to form a tapered section of a few cm lengths (Birks et al., 2000). When the diameter of the taper becomes comparable to the core diameter of the microstructured fibers, pretty much the same properties are observed. In the tapered section the action of the fiber core is taken over by the whole fiber. The original fiber core then is much too small to have any influence on the light propagation. The fraction of evanescent field around the taper and along with it the dispersion characteristics can be adjusted by choosing a suitable taper diameter.

The peak intensity that can be reached with a given mode-locked laser does not only depend on the pulse duration but also inversely on the repetition rate (for fixed average power). Lower repetition rates concentrate the available power into fewer pulses per second. Comparing different pulsed lasers, the repetition rates cover 13 orders of magnitude from the largest pulse energy (one laser shot per 1000 sec) to

---

[19] Some authors refer to these fibers as photonic crystal fibers that need to be distinguished from photonic bandgap fibers. The latter use Bragg diffraction to guide the light, while the fibers discussed here use the traditional index step, with the refractive index determined by the air filling factor.

[20] This term is frequently used for very broad spectra obtained by self-phase modulation (as well as "white light continuum"), even in the context of frequency combs that are clearly not a continuum.

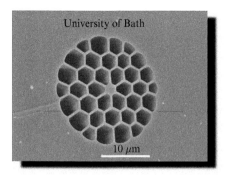 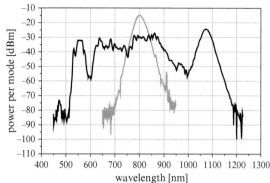

Fig. 4.12 Left: SEM image of the core of a microstructured fiber made at the University of Bath, UK (Knight et al., 1996). The light is guided in the central part but the evanescent part of the wave penetrates into the air holes that run parallel to the fiber core and lower the effective refractive index without any doping. The guiding mechanism is the same as in a conventional single mode fiber. **Right:** Power per mode on a logarithmic scale (0 dBm = 1mW). The lighter 30 nm (14 THz, measured at -3 dB) wide spectrum displays the laser intensity and the darker octave-spanning spectrum (532 nm through 1064 nm) is observed after the microstructured fiber that was 30 cm long. The laser was operated at $\omega_r = 2\pi \times$ 750 MHz (modes not resolved) with 25 fs pulse duration. An average power of 180 mW was coupled through the microstructured fiber (Holzwarth et al., 2001b).

highly repetitive lasers with $\omega_r = 2\pi \times 10$ GHz. It has been long known that with enough peak intensity one can produce very wide spectra that were initially called "white light continuum". Unfortunately, for a long time this was only possible at a repetition rate of around 1 kHz that indeed justifies the name: the generated spectrum could be called a "continuum" as there was not much hope to resolve the modes in any way for self-referencing (see next section) or by a beat note with another continuous wave (cw) laser. Because of its high efficiency, the microstructured fiber allowed to generate an octave wide spectra with repetition rates up to 10 GHz (Bartels et al., 2009), that conveniently allowed the beat notes to be separated. In addition, a large mode spacing puts more power in each mode improving the signal-to-noise ratio of the beat notes.

The laser that was in wide spread use in the early days of optical frequency combs was a rather compact titanium:sapphire ring laser with typical repetition rates of 500 MHz to 1 GHz (Bartels et al., 1999). The ring design solved another problem that is frequently encountered when coupling a laser into an optical fiber. Optical feedback from the fiber may disturb the laser operation and even prevent mode locking in some cases. The standard solution to this problem would be to place an optical isolator between the fiber and the laser. In this case however one will have to compensate the devices group velocity dispersion, otherwise it will most likely prevent subsequent spectral broadening. In a ring laser the pulses reflected back from the fiber travel in the opposite direction and do not talk to the laser pulses unless they meet inside the laser crystal. The latter can be prevented by observing the distance of the fiber from the laser. Disadvantages of these lasers are that they are not easy to align and have so far not become turn key systems that can be operated unattended for a long time, say in an optical clock.

Even though microstructured fibers have allowed the simple $f - 2f$ self–referencing (see next section) for the first time, they also have some drawbacks. To achieve the desired properties, the microstructured fibers need to have a rather tiny core. The coupling to this core causes problems due to mechanical instabilities and temperature drifts, even with low level and stable mounts. Another problem may be the strong polarization dependence of the fibers broadening action, that lead to slow drifts of the spectral envelope. Of course servo systems can be installed to keep all these parameters at optimum values. However, current frequency comb systems that need to operate continuously are based on mode-locked fiber lasers. The fiber lasers are somewhat more noisy, in particular broad band noise floor exists between the modes. As explained in Sec. 4.1.2, the additional phase fluctuations quickly average away for applications in optical frequency metrology. With the proper design fiber lasers can avoid the mechanical and temperature related drifts and can stay phase-locked for months without human attention. Microstructured fibers or so-called "highly nonlinear fibers", that are commercially available, can be spliced directly to a mode-locked fiber laser, virtually eliminating the alignment sensitive parts.

A problem with spectral broadening by self-phase modulation in general is an excess noise level of the beat notes well above the shot noise limit (Holzwarth et al., 2000; Corwin et al., 2009; Holman et al., 2003; Wetzel et al., 2012). It is believed that modulation instability is causing this problem by providing optical gain in the fiber that may not be seeded by the laser if its spectral envelope is too narrow (Dudley and Coen, 2002). In some cases the noise completely takes over and even washes out the modes from the laser so that one obtains a genuine continuum (Holzwarth, 2001) instead of a comb. As a rule of thumb, it seems that the initial pulses should not be too long to avoid this. Lasers that reach an octave-spanning spectrum without using any external self-phase modulation can solve this problem (Ell et al., 2001; Fortier et al., 2001; Matos et al., 2004, 2006). These lasers still seem to be delicate to handle and are not in wide spread use for frequency comb applications. Other self-referencing schemes do not require an octave-spanning spectrum as detailed in the next section.

4.1.3.6   Stabilizing the frequency comb

The measurement of ω_{CE} fixes the position of the whole frequency comb and is called self-referencing. The method relies on measuring the frequency gap between *different* harmonics derived from the *same* laser or frequency comb. The first crude demonstration (Hänsch et al., 1999) employed the 4th and the 3.5th harmonic of a $f = 88.4$ THz (3.39 μm) methane stabilized HeNe laser to determine ω_{CE} according to $4\omega_n - 3.5\omega_{n'} = (4n - 3.5n')\omega_r + 0.5\omega_{CE} = 0.5\omega_{CE}$ with $4n - 3.5n' = 0$. To achieve the condition of the latter equation both n and n' have to be active modes of the frequency comb. The required bandwidth is $0.5f = 44.2$ THz which is what the 73 fs laser together with a single mode fiber as discussed in the previous chapter could generate.

A much simpler approach is to fix the absolute position of the frequency comb by measuring the gap between ω_n and ω_{2n} which is the $f - 2f$ method[21] already mentioned in Sec. 4.1.2. Most conveniently one would use the modes from the frequency comb (Reichert et al., 2000; Diddams et al., 2000; Jones et al., 2000) rather than using the comb to measure the frequency difference between a continuous wave laser and its own second harmonic (Holzwarth et al., 2000). In this case the carrier-envelope offset frequency ω_{CE} is directly produced by beating the frequency doubled[22] red wing of the comb $2\omega_n$ with the blue side of the comb at ω_{2n}: $2\omega_n - \omega_{n'} = (2n - n')\omega_r + \omega_{CE} = \omega_{CE}$ where again the mode numbers n and n' are chosen such that $(2n - n') = 0$. This approach requires an octave-spanning comb, i.e. a spectral width of $(2/3) \times 377$ THz $= 251$ THz if centered at the titanium:sapphire gain maximum at 795 nm (377 THz).

Fig. 4.13 sketches the $f-2f$ self-referencing method. The spectrum of a titanium:sapphire mode-locked laser is first broadened to more than one optical octave with a microstructured fiber. A broad band $\lambda/2$ wave plate allows to choose the polarization with the most efficient spectral broadening. After the fiber a dichroic mirror separates the infrared ("red") part from the green ("blue"). The former is frequency doubled in a nonlinear crystal and reunited with the green part to create a wealth of beat notes, all at ω_{CE}. These beat notes emerge as frequency differences between $2\omega_n - \omega_{2n}$ according to Eq. 4.14 for various values of n. The number of contributing modes is given by the phase matching bandwidth $\Delta\nu_{pm}$ of the doubling crystal and can easily exceed 1 THz. To bring all these beat notes at ω_{CE} in phase, so that they all add constructively, an adjustable delay for example in form of a pair of glass wedges or corner cubes may be used. It is straight forward to show that the condition for a common phase of all these beat notes is that the green and the doubled infrared pulse reach the photo detector at the same time. The adjustable delay allows to compensate for different group delays, including the fiber. In practice the delay needs to be correct within $c\Delta\nu_{pm}$ which is 300 μm for $\Delta\nu_{pm} = 1$ THz. Outside this range a beat note at ω_{CE} is usually not detectable.

Whereas the half wave plates in the two interferometer arms are used to adjust for whatever polarization exits the microstructured fiber, the half wave plate between the two polarizing beam splitters helps to find the optimum relative intensity of the two beating pulses. As shown in Sec. 4.1.2, the maximum signal–to–noise ratio is obtained for equal intensities reaching the detector within the optical bandwidth that contributes to the beat note (Reichert et al., 1999). Experimentally this condition is most conveniently adjusted by observing the signal-to-noise ratio of the ω_{CE} beat note with a radio frequency spectrum analyzer.

A grating is used to prevent the extra optical power, that does not contribute to the signal but adds to the noise level, from reaching the detector. Typically, only a large relative bandwidth of say 1 THz/377 THz needs to be selected so that a very

---

[21] A related and almost as simple method is the $2f - 3f$ self-referencing (Morgner et al., 2001; Diddams et al., 2003)

[22] It should be noted that this does not simply mean that the frequency of each individual mode is doubled, but the sum frequencies of all mode combinations are generated. Otherwise the mode spacing, and therefore the repetition rate, would be doubled as well.

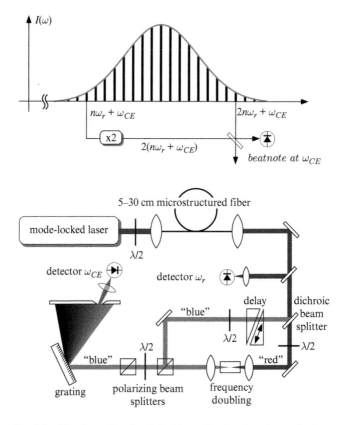

Fig. 4.13 Top: $f - 2f$ self-referencing by detecting a beat note at ω_{CE} between the frequency doubled "red" wing $2(n\omega_r + \omega_{CE})$ of the frequency comb and the "blue" modes at $2n\omega_r + \omega_{CE}$. **Bottom:** Layout of the self-referencing scheme. See text for details.

moderate resolution illuminating 377 lines is sufficient. For this reason it is usually not necessary to use a slit between the grating and the photo detector. Sufficient resolution can be reached with a small 1200 lines per mm grating illuminated with a beam collimated with $\times 10$ microscope objective out of the microstructured fiber.

It should be noted that the layout in the lower part of Fig. 4.13 is not necessary the best in practice. It was the one that was used in the early days and nicely shows the principle. Meanwhile other layouts have emerged that, for example, use mostly common paths (see for example (Koke et al., 2009)). This makes the set-up much less sensitive to vibrations. Selecting the spectral region to detect ω_{CE} may be done with an interference filter instead of a grating. For fiber lasers one can also use a fiber Bragg grating and so on.

When detecting the beat note as described above, more than one frequency component is obtained for two reasons. First of all any beat note, even between two cw lasers, generates two components because the radio frequency domain cannot decide which of the two optical frequencies is larger than the other (negative and positive beat frequencies). Secondly, observing the beat notes between frequency combs, not

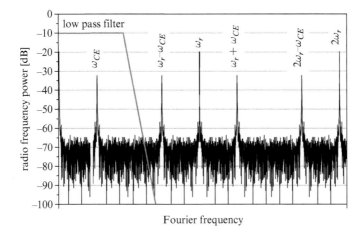

Fig. 4.14 Radio frequency spectrum produced by a self-referencing nonlinear interferometer such as the $f - 2f$ interferometer shown in Fig. 4.13. A low pass filter with a cut-off at $0.5\omega_r$ selects the component at $\pm \omega_{CE}$.

only the desired component $k = 2n - n' = 0$ is registered, but all integer values of k, positive and negative contribute, up to the bandwidth of the photo detector. This leads to a set of radio frequency beat notes at $k\omega_r \pm \omega_{CE}$ for $k = \ldots - 1, 0,$ $+1\ldots$. In addition, the repetition rate, including its harmonics will most likely give the strongest components. After carefully adjusting the nonlinear interferometer, spatially and spectrally, and scanning the delay line for the proper pulse arrival times, the radio frequency spectrum may look like the one shown in Fig. 4.14. A low pass filter with a cut-off frequency of $0.5\omega_r$ selects exactly one beat note at $\pm\omega_{CE}$. No optical resolution to separate a mode is required. The design of the proper low pass filter may be tricky, depending on how much stronger the repetition rate signal exceeds the beat note at ω_{CE}. The sketch in Fig. 4.14 gives a feeling on how steep this filter needs to be at the cut-off in order to suppress the unwanted components below the noise level. Such a suppression is required for taking the full advantage of the signal-to-noise ratio. For this reason it is desirable to work at higher repetition rates. At ω_r around $2\pi \times$ 800 MHz, as used mostly for the ring titanium:sapphire lasers described above, the filter requirements are much more relaxed than say at 80 MHz. In addition, a larger repetition rate concentrates more power in each mode further improving the beat notes with the frequency comb. It should be noted though, that with larger repetition rates it may be difficult to achieve the necessary peak power for spectral broadening (see Sec. 4.1.3).

As described, both degrees of freedom ω_r and ω_{CE} of the frequency comb can be measured up to a sign in ω_{CE} that will be discussed below. For stabilization of these frequencies, say relative to a radio frequency reference, it is necessary to be able to control them. Again the repetition rate turns out to be simpler. Mounting one of the lasers cavity mirrors on a piezo electric transducer as shown in Fig. 4.7, allows to control the pulse round trip time. Another option is offered by mode-locked lasers that use prism pairs to compensate the intracavity group velocity dispersion.

In this case tipping the mirror at the dispersive end where the cavity modes are spatially separated, changes the relative cavity lengths of the individual modes and thereby the mode spacing in frequency space (Udem et al., 1999b; Holzwarth et al., 2000). Locking the repetition rate of fiber lasers works pretty much the same way by changing the cavity length with a fiber stretcher adjusting a distance in a free-space section if there is one. In practice the detected repetition frequency is mixed with the radio frequency reference, i.e. the frequency difference is generated, lowpass filtered and with appropriate gain send back to the piezo electric transducer. The laser acts as the voltage controlled oscillator in the phase-locked loop discussed in Sec. 4.1.2.

Setting up a phase-locked loop for the repetition rate therefore seems rather straightforward. However, some caution concerning the servo bandwidth needs to be observed. It turns out that the large frequency multiplication factor n in Eq. 4.14 may also multiply the noise of the reference oscillator. The phase noise power for direct frequency multiplication by n increases proportional to n^2 (Walls and DeMarchi, 1975), so that a factor of $n = 10^6$, that would take us from a 100 MHz radio frequency signal to a 100 THz optical signal, increases the noise by 120 dB. On this basis it has been predicted that, using even the best available reference oscillator, it is impossible to multiply in a single step from the radio frequency domain into the optical (Telle, 1996). The frequency comb does just that, but avoids the predicted "carrier collapse". This is because the laser acts as a flywheel in the optical domain that does not follow the fast phase fluctuations of the reference oscillator instantaneously but averages them out, unless the servo bandwidth is too large (see Sec. 4.1.2). The n^2 multiplication law does not apply, because it assumes a phase stiff frequency multiplication. Fortunately, a decent free running mode-locked laser shows very good phase stability of the pulse train on its own. For averaging times shorter than typical acoustic vibrations of several ms period, such a laser shows better phase stability than a high quality synthesizer (Hollberg et al., 2001). A small servo bandwidth may be implemented electronically by appropriate filtering or mechanically by using larger masses than the usual tiny mirrors mounted on piezo transducers for high servo speed. The mechanical low pass filter has the advantage that it is essentially free of additional noise.

Controlling the carrier envelope frequency requires more effort. Experimentally it turned out that the energy of the pulse stored inside the mode-locked laser has a strong influence on ω_{CE}. Conventional soliton theory (Hasegawa and Tappert, 1973) predicts a dependence of the phase velocity but no dependence of the group velocity on the pulse peak intensity such that Eq. 4.16 predicts the wrong sign of this effect. In addition the effect on ω_{CE} would be very small. In some cases one even observes sign reversal as the intracavity pulse energy is changed (Holman et al., 2003). Higher-order terms that need to be added to the nonlinear Schrödinger equation are responsible for these observed deviations (Haus and Ippen, 2001). In most cases the carrier-envelope frequency is phase-locked using the power of the pump laser that is either controlled directly or with an external modulator. An acousto-optic modulator, that drains an adjustable part of the pump laser power has been used mostly with titanium:sapphire lasers. An electro-optic intensity modulator may also be used, but could be less effective because its sinusoidal dependence of the transmission on the applied voltage may require a larger bias. Fiber lasers can be conveniently controlled via the injection current of the semiconductor pump lasers. While the repetition rate should not be locked too tight for the reason given above, no such restriction exists for the carrier-envelope frequency. It is only added to the optical frequency via Eq. 4.14 and hence does not easily lead to a collapse of the optical carrier frequencies.

Mode-locked lasers that use pairs of prisms to compensate for group velocity dispersion generally show a larger carrier envelope frequency noise. This is because intensity fluctuations slightly change the pulse round trip path due to the intensity dependent refraction of the titanium:sapphire crystal (Helbing et al., 2002). Small variations of the beam pointing results in a varying prism intersection. It should be noted that already 50 μm of extra BK7 glass in the path shifts the carrier envelope phase by 2π and the carrier envelope frequency by ω_r. In many other cases the carrier envelope frequency fluctuations seem to be dominated by the pump laser noise, so that stabilizing ω_{CE} with a modulator as described above even reduces this noise. Fiber-based mode-locked lasers typically come with significantly larger noise of the carrier-envelope frequency than titanium:sapphire lasers. Careful stabilization of the semiconductor pump lasers is recommended to reduce this problem (McFerran et al., 2006). Whereas the pump laser power can be controlled with very large bandwidth, for example with external modulators, the intracavity pulse energy may not react instantaneously. The latter is responsible for adjusting ω_{CE}. The common fiber lasers gain material, erbium and ytterbium, have upper state lifetimes of milliseconds. This puts a severe low pass filter on the control bandwidth of the carrier-envelope frequency. The titanium:sapphire laser does not suffer from this problem, but other ytterbium doped lasers like thin disk lasers do. One way around this is to control the intracavity pulse energy with an additional laser that quickly quenches the upper laser level or by controlling the intracity loss, for example with an acousto-optic modulator (Lee et al., 2012; Pronin et al., 2015). Intracavity electro-optic phase modulators in combination with an "integer waveplate" can also be used for fast ω_{CE} control (Hänsel et al., 2017). However, for the standard frequency metrology application, that locks ω_r and ω_{CE} in order to measure ω_n via Eq. 4.14, the carrier-envelope frequency does not have to be tightly locked. Large range phase detectors may be used instead of trying to increase the control bandwidth (see Sec. 4.1.2). The best strategy also depends on the intended application.

To avoid having to stabilize the carrier-envelope frequency, one can employ difference frequency generation in a nonlinear crystal. Subtracting the frequencies of some comb lines Eq. (4.14) from the frequencies of other comb lines cancels ω_{CE}. A properly phase-matched doubling crystal used for $f - 2f$ self-referencing (Fig. 4.13) also generates the frequency difference at f (Zimmermann et al., 2004). Without additional measures the resulting difference frequency comb is rather narrow, a disadvantage that may be ignored if the optical target frequency is covered. However, broadband difference frequency schemes exist and very wide self–stabilized combs have been generated in this way (Fuji et al., 2005).

Measuring the frequency of an unknown continuous wave laser at ω_L with a stabilized frequency comb, involves the creation of yet another beat note ω_b with the comb. For this purpose the beam of this laser is matched with the beam that contains the frequency comb, say with similar optics components as used for creating the carrier envelope beat note. A dichroic beam splitter, just before the grating in Fig. 4.13, could be used to reflect out the spectral region of the frequency comb around ω_L without affecting the beat note at ω_{CE}. This beam could then be fed into another set-up consisting of two polarizing beam splitters, one half wave plate, a spectral filter such as a grating and a photo detector for an optimum signal-to-noise ratio. The frequency of the continuous wave laser is then given by

$$\omega_L = n\omega_r \pm \omega_{CE} \pm \omega_b \qquad (4.33)$$

where the same considerations as above apply for the sign of the beat note ω_b. These signs may be determined by introducing small changes to one of the frequencies (with a known sign) while observing the sign of changes in another frequency. For example the repetition rate could be increased by picking a slightly different frequency of the reference oscillator. If ω_L stays constant we expect ω_b to decrease (increase) if the "+" sign ("−" sign) is correct.

The last quantity that needs to be determined is the mode number n. If the optical frequency ω_L is already known to a precision better than the mode spacing, the mode number can simply be determined by solving the corresponding Eq. 4.33 for n and allowing for an integer solution only. A coarse measurement could be provided by a wave meter for example if its resolution and accuracy is trusted to be better than the mode spacing of the frequency comb. If this is not possible, at least two measurements of ω_L with two different and properly chosen repetition rates may leave only one physically meaningful value for ω_L (Holzwarth et al., 2001a).

Rather than phase-locking ω_r and ω_{CE} to a known reference derived from a radio frequency atomic clock, one can phase-lock one of the optical modes to a stable optical reference, say from an optical frequency standard. The carrier-envelope frequency can be locked to a reference frequency derived from the repetition rate (Diddams et al., 2001). In this case the optical frequency comb uses no radio frequency input and acts as a divider for the optical input frequency. The repetition rate becomes the countable clock output and may be used as a time base. The optical frequency reference together with the frequency comb (the optical clockwork mechanism) form an optical clock (for a review see (Ludlow et al., 2015)).

Operating the frequency comb with a repetition rate that is close to an integer multiple of 10 MHz one can use a prescaler to generate a signal that is commonly used as a reference for counters, synthesizers, spectrum analyzers, and so on. In this way every radio frequency that appears in the set-up is proportional to ω_r. In particular this means $\omega_{CE} = \alpha\omega_r$ where α is determined by the local oscillator used for phase-locking ω_{CE}. A frequency comb stabilized in this way may be used to measure another optical frequency by recording the beat note $\omega_b = \beta\omega_r$ where β is derived from the ω_r-referenced frequency counter. The ratio of the two optical frequencies is then given by $(n_2 + \alpha + \beta)/(n_1 + \alpha)$ where n_1 and n_2 are the corresponding mode numbers (Diddams et al., 2001). It is important to note that the measurement of this ratio is not limited in accuracy by a cesium standard so that optical frequency standards with significantly lower uncertainty can be compared (Rosenband et al., 2008). Simpler, but maybe less elegant is it to measure the two optical frequencies simultaneously with a radio frequency referenced comb that drops out in the data evaluation

(Nemitz et al., 2016). Another method to measure an optical frequency ratio ω_1/ω_2 uses direct-digital-synthesis of a signal $(\omega_{CE} + \omega_{b2})n_1/n_2$ where n_1 and n_2 are the mode numbers of the two optical signals beating with the comb at ω_{b1} and ω_{b2}. Subtracting $\omega_{CE} + \omega_{b1}$, that is like $\omega_{CE} + \omega_{b2}$ generated with a mixer, yields the radio frequency deviation from the ratio $\omega_1 - \omega_2 n_1/n_2$ (Stenger et al., 2002). The interesting feature is that the noise of ω_r cancels out and does not even have to be phase-locked. To test this method, the frequency ratio of a laser with its own second harmonic was measured and found to deviate by not more than $(1 \pm 7) \times 10^{-19}$ from 2.

Instead of operating an optical clock that uses a well defined optical reference frequency, one might stabilize a frequency comb to a high finesse cavity. Sub-hertz linewidths in the optical region are possible (see for example (Kessler et al., 2015)). This corresponds to a spectral purity that is much better than anything that can be obtained in the radio frequency domain. One may use the frequency comb to convert this linewidth to another (optical) spectral region (Schibli et al., 2008; Fang et al., 2013) or to the radio frequency domain to reach new levels of stability (Xie et al., 2016). The afore mentioned quadratic noise multiplication process is reversed in the division. In contrast to the phase-locked loops above, these applications require the largest possible control bandwidths, for both ω_n and ω_{CE}. Controlling the cavity length as usual with a piezo transducer is too slow and hence not appropriate for this application. The above mentioned intracavity electro-optic phase modulators in combination with "integer waveplates" can provide fast control for both ω_n and ω_{CE} (Hänsel et al., 2017).

4.1.3.7   Comb fixed points

Controlling the frequency comb is done by changing some operational parameters (see Fig. 4.7). In general this means that a linear combination of the carrier-envelope frequency ω_{CE} and the mode spacing ω_r is affected. If we assume that the defining comb property (equidistant modes) is maintained, it follows that there must be a fixed point to each control, i.e. a frequency ω_f that is invariant under its action (Haverkamp et al., 2004; Telle et al., 2002). One may even renumber the modes[23] counting from the closest mode to the fixed point. This fixed point does not have to be within the spectral envelope of the comb. As an example consider the cavity lengths that will change all mode frequencies ω_n in proportion to that frequency. The fixed point of the cavity length control is at $\omega_f = 0$. All modes appear as if they were attached to an elastic tape that is fixed at the origin. As such the cavity length control mostly changes ω_r because $\omega_{CE} \approx \omega_f$. However, this statement depends on how the modes are numbered, as in Eq. 4.14 or Eq. 4.22 for example. Rather than saying which frequency varies, it is more unambiguous to state which frequency is fixed under a certain action.

Tilting the mirror at the dispersive end of a prism compensated mode-locked laser as described in the previous section has its fixed point at the pivot point, i.e. at the mode for which the cavity length does not change. As these types of lasers are no longer in wide-spread use, the intracavity pulse energy may be employed which has a similar effect. Its fixed point is close to the center of the comb's spectral envelope

---

[23] The numbering that assigns ω_{CE} to be 0-th mode is not the only obvious choice: see Eq. 4.22.

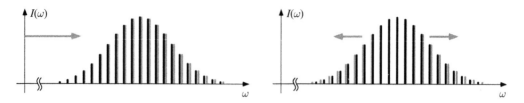

Fig. 4.15 Fixed-point concept to illustrate the modification of the frequency comb upon changing some operational parameters. **Left:** Changing the cavity lengths leaves the (virtual) mode at the origin fixed while all other modes move es if they were attached to an elastic tape. **Right:** Changing the pump power stretches the comb roughly about its center like in an accordion.

(Haverkamp et al., 2004). Using this control causes the comb's to stretch or contract like an accordion. The remarkable thing about this fixed point is that it seems to be independent of the type of laser and it even tunes along when tuning the combs spectral envelope (Walker et al., 2007). Fig. 4.15 sketches the fixed point concept for the two common cases.

The fixed-point concept is very useful to quickly grasp which control may work to stabilize which degree of freedom. For example, the cavity length will not be useful to change the carrier-envelope frequency. On the other hand, the intracavity pulse energy is very effective for that purpose, but useless to lock a mode close to the center of the comb to another optical frequency reference. In this context the parametrization of Eq. 4.14 could be confusing because the "accordion control" almost exclusively changes the mode spacing ω_r in a sense, but is nevertheless very effective for ω_{CE}. With the fixed-point concept we can easily understand this. Likewise it is not clear what would be a good control that would change ω_{CE} exclusively in Eq. 4.14. Its fixed-point would be at $\omega_f = \pm\infty$. Hence, with the fixed-point concept it is much better to understand why it is difficult to find a comb control that varies only ω_{CE}.

Locking a mode of the frequency comb to a stable optical reference, as discussed in the previous section, requires a fast control of that mode. In a standard mode-locked laser one has only the cavity length and the pump power to control. Unfortunately, the pump power may be fast but is not very useful to lock an optical mode because the fixed point is close. The cavity length on the other hand would be very effective for that purpose because of its remote fixed point. However, the cavity length control is usually slow, at least when it is driven by a piezo transducer. The solution is an intracavity electro-optic phase modulator that has its fixed point close to zero frequency.

Besides knowing which control can be used for locking, the fixed-point concept is also useful to estimate how noise enters the frequency comb. The noise fixed point can be found by measuring the correlation of the noise of ω_{CE} and ω_r which then gives a hint on the noise source. A noisy pump laser for example is expected to give rise to "accordion noise" with the fixed point close to the carrier frequency. Acoustic perturbations of the cavity length are usually small in relative units, such that variations of ω_r, ω_{CE}, and ω_n are also small in relative units. Beating a mode of the comb with a stable reference laser will nevertheless reveal its noise in absolute units.

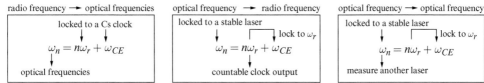

Fig. 4.16 Frequency conversions possible with the frequency comb. **Left:** by locking to radio frequencies ω_r and ω_{CE} of the comb to a Cs atomic clock, the optical modes ω_n may be used to determine atomic transition frequencies in units of Hz. **Center:** In reverse operation the comb is referenced to an optical transition extracting a pre-scaled radio frequency copy. **Right:** Phase coherent optical-optical frequency ratios are also measured.

With the mixing scheme that involves direct-digital-synthesis as described above, it is possible to effectively cancel out the noise from the frequency comb, provided the fixed point of the noise source is known. With a free running repetition rate it is reasonable to assume that the main noise is due to cavity length fluctuations with the fixed point at $\omega_f = 0$. By mixing the proper small integer fractions of ω_{CE} and ω_b with a harmonic of ω_r, the noise of the repetition rate is cancelled out from the detected beat notes (Telle et al., 2002).

4.1.3.8 *Frequency comb summary*

To conclude the first part, we summarize the combs applications for frequency metrology. By virtue of Eq. 4.14 the frequency comb can be used to phase coherently connect the radio frequency domain with the optical domain. This is enabled by the large integer n ($\approx 10^5 \ldots 10^6$) that the pulse repetition frequency is multiplied with. As sketched in Fig. 4.16, three interesting frequency conversions are possible with the frequency comb. Locking the radio frequencies ω_r and ω_{CE} to a Cs clock references the optical modes to the SI second in order to determine optical transition frequencies of interest with the utmost precision. Using the repetition rate as a reference to stabilize the carrier envelope frequency allows to divide an optical frequency down to the radio frequency domain or to directly measure optical frequency ratios.

4.2 Atomic hydrogen

4.2.1 Introduction

Atomic hydrogen has been the Rosetta Stone for quantum physics. Through the successive refinement of its theoretical description, starting from the phenomenological description of J. J. Balmer, it provided a concise argument for the wave nature of matter. N. Bohr's quantization requirement, that the electrons angular momentum must be an integer multiple of $\hbar$, can be understood such that the corresponding de Broglie wave forms a standing wave. E. Schrödinger formulated the appropriate wave equation for matter waves. With the equation that bears his name, he put the emerging field of quantum mechanics on solid grounds. Later P. A. M. Dirac figured out how to include relativity. These refinements were introduced to keep up with increasing experimental accuracy, and to fix the resulting inconsistencies, but continuously complicated the theoretical picture. W. E. Lamb and R. C. Retherford discovered a discrepancy between experimental observations and the Dirac theory in 1947, which is now commonly referred to as the "Lamb-shift". It turned out that this was not due to a failure of quantum mechanics but because it had not been applied in the most rigourous way. In particular, effects of the quantum vacuum, both the electronic and photonic part, had been neglected up to then. Their inclusion led to the development of Quantum Electrodynamics (QED) by E. A. Ühling, H. A. Bethe, R. P. Feynman, and others. Unlike any of the predecessors, QED survived about six orders of magnitude improvement of the experimental accuracy and is still the current description today. Of course, several discrepancies between experiment and theory have shown up in the last sixty-five years, but thus far they could all be traced back to errors in the computation of QED terms, neglection of higher-order terms or to underestimation of experimental errors. In that sense, QED is probably the most successful theory in all of physics. As such it has served as a template for all subsequent quantum field theories.

Experimentally precise determinations of the energy levels of simple atomic systems are required for testing QED. Differences of energy levels can be determined by measuring transition frequencies between two levels. The sharpest transition in atomic hydrogen occurs between the 1S ground state and the metastable 2S state. Its transition frequency has now been measured with almost 15 digits accuracy using an optical frequency comb and a cesium atomic clock as a reference (Parthey et al., 2011). Unfortunately, any of the other metrologically relevant transitions have linewidths of several MHz.[24] At least one large linewidth transition is required to test QED (see next section to find out why), so that a very good understanding of the observed line shape is required (see Sec. 4.2.3).

While the existing experimental hydrogen data has been consistent with QED until very recently, a measurement of the $2S - 2P_{1/2}$ transition frequency in muonic hydrogen performed in 2010 was found to be in significant contradiction with it (Pohl et al., 2010). This has subsequently been confirmed with the $2S - 2P_{3/2}$ transition frequency in muonic hydrogen (Antognini et al., 2013). This problem is the so-called "proton size puzzle" because muonic hydrogen is mostly sensitive to the proton charge radius (see next section). At this time, it is not clear whether or not this is due to a failure of theory, its application or due to a measurement problem (or all of the above).

More data is required to shed light on this issue. The only route for progress on the hydrogen experiments is to improve the measurement of one of the broader lines. For that reason we have been working for five years on improving the accuracy of the 2S–4P transitions in regular hydrogen. The recent result delivers a value for the proton radius that is as accurate as from the combined previous hydrogen data. It supports the muonic value. This means that the data from regular hydrogen is no longer self-consistent and the puzzle is reinforced. Even more high-precision data is

---

[24] The Rydberg levels do also have a long lifetime and hence could contribute with narrow linewidth transition. However, these levels are extremely sensitive to stray electric fields such that so far no precision measurements could be performed.

required to solve this problem or to discover new theoretical ingredients. This justifies our passion for precision. The frequency comb is the most important tool for that.

4.2.2 Theory

The current state of the QED theory of atomic hydrogen includes a large number of terms and a couple of constants that cannot be computed from first principles. It is usually expressed in a power series of the fine structure constant α and $\ln(\alpha)$:

$$E = R_\infty \left(-\frac{1}{n^2} + X_{20}\alpha^2 + X_{30}\alpha^3 + X_{31}\alpha^3 \ln(\alpha) + X_{40}\alpha^4 + \ldots + \delta_{l,0} \frac{16\pi^2 m_e^2 c^2 \alpha^2}{3n^3 h^2} r_p^2 \right).$$

$$(4.34)$$

The QED expression in brackets actually contains a much larger number of terms than this way of writing suggests. The leading order is given by the well known result of Bohr and Schrödinger. Some of the coefficients X_{mn} are themself lengthy expressions that depend on constants like the electron to proton mass ratio m_e/m_p to take the finite mass nuclear recoil into account, as well as on more quantum numbers. The last term describes the effect of the finite size of the nucleus in terms of the r.m.s. proton charge radius r_p for S-states ($l = 0$). For hydrogen, this term contributes very little to the total energy,[25] but adds the largest part to the uncertainty, at least if r_p is obtained from electron-proton scattering (see for example table 2 of (Biraben, 2009)). For non-S states, this term contributes even less and is understood to be included in the X_{mn} series. A complete list of terms is presented and regularly updated by the CODATA team (Mohr et al., 2016).

I should be noted that Eq. 4.34 is just one way of summarizing the QED terms for atomic hydrogen. Sometimes this is separated into the Dirac term (with some recoil corrections also including the leading $-1/n^2$ term) and the remaining terms called the "Lamb-shift". Even though is was very important to discover it, from the modern perspective the "Lamb-shift" is merely a deviation from an incomplete theory. The various contributions to the energy levels are sketched in Fig. 4.17.

Like any other theory, QED can deliver only pure numbers. If it could generate results in SI units, it would know about our somewhat arbitrary choice of these units. No fundamental theory can possibly know that. Hence we need to multiply the QED expression with the Rydberg constant R_∞, if we decide to measure in SI units instead of atomic units. Measuring in atomic units would mean that we would determine ratios of transition frequencies (in fact that is all the theory can predict). The Rydberg constant, or any other unit converter[26] would drop out in that

---

[25] For that reason the additional constants that appear there can be taken with a few digits accuracy from other experiments.

[26] Sometimes the Rydberg constant is considered to be fundamental. That however depends on how fundamentality is defined. Here we would argue that a fundamental constant needs to be dimensionless, otherwise we can make it go away by picking the right units. How can something that is disposable be fundamental? On the other hand, what makes the Rydberg constant important, is the fact that is accurately known *and* that it is made up of other constants: $R_\infty = m_e c^2 \alpha^2/2h$. For this reason it is an important corner stone for choosing the best values for the other constants (Mohr et al., 2016). However, importance does not necessarily make it fundamental.

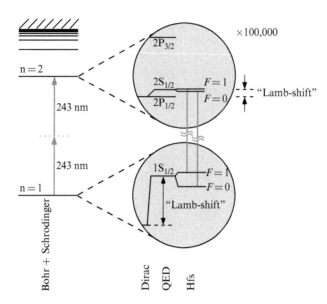

Fig. 4.17 Various contributions to the energy levels of atomic hydrogen. The result of the Bohr–Schrödinger theory is given by the first term in Eq. 4.34. Adding relativity in form of the Dirac theory leads to a modification and a splitting of terms with different electron angular momenta. Including the electronic and photonic vacuum, in form of vacuum polarization and self-energy respectively, leads to the so-called "Lamb-shift", that is actually not a shift in the sense that it could be turned on or off. These vacuum effects lift the degeneracy of the same electron angular momentum states. The last correction is due to the magnetic interaction of the electron with the nucleus and is called hyperfine structure (HFS). It is extremely precisely measured, so that spectroscopic results are usually corrected for it. It is therefore not discussed any further here. The sharpest transition takes place between the 1S ground state and the 2S metastable state with two photons at about 243 nm. The two-photon transition can be arranged such that the Doppler effect is cancelled out in first order (see Sec. 4.2.3). The 1S–2S transition frequency has therefore been measured with by far the smallest uncertainty.

ratio. These measurements are possible and have been done in the past (Berkeland et al., 1995; Weitz et al., 1995; Bourzeix et al., 1996), but requires to operate two spectroscopic experiments simultaneously. This was a powerful method before the frequency comb made it possible to measure optical frequencies in SI units (see Sec. 4.1 of this review). Nowadays it is better to measure transition frequencies separately and determine the Rydberg constant from them. To avoid further unit conversions we think of the Rydberg constant in Eq. 4.34, as well as the energies, given there in Hz. Conversion to commonly used wavenumber units can be done without loss of accuracy[27] because the speed of light has a defined value within the SI. The differences of energy levels directly give the associated transition frequency.

---

[27] This is not so for converting to Joules until Planck's constant will also be assigned an exact value in the next revision of the SI that is scheduled for 2018.

The constants that enter Eq. 4.34 are playing an important role. In principle every constant that has a unit can be removed simply by promoting them to be a unit.[28] Measuring in SI units we do not want to do that for R_∞, even if we could. The fine structure constant and the electron to proton mass ratio are dimensionless and hence cannot be removed in this way. They (hopefully) have the same values even at a distant planet, where the inhabitants do not use our SI units. In that sense, these constants are really fundamental, unless somebody figures out how to compute them (Leuchs and Sánchez-Soto, 2013). Every fundamental constant that enters a theory reduces its predictive power. In the limit of equal number of constants and computable numbers, a theory would loose all its predictive power and becomes a tautology.

Therefore, an inventory of the number of constants in the QED description of atomic hydrogen is essential: So far we have the fine structure constant α, the electron to proton mass ratio m_e/m_p, the proton charge radius r_p, and the Rydberg constant R_∞. In practice, the first two are better taken from other experiments. The fine structure constant is obtained with a relative uncertainty of 3.7 parts in 10^{10} by comparing the measured value of the electron g-factor with the corresponding QED expression (Hanneke et al., 2008; Aoyama et al., 2012). In contrast to Eq. 4.34 this expression is linear in α, so that it is about 137 times more sensitive to α. The other constant that is advantageously taken from another experiment, given the accuracy of currently available data, is the electron to proton mass ratio. Similar to the fine structure constant, the hydrogen energy levels are not sensitive to this quantity in first order. The simplest non-relativistic recoil correction is obtained by replacing the electron mass that enters R_∞ with its reduced mass $m_e m_p/(m_e + m_p) \approx m_e(1 - m_e/m_p)$. On the other hand, mass ratios of particles that can simultaneously be held in a Penning trap, can be determined with relative uncertainties in the 10^{-10} range, by comparing their cyclotron frequencies (Rainville et al., 2004). In this way the electron to proton mass ratio has been determined (Heiße et al., 2017; Mohr et al., 2016) with an uncertainty lower than with extraction from Eq. 4.34.

This leaves us with two constants r_p and R_∞ that we treat as adjustable parameters to be determined from the measured hydrogen transition frequencies. The accuracy with which these parameters can be determined is given by the *second* precise measurement. In the CODATA adjustment (Mohr et al., 2016) 15 measured transition frequencies are used. Whereas the 1S–2S transition frequency has been measured with almost 15 digits of accuracy, all other listed measurements have accuracies of around 12 digits. It is always possible to use two transition frequencies to fix the parameters to make the theory consistent with these two transition frequencies. For a real QED test we need to use at least three measurements. In other words, one uses an overdetermined system (more measurements than parameters) to test whether or not one finds the same values for the parameters for different combinations of measurements. The best test will use all of the most accurately measured hydrogen transitions, i.e. the ones listed by CODATA, to determine 14 values of the proton charge radius and 14 values for the Rydberg constant. The consistency of these values

---

[28] In case two ore more constants enter with the same units, some of them can be replaced with dimensionless ratios so that we always end up with the same number of constants and units.

then tests the consistency of QED with observations. Again such a test is currently limited to about 12 digits by the available experimental data.

Ending up with two parameters brings up the question if one can obtain those from other experiments like it is done for the fine structure constant and the electron to proton mass ratio. There is little hope that this can be done for the Rydberg constant even though it scales all atomic and molecular transition frequencies by an expression analogous to Eq. 4.34. Assuming infinitely precise measurements of transition frequencies, the Rydberg constant can only be determined as accurate as the expression in brackets. There is nothing that can compete with atomic hydrogen and hydrogen-like systems in that respect.

The situation is quite different for the proton charge radius. Using muonic hydrogen instead of regular hydrogen with a muon replacing the electron, the charge radius term in Eq. 4.34 gets much larger because the muon possesses a ≈ 207 times larger mass. The finite size term increases by more than 4 orders of magnitude relative to the other terms and by about 7 orders of magnitude in absolute terms, as the Rydberg constant is proportional to the orbiting lepton mass as well. Therefore, the measurement of transition frequencies in muonic hydrogen provides a much more sensitive method to determine the proton charge radius compared to regular hydrogen. A rather moderate measurement precision is sufficient to obtain a precise value for r_p. Measurements on this system have found a value for the proton charge radius that is in significant contradiction with the value obtained from regular hydrogen (the "proton size puzzle"). The current situation is summarized in Fig. 4.18.

4.2.3 Experiment

Inspecting Fig. 4.18 it seems obvious that the "proton size puzzle" could be due to a yet unknown systematic uncertainty in one of the measurements. If the 1S–2S transition frequency would be wrong for example, all hydrogen values for r_p would shift in this plot. To explain the "proton size puzzle", however, the 1S–2S frequency must be wrong by 4000 standard deviations (or 40 observational linewidths). It seems unlikely that such a large systematic has yet been undiscovered. Similarly, the muonic 2S–2P transition needs to be off by 100 standard deviations (or 4 observational linewidths) to solve the puzzle.

If the problem is really on the experimental side, it seems more likely that it is due to one of the broad linewidth transitions, i.e. not the 1S–2S transition. The latter has a natural linewidth of only 1.3 Hz but an observational linewidth (mostly due to the time of flight, ionization, and AC Stark) of about 1 kHz (Kolachevsky et al., 2006). The other lines used in Fig. 4.18 have natural linewidths between 0.1 and 10 MHz. For a well designed spectrometer the observational linewidth is close to the natural linewidth in these cases. While one can readily convert the measured transition frequencies and compare them with the energy levels given in Eq. 4.34, it is not so trivial to extract these transition frequencies from the measured laser frequencies and the recorded broad lineshapes. The "proton size puzzle" measures about 10 kHz at the 2S–4P transition with a natural linewidth of 12.9 MHz. Tiny line distortions are important and may lead to significant spurious shifts, depending on the line fitting function. This issue is further discussed in Sec. 4.2.3.

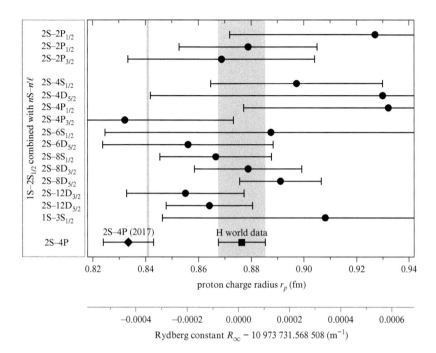

Fig. 4.18 The proton r.m.s. charge radius r_p and the Rydberg constant R_∞ (lower horizontal scale) obtained by pairs of measured hydrogen transition frequencies. Since the 1S–2S transition is by far the most precise, we combine this one with the other transitions listed in the left box. The upper three values are radio frequency measurements. All others are in the optical regime. The right-hand shaded range is obtained when combining all hydrogen data (except for the new 2S–4P) through a weighted mean (square labeled "H world data"). The left vertical gray region marks the result from muonic hydrogen, which deviates by 4.0 combined standard deviations from the weighted hydrogen value. This discrepancy is called the "proton size puzzle". A recent remeasurement of the 2S–4P transitions (Beyer et al., 2017) is in agreement with the muonic value (diamond labeled "2S–4P (2017)"). With the latter value, the hydrogen data that used to be consistent with QED, no longer is. The better way to analyze overdetermined data of this kind is to perform a least squares adjustment (Crowe et al., 1957). However, such an analysis makes sense only for consistent data. The purpose of this plot is to show that the data (and QED) are not consistent.

4.2.3.1 The 1S–2S transition

The possibility to readily count optical frequencies has set the stage for new improved experiments to test the predictions of QED. At the Max-Planck-Institute of Quantum Optics (MPQ) in Garching the uncertainty of the 1S–2S transitions frequency at around 2466 THz has been continuously reduced over many years (Udem et al., 1997; Niering et al., 2000; Fischer et al., 2004; Parthey et al., 2011). The spectrometer used for that purpose is sketched in Fig. 4.19. It consists of a highly stable laser at 972 nm that is frequency doubled twice to 243 nm. This radiation is used to drive the 1S–2S transition (see Fig. 4.17) in an enhancement resonator that is collinear with a cold atomic hydrogen beam. The emission linewidth of the laser is narrowed to

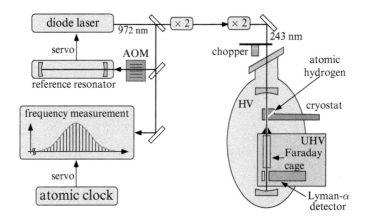

Fig. 4.19 Exciting the hydrogen 1S–2S transition with two counter propagating photons in a standing wave field at 243 nm that is obtained by frequency doubling a diode laser at 972 nm twice. The laser is stabilized to a high finesse cavity to reduce its linewidth and drift rate. It is scanned using an acousto-optic modulator (AOM) while its frequency is measured with a frequency comb that is stabilized to a cesium atomic clock. Hydrogen atoms are generated in a gas discharge (not shown) that breaks up hydrogen molecules. The copper nozzle at 5.8 K is used to thermalize the hydrogen atoms that form a beam inside the vacuum system. The chopper wheel is used to selectively excite the slower atoms. As the laser is scanned across the resonance, the 2S atoms are detected with the Lyman-α detector (electric field + photo-multiplier). The latest value of the 1S–2S resonance frequency is 2 466 061 413 187 035 (10) Hz (hyperfine centroid) (Parthey et al., 2011).

below 1 Hz by stabilizing it to an external reference cavity (Alnis et al., 2008). The stabilization also reduces the drift rate to about 0.1 Hz per second.

Hydrogen atoms are produced in a gas discharge and ejected from a copper nozzle kept at a temperature of 5.8 K.[29] The r.m.s. mean thermal velocity of atoms is $\sqrt{3KT/m} = 380$ m/s, or roughly $10^{-6}c$. This would also be the fractional frequency shift due to the first order Doppler effect: $\Delta\omega = \vec{k}\cdot\vec{v}$ or $\Delta\omega/\omega = v/c$. However, a linear enhancement cavity ensures that the exciting light field is made of two counter propagating laser fields, so that the Doppler effect can be cancelled to first order (Cagnac et al., 1973; Grynberg et al., 1977). As sketched in Fig. 4.17, two photons are required to drive the transition. If these photons are counter propagating, i.e. with wave vectors $+\vec{k}$ and $-\vec{k}$, their total Doppler shift adds to zero. The matrix elements for absorbing two photons from the same direction have the same magnitude (for an S → S transition and linear laser polarization). This means that the Doppler free line sits on a Doppler broadened line with the same integrated area. The Doppler broadened component is roughly 1 GHz wide, i.e. 6 orders of magnitude wider than the Doppler free component, and hence its amplitude is 6 orders of magnitude smaller. In practice we do not observe the Doppler broadened

---

[29] This temperature is an optimum with respect to the complex dynamics of freezing out hydrogen, recombination to molecular hydrogen and thermalization.

component. Another very important advantage of the two-photon excitation of the hydrogen 1S–2S transition is that it does not require light at 121.5 nm. Generating laser radiation at this wavelength is not impossible but very difficult (Kolbe et al., 2012).

After travelling about 13 cm, the hydrogen atoms that are excited to the 2S state are recorded as a function of laser frequency. To detect the 2S atoms, a small electric field is generated in front of a photo-multiplier tube. Perturbation theory tells us that this mixes the 2P state into the 2S state and causes its immediate decay to the 1S ground state.[30] The photo-multiplier tube picks up the Lyman-α photon (121.5 nm) that is emitted upon this decay.

There are two main systematic effects of almost the same magnitude. The first one is the second-order Doppler effect. While the first order is of geometric origin,[31] the second-order is due to time dilation and is given by $\sqrt{1-(v/c)^2} \approx 1 - v^2/2c^2$. With the numbers given above, the second-order Doppler effect is 8×10^{-13} in relative units. To further reduce this effect we use the chopper wheel shown in Fig. 4.19 that periodically interrupts the exciting laser. By accepting Lyman-α photons only after a delay τ that the light field has been turned off, we restrict the speed of the atoms to a maximum of $v < L/\tau$, where the length $L = 13$ cm is the distance from the copper nozzle to the Lyman-α detector. The delayed lines are shown at the left-hand side of Fig. 4.20. For the analysis we used delays between 0.8 ms and 1.2 ms, reducing the second-order Doppler effect to roughly 4×10^{-14}. By modelling we can further correct our data to within 10% of the remaining second-order Doppler effect.

The second large systematic effect is the AC Stark or light shift. This is very common for two-photon transitions. Because the matrix elements are much smaller than for dipole allowed transitions, a large laser intensity is required. A pure two-level system would not be subject to the AC Stark shift of the line center. It would merely be a detuning dependent line shift that is symmetric about the line center which is described as power broadening. The presence of other levels allows the two-photon transition and gives rise to a different AC Stark effect. It also shifts the line when the laser is on resonance. The way to compensate for it is to measure at different laser intensities and extrapolate the line center position to zero as shown at the right-hand side of Fig. 4.20. This procedure gives the final uncertainty of 4×10^{-15} as stated in Fig. 4.19.

4.2.3.2 *The 2S–4P transitions*

To provide another precisely measured transition in atomic hydrogen, we are using the 1S–2S apparatus now as a source of cold 2S atoms. Almost all measurements used to test QED use the 2S state as the initial state (see Fig. 4.18). So far all previous experiments have employed electron impact excitation to populate the 2S state. This method severely heats the atomic beam and contributes to deviations from the Maxwellian velocity distribution. This complicates a detailed line shape analysis

---

[30] The lifetime of the 2P state is only 1.6 ns.

[31] Adding or subtracting the number of extra wavefronts an atom sees per unit of time as it flies through a laser field is causing a frequency shift of $\Delta\omega = kv = 2\pi \times v/\lambda$.

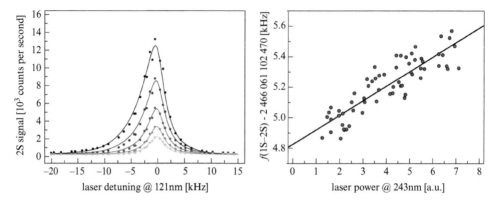

Fig. 4.20 Left: Detecting the 2S state with the spectrometer in Fig. 4.19 with delay times $\tau = 0.1, 0.2, 0.3, 0.4$ ms (top to bottom) measured after the exciting laser has been interrupted by the chopper wheel. With this method we let the faster atoms escape and reduce the second-order Doppler effect, and along with the signal. The second-order Doppler effect gives rise to an asymmetric line deformation, because it shifts only with one sign. For the short delay times this can be clearly seen, while for the longer delays the lines are sufficiently symmetric. From Monte Carlo simulations we know that we can safely fit the longer delayed lines with a Lorentzian, causing only a negligible bias to the line center. Experimentally, the various delays are recorded simultaneously by sorting all photon detection events into delay time bins. **Right:** Line centers as a function of laser power. The AC Stark shift is extracted out by taking the limit of zero laser power.

which is essential to determine the line center much better than the linewidth. Another complication arises from the unresolved hyperfine structure of the P states. Previous experiments have relied on selecting the hyperfine components by laser polarization. Thanks to the well resolved hyperfine splitting of the 1S–2S transition we can now selectively populate the 2S$(F = 0)$ hyperfine state with the 243 nm laser. As the $F = 0 \rightarrow F = 0$ transition is strictly forbidden (independent of polarization) for dipole transitions, only one hyperfine component is left in the 2S–4P$_{1/2}$ manifold. Likewise, for the 2S–4P$_{3/2}$ transition only the $F = 0 \rightarrow F = 1$ can be excited. Avoiding the electron beam also helps to reduce the DC Stark effect by introducing fewer free charges into the system.

The apparatus that is sketched in Fig. 4.21 uses two ultra-stable laser beams at 972 nm. One of them operates at a fixed frequency. Its 4th harmonic runs collinearly with the atomic beam and drives the 1S–2S two-photon transition as in Fig. 4.19. From the previous 1S–2S experiments we know that the velocities of the resulting 2S atoms are essentially Maxwellian distributed, with a temperature that is very close to 5.8 K, i.e. the temperature of the nozzle. The second laser at 972 nm is detuned by about 2.3 GHz (3.6 GHz) and scanned in frequency using an acousto-optic modulator (AOM) such that its second harmonic probes the 2S–4P$_{1/2}$ (2S–4P$_{3/2}$) transition at 486 nm. Transitions to the 4P levels are detected via their decay to the 1S ground state upon which they emit a Lyman-γ photon at 97 nm. These photons are detected via photo electrons that they generate inside two graphite coated cylinders. The photo electrons are guided to channel electron multipliers (CEM) with a suitable

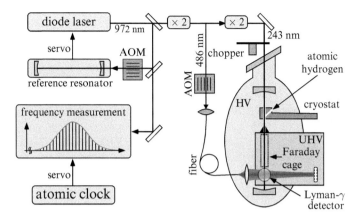

Fig. 4.21 The 1S–2S apparatus of Fig. 4.19 (in light gray) completed with a second identical laser at 972 nm (drawn as one) that is frequency doubled only once to drive the 2S–4P transition at 486 nm. The leading order Bohr–Schrödinger term in the energy levels Eq. 4.34 predicts the frequency ratio of the 1S–2S and 2S–4P to be exactly 4, so that in reality it is close to that value. The 2S–4P laser is illuminating the atomic beam at an angle as close as possible to 90°. An actively stabilized highly reflecting mirror ensures that the wavefronts of the beam coupled back into the fiber closely match those of the forward wave. This suppresses the shift resulting from the first-order Doppler effect to a high degree.

electric field configuration. With this detector we obtain a large solid acceptance angle (Matveev et al., 2013), while the large work function of graphite (≈ 5 eV) suppresses the background from the 486 nm laser. The photo electron detector is sketched at the left-hand side of Fig. 4.22.

The largest systematic uncertainty, that needs to be taken care of, is the first-order Doppler shift that is easily avoided in two-photon spectroscopy by using counter propagating laser beams. The full size of this effect would be $\Delta\omega/\omega = (v/c)\cos(\vartheta) \approx \pm 10^{-6}\cos(\vartheta)$ at a temperature of $T = 5.8$ K in relative units. There is no hope that the angle ϑ can be adjusted close enough to 90° to reach the accuracy goal of $\approx 10^{-12}$, also because both the atomic beam and the laser beam are slightly divergent. Similarly to the two-photon case, two counter propagating waves may be used that ideally give rise to two lines that are equal in amplitude and shape and are symmetrically separated by $2\Delta\omega$. The condition is that at each point in space the two laser waves must have the same intensity and the wave vectors must be anti-parallel. Corner cube reflectors have been used in the past for this purpose but fail to meet the requirements here. Even though the angular deviations can be controlled to better than 10^{-6} rad, they rotate the polarization in a complicated way, cause a beam displacement, and are rather lossy. Instead, we generate a high quality Gaussian beam using spatial mode cleaning of a single mode fiber and carefully designing a system of collimating lenses with as low as possible spherical aberrations. The waist of the collimated beam is placed on a flat high reflecting mirror (99.995%) and coupled back into the same fiber with a measured efficiency of 100.4(9)% (see right hand side of Fig. 4.22). In addition, we adjust the angle ϑ close to 90° (within about 0.08°) by minimizing the observational linewidth of the transition (Berkeland et al., 1995).

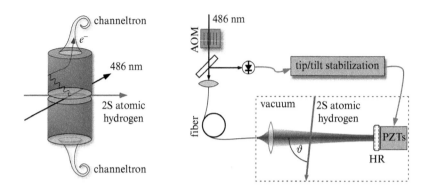

Fig. 4.22 Details of the 2S–4P spectrometer of Fig. 4.19. **Left:** Lyman-γ detector. Atoms in the 4P states quickly release a Lyman-γ photon at 97 nm. To obtain a large solid angle of detection, photons are not detected directly. Instead we detect photo-electrons released from a cylindrical graphite coated surface. The photo-electrons are guided by electric fields to two channel electron multiplier (CEM) for detection. The atoms are shielded from the electric fields of the CEMs by a Faraday cage (not shown). **Right:** Doppler cancellation with a standing wave. A single mode fiber and a low aberration collimator lens generate a high quality Gaussian beam that is back coupled into the fiber with an efficiency close to 100%. Two dither locks, operating with two lock-in amplifiers at two distinct modulation frequencies, are used for the tip/tilt stabilization. In addition the lens position is controlled to place the waist at the flat high reflecting (HR) mirror. Not shown for clarity is a second mode cleaning fiber followed by a power stabilisation.

This strategy of reducing the first-order Doppler shift is only a qualitative one. We do not want to rely on having generated a good enough laser beam without quantitative verification. Fortunately, we can once again employ the laser generated 2S atoms for that purpose. Similar to the delayed detection of Lyman-α photons discussed in Sec. 4.2.3, we detect the Lyman-γ photons after the 2S atoms generating laser beam has been turned off. From Monte Carlo simulations we find that the mean velocity of the 2S atoms that are excited to the 4P state[32] can be varied with this method between 85(10) m/s and 295(40) m/s. Any residual first-order Doppler shift is proportional to velocity. With the current apparatus we could set an upper limit on this systematic effect of 3 kHz or $\Delta\omega/\omega = 5 \times 10^{-12}$ (Beyer et al., 2017) with room for future improvement.

Another effect that can lead to systematic shifts on the order of the proton size discrepancy is due to line pulling effects by the so-called quantum interference[33] between different excitation and decay paths (Horbatsch and Hessels, 2010). Despite its name, this effect can be understood with two[34] phased classical dipole emitters that have resonance frequencies at ω_0 and $\omega_0 + \Delta$. For simplicity and adapted to the

---

[32] Note that the Maxwellian velocity distribution of the 1S atoms is not identical to the velocity distribution of the 2S atoms, because the 1S–2S excitation probability depends on velocity. In addition, the 2S–4P excitation is also velocity dependent.

[33] Sometimes also called cross damping.

[34] Generalization to more than two resonances is done in an analogous way.

2S–4P multiplet, we assume that both resonances have the same damping constant Γ. Furthermore the driven dipoles are given by the vectors $\vec{D}_0$ and $\vec{D}_1$ that do not need to be parallel. With the usual susceptibilities the total dipole moment is given by:

$$\vec{D}(\omega) = \frac{\vec{D}_0}{(\omega_0 - \omega) + i\Gamma/2} + \frac{\vec{D}_1}{(\omega_0 + \Delta - \omega) + i\Gamma/2} \tag{4.35}$$

The quantum mechanical description is analogous to Eq. 4.35, except for the rule to compute the dipole moments $\vec{D}_0$ and $\vec{D}_1$. According to the Kramers–Heisenberg formula, they are given as products of the driving laser field, the absorbing dipole matrix element (between the initial and excited state) and the emitting dipole matrix element (between the excited and the final state). In case the same initial and final states are connected via different indistinguishable excited states, interference can take place. This effect has first been described with doubly excited autoionizing states by U. Fano (Fano, 1935). The corresponding line shape that he derived is referred to as Fano resonance.

For the qualitative discussion here, we can ignore the exact mechanism that generates the dipoles. If we placed them at the origin, they generate a field $\vec{E}(\vec{r})$ at position $\vec{r}$ that is given by:

$$\vec{E}(\vec{r}) \propto \left(\vec{r} \times \vec{D}(\omega)\right) \times \vec{r}/r^3 \tag{4.36}$$

The power spectrum $P(\omega)$, i.e. the line shape, is proportional to the square modulus $\vec{D}(\omega)$. It consists of two real valued Lorentzians and a non-Lorentzian cross term. The latter depends not only on the relative orientation of $\vec{D}_0$ and $\vec{D}_1$ but also on the direction of the emitted radiation relative to the orientation of the dipoles. Because the orientation of the dipoles is itself a function of the laser polarization, the observed cross term will effectively depend on the orientation of the laser polarization relative to detection direction. This means that we must deal with deviations from a Lorentzian that depend on the geometry of excitation and detection. If the detection is not point-like, which is better for large quantum efficiency, the line distortions also depend on the geometry of the detector. However, it can be shown that in the limit of a 4π detection solid angle, the cross term disappears. Hence we have tried to get close to this situation with the Lyman-γ photo electron detector shown at the left-hand side of fig. 4.22.

For a sufficiently large spectral separation of the two resonances ($\Delta/\Gamma \gg 1$), the second resonance at $\omega_0 + \Delta$ can be treated as a perturbation to the resonance at ω_0 and the full line shape $P(\omega)$ can be expanded around the resonance at ω_0 (Jentschura and Mohr, 2002):

$$P(\omega) \approx \frac{C}{(\omega - \omega_0)^2 + (\Gamma/2)^2} + a(\omega - \omega_0) + \frac{b(\omega - \omega_0)}{(\omega - \omega_0)^2 + (\Gamma/2)^2}. \tag{4.37}$$

The first term in Eq. 4.37 is the unperturbed resonance at $\omega = \omega_0$ with strength C. The second term takes into account that the main resonance sits on the pedestal of

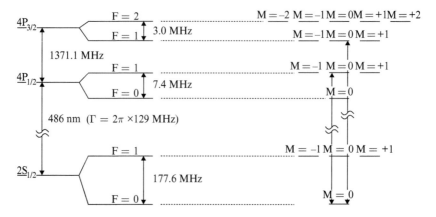

Fig. 4.23 Level scheme of the 2S–4P multiplett. **Left:** The fine structure splitting of the 4P state is around 1.4 GHz while the hyperfine splittings are 7.4 and 3.0 MHz. **Right:** Preparing the atoms in the 2S($F = 0$) state, there are only two allowed Zeeman components that are spectrally separated by about 1.4 GHz, i.e. $\Delta/\Gamma \approx 100$. The cross term should be taken into account if one wants to find the line center better than to 1% of the linewidth, i.e. better than 13 kHz (Horbatsch and Hessels, 2010).

the perturbing resonance, which appears as a linear background. The coefficient a is easily obtained from linearizing the distant resonance. The last term is due to the cross term and has a dispersive shape. The usefulness of this expansion lies in the fact that the full and complicated geometry dependence is described by only one additional coefficient b, that is then used as a fit parameter. It turns out that the latter has the largest influence on the line pulling but is often ignored, whereas the second term has a smaller effect but is regularly included when dealing with double resonances. Fig. 4.24 shows the relative importance of the two terms.

The magnitude of the line pulling depends on the way the line center is determined. A symmetric line may be fitted with any other symmetric line without an apparent line pulling, so that we need to be concerned only with asymmetric line distortions like the correction terms in Eq. 4.37. In this case it is important to agree upon the way the line pulling is measured. Especially when it comes to correcting them by modelling, one needs to be sure that the definition used in the model is the same as the one employed in analyzing experimental data. One might define the line pulling as the shift of the maximum of the line shape (Jentschura and Mohr, 2002). However, this definition will not work when dealing with experimental data that is subject to noise. An operational way that is adapted to the experiment, is to fit a Lorentzian to the slightly distorted line shape and determine the line pulling by fitting the *same* wrong line shape to artificial data obtained from modelling. This definition gives a line pulling of the line shape derived from Eq. 4.36 that is a factor of 2 different from the maximum method. Of course the operational method is only reasonable if the line pullings are small (in some sense). Because the asymmetric distortions are the ones that need to be corrected, it is important to use the same span and sampling on the frequency axis in the model and the experiment.

From Fig. 4.23 and Fig. 4.24 it can be derived that the second term in Eq. 4.37 has a much smaller contribution to the line pulling (about $3 \times 10^{-10}\Gamma = 4$ mHz) and can be safely ignored. This simplifies the fitting function further. There is however another effect that is not yet included in Eq. 4.37: the atomic beam that is sketched as a straight line in Fig. 4.21 is actually slightly divergent so that there is a range of possible angles ϑ (see also right-hand side of Fig. 4.22). As explained above, the line center is independent of this angle but the width is not. Assuming the transversal velocity distribution to be Maxwellian, we fold Eq. 4.37 with a Gaussian with a full width at half maximum of Γ_G. For the first term this leads to the well known Voigt profile, the second term we ignore and the last term can be expressed as the real part of a complex Lorentzian (see Eq. 4.35). The fitting function can therefore be expressed through a complex Voigt profile, i.e. folding a Gaussian with a complex Lorentzian. The observed line shape is then given by the real part of the complex Voigt profile adding a (hopefully) small correction in terms of the imaginary part of the complex Voigt profile (Schippers, 2016; Beyer et al., 2017). For computational reasons the Voigt function is usually expressed in terms of the Faddeeva function (also known as

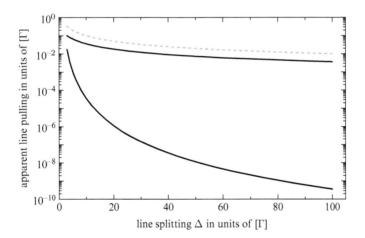

Fig. 4.24 Line shifts obtained when line distortions due to quantum interference are not properly taken into account. To quantify this error, samples are generated using the full line shape model given by Eq. 4.36 with two lines separated by Δ of identical intensity and linewidth Γ. By fitting within a region of $\omega = (\omega_0 - 2\Gamma, \omega_0 + 2\Gamma)$ two Lorentzians (separated by Δ) the upper black curve is obtained, which measures the apparent line shift as a function of line separation (both in units of Γ). For the 2S–4P experiment with $\Delta/\Gamma \approx$ 100 a line pulling of several % of the linewidth takes place. Very similar curves are obtained when fitting with the first and second term in Eq. 4.37. This is another way of including the sloping pedestal of a perturbing remote line. On the other hand fitting with the first and last term of Eq. 4.37 reduces this error by many orders of magnitude (lower black curve; with small modifications with the quantum mechanical treatment). The non-Lorentzian cross term is more important than including the remote line. The gray dashed curve gives an estimated upper limit of this effect according to a rule of thumb derived by (Horbatsch and Hessels, 2010): "The cross term may be ignored, if the accuracy goal for finding the line center is less than Γ/N with the next line separated by at least $N\Gamma$."

the complex error function) (Abramowitz and Stegun, 1972) $w(z)$ for which fast and precise computer routines exist:

$$\mathcal{F}(\omega) = A\left\{\Re[w(z)] + 2\eta\Im[w(z)]\right\} \quad \text{with} \quad z = 2\sqrt{\ln(2)}\left(\frac{\omega - \omega_0}{\Gamma_G} + i\frac{\Gamma}{2\Gamma_G}\right). \quad (4.38)$$

This is the fit function used for the data analysis (Beyer et al., 2017). It involves 6 free parameters, the center frequency ω_0, the amplitude A, a constant background, the Lorentzian and Gaussian widths (Γ and Γ_G) and the asymmetry parameter $\eta = b\Gamma/4C$ (see Eq. 4.37). Given a not so small detection solid angle, the quantum interference shift (due to using the wrong fit function) is small enough such that it can be taken out by the extra adjustable parameter η with sufficient accuracy.

Another option to avoid this problem would be to detect the surviving 2S atoms (like in the 1S–2S set-up) instead of the 2P–1S Lyman-γ emission. This signal is almost immune to the quantum interference effect. Most of the measurements presented in Fig. 4.18 have used that method and the others do not show a large line distortion due to quantum interference. Therefore, this effect cannot explain the "proton size puzzle". The disadvantage of the 2S detection scheme is that one needs to observe a dip on a large and noisy background. For that reason we decided to rather use the Lyman-γ emission and deal with the quantum interference effect. It should be noted that this is a rare case where improving the statistics (larger detection solid angle) also reduces the systematics.

The last systematic line shift that will be mentioned here is due to light forces on the atomic trajectories. For this discussion the laser field is described very well as a plane standing wave with a distance of $\lambda/2 = 243$ nm between adjacent nodes or anti-nodes. The first (but incomplete) idea would be that the atoms are pulled into the nodes when the laser is blue detuned (low field seekers) and into the anti-nodes when it is red detuned (high field seekers). In that case one would expect that the line strength is somewhat reduced at its blue side and enhanced at the red side (Minardi et al., 1999). However, the situation is not that simple. A mechanical pull does not necessary mean that the atoms will predominantly stay at their potential minimum without damping. An oscillating trajectory may be present instead. In our case also this approach is not appropriate, because the thermal de Broglie wavelength $h/\sqrt{2\pi mKT}$ gives about $\lambda/2$ for a temperature that is associated with the width[35] of the transverse velocity distribution ($T \approx 10^{-3} \times 5.8$ K). This means that the atoms are transversally more delocalized than the structure size of the potential. Without being able to give the position of the atom with an uncertainty much smaller than $\lambda/2$, one cannot assign a force. Another delocalization of the atoms takes place, even at large temperatures because of the photon recoil. To give a simple example, consider an atom which experiences a $\pi/2$–pulse at the entrance of the laser interaction zone. After that the 2S part of the wave function will separate from the 4P part. By how

---

[35] Often the thermal de Broglie wavelength is confused with de Broglie wavelength. The former is actually not the wavelength of the matter wave but its coherence length. The reason why one can usually use them interchangeably without obtaining wrong results is that the width of the thermal velocity distribution is numerically almost identical to its mean.

much depends on the longitudinal velocity. In our case the atom delocalizes because of this effect by $40 \times \lambda/2$. These and more conditions for quantum paths or classical trajectories are given in (Cohen-Tannoudji, 1977). For our 2S-4P set-up we need to describe the atomic motion, at least in the direction of the laser beams, with wave mechanics (Chang, 2018). In this way we find that the light force induced frequency shift is very small and adds an uncertainty of 300 Hz. Applying a few other small corrections yields for the 2S–4P fine structure centroid 616 520 931 626.8(2.3) kHz and a value for the proton charge radius compatible with the muonic value (see "2S–4P (2017)" in Fig. 4.18).

4.2.3.3 The 1S–3S transition

Another transition that is currently under investigation in the group of F. Biraben at the Laboratoire Kastler Brossel in Paris and in our group is the 1S–3S two-photon transition. Its frequency has been measured by the Paris team with an uncertainty of 13 kHz (Arnoult et al., 2010). Among the challenges here is the required wavelength of 205 nm, which is difficult to produce as a continuous wave (cw). The only commercially available crystal that can generate it as a second harmonic is beta barium borate (BBO). Not only does the required phase matching angle lie very close to 90° for that crystal, it also suffers from photochemical reactions on the surfaces and from the photo-refractive effect. Sum frequency generation (Galtier et al., 2014) instead of second harmonic generation (Bourzeix et al., 1997) seems to mitigate the latter problem. By cooling the crystal down to -10° C the phase matching angle shifts by about 1° and effective nonlinearity increases (Peters et al., 2009). To deal with these crystal issues and to test a new method for high resolution spectroscopy, we are employing a mode-locked laser rather than a cw laser, whose spectrum can be described as a frequency comb source (Peters et al., 2013). It features a high peak intensity and therefore greater nonlinear interactions and does not induce the photo-refractive effect. Thus, a deep ultraviolet frequency comb can be generated more efficiently than cw radiation.

As detailed in Sec. 4.1.3 and 4.1.3 the comb modes can be expressed as $\omega_n = n\omega_r + \omega_{CE}$ (see Eq. 4.14) with the mode number n, the pulse repetition frequency ω_r and the carrier-envelope offset frequency ω_{CE}. To drive a two-photon transition with a frequency comb, the modes add pairwise to produce the transition energy $\hbar\omega_{eg}$. By tuning ω_{CE} such that a comb mode ω_n corresponds to half the transition frequency, all mode pairs m which satisfy the relation $\omega_{eg} = \omega_{n-m} + \omega_{n+m}$ contribute to the excitation rate. This is sketched for $m = -1, 0, +1$ at the left side of Fig. 4.25. The two-photon resonance condition is also satisfied when the transition occurs exactly between two comb modes. Therefore, the spectroscopic signal repeats with half the repetition rate when scanning ω_{CE}. Quite commonly, the transition frequency of two-photon transitions is given as "atomic frequency", i.e. twice the frequency of the exciting laser. On this scale the recorded spectrum repeats itself with the pulse repetition rate.

For Fourier-limited pulses, all modes of the frequency comb are properly phased and the excitation paths add up coherently. In this way a pulsed laser drives a two-photon transition as efficiently as a cw laser with the same average intensity, while the spectral linewidth is limited by the width of a single comb mode rather than by the

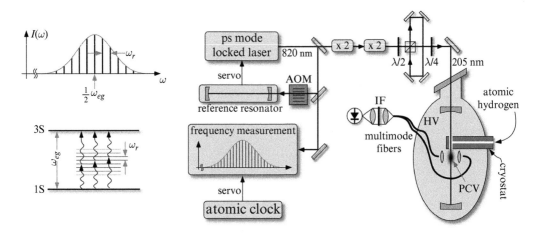

Fig. 4.25 Left: Spectral envelope of the frequency comb tuned to excite a two-photon transition at ω_{eg}. On resonance, pairwise addition of properly phased modes provides an efficient excitation of the atoms. Three such mode pairs are shown. **Right:** A mode-locked titanium:sapphire laser provides Fourier-limited pulses of 1.3 ps pulse duration at $\omega_r = 2\pi \times 78.75$ MHz repetition frequency. The repetition rate is doubled by coupling into an enhancement cavity with a free spectral range of 157.5 MHz for frequency doubling (first "× 2"). After the second doubling stage, we obtain 50 mW of 205-nm radiation with a pulse duration of 2 ps. A third cavity inside a vacuum chamber (free spectral range 157.5 MHz) enhances the power to 500 mW for spectroscopy. Hydrogen atoms are escaping from a cold T-shaped copper nozzle, similar to the 1S-2S and 2S–4P set-ups. An interferometer allows adjustment of the polarization and the pulse collision volume (PVC) where the Doppler-free excitation of the 1S–3S transition takes place. With a quarter wave plate ($\lambda/4$) we generate colliding pulses with σ_- and σ_+ polarization. Four (only two shown) multimode fibers are used to guide the emitted Balmer-α light (3S → 2P decay) through an interference filter (IF) onto a photo detector. Additional fibers (not shown) pick up Balmer-α light at the other side of the nozzle. The Doppler broadened signal obtained outside the PCV is independent of the laser frequency and used to normalize the signal for fluctuations of the flow of hydrogen atoms and laser power.

spectral envelope of the comb (Baklanov and Chebotayev, 1977). Furthermore, the AC Stark shift is due to the average power rather than due to the peak power (Fendel et al., 2007). In fact, the first application of a frequency comb was for two-photon direct frequency comb spectroscopy (Eckstein et al., 1978).

The two-photon direct frequency comb spectroscopy is not exactly free of the first-order Doppler effect. In the frequency domain one would argue that the contributing mode pairs do not have the same frequency. However, for each pair m there is a pair $-m$ such that the Doppler shift is balanced, provided that the spectral envelopes of the counter propagating pulses are identical. This condition is closely matched with an enhancement cavity. In the time domain one would argue that the finite excitation region leads to a time of flight broadening. In both cases the first-order Doppler effect gives rise to a line broadening rather than to a line shift. At a thermal velocity of about $v = 380$ m/s (5.8 K) a pulse duration of $\tau_{1/2} = 2$ ps (FWHM) gives rise to a time of flight (or Doppler) broadening of $\Delta\omega = 8\ln(2)v/(c\tau_{1/2}) = 2\pi \times 574$ kHz. Taking the velocity dependence of the excitation rate into account, the time of flight broadening is even smaller. The pulse duration is therefore well adapted to the natural linewidth of 1.0 MHz of the 1S–3S transition.

The experimental realization (Yost et al., 2016), that is sketched at the right hand side of Fig. 4.25, uses a ps titanium:sapphire mode-locked laser[36] at 820 nm that is frequency doubled twice to 205 nm using two enhancement cavities (Peters et al., 2009). The signal is obtained by detecting Balmer-α photons from the decay of the 3S/3D to the 2P state at 656 nm. As shown in Fig. 4.25, these photons are collected with lenses and coupled into multimode fibers that are connected to a photo detector.

A delay line generates double pulses that meet twice per round trip inside the enhancement cavity inside the vacuum chamber. This effectively doubles the repetition rate again[37] such that the spectrum shown at the right-hand side of Fig. 4.26 repeats itself with a period of 315 MHz. The delay line superimposes the two pulse trains

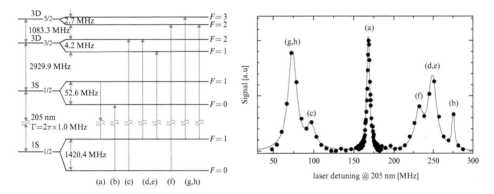

Fig. 4.26 Left: Level scheme of the 1S–3S/D multiplet and transitions allowed with colliding pulses of σ_- and σ_+ polarization respectively. Comparing the energy separation of the levels with the natural linewidth of the 1S–3S transition (1 MHz), we estimate the quantum interference effect to be important only when the accuracy goal is below 1 kHz. **Right:** With two-photon direct frequency comb spectroscopy the observed spectrum repeats with the pulse repetition rate. This means that the separation of the line components appear modulo w_r. To obtain a less crowed spectrum we effectively double repetition rate from the laser twice to $w_r = 2\pi \times 315$ MHz. By fine tuning we obtain a well isolated 1S($F = 1$)–3S ($F = 1$) component (a). The components (g,h) and (d,e) are not resolved in this preliminary measurement that was taken with a room temperature nozzle and therefore with an increased time of flight broadening (Yost et al., 2016).

[36] One may argue if a ps mode-locked laser should be called a frequency comb because it cannot be directly self-referenced. Nevertheless, for simplicity we are referring to this type of spectroscopy as direct frequency comb spectroscopy.

[37] In the frequency domain this can be understood by the shift theorem (see Eq. 4.25). Delaying the pulse by half the repetition time means in the frequency domain that every second mode is phase shifted by π and hence destructively interferers with the spectrum of the undelayed pulse train. In the case discussed here, this takes place for the atoms even with orthogonal circular polarization, because the Doppler free two-photon transition rate is proportional to the product (not the scalar product) of the two counter propagating fields.

with σ_- and σ_+ polarizations respectively. This has two advantages: first it avoids loosing half the power in comparison with a delay line based on a Michelson or Mach–Zehnder interferometer. Secondly, it allows one to reduce the excitation of the 1S–3D transitions with respect to the 1S–3S transitions using the more complex selection rules for two-photon transitions (Grynberg et al., 1977).[38]

Similar to the 2S–4P transition, the 1S–3S has a fine and hyperfine structure. Because one selection rule that applies for two-photon transitions dictates $\Delta l = (0, 2)$, the 1S–3D$_{3/2}$ and 1S–3D$_{5/2}$ are also allowed. The left-hand side of Fig. 4.26 shows all the involved levels (except for the Zeeman structure) and transitions that are allowed with the given polarizations.

The set of systematic effects, that are important for the pulsed 1S–3S spectroscopy, is essentially the same as already discussed for the other transitions. Therefore, we only consider the additions to this list here. To model the quantum interference for two-photon transitions one needs to expand the usual Kramers–Heisenberg formula for the three involved photons, two for excitation and one emitted photon. A detailed study of this mechanism has been done and finds that the systematic shift is ≈ 1 kHz if the associated line distortion is ignored by simply fitting a sum of Lorentzians to the data (Yost et al., 2014). This is compatible with the estimation rule for one photon transitions of Horbatsch and Hessels (Horbatsch and Hessels, 2010). The "proton size puzzle" measures 7 kHz at the 1S–3S transition frequency so that quantum interference should not prevent us from making a statement about that problem.

When it comes to spectroscopy, the usage of a frequency comb to excite the two-photon transition seems to have no real advantage over the usage of a cw laser, unless this laser radiation is difficult or impossible (see next section) to build. For the 1S–3S spectroscopy, however, it can improve the signal-to-noise ratio because the Doppler free signal is generated in a small volume where the counter propagating pulses collide. Light collection optics adapted to this volume may be used. Since this volume is easily translated spatially, it allows to check for possible collision shifts in a simple way as the atomic density reduces roughly with the square of the distance to the nozzle.

These advantages are contrasted with problems that are due to a possible chirp of the laser pulses in certain geometries. In the frequency domain, a chirp is described through a spectral phase (see Sec. 4.1.3). At first glance one would argue that adding a mode number dependent phase to a frequency comb does not change the frequencies of the modes. In fact this is true for atoms that do not move relative to the pulse collision volume. In this case a chirp merely leads to a reduced transition rate, because the various paths that lead from the ground to the excited state no longer all interfere constructively (see left-hand side of Fig. 4.25). For atoms at rest this reduction is proportional to the time-bandwidth-product (Reinhardt et al., 2010). This rule may be applied to the spectroscopy of single trapped ions and/or to pulses that are chirped without changing their duration (by self-phase modulation). Chirping pulses by group velocity dispersion leads to a larger pulse collision volume and hence to a compensation of the signal reduction for example in a gas sample (Ozawa and Kobayashi, 2012).

---

[38] The two-photon operator has a rank 0 and a rank 2 component. In addition, the two photons may have differen polarizations, like in the case that is discussed here.

The pulses that are generated by our laser are close to the transform limit. Frequency doubling may lead to a reduction of the spectral width if the phase matching bandwidth is smaller than the pulse bandwidth. However, this does lead to a corresponding extension in the time domain and will not induce a chirp, at least for the ideal (but bandwidth limited) doubling crystal (see Sec. 3.4.1 in (Diels and Rudolph, 2006)). However, self phase modulation within the doubling crystals can play a role (Wang and Weiner, 2003). Experimentally, such a chirp can be varied for example by changing the crystal position, possible by sacrificing efficiency.

Unfortunately, it turns out that the chirp may also give rise to a residual first-order Doppler shift when the atoms form a diverging beam. To describe this effect we write the laser pulse trains similar to Sec. 4.1.3 (Yost et al., 2016):

$$\mathcal{E}_\pm = A_0 \sum_n e^{-(1+ib)(t-nT\pm z/c)^2/\tau^2 - r^2/w^2 - i(\omega_c t \pm \omega_c z/c)}, \tag{4.39}$$

with the repetition time $T = 2\pi/\omega_r$, the pulse duration τ and the optical carrier frequency ω_c. The pulse trains travel along the $\pm z$-direction with a field amplitude that falls off with the radial coordinate $r = \sqrt{x^2+y^2}$ (FWHM intensity $= 2\sqrt{\ln(2)}w$). This form is valid in the weak-focusing (plane wave) approximation. The chirp of the pulses is included with the parameter b. The Doppler free two-photon signal will occur only at the position where the pulses collide. The corresponding two-photon Rabi frequency is proportional to the product of the fields of the counter propagating pulses:[39]

$$\mathcal{E}_-\mathcal{E}_+ \sim \sum_n e^{-2(1+ib)((t-nT)^2+z^2/c^2)/\tau^2 - 2r^2/w^2 - 2i\omega_c t}. \tag{4.40}$$

Assuming a single atom that moves along a well defined trajectory $\vec{r} = (x_0 + v_x t, y_0 + v_y t, z_0 + v_z t)$ through the pulse collision volume, the Doppler shift can be obtained from the time dependent phase of the complex Rabi frequency. The situation here is more complicated than with a simple spatial phase factor of the type $\exp(+i\vec{k}\cdot\vec{r})$, that would give a constant frequency offset. Therefore, we compute the power spectral density of the Rabi frequency, in the rotating wave approximation without the term $\exp(-2i\omega_c t)$, and obtain a comb-like spectrum.[40] The individual comb lines are broadened by the time of flight effect and shifted in the limit $v_z \ll c$ by

$$\Delta\omega = b\frac{4v_z\left(v_x v_z x_0 + v_y v_z y_0 - (v_x^2+v_y^2)z_0\right)}{v_z^2 w^2 + (v_x^2+v_y^2)c^2\tau^2}. \tag{4.41}$$

In an isotropic gas this shift disappears when averaging over trajectories. In a directed atomic beam this is not the case, at least when this beam is diverging. At least v_z and one of the terms $v_x x_0$ or $v_y y_0$ may not average to zero.

---

[39] The chirp parameter b keeps its sign upon reflection to preserve the color sequence of the pulse.
[40] An analytic but lengthy expression exists.

We can further simplify this expression for the geometry discussed here with $\vec{r} = (x_0 + v_x t, 0, v_z t)$. Considering the divergence of the atomic beam in one plane only is sufficient to obtain an approximate expression in the limit $v_z \gg v_x$ (Yost et al., 2016):

$$\Delta\omega \approx b\frac{4v_x x_0}{w^2} \quad \overrightarrow{\text{average atomic trajectories}} \quad \langle\Delta\omega\rangle \approx b\frac{v}{L}. \quad (4.42)$$

The expression at the left would average out because both v_x and x_0 appear with both signs, unless these quantities are correlated. This is the case in a diverging atomic beam. The expression at the right-hand side is obtained by averaging over all trajectories possible with a nozzle radius R_n that is positioned at a distance L from the center of the pulse collision volume with a weighting that takes the excitation probability into account. In addition we adapted to our geometry with the hierarchy $L \gg R_n \gg w \gg c\tau$. With that and $v = 380$ m/s, $L = 10$ mm, and $b = 0.5$, we obtain a small systematic shift of $\Delta\omega \approx 2\pi \times 3.0$ kHz. Further reduction is possible by varying the chirp, for example by changing the position of the doubling crystals, and extrapolating to zero.

4.2.4   XUV direct comb spectroscopy

Direct frequency comb spectroscopy may be pushed to even shorter wavelengths like the extreme ultraviolet (XUV) from 124 nm down to 10 nm. Laser-like pulsed emission at very short wavelengths can be produced with a method referred to as high harmonic generation (HHG) (Eden, 2004). For this purpose, infrared femtosecond pulses are focused in a gas target employing its extreme nonlinear response when subject to intensities exceeding $\approx 5 \times 10^{12}$ W/cm^2. In most HHG experiments a nozzle emitting a jet of noble gas atoms is placed near the focus of an amplified femtosecond pulse train. A collimated beam of high harmonics, that is collinear with the driving laser beam emerges, underlining their spatial coherence. Wavelengths down to the soft x-ray regime have been produced (Popmintchev et al., 2012). For harmonics up to 62 nm the temporal coherence has been verified (Benko et al., 2014). The XUV frequency comb that is sketched in Fig. 4.27, may allow high resolution laser spectroscopy in this spectral region for the first time with the method described in the previous section.

Besides the two-photon direct frequency comb spectroscopy, other types of direct comb spectroscopy may be used (see introduction). It is possible to excite a dipole allowed single-photon transition with a frequency comb (Gerginov et al., 2005). However, only one mode of the comb contributes to the transition rate. On the other hand, dipole allowed transitions usually have much larger matrix elements, so that the total transition rate could be even larger than with a two-photon transition. XUV direct frequency comb spectroscopy that has been demonstrated so far is of this type (Kandula et al., 2010; Cingöz et al., 2012; Ozawa and Kobayashi, 2013). If one wants to use this new technique for testing QED, a transition in a hydrogen-like or otherwise calculable system has to be addressed. For good measurement accuracy a narrow

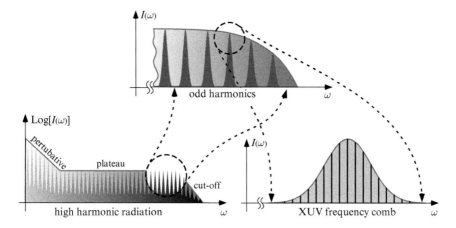

Fig. 4.27 Hierarchy of the laser spectrum that is obtained with high harmonic generation. Three regimes are distinguished, the lower order pertubative harmonics, the plateau, and the cut-off. Due to the inversion symmetry of the nonlinear gas medium, only odd harmonics of the driving laser are generated. Each of those harmonics consists of a frequency comb with a mode spacing given by the repetition rate of the driving laser and an odd harmonic of the carrier offset frequency.

linewidth is desirable. The 1S–2S transition in hydrogen-like helium is the perfect candidate in that respect (see below).

A condition for direct frequency comb spectroscopy is that the mode spacing is larger than the observational linewidth. Otherwise the atom or ion will not resolve the comb structure. This requirement is contradicting the requirement for high peak intensities necessary to drive the HHG. A large pulse repetition rate distributes the limited average power among many pulses per second. One solution to this problem is to employ a high power laser (Hädrich et al., 2014). Another solution is to generate HHGs inside an enhancement resonator (Gohle et al., 2005; Jones et al., 2005). Repetition rates on the order of 100 MHz can be handled in a straightforward manner. However, for metrologically relevant narrow transitions this is actually much larger than required. Reducing the repetition rate for higher HHG yield in this way requires very large enhancement resonators (Ozawa et al., 2008, 2015) that may give rise to mechanical instabilities. Nevertheless, XUV powers of several μW have been generated at large repetition rates (Pupeza et al., 2013).

Another problem with the intracavity approach is to extract the XUV radiation which is collinear with the driving field and, due to its spatial coherence and shorter wavelength, even comes with a smaller diffraction angle. No transmitting optics is available at this wavelength. Several solutions for this problem have been worked out and demonstrated over the years. An uncoated Brewster plate can be used that has very low loss for the fundamental laser. At the same time, due to the large imaginary and small real part of the refractive index, this plate leads to a significant Fresnel reflectivity for XUV radiation. The specific problems with this method are the unwanted dispersion that can limit the number of modes that could be simultaneously resonant. This is a particular problem for high finesse cavities. At high fundamental powers the Brewster output coupler can be subject to thermal lensing and can give rise to a nonlinear contribution to the group velocity dispersion. An alternative method uses a cavity with a small hole in one of its mirrors to extract the XUV radiation (Pupeza et al., 2013). While this method works best for higher-order harmonics, it may not be optimal for the lower order harmonics, because their diffraction angle is larger. Rather than accepting the loss due to the hole one can design a higher-order transverse cavity mode that avoids this hole, while the generated XUV radiation exits the cavity through it (Weitenberg et al., 2011). This method is more advantages for the lower order harmonics with larger divergence (Pupeza et al., 2014). Yet another method uses a diffraction grating etched on one of the cavity mirrors with a ruling too small to perturb the fundamental wave but large enough to extract the XUV radiation in first diffraction order (Yost et al., 2008). A specific feature of this method is the spatial separation of the various wavelengths that make up the high harmonics. This may or may not be desirable depending on the application. To make use of the full power for two-photon direct frequency comb spectroscopy, it is required that all modes from a certain HHG order focus at the same spot, so that the grating output coupler does not seem to be the optimal solution for that purpose.

So far, all high resolution spectroscopy experiments were performed with continuous wave lasers that reach wavelengths up to the near ultraviolet (see Sec. 4.2.3). The spectral region beyond has not been accessible for high resolution laser spectroscopy. This is an unfortunate situation because all hydrogen-like ions have their

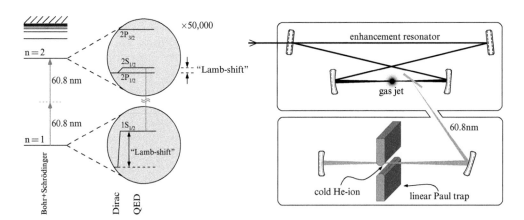

Fig. 4.28 Left: Energy levels of He$^+$ analogous to Fig. 4.17 but scaled by a factor $Z^2 = 4$. Therefore, the 1S–2S transition occurs at a laser wavelength of 60.8 nm rather than 243 nm. The leading order QED effects even scale with $Z^3 = 16$ (magnification $\times$ 50, 000 instead of $\times$ 100, 000) which makes this system more sensitive to higher-order theoretical contributions. In contrast to hydrogen, ^{4}He$^+$ has no hyperfine structure. **Right:** Sketch of the set-up being built to excite the 1S–2S two-photon transition. Intracavity high harmonics are generated in a gas jet and then coupled out, for example with a Brewster plate, to another vacuum chamber that houses an ion trap for He$^+$ and another ion for sympathetic cooling. Mirrors at 60.8 nm are not very effective so that it will be difficult to deliver a large fraction of power to the ion or even reflect it back. However, the two-photon transition does not have to be Doppler free in this case because the ion is essentially at rest.

metrologically relevant transitions there. With the XUV frequency combs it may now become feasible to excite the 1S–2S transition in hydrogen-like He^+ at 60.8 nm (Herrmann et al., 2009). Besides providing another independent QED test, this system has several advantages: as the power series in Eq. 4.34 is written for hydrogen, α has to be replaced by $Z\alpha$ for hydrogen-like systems with nuclear charge $Z > 1$. Therefore, He^+ provides a more sensitive test of the higher-order terms, in particular for yet uncalculated terms of order $(Z\alpha)^{6\cdots 7}$. On the experimental side, He^+ holds the advantage that it can be held in an ion trap and sympathetically cooled by another ion with an easy to generate cooling wavelength. In contrast, cooling and trapping neutral hydrogen is much more challenging because the cooling wavelength of 121 nm is difficult to generate and neutral atom traps are much more shallow. Both of the leading order systematic effects of the hydrogen 1S–2S transition are due to the atomic motion which is absent for a trapped ion. The AC Stark effect can be significantly reduced by illuminating longer with lower intensity. The helium nuclear charge radius has been measured using muonic He^+ (Nebel et al., 2012) and is theoretically better understood than the proton radius. The apparatus that is currently being set-up in our laboratory is sketched along with the level scheme in Fig. 4.28.

Acknowledgments

I would like to acknowledge the hard work of generations of PhD students, postdocs, and senior scientists that worked on high resolution laser spectroscopy at MPQ. The current crew on hydrogen and hydrogen-like systems is Lothar Maisenbacher, Alexey Grinin, Vitaly Andreev, Fabian Schmid, Arthur Matveev, Johannes Weitenberg, and Akira Ozawa. Many thanks to them also for proof-reading the manuscript.

References

Abramowitz, M., and Stegun, I. (1972). Handbook of Mathematical Functions with Formulas, Graphs, and Mathematical Tables (10th edn). Washington, D.C.

Agrawal, G. P. (2013). Nonlinear Fiber Optics (5th edn). Academic Press, Oxford.

Alnis, J. et al. (2008). Subhertz linewidth diode lasers by stabilization to vibrationally and thermally compensated ultra low-expansion glass fabry–pérot cavities. *Physical Review A*, 77, 053809.

Antognini, A. et al. (2013). Proton structure from the measurement of 2s-2p transition frequencies of muonic hydrogen. *Science*, 339, 417–20.

Aoyama, T. et al. (2012). Tenth–order qed contribution to the electron g-2 and an improved value of the fine structure constant. *Physical Review Letters*, 109, 111807.

Arnoult, O. et al. (2010). Two-photon frequency comb spectroscopy of the 6s–8s transition in cesium. *Euro Physics Journal D*, 60, 243–56.

Baklanov, Ye. V., and Chebotayev, V. P. (1977). Narrow resonances of two-photon absorption of super-narrow pulses in a gas. *Applied Physics*, 12, 97–9.

Baltuška, A. et al. (2003). Attosecond control of electronic processes by intense light fields. *Nature*, 421, 611–15.

Bartels, A., Dekorsy, T., and Kurz, H. (1999). Femtosecond ti:sapphire ring laser with a 2-ghz repetition rate and its application in time-resolved spectroscopy. *Optics Letters*, 24, 996–8.

Bartels, A., Heinecke, D., and Diddams, S. A. (2009). 10-ghz self-referenced optical frequency comb. *Science*, 326, 681.

Bartels, A. et al. (2004). Stabilization of femtosecond laser frequency combs with subhertz residual linewidths. *Optics Letters*, 29, 1081–3.

Benko, C. et al. (2014). Extreme ultraviolet radiation with coherence time greater than 1 s. *Nature Photonics*, 9, 530–6.

Berkeland, D. J., Hinds, E. A., and Boshier, M. G. (1995). Precise optical measurement of lamb shifts in atomic hydrogen. *Physical Review Letters*, 75, 2470–3.

Bernhardt, B. et al. (2010). Cavity-enhanced dual-comb spectroscopy. *Nature Photonics*, 4, 55–7.

Beyer, A. et al. (2017). The rydberg constant and proton size from atomic hydrogen. *Science*, 385, 79–85.

Biraben, F. (2009). Spectroscopy of atomic hydrogen: how is the rydberg constant determined? *European Physical Journal Special Topics*, 172, 109–19.

Birks, T. A., Wadsworth, W. J., and Russell, P. St. J. (2000). Coherence properties of supercontinuum spectra generated in photonic crystal and tapered optical fibers. *Optics Letters*, 25, 1415–17.

Bourzeix, S. et al. (1997). Ultra-violet light generation at 205 nm by two frequency doubling steps of a cw titanium–sapphire laser. *Optics Communications*, 133, 239–44.

Bourzeix, S. et al. (1996). High resolution spectroscopy of the hydrogen atom: determination of the 1s lamb shift. *Physical Review Letters*, 76, 384–7.

Bramwell, S. R., Kane, D. M., and Ferguson, A. I. (1985). Frequency offset locking of a synchronously pumped mode—locked dye laser. *Optics Communications*, 56, 112–16.

Cagnac, B., Grynberg, G., and Biraben, F. (1973). Spectroscopie d'absorption multiphotonique sans effet doppler. *Journal de Physique*, 30, 845–58.

Chang, Y. (2018). Light force induced line distortions in laser spectroscopy. To be published.

Chou, C. W. et al. (2010). Optical clocks and relativity. *Science*, 329, 1630–3.

Cingöz, A. et al. (2012). Direct frequency comb spectroscopy in the extreme ultraviolet. *Nature*, 482, 68–71.

Cohen-Tannoudji, C. (1977). Frontiers in Laser Spectroscopy, Les Houches, Session XXVII. North–Holland Publishing Company, Amsterdam.

Corwin, K. L. et al. (2009). Fundamental noise limitations to supercontinuum generation in microstructure fiber. *Physical Review Letters*, 90, 113904.

Crowe, E. R., Cohen, K. M., and Dumond, J. W. M. (1957). The Fundamental Constants of Physics (1st edn). Interscience Publishers, New York.

Dehmelt, H. G. (1982). Mono-ion oscillator as potential ultimate laser frequency standard. *IEEE Transactions on Instrumentation and Measurement*, 31, 83–7.

Diddams, S. A. et al. (2002). Femtosecond–laser-based optical clockwork with instability $\leq 6.3 \times 10^{16}$ in 1 s. *Optics Letters*, 27, 58–60.

Diddams, S. A. et al. (2000). Direct link between microwave and optical frequencies with a 300 thz femtosecond laser comb. *Physical Review Letters*, 84, 5102–5.

Diddams, S. A. et al. (2003). Design and control of femtosecond lasers for optical clocks and the synthesis of low-noise optical and microwave signals. *IEEE Journal of Quantum Electronics*, 9, 1072–80.

Diddams, S. A. et al. (2001). An optical clock based on a single trapped ^{199}Hg^{+}ion. *Science*, 293, 825–8.

Diels, J. C., and Rudolph, W. (2006). Ultrashort Laser Pulse Phenomena (2nd edn). Elsevier, New York.

Dudley, J. M., and Coen, S. (2002). Coherence properties of supercontinuum spectra generated in photonic crystal and tapered optical fibers. *Optics Letters*, 27, 1180–2.

Dudley, J. M., Genty, G., and Coen, S. (2006). Supercontinuum generation in photonic crystal fiber. *Reviews of Modern Physics*, 78, 1135–84.

Eckstein, J. N. (1978). PhD thesis. Stanford.

Eckstein, J. N., Ferguson, A. I., and Hänsch, T. W. (1978). High-resolution two-photon spectroscopy with picosecond light pulses. *Physical Review Letters*, 40, 847–50.

Eden, J. G. (2004). High-order harmonic generation and other intense optical field—matter interactions: review of recent experimental and theoretical advances. *Progress in Quantum Electronics*, 28, 197–246.

Ell, R. et al. (2001). Generation of 5-fs pulses and octave-spanning spectra directly from a ti:sapphire laser. *Optics Letters*, 26, 373–75.

Evenson, K. M. et al. (1973). Accurate frequencies of molecular transitions used in laser stabilization: the 3.39μm transition in ch4 and the 9.33μ and 10.18μm transitions in co$_2$. *Applied Physics Letters*, 22, 192–5.

Fang, S. et al. (2013). Optical frequency comb with an absolute linewidth of 0.6 hz–1.2 hz over an octave spectrum. *Applied Physics Letters*, 102, 231118.

Fano, U. (1935). Sullo spettro di assorbimento dei gas nobili presso il limite dello spettro d'arco. *Nuovo Cimento*, 12, 154–61.

Fendel, P. et al. (2007). Two-photon frequency comb spectroscopy of the 6s–8s transition in cesium. *Optics Letters*, 32, 701–99.

Fischer, M. et al. (2004). New limits on the drift of fundamental constants from laboratory measurements. *Physical Review Letters*, 92, 230802.

Fortier, T. M., Jones, D. J., and Cundiff, S. T. (2001). Phase stabilization of an octave–spanning ti:sapphire laser. *Optics Letters*, 28, 2198–200.

Fuji, T. et al. (2005). Monolithic carrier-envelope phase-stabilization scheme. *Optics Letters*, 30, 332–4.

Galtier, S. et al. (2014). Ultraviolet continuous-wave laser source at 205 nm for hydrogen spectroscopy. *Optics Communications*, 324, 34–7.

Gardner, F. M. (2005). Phaselock Techniques (3rd edn). John Wiley & Sons, New York.

Gerginov, V. et al. (2005). High-resolution spectroscopy with a femtosecond laser frequency comb. *Optics Letters*, 30, 1734–6.

Glauber, R. J. (1963). Coherent and incoherent states of radiation field. *Physical Review*, 131, 2766–88.

Gohle, C. et al. (2005). A frequency comb in the extreme ultraviolet. *Nature*, 436, 234–7.

Goulielmakis, E. et al. (2008). Single-cycle nonlinear optics. *Science*, 320, 1614–17.

Grynberg, G. et al. (1977). Doppler-free two-photon spectroscopy of neon. ii. line intensities. *Journal de Physique*, 38, 629–40.

Hädrich, S. et al. (2014). High photon flux table-top coherent extreme-ultraviolet source. *Nature Photonics*, 8, 779–83.

Hall, J. L. (2000). Optical frequency measurement: 40 years of technology revolutions. *IEEE Journal of Selected Topics in Quantum Electronics*, 6, 1136–44.

Hanneke, D., Fogwell, S., and Gabrielse, G. (2008). Spectroscopy of atomic hydrogen: how is the rydberg constant determined? *Physical Review Letters*, 100, 120801.

Hänsch, T. W. (1989). The Hydrogen Atom (1st edn). Springer, New York.

Hänsch, T. W. et al. (1999). Measuring the frequency of light with ultrashort pulses. Proceedings of the 1999 Joint Meeting of the European Frequency and Time Forum and the IEEE International Frequency Control Symposium, 602–25.

Hänsel, W. et al. (2017). Electro-optic modulator for rapid control of the carrier-envelope offset frequency. *CLEO*, SF1C.5.

Hasegawa, A., and Tappert, F. (1973). Transmission of stationary nonlinear optical pulses in dispersive dielectric fibers. i. anomalous dispersion. *Applied Physics Letters*, 23, 142–4.

Haus, H. A., and Ippen, E. P. (2001). Group velocity of solitons. *Optics Letters*, 26, 1654–6.

Haverkamp, N. et al. (2004). Frequency stabilization of mode-locked erbium fiber lasers using pump power control. *Applied Physics B*, 78, 321–4.

Heiße, F. et al. (2017). High-precision measurement of the proton's atomic mass. *Physical Review Letters*, 119, 033001.

Helbing, F. W. et al. (2002). Carrier–envelope-offset dynamics and stabilization of femtosecond pulses. *Applied Physics B*, 74, S35–S42.

Herrmann, M. et al. (2009). Feasibility of coherent xuv spectroscopy on the 1s-2s transition in singly ionized helium. *Physical Review A*, 79, 052505.

Hocker, L. O., Javan, A., and Rao, D. R. (1967). Absolute frequency measurement and spectroscopy of a gas laser transition in the far infrared. *Applied Physics Letters*, 10, 147–9.

Hocker, L. O. et al. (1968). Frequency mixing in the infrared and far-infrared using a metal-to-metal point contact diode. *Applied Physics Letters*, 12, 401–2.

Hollberg, L. et al. (2001). Optical frequency standards and measurements. *IEEE Journal of Quantum Electronics*, 37, 1502–13.

Holman, K. W. et al. (2003). Intensity-related dynamics of femtosecond frequency combs. *Optics Letters*, 28, 851–3.

Holzwarth, R. (2001). PhD thesis. Ludwig-Maximilians-Universität Munich, Germany.

Holzwarth, R. et al. (2001a). Absolute frequency measurement of iodine lines with a femtosecond optical synthesizer. *Applied Physics B*, 73, 269–71.

Holzwarth, R. et al. (2001b). A new type of frequency chain and its application to optical frequency metrology. *Laser Physics*, 11, 1100–9.

Holzwarth, R. et al. (2000). Optical frequency synthesizer for precision spectroscopy. *Physical Review Letters*, 85, 2264–7.

Horbatsch, M., and Hessels, E. A. (2010). Shifts from a distant neighboring resonance. *Physical Review A*, 82, 052519.

Huber, A. et al. (1998). Hydrogen-deuterium 1s-2s isotope shift and the structure of the deuteron. *Physical Review Letters*, 80, 468–71.

Ikegami, T. et al. (1996). Accuracy of an optical parametric oscillator as an optical frequency divider. *Optics Communications*, 127, 69–72.

Imai, K., Kourogi, M., and Ohtsu, M. (1998). 30-thz span optical frequency comb generation by self-phase modulation in an optical fiber. *IEEE Journal of Quantum Electronics*, 34, 54–60.

Javan, A., Ballika, E. A., and Bond, W. L. (1962). Frequency characteristics of a continuous-wave He-Ne optical maser. *Journal of the Optical Society of America*, 52, 96–8.

Javan, A., Herriott, D. R., and Bennett, W. R. (1961). Population inversion and continous optical maser oscillation in a gas discharge containing a He-Ne mixture. *Physical Review Letters*, 6, 106–10.

Jentschura, U. D., and Mohr, P. J. (2002). Nonresonant effects in one- and two-photon transitions. *Canadian Journal of Physics*, 80, 633–44.

Johnson, L. A. M., Gill, P., and Margolis, H. S. (2015). Evaluating the performance of the npl femtosecond frequency combs: agreement at the 10^{21} level. *Metrologia*, 52, 62–71.

Jones, D. J. et al. (2000). Carrier-envelope phase control of femtosecond mode-locked lasers and direct optical frequency synthesis. *Science*, 288, 635–9.

Jones, R. J. et al. (2005). Phase-coherent frequency combs in the vacuum ultraviolet via high-harmonic generation inside a femtosecond enhancement cavity. *Physical Review Letters*, 94, 193201.

Kandula, D. Z. et al. (2010). Extreme ultraviolet frequency comb metrology. *Physical Review Letters*, 105, 063001.

Kessler, T. et al. (2015). A sub-40-mhz-linewidth laser based on a silicon single-crystal optical cavity. *Nature Photonics*, 6, 687–92.

Kingston, R. H. (1978). Detection of Optical and Infrared Radiation (1st edn). Springer-Verlag, Berlin.

Kippenberg, T. J., Holzwarth, R., and Diddams, S. A. (2011). Microresonator-based optical frequency combs. *Science*, 332, 555–9.

Knight, J. C. et al. (1996). All-silica single-mode optical fiber with photonic crystal cladding. *Optics Letters*, 21, 1547–9.

Koke, C. et al. (2009). Performance comparison of interferometer topologies for carrier-envelope phase detection. *Applied Physics B*, 95, 81–4.

Kolachevsky, N. et al. (2006). Photoionization broadening of the 1s–2s transition in a beam of atomic hydrogen. *Physical Review A*, 74, 052504.

Kolbe, D., Scheid, M., and Walz, J. (2012). Triple resonant four-wave mixing boosts the yield of continuous coherent vacuum ultraviolet generation. *Physical Review Letters*, 109, 063901.

Kourogi, M., Nakagawa, K., and Ohtsu, M. (1993). Wide-span optical frequency comb generator for accurate optical frequency difference measurement. *IEEE Journal of Quantum Electronics*, 29, 2693–701.

Krausz, F. et al. (1992). Femtosecond solid-state lasers. *IEEE Journal of Quantum Electronics*, 28, 2097–122.

Lee, C. C. et al. (2012). Frequency comb stabilization with bandwidth beyond the limit of gain lifetime by an intracavity graphene electro-optic modulator. *Optics Letters*, 37, 3084–6.

Leuchs, G., and Sánchez-Soto, L. L. (2013). A sum rule for charged elementary particles. *European Physical Journal D*, 67, 57.

Lezius, M. et al. (2016). Space-borne frequency comb metrology. *Optica*, 3, 1381–7.

Ludlow, A. D. et al. (2015). Optical atomic clocks. *Reviews of Modern Physics*, 87, 637–701.

Ma, L. S. et al. (2004). Optical frequency synthesis and comparison with uncertainty at the 10^{-19} level. *Science*, 303, 1843–5.

Maiman, T. H. (1960). Stimulated optical radiation in ruby. *Nature*, 187, 493–4.

Matos, L. et al. (2004). Direct frequency comb generation from an octave-spanning, prismless ti:sapphire laser. *Optics Letters*, 29, 1683–5.

Matos, L. et al. (2006). Octave-spanning ti:sapphire laser with a repetition rate ¿ 1 ghz for optical frequency measurements and comparisons. *Optics Letters*, 31, 1011–13.

Matveev, A. et al. (2013). Precision measurement of the hydrogen 1s–2s frequency via a 920-km fiber link. *Physical Review Letters*, 110, 230801.

McFerran, J. J. et al. (2006). Elimination of pump-induced frequency jitter on fiber-laser frequency combs. *Optics Letters*, 31, 1997–2000.

Minardi, F. et al. (1999). Frequency shift in saturation spectroscopy induced by mechanical effects of light. *Physical Review A*, 60, 4164–7.

Mohr, P. J., Newell, D. B., and Taylor, B. N. (2016). Codata recommended values of the fundamental physical constants: 2014. *Reviews of Modern Physics*, 88, 035009.

Morgner, U. et al. (2001). Nonlinear optics with phase-controlled pulses in the sub-two-cycle regime. *Physical Review Letters*, 86, 5462–5.

Nebel, T. et al. (2012). The lamb-shift experiment in muonic helium. *Hyperfine Interactions*, 212, 195–201.

Nemitz, N. et al. (2016). Optical frequency synthesis and comparison with uncertainty at the 10−19 level. *Nature Photonics*, 10, 258–1845.

Niering, M. et al. (2000). Measurement of the hydrogen 1s–2s transition frequency by phase coherent comparison with a microwave cesium fountain clock. *Physical Review Letters*, 84, 5496–9.

Ozawa, A., and Kobayashi, Y. (2012). Chirped-pulse direct frequency-comb spectroscopy of two-photon transitions. *Physical Review A*, 86, 022514.

Ozawa, A., and Kobayashi, Y. (2013). Vuv frequency comb spectroscopy of atomic xenon. *Physical Review A*, 87, 022507.

Ozawa, A. et al. (2008). High harmonic frequency combs for high resolution spectroscopy. *Physical Review Letters*, 100, 253901.

Ozawa, A. et al. (2015). High average power coherent vuv generation at 10 mhz repetition frequency by intracavity high harmonic generation. *Optics Express*, 23, 15107–18.

Parthey, C. G. et al. (2011). Improved measurement of the hydrogen 1s–2s transition frequency. *Physical Review Letters*, 107, 203001.

Peters, E. et al. (2009). A deep-uv optical frequency comb at 205 nm. *Optics Express*, 17, 9183–90.

Peters, E. et al. (2013). Frequency-comb spectroscopy of the hydrogen 1S-3S and 1S-3D transitions. *Annalen der Physik*, 525, L29–L34.

Pohl, R. et al. (2010). The size of the proton. *Nature*, 466, 213–16.

Popmintchev, T. et al. (2012). Bright coherent ultrahigh harmonics in the kev x-ray regime from mid-infrared femtosecond lasers. *Science*, 336, 1287–91.

Prevedelli, M., Freegarde, T., and Hänsch, T. W. (1995). Phase locking of grating-tuned diode lasers. *Applied Physics B*, 60, S241–S248.

Pronin, O. et al. (2015). High-power multi-megahertz source of waveform-stabilized few-cycle light. *Nature Communications*, 6, 6988.

Protopopov, V. V. (2009). Laser Heterodyning (1st edn). Springer Series in Optical Sciences, New York.

Pupeza et al. (2014). Cavity-enhanced high-harmonic generation with spatially tailored driving fields. *Physical Review Letters*, 112, 103902.

Pupeza et al. (2013). Compact high-repetition-rate source of coherent 100 ev radiation. *Nature Photonics*, 7, 608–12.

Rainville, S., Thompson, J. K., and Pritchard, D. E. (2004). An ion balance for ultra-high-precision atomic mass measurements. *Science*, 303, 334–8.

Ranka, J. K., Windeler, R. S., and Stentz, A. J. (2000). Visible continuum generation in air–silica microstructure optical fibers with anomalous dispersion at 800 nm. *Optics Letters*, 25, 25–7.

Reichert, J. et al. (1999). Measuring the frequency of light with mode-locked lasers. *Optics Communications*, 172, 59–68.

Reichert, J. et al. (2000). Phase coherent vacuum-ultraviolet to radio frequency comparison with a mode-locked laser. *Physical Review Letters*, 84, 3232–5.

Reinhardt, S. et al. (2007). Test of relativistic time dilation with fast optical atomic clocks at different velocities. *Nature Physics*, 3, 861-4.

Reinhardt, S. et al. (2010). Two-photon direct frequency comb spectroscopy with chirped pulses. *Physical Review A*, 81, 033427.

Rosenband, T. et al. (2008). Frequency ratio of Al$^+$ and Hg$^+$ single-ion optical clocks; metrology at the 17th decimal place. *Science*, 319, 1808–12.

Rush, D. W., Ho, P. T., and Burdge, G. L. (1984). The coherence time of a mode—locked pulse train. *Optics Communications*, 52, 41–5.

Russell, P. St. J. (2003). Photonic crystal fibers. *Science*, 299, 358–62.

Saitoh, T., Kourogi, M., and Ohtsu, M. (1995). A waveguide type optical frequency comb generator. *IEEE Photonics Letters*, 7, 197–9.

Schibli, T. R. et al. (2008). Optical frequency comb with submillihertz linewidth and more than 10w average power. *Nature Photonics*, 2, 355–9.

Schippers, S. (2016). Analytical expression for the convolution of a fano line profile with a gaussian. arXiv:[physics.atom-ph], 1203.4281v2.

Schnatz, H. et al. (1996). First phase-coherent frequency measurement of visible radiation. *Physics Review Letters*, 79, 18–21.

Schumaker, B. L. (1987). Noise in homodyne detection. *Optics Letters*, 9, 189–91.

Siegman, A. E. (1986). Lasers (1st edn). University Science Books.

Siegman, E. A. (1966). The antenna properties of optical heterodyne receivers. *Applied Optics*, 5, 1588–94.

Spence, D. E., Kean, P. N., and Sibbett, W. (1991). 60-fsec pulse generation from a self-mode-locked ti:sapphire laser. *Optics Letters*, 16, 42–4.

Stenger, J. et al. (2002). Ultraprecise measurement of optical frequency ratios. *Physical Review Letters*, 88, 073601.

Stenger, J., and Telle, H. R. (2000). Intensity-induced mode shift in a femtosecond laser by a change in the nonlinear index of refraction. *Optics Letters*, 25, 1553–5.

Sutter, D. H. et al. (1999a). Semiconductor saturable–absorber mirror—assisted kerr–lens mode–locked ti:sapphire laser producing pulses in the two–cycle regime. *Optics Letters*, 24, 631–3.

Sutter, D. H. et al. (1999b). Sub-two-cycle pulses from a kerr-lens mode-locked ti:sapphire laser. *Optics Letters*, 24, 411–13.

Swann, W. C. et al. (2006). Fiber-laser frequency combs with subhertz relative linewidths. *Optics Letters*, 31, 3046–8.

Takano, T. et al. (2016). Geopotential measurements with synchronously linked optical lattice clocks. *Nature Photonics*, 10, 662–6.

Telle, H. R. (1996). Frequency Control of Semiconductor Lasers. Wiley, New York.

Telle, H. R., Mesched, D., and Hänsch, T. W. (1990). Realization of a new concept for visible frequency division: phase-locking of harmonic and sum frequencies. *Optics Letters*, 15, 532–4.

Telle, H. R., Lipphardt, B., and Stenger, J. (2002). Kerr-lens, mode-locked lasers as transfer oscillators for optical frequency measurements. *Applied Physics B*, 74, 1–6.

Udem, Th. (1997). PhD thesis. Ludwig–Maximilians–Universität München, Germany.

Udem, Th., Holzwarth, R., and Hänsch, T. W. (2004). Optical frequency metrology. *Nature*, 416, 233–7.

Udem, Th. et al. (1997). Phase-coherent measurement of the hydrogen 1s-2s transition frequency with an optical frequency interval divider chain. *Physics Review Letters*, 79, 2646–9.

Udem, Th. et al. (1998). Accuracy of optical frequency comb generators and optical frequency interval divider chains. *Optics Letters*, 23, 1387–9.

Udem, Th. et al. (1999a). Accurate measurement of large optical frequency differences with a mode-locked laser. *Optics Letters*, 24, 881–3.

Udem, Th. et al. (1999b). RF spectrum of a signal after frequency multiplication; measurement and comparison with a simple calculation. *Physical Review Letters*, 82, 3568–71.

Walker, D. R. et al. (2007). Frequency dependence of the fixed point in a fluctuating frequency comb. *Applied Physics B*, 89, 535–8.

Walls, F. L., and DeMarchi, A. (1975). RF spectrum of a signal after frequency multiplication; measurement and comparison with a simple calculation. *IEEE Transactions on Instrumentation and Measurement*, 24, 210–7.

Wang, H., and Weiner, A. M. (2003). Efficiency of short-pulse type-i second-harmonic generation with simultaneous spatial walk-off, temporal walk-off, and pump depletion. *IEEE Journal on Quantum Electronics*, 39, 1600–18.

Weitenberg, J. et al. (2011). Transverse mode tailoring in a quasi-imaging high-finesse femtosecond enhancement cavity. *Optics Express*, 19, 9551–61.

Weitz, M. et al. (1995). Precision measurement of the ls ground-state lamb shift in atomic hydrogen and deuterium by frequency comparison. *Physical Review A*, 52, 2664–81.

Wetzel, B. et al. (2012). Real-time full bandwidth measurement of spectral noise in supercontinuum generation. Scientific Reports, 2, 882.

Wilken, T. et al. (2012). A spectrograph for exoplanet observations calibrated at the centimetre-per-second level. Nature, 485, 611–14.

Wineland, D. J. et al. (1989). The Hydrogen Atom (1st edn). Springer.

Wong, N. C. (1992). Optical frequency counting from the uv to the near ir. Optics Letters, 17, 1155–7.

Xie, X. et al. (2016). Photonic microwave signals with zeptosecond-level absolute timing noise. Nature Photonics, 11, 44–8.

Xu, L. et al. (1996). Route to phase control of ultrashort light pulses. Optics Letters, 21, 2008–10.

Yost, D. C. et al. (2016). Spectroscopy of the hydrogen 1s–3s transition with chirped laser pulses. Physical Review A, 93, 042509.

Yost, D. C. et al. (2014). Quantum interference in two-photon frequency-comb spectroscopy. Physical Review A, 90, 012512.

Yost, D. C., Schibli, T. R., and Ye, J. (2008). Efficient output coupling of intracavity high-harmonic generation. Optics Letters, 33, 1099–1101.

Zimmermann, M. et al. (2004). Optical clockwork with an offset-free difference-frequency comb: Accuracy of sum- and difference-frequency generation. Optics Letters, 29, 310–12.

5

Collective effects in quantum systems

Thierry Giamarchi

Department of Quantum Matter Physics, University of Geneva
1211 Genève, Switzerland

Giamarchi, T. "Collective effects in quantum systems." In *Current Trends in Atomic Physics*.
Edited by Antoine Browaeys, Thierry Lahaye, Trey Porto, Charles S. Adams, Matthias
Weidemüller, and Leticia F. Cugliandolo. Oxford University Press (2019).
© Oxford University Press. DOI: 10.1093/oso/9780198837190.003.0005

Chapter Contents

5.1 Introduction

The goal of the course was to introduce the methods allowing the treatment of quantum systems of interacting particles and discuss the resulting physics.

Treating such systems is one of the most challenging problems of quantum physics, with important consequences in connection with novel properties of materials. Indeed interactions can change drastically the properties compared to non-interacting systems, and allow novel effects to appear. One such example, but not the only one, is the occurrence of superconductivity and superfluidity. However dealing with such problems is highly non-trivial since the Hilbert space is humongous (of the order of 2^N for spins $1/2$ on N sites). The entanglement of the wavefunction (totally symmetric for bosons, totally antisymmetric for fermions) is an additional important ingredient that makes the problem also difficult to tackle numerically.

Recently cold atoms have proven to be remarkable systems to realize quantum simulators, i.e. experimental systems realizing near perfect realizations of models of such interacting systems. Condensed matter and cold atomic physics have thus moved hand in hand to make progress on these problems.

The course thus introduced the basic methods of second quantization, and discussed various models (tight binding, Hubbard model, Heisenberg model, etc.). These are the simplest realizations of such interacting systems with direct realization in cold atomic systems.

5.2 Basic methods and concepts

Unfortunately time constraints prevent the author from writing a fully redacted version of the course in these notes. In addition the material is already well represented in the literature. I thus give in these notes just some links or references where the reader can find the material presented in the course.

The note of the course done in Les Houches itself, together with the slides complementing the "blackboard" notes can be found on https://www.dropbox.com/sh/oaq4uy2qogvz6t7/AAC4OfNBgBlLdtPOR_SUuwL1a?dl=0

In a more redacted form the basic concepts of solid-state physics and interacting systems can be found in the litterature. My personal favorite for an introduction to condensed matter is (Ziman, 1972) even if there are clearly more modern textbooks. Many-body concepts and methods are well explained in e.g. (Mahan, 1981). In addition most of the introductory material can be found in unpublished notes of a master course on solid-state systems https://dqmp.unige.ch/giamarchi/local/people/thierry.giamarchi/pdf/many-body.pdf

These notes cover most of the introductory material with a level suitable for beginners in the field but going quite into details on the physics of interacting systems (Hubbard model, Mott transition, quantum magnetism, etc.). An excellent book on quantum magnetism is also (Auerbach, 1998).

5.3   Correlated systems and cold atoms

The discussion on how cold atomic systems are connected to this physics of interacting quantum systems was already discussed in several previous courses. I give here personal references in which the reader can find more extensive discussions and references. The course (Giamarchi, 2011) and (Georges and Giamarchi, 2012) discuss the physics of such interacting systems in connection with cold atoms. In addition to covering the basic material they also touch on the physics of one-dimensional systems that will be discussed more in detail in the next section.

5.4   One-dimensional quantum systems

One important class of systems that could be realized with cold atomic systems are systems of low dimensionality. In particular cold atoms allowed very controlled realizations of one- and quasi-one-dimensional quantum systems. This is particularly interesting since the physics of such systems can be a priori very different from their high dimensional counterparts. Courses that address the physics of such systems in direct connection with cold atoms (again personal references in which more references and details can be found) are (Giamarchi, 2006) and (Giamarchi, 2016). Quite complete references also exist for one-dimensional systems such as the book (Giamarchi, 2004) or the most recent review for 1D bosons (Cazalilla et al., 2011)

5.5   Since the course

It is the hope of the author (despite the very sketchy form of those notes) that these few references will help the reader find their way in the maze of the literature on strong correlations and the contact with cold atoms. Since the course spectacular progress has been accomplished on the fermionic side, especially with the use of "fermion microscopes" and much more is certainly to come. Good theories and theorists are certainly needed in this active field.

References

Auerbach, A. (1998). *Interacting Electrons and Quantum Magnetism* (2nd edn). Springer, Berlin.

Cazalilla, M. A. et al. (2011). One dimensional bosons: from condensed matter systems to ultracold gases. *Reviews of Modern Physics*, **83**, 1405.

Georges, A., and Giamarchi, T. (2012). Strongly correlated bosons and fermions in optical lattices. In *Many-Body Physics with Ultracold Gases* (ed. C. Salomon, G. Shlyapnikov and L. F. Cugliandolo), Volume XCXIV, Les Houches 2010. Oxford. arXiv:1308.2684.

Giamarchi, T. (2004). *Quantum Physics in One Dimension*. Volume 121, International series of monographs on physics. Oxford University Press, Oxford.

Giamarchi, T. (2006). Strong correlations in low dimensional systems. In *Lectures on the Physics of Highly Correlated Electron Systems X*, p. 94. AIP conference proceedings. arXiv:cond-mat/0605472.

Giamarchi, T. (2011). Interactions in quantum fluids. In *Ultracold Gases and Quantum Information* (ed. C. Miniatura et al.), Volume XCI, Les Houches 2009, p. 395. Oxford. arXiv:1007.1030.

Giamarchi, T. (2016). Clean and dirty one dimensional systems. In *Quantum Matter at Ultralow Temperatures* (ed. S. I. di Fisica), Amsterdam. IOS Press.

Mahan, G. D. (1981). *Many Particle Physics*. Plenum, New York.

Ziman, J. M. (1972). *Principles of the Theory of Solids*. Cambridge University Press, Cambridge.

6
Macroscopic scale atom interferometers: introduction, techniques, and applications

TIM KOVACHY, ALEX SUGARBAKER, REMY NOTERMANS, PETER ASENBAUM, CHRIS OVERSTREET, JASON M. HOGAN, AND MARK A. KASEVICH

Department of Physics, Stanford University
Stanford, CA 94305 USA

Kovachy, T., et al. "Macroscopic scale atom interferometers: introduction, techniques, and applications." In *Current Trends in Atomic Physics*. Edited by Antoine Browaeys, Thierry Lahaye, Trey Porto, Charles S. Adams, Matthias Weidemüller, and Leticia F. Cugliandolo. Oxford University Press (2019). © Oxford University Press.
DOI: 10.1093/oso/9780198837190.003.0006

Chapter Contents

6.1 Overview of these notes

These lecture notes provide an introduction to light-pulse atom interferometry. Atom interferometers are one of the most sensitive ways to measure inertial forces, making them a valuable tool for a broad set of practical applications and fundamental physics tests (Cronin et al., 2009). An overview of atom optics techniques and interferometer phase shift calculations is given. Recent advances have enabled atom interferometers that cover macroscopic scales in space (tens of centimeters) and in time (multiple seconds), dramatically improving interferometer sensitivity in a wide range of applications. We review these advances and recent experiments performed with macroscopic scale atom interferometers in the 10 m tall atomic fountain at Stanford. The notes are organized as follows. Sec. 6.2 provides an introduction to the concepts of atom optics and atom interferometry. Sec. 6.3 describes Large Momentum Transfer atom optics and their use in macroscopic scale interferometers. Associated experimental challenges and techniques to overcome them are discussed. Sec. 6.4 gives an overview of the 10 m atomic fountain at Stanford, in which the macroscopic scale atom interferometers have been implemented. Sec. 6.5 discusses some of the applications of atom interferometry, with an emphasis on our recent work in the 10 m fountain. Sec. 6.6 reviews the idea of matter wave lensing, which is used to generate the ultra-low-velocity-spread atom sources required for long duration atom interferometry. Sec. 6.7 describes the use of an optical lattice to efficiently launch atoms at 13 m/s into the 10 m fountain. A notable result is the observed suppression of the spontaneous emission scattering rate for a blue-detuned vs. a red-detuned lattice. Sec. 6.8 provides an analytical treatment of the phase shifts induced by matter wave beam splitters and waveguides based on optical lattice manipulations. Due to their exceptionally high transfer efficiency, lattice-based atom optics are a promising candidate for next generation macroscopic scale interferometers.

6.2 Introduction to atom interferometry

6.2.1 Beam splitters and mirrors for atoms: a simplified treatment using the two level atom

The quantum mechanical wavelike nature of atoms can be exploited to perform experiments typically associated with light. These quantum mechanical matter waves can be manipulated with the atomic equivalents of lenses, beam splitters, and mirrors. Atom interferometry is conceptually analogous to optical interferometry. In both cases, an incident wave is split into two paths by a beam splitter. The two paths are later redirected back toward each other with mirrors and overlapped on a final beam splitter to produce an interference pattern. It has long been known how to make beam splitters and mirrors for optical beams. Work to develop and improve beam splitters and mirrors for atoms has been underway for the past several decades. The two most widely used techniques for atom interferometry respectively rely on material gratings and optical pulses to manipulate the atomic wavefunctions (Cronin et al., 2009). Our work has focused on the optical pulse technique, often referred to as light-pulse atom interferometry, so we will focus the discussion on this method.

A discussion of atom optics for light-pulse atom interferometry naturally begins with consideration of the standard quantum mechanical treatment of the two level atom. Although the two level atom is a simplified picture, it captures much of the essential physics. We consider an atom with two levels $|1\rangle$ and $|2\rangle$ that have frequency difference $\omega_0 \equiv \omega_2 - \omega_1$. We will analyze the effect of the electric dipole interaction (Metcalf and van der Straten, 1999) of the atom with an electric field $\mathbf{E} = \mathbf{E_0}\cos(\phi - \omega_0 t)$ at the location of the atom (ϕ is the local phase of the field). Under the electric dipole approximation, we neglect effects arising from variation of the field over the spatial extent of the atom's electron orbitals, as this spatial extent is typically much smaller than an optical wavelength. For simplicity, we have assumed that the field oscillates on resonance with the atomic transition. The electric dipole interaction is characterized by an interaction Hamiltonian $H_{\text{int}} = -\hat{\boldsymbol{\mu}} \cdot \mathbf{E}$, where $\hat{\boldsymbol{\mu}}$ is the electric dipole operator for the atom. We write the state of the atom in terms of amplitudes of the two states:

$$|\psi(t)\rangle = c_1(t)\,|1\rangle\,e^{-i\omega_1 t} + c_2(t)\,|2\rangle\,e^{-i\omega_2 t}. \tag{6.1}$$

Defining a Rabi frequency $\Omega \equiv -\frac{1}{\hbar}\langle 2|\,\hat{\boldsymbol{\mu}}\cdot\mathbf{E}\,|1\rangle$, the dynamical equations for the two amplitudes are

$$\frac{dc_1(t)}{dt} = -ic_2(t)\Omega\cos(\phi - \omega_0 t)e^{-i\omega_0 t} = -ic_2(t)\frac{\Omega}{2}\left(e^{-i\phi} + e^{-i(2\omega_0 t - \phi)}\right)$$

$$\frac{dc_2(t)}{dt} = -ic_1(t)\Omega\cos(\phi - \omega_0 t)e^{i\omega_0 t} = -ic_1(t)\frac{\Omega}{2}\left(e^{i\phi} + e^{i(2\omega_0 t - \phi)}\right). \tag{6.2}$$

We can simplify these equations by making the rotating wave approximation, dropping the rapidly oscillating terms at frequency $2\omega_0$ (Metcalf and van der Straten, 1999). This approximation assumes that $\omega_0 \gg \Omega$, which is typically the case for atomic systems. The dynamical equations then reduce to

$$\frac{dc_1(t)}{dt} = -i\frac{\Omega}{2}e^{-i\phi}c_2(t)$$

$$\frac{dc_2(t)}{dt} = -i\frac{\Omega}{2}e^{i\phi}c_1(t), \tag{6.3}$$

which admit the simple solution

$$c_1(t) = c_1(0)\cos\left(\frac{\Omega t}{2}\right) - ie^{-i\phi}c_2(0)\sin\left(\frac{\Omega t}{2}\right)$$

$$c_2(t) = -ie^{i\phi}c_1(0)\sin\left(\frac{\Omega t}{2}\right) + c_2(0)\cos\left(\frac{\Omega t}{2}\right). \tag{6.4}$$

These equations correspond to sinusoidal oscillation of population between the two states (see Fig. 6.1b). This can be quickly seen by considering the case where all the population begins in $|1\rangle$, so that $c_1(0) = 1$ and $c_2(0) = 0$. The populations $p_1(t) = |c_1(t)|^2$ and $p_2(t) = |c_2(t)|^2$ then evolve as

(a)

(b)

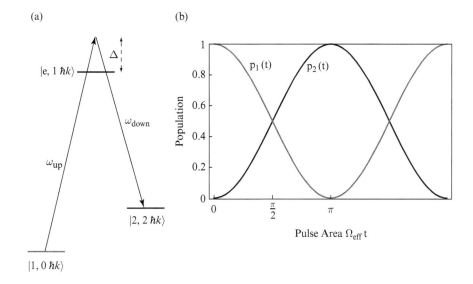

Fig. 6.1 Diagram of a Raman transition. (a) The Raman transition couples two hyperfine ground states $|1\rangle$ and $|2\rangle$ through an intermediate optically excited state $|e\rangle$. Atomic population initially in $|1\rangle$ with upward momentum $0\hbar k$ undergoes stimulated absorption of a photon from an upward propagating laser beam with frequency ω_{up} followed by stimulated emission of a photon into a downward propagating laser beam with frequency ω_{down}, yielding a total upward momentum kick of $2\hbar k$ in addition to the internal state change from $|1\rangle$ to $|2\rangle$. (b) The atomic population undergoes a sinusoidal oscillation between the states $|1, 0\hbar k\rangle$ and $|2, 2\hbar k\rangle$ as the pulse area $\Omega_{\text{eff}}t$ is increased. For $\Omega_{\text{eff}}t = \pi/2$, the Raman transition acts as a 50/50 beam splitter for the two states. For $\Omega_{\text{eff}}t = \pi$, the Raman transition acts as a mirror, interchanging population between the two states.

$$p_1(t) = \cos^2\left(\frac{\Omega t}{2}\right)$$
$$p_2(t) = \sin^2\left(\frac{\Omega t}{2}\right). \tag{6.5}$$

There are two especially interesting times t with respect to Eq. 6.5. If the pulse area Ωt is equal to $\pi/2$ (we call this a $\pi/2$-pulse), then the population is evenly divided between $|1\rangle$ and $|2\rangle$. This corresponds to the operation

$$|1\rangle \longrightarrow \frac{1}{\sqrt{2}}\left(|1\rangle - ie^{i\phi}|2\rangle\right)$$
$$|2\rangle \longrightarrow \frac{1}{\sqrt{2}}\left(-ie^{-i\phi}|1\rangle + |2\rangle\right). \tag{6.6}$$

Until this point, we have not paid attention to the external degrees of freedom of the atom. It is now important to note that the excitation of atomic population from $|1\rangle$

to $|2\rangle$ via an electric dipole interaction corresponds to the stimulated absorption of a photon from the driving field (Metcalf and van der Straten, 1999). By conservation of momentum, the photon's momentum is necessarily transferred to the atom. For example, if the field arises from a laser beam propagating in a particular direction, then the atom will receive a single-photon momentum kick in that direction. Where k is the laser wave number, the momentum kick has magnitude $\hbar k$. For the Rb atoms that we work with, and laser light near the 780 nm D$_2$ transition, the corresponding recoil velocity ($\hbar k/m$ for atomic mass m) is about 6 mm/s (for Rb atoms, 6 mm/s corresponds to an effective temperature of 400 nK). State $|2\rangle$ should be associated with this additional recoil velocity. The $\pi/2$-pulse therefore divides the atomic state into two parts with different velocities, making it a beam splitter. The de-excitation of population from $|2\rangle$ to $|1\rangle$ corresponds to stimulated emission of a photon into the field. The atom then receives a momentum kick $\hbar k$ in the direction opposite to the laser beam propagation, losing the momentum that it had gained upon initial excitation.

The second interesting time t occurs when $\Omega t = \pi$ (a π-pulse). A π-pulse yields the operation

$$|1\rangle \longrightarrow -ie^{i\phi}|2\rangle$$
$$|2\rangle \longrightarrow -ie^{-i\phi}|1\rangle, \tag{6.7}$$

fully swapping the population between the two states. We describe this as a mirror operation, as population is 'reflected' from one state to the other, along with the corresponding momentum kick.

Note that the transfer of population from one state to the other is associated with a phase factor: $e^{i\phi}$ for $|1\rangle \longrightarrow |2\rangle$ and $e^{-i\phi}$ for $|2\rangle \longrightarrow |1\rangle$. We can interpret the phase factor as follows. When the atom absorbs a photon from the field ($|1\rangle \longrightarrow |2\rangle$), the local phase ϕ of the field is imprinted onto the atom's state. When the atom emits a photon into the field, the atom receives the conjugate phase factor. These imprinted phases play a crucial role in determining the response of the interferometer phase shift to inertial effects.

6.2.2 Beam splitters and mirrors for atoms: the three level atom

The example of the two level atom discussed in Sec. 6.2.1 illustrates many of the key features of light-pulse beam splitters and mirrors. However, for the alkali atoms that have so far been most widely used in atom interferometry, there are some additional details that are important to consider.[1] For the Rb atoms that we use, population in optically excited states rapidly decays via spontaneous emission (the decay rate is

---

[1] There has been recent interest in using atomic species that have narrow optical clock transitions, such as Sr, for atom interferometry with reduced sensitivity to laser phase noise, with the clock transition being used for atom optics. In this case, the two level atom would be the relevant description, with the two states being the lower and excited optical clock states (Graham et al., 2013).

$\Gamma = 2\pi \times 6$ MHz for the D_2 line transitions that we use). An atom that undergoes spontaneous emission does not contribute to the interference signal—it receives a photon recoil momentum kick in a random direction that typically removes it from an ultracold atom ensemble that is well below the recoil temperature (Metcalf and van der Straten, 1999). Therefore, we want to avoid using an optically excited state as one of the states in a two level system.

Instead, we turn to the quantum mechanical problem of the three level atom. We consider an optically excited state $|e\rangle$ and two long-lived, lower energy states $|1\rangle$ and $|2\rangle$ with a comparatively small frequency difference. For Rb, states $|1\rangle$ and $|2\rangle$ could correspond to the two ground state hyperfine levels, which have a microwave frequency splitting. For appropriate parameters, it is possible to create an effective two level system from the three level atom, with $|1\rangle$ and $|2\rangle$ coupled by two-photon optical transitions and $|e\rangle$ acting as an intermediary state that contains only a small amount of atomic population (Berman, 1997).

To be concrete, let us say that an upward propagating laser beam with frequency ω_{up} and local phase ϕ_{up} couples $|1\rangle$ and $|e\rangle$ with Rabi frequency Ω_1 and detuning from resonance $\Delta \equiv \omega_{\text{up}} - (\omega_e - \omega_1)$, and a downward propagating laser beam with frequency ω_{down} and local phase ϕ_{down} couples $|2\rangle$ and $|e\rangle$ with Rabi frequency Ω_2 and a similar detuning approximately equal to Δ (i.e., $\omega_{\text{down}} - (\omega_e - \omega_2) \approx \Delta$). In the limit where $\Delta \gg \Omega_1, \Omega_2, \Gamma$, and if the frequency difference $\omega_{\text{up}} - \omega_{\text{down}}$ is on resonance with the frequency difference between states $|1\rangle$ and $|2\rangle$, then within good approximation the dynamics proceed as for the resonantly driven two level atom described in Sec. 6.2.1 (Berman, 1997). The system is described by an analogue to Eq. 6.4, with an effective Rabi frequency $\Omega_{\text{eff}} \equiv \frac{\Omega_1 \Omega_2}{2\Delta}$ and a phase factor equal to the difference of the local phases ϕ_{up} and ϕ_{down} of the upward and downward propagating beams:[2]

$$c_1(t) = c_1(0) \cos\left(\frac{\Omega_{\text{eff}} t}{2}\right) - i e^{-i(\phi_{\text{up}} - \phi_{\text{down}})} c_2(0) \sin\left(\frac{\Omega_{\text{eff}} t}{2}\right)$$

$$c_2(t) = -i e^{i(\phi_{\text{up}} - \phi_{\text{down}})} c_1(0) \sin\left(\frac{\Omega_{\text{eff}} t}{2}\right) + c_2(0) \cos\left(\frac{\Omega_{\text{eff}} t}{2}\right). \tag{6.8}$$

As for the two level atom, beam splitters and mirrors can be realized by $\pi/2$-pulses and π-pulses, respectively, with the pulse area defined in terms of the effective Rabi frequency Ω_{eff}. The transfer of population from one state to the other is now associated with the modified phase factors $e^{i(\phi_{\text{up}} - \phi_{\text{down}})}$ for $|1\rangle \longrightarrow |2\rangle$ and $e^{-i(\phi_{\text{up}} - \phi_{\text{down}})}$ for $|2\rangle \longrightarrow |1\rangle$. These phase factors can be understood as arising from two-photon transitions that couple $|1\rangle$ and $|2\rangle$. For transfer from $|1\rangle \longrightarrow |2\rangle$, population is transferred from $|1\rangle \longrightarrow |e\rangle$ by stimulated absorption of a photon from the upward propagating beam, imprinting phase ϕ_{up}. Stimulated emission of a photon into the downward propagating beam transfers the population from $|e\rangle \longrightarrow |2\rangle$, imprinting phase $-\phi_{\text{down}}$, and yielding a total imprinted phase of $(\phi_{\text{up}} - \phi_{\text{down}})$. The inverse of this process yields a total imprinted phase of $-(\phi_{\text{up}} - \phi_{\text{down}})$ for transfer from $|2\rangle \longrightarrow |1\rangle$.

---

[2] We ignore light shifts for the moment. We will discuss them in detail in Sec. 6.3.2.

In an actual multilevel atom, there are typically several closely spaced excited states $|e\rangle$ that contribute to the coupling of states $|1\rangle$ and $|2\rangle$. The observed effective Rabi frequency then consists of a sum of the individual contributions from this collection of optically excited states.

The fact that we are now considering two-photon transitions as opposed to the single-photon transitions considered in Sec. 6.2.1 affects the momentum kicks associated with the two states. For transfer from $|1\rangle \longrightarrow |2\rangle$, the absorption of the photon from the upward propagating beam gives the atom a momentum kick $\hbar k$ in the upward direction. The stimulated emission of a photon into the downward propagating beam gives the atom another upward kick of $\hbar k$. So state $|2\rangle$ should be associated with an upward momentum kick of $2\hbar k$.

If, as suggested above, states $|1\rangle$ and $|2\rangle$ correspond to the two ground state hyperfine levels of our Rb atoms, these two-photon transitions are known as Raman transitions. The use of Raman transitions for beam splitters and mirrors in an atom interferometer (with Na atoms) was first demonstrated in 1991 (Kasevich and Chu, 1991), and Raman transitions remain a widely used and versatile tool for atom optics to this day. A schematic of a Raman transition is illustrated in Fig. 6.1.

Another possibility is that $|1\rangle$ and $|2\rangle$ do not correspond to different internal atomic states, but only to different external momentum states differing by momentum $2\hbar k$ (see Fig. 6.3). This situation is know as a Bragg transition (Kozuma et al., 1999). For a Bragg transition, it makes sense to relabel states $|1\rangle$ and $|2\rangle$ as $|p_0\rangle$ and $|p_0 + 2\hbar k\rangle$, where p_0 is the initial vertical momentum of the atom. Conservation of energy provides a resonance condition for $\omega_{\text{up}} - \omega_{\text{down}}$ in the case of a Bragg transition. When atomic population is transferred from $|p_0\rangle$ to $|p_0 + 2\hbar k\rangle$, there is an external kinetic energy gain of

$$\Delta E = \frac{(p_0 + 2\hbar k)^2}{2m} - \frac{p_0^2}{2m} = 2p_0\frac{\hbar k}{m} + 2\frac{\hbar^2 k^2}{m} = 2p_0 v_{\text{r}} + 4\hbar\omega_{\text{r}}, \tag{6.9}$$

where $v_{\text{r}} \equiv \frac{\hbar k}{m}$ and $\omega_{\text{r}} \equiv \frac{\hbar k^2}{2m}$ are respectively the recoil velocity and the recoil frequency associated with a single-photon momentum kick. Meanwhile, the atom gains energy $\hbar(\omega_{\text{up}} - \omega_{\text{down}})$ from the laser fields. Since the internal state of the atom does not change, all of the energy gained from the laser fields necessarily goes into the kinetic energy gain ΔE. Therefore,

$$\omega_{\text{up}} - \omega_{\text{down}} = \frac{\Delta E}{\hbar} = 2\frac{p_0}{\hbar}v_{\text{r}} + 4\omega_{\text{r}}. \tag{6.10}$$

For both the Raman and Bragg cases, it is important to consider the spontaneous emission loss rate arising from residual atomic population in $|e\rangle$. For a laser beam spectral component with single-photon Rabi frequency Ω and detuning Δ, the induced excited state population for $\Delta \gg \Omega, \Gamma$ is (Metcalf and van der Straten, 1999)

$$P_{\text{e}} = \frac{\Omega^2}{4\Delta^2}. \tag{6.11}$$

The corresponding spontaneous emission loss rate is

$$R = \Gamma P_e = \Gamma \frac{\Omega^2}{4\Delta^2}. \tag{6.12}$$

If this spectral component pairs with a second spectral component of comparable Rabi frequency to drive a two-photon transition, as discussed above the effective Rabi frequency is $\Omega_{\text{eff}} = \frac{\Omega^2}{2\Delta}$. To drive a π-pulse, the light must be pulsed on for a time $t_\pi = \frac{\pi}{\Omega_{\text{eff}}}$. This leads to a fraction of the atoms $L_\pi \sim R t_\pi$ being lost to spontaneous emission during a π-pulse. The loss scales as

$$L_\pi \sim \Gamma \frac{\Omega^2}{4\Delta^2} \frac{\pi}{\frac{\Omega^2}{2\Delta}} \sim \frac{\Gamma}{\Delta}. \tag{6.13}$$

This suggests that in order to minimize spontaneous emission losses, the detuning Δ should be made as large as possible. There are additional constraints that make it undesirable to have Δ be arbitrarily large. The single-photon Rabi frequency Ω is proportional to the electric field amplitude, so Ω_{eff} is proportional to the laser intensity and inversely proportional to Δ. If arbitrarily high laser power were available, a fixed t_π could be maintained as Δ is increased by turning up the laser power.[3] We want t_π to be short enough so that the pulse is not excessively Doppler selective. From fundamental properties of the Fourier transform, the width of the range of detunings over which a π-pulse can maintain high transfer efficiency is inversely proportional to t_π. If this width becomes too narrow, the majority of the atoms will be not be efficiently transferred due to the finite velocity spread of the cloud, which leads to Doppler shifts away from resonance. For our typical atom sources, we do not want t_π to be longer than ~ 100 μs. In practice, the laser power is limited, constraining the maximum detuning that can be used.

This discussion indicates that higher laser power is generally helpful for atom optics, as it allows for the reduction of spontaneous emission losses while maintaining the same effective Rabi frequency. We have therefore expended significant effort to build a high power, frequency doubled laser system. Chiow et al. 2012 discusses the demonstration of a 43 W laser source at 780 nm (the wavelength of the Rb D_2 line) that is suitable for atom optics.

6.2.3 The phase shift of an atom interferometer: simple example using a perturbative treatment

In order to assess possible applications for atom interferometry, it is necessary to know how the interferometer phase shift depends on various quantities of interest, such as inertial forces. As a start, we will perform a simple phase shift calculation

---

[3] Using a smaller laser beam waist can also increase the intensity, but it is typically desirable not to make the waist too small to avoid laser intensity inhomogeneities across the atom cloud.

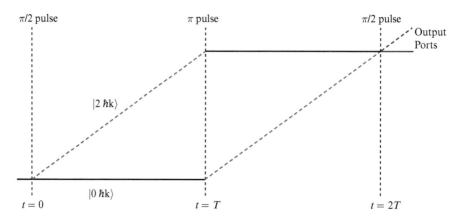

Fig. 6.2 Spacetime diagram of a three pulse Mach-Zehnder interferometer with time between pulses (pulse spacing) T.

for a spatially and temporally constant gravitational acceleration g. In Sec. 6.2.4, we will give an overview of the more formal method for calculating phase shifts in more general situations.

We consider a Mach-Zehnder interferometer with time between pulses (pulse spacing) T (see Fig. 6.2). At time $t = 0$, a beam splitter ($\pi/2$-pulse) splits the initial matter waves into a superposition of two paths with vertical momentum differing by $2\hbar k$. These two paths drift apart for time T. At time $t = T$, a mirror (π-pulse) interchanges the momenta of the two interferometer arms, so that they start to move back toward each other. At time $t = 2T$, the two arms spatially overlap and are interfered by a final beam splitter. The normalized populations P_1 and P_2 in the two interferometer output ports depend on the interferometer phase shift $\Delta\phi$ as

$$P_1 = \frac{1}{2} + \frac{1}{2}C\cos(\Delta\phi)$$
$$P_2 = \frac{1}{2} - \frac{1}{2}C\cos(\Delta\phi), \tag{6.14}$$

where C is the interferometer contrast, which ranges between zero and one depending on the quality of the interference. Fig. 6.2 shows a schematic of such a Mach-Zehnder interferometer.

The atoms move in a potential $U(z) = mgz$, where m is the atomic mass. To perform a 'back-of-the-envelope' phase shift calculation, we will neglect the effect of this potential on the atomic trajectories and assume that the interferometer phase shift comes entirely from integrating the Lagrangian over the two interferometer arms. This method is motivated by the Feynman path integral approach to quantum mechanics, and it turns out that our 'back-of-the-envelope' approximation is formally justified as a perturbative method that gives exactly the correct answer for the simple situation considered here (Storey and Cohen-Tannoudji, 1994). The Lagrangian is given by $\mathcal{L} = \frac{1}{2}m\dot{z}^2 - mgz$. Due to the symmetry of the interferometer, the integral of the kinetic energy term is the same over both paths and cancels from the interferometer phase shift. The phase shift is then determined by the difference between the gravitational potential term integrated over the upper versus the lower path:

$$\Delta\phi = \phi_{\text{upper}} - \phi_{\text{lower}} = \left(-\frac{1}{\hbar}\int_{\text{upper path}} mgz\right) + \left(\frac{1}{\hbar}\int_{\text{lower path}} mgz\right) \tag{6.15}$$

$$= -\frac{m}{\hbar}g\int_0^{2T} \Delta z(t)dt, \tag{6.16}$$

where $\Delta z(t) \equiv z_{\text{upper}}(t) - z_{\text{lower}}(t)$ is the spatial separation of the upper and lower paths.

The phase shift is proportional to the enclosed spacetime area $\int_0^{2T} \Delta z(t)dt$ of the interferometer. For the interferometer geometry we are considering, the enclosed spacetime area is equal to $\Delta z_{\text{max}} T$, where Δz_{max} is the maximum path separation reached during the interferometer. Therefore,

$$\Delta\phi = -\frac{m}{\hbar}g\Delta z_{\text{max}}T. \tag{6.17}$$

Since $\Delta z_{\text{max}} = \frac{2\hbar k}{m}T$, we can alternatively write

$$\Delta\phi = -2kgT^2. \tag{6.18}$$

This phase shift calculation demonstrates that to increase the acceleration sensitivity of an atom interferometer, the enclosed spacetime area should be increased. This can be achieved by increasing the pulse spacing T or by implementing beam splitters that transfer many photon recoils of momentum instead of just two. We will discuss large area interferometers in detail in Sec. 6.3.

6.2.4 The phase shift of an atom interferometer: formal treatment

In Sec. 6.2.3, the phase shift from a gravitational acceleration g was calculated using a simple perturbative method that ignored the effect of g on the atomic trajectories. To calculate phase shifts in a general and formal framework, it is necessary to explicitly account for modifications to the atomic trajectories from external forces. The discussion here will provide a summary of a formal method for calculating phase shifts, following the treatment in Storey and Cohen-Tannoudji 1994 and in Hogan et al. 2009.

Generally speaking, contributions to the interferometer phase shift can arise from several sources. The first class of contributions, familiar from Sec. 6.2.3, comes from taking the difference of the Lagrangian integrated over the two interferometer arms.

This is called the propagation phase, which we denote $\Delta\phi_{\text{prop}}$:

$$\Delta\phi_{\text{prop}} = \left(\frac{1}{\hbar}\int_{\text{upper path}}\mathcal{L}\right) - \left(\frac{1}{\hbar}\int_{\text{lower path}}\mathcal{L}\right). \tag{6.19}$$

Second, recall from Sec. 6.2.2 that the local phase of the laser field at the location of the atom is imprinted on the atom's wavefunction when momentum is transferred during beam splitter or mirror pulses. The difference of the net imprinted phase for the two interferometer arms is called the laser phase, $\Delta\phi_{\text{laser}}$. For each interferometer path, the local phase is evaluated at the classical center of mass position of that path at the time when a given atom optics pulse occurs. Where $\phi_{\text{imprint}}\left(\mathbf{r}_i^{\text{upper}}\right)$ is the imprinted phase on the upper path from atom optics pulse i and $\phi_{\text{imprint}}\left(\mathbf{r}_i^{\text{lower}}\right)$ is the imprinted phase on the lower path from pulse i ($\mathbf{r}_i{}^{\text{upper}}$ and $\mathbf{r}_i{}^{\text{lower}}$ are the positions along the upper and lower paths at which the local phase is evaluated for pulse i), the laser phase is

$$\Delta\phi_{\text{laser}} = \left(\sum_i \phi_{\text{imprint}}\left(\mathbf{r}_i^{\text{upper}}\right)\right) - \left(\sum_i \phi_{\text{imprint}}\left(\mathbf{r}_i^{\text{lower}}\right)\right). \tag{6.20}$$

Finally, if the two paths do not perfectly overlap at the end of the interferometer, there is an additional contribution called the separation phase, $\Delta\phi_{\text{sep}}$. Where $\bar{\mathbf{p}}_{\text{final}}$ is the mean canonical momentum of the two interferometer paths in a given output port[4] after the final beam splitter and $\mathbf{r}_{\text{final}}^{\text{upper}} - \mathbf{r}_{\text{final}}^{\text{lower}}$ is the separation between the upper and lower paths when the final beam splitter is applied, the separation phase is

$$\Delta\phi_{\text{sep}} = -\frac{1}{\hbar}\bar{\mathbf{p}}_{\text{final}} \cdot \left(\mathbf{r}_{\text{final}}^{\text{upper}} - \mathbf{r}_{\text{final}}^{\text{lower}}\right). \tag{6.21}$$

The total phase shift is the sum of these three contributions:

$$\Delta\phi = \Delta\phi_{\text{prop}} + \Delta\phi_{\text{laser}} + \Delta\phi_{\text{sep}}. \tag{6.22}$$

We can now calculate the phase shift for the example interferometer considered in Sec. 6.2.3 using this formal method. When the effect of g on the atomic trajectories is taken into account, the kinetic energy term in the propagation phase is no longer zero. In fact, it cancels with the potential energy term to yield $\Delta\phi_{\text{prop}} = 0$. The two interferometer arms overlap perfectly at the end, so $\Delta\phi_{\text{sep}} = 0$. The entire phase shift therefore comes from the laser phase. Accounting for the positions of the freely falling trajectories of the interferometer arms at each of the three atom optics pulses, the laser phase, and hence the full phase shift, is

$$\Delta\phi = -2kgT^2, \tag{6.23}$$

---

[4] In general, $\Delta\phi_{\text{laser}}$ and $\Delta\phi_{\text{sep}}$ can individually differ depending on which interferometer output port is considered. However, their sum $\Delta\phi_{\text{laser}} + \Delta\phi_{\text{sep}}$ is the same for both output ports (Hogan et al., 2009).

in agreement with the result from the treatment in Sec. 6.2.3. It is interesting to note that in the perturbative treatment of Sec. 6.2.3, the phase shift comes from the propagation phase calculated using the unmodified trajectories, while for the full treatment this same phase shift instead enters through the laser phase. This is an example of a more general method of calculating phase shifts from perturbing potentials—in perturbation theory, the change in the phase shift is calculated by integrating the perturbing potential over the unperturbed interferometer trajectories (Storey and Cohen-Tannoudji, 1994). In general, there will also be terms in the formal phase shift calculation that enter through the propagation phase.

6.3   Large Momentum Transfer (LMT) atom optics

6.3.1   Motivation for LMT atom optics

For atom interferometry to reach its ultimate potential, it is desirable to realize ultra-sensitive atom interferometers via large enclosed spacetime area $\Delta z_{max} T$, where Δz_{max} is the maximum path separation reached during the interferometer and T is the interferometer pulse spacing. As shown in Sec. 6.2.3, the acceleration sensitivity scales proportionally with the enclosed spacetime area. One way to increase the enclosed spacetime area is to implement beam splitters that lead to larger momentum splittings between the two interferometer arms than the $2\hbar k$ momentum splitting of standard light-pulse atom optics. Beam splitters and mirrors that deliver $n\hbar k$ momentum kicks to the atoms, with $n > 2$, are commonly called Large Momentum Transfer (LMT) atom optics. The path separation Δz_{max}, and hence the interferometer area, is proportional to n, as a larger momentum splitting allows the two interferometer paths to separate by a larger distance over a given time T.

Since Δz_{max} is proportional to T, the interferometer area is proportional to T^2. In addition to implementing interferometers with LMT atom optics, we would therefore like to make T as large as possible. For terrestrial experiments, T is limited by the height over which the atoms can freely fall before hitting the bottom of the apparatus. Maximizing T has motivated us to construct a 10 m atomic fountain for atom interferometry and others to perform atom interferometry in a freely falling capsule that takes advantage of the microgravity environment provided by the 146 m tall drop tower at the Center of Applied Space Technology and Microgravity (ZARM) in Bremen (Müntinga et al., 2013). Our 10 m fountain apparatus is discussed in Sec. 6.4.

One of the major goals of our work was to demonstrate atom interferometers that have both very large momentum transfer and very long interrogation time, so that Δz_{max} and T reach macroscopic scales in space (tens of centimeters) and in time (~ 1 second), respectively (Kovachy et al., 2015b; Asenbaum et al., 2017). There are a number of technical challenges associated with these macroscopic scale interferometers. Even for short T, it takes significant care to implement LMT atom optics that efficiently transfer many momentum kicks to the atoms without washing out the interference signal due to dephasing from the atom-light interaction. Sources of dephasing include laser-intensity- and velocity-dependent atom-optics-induced phase shifts (the laser-intensity-dependent phase shifts couple to laser intensity variations across the atom cloud) and wavefront imperfections in the atom optics laser beams. As discussed in Sec. 6.3.3, it becomes significantly more difficult to mitigate the harmful effects of such technical imperfections as n becomes larger and T becomes longer.

6.3.2   Methods for LMT atom optics

There are a number of different methods for LMT atom optics, each with its own set of advantages and disadvantages. It is useful to have this variety of methods available, as there are many applications of atom interferometry that can benefit from LMT atom optics, and the optimal method may depend on the specific constraints imposed by a particular situation. We will now discuss several of these methods.

6.3.2.1   *Sequential raman transitions*

One way to create an LMT beam splitter involves a sequence of Raman transitions (McGuirk et al., 2000). One implementation of a sequential Raman LMT beam splitter, which we realized in the 10 m atomic fountain apparatus (Dickerson, 2014; Sugarbaker, 2014), consists of an initial $\pi/2$-pulse that splits the interferometer arms in momentum by $2\hbar k$, and a subsequent sequence of π-pulses that increase this momentum splitting by selectively delivering additional $2\hbar k$ momentum kicks to one of the arms. As in Sec. 6.2.2, we denote the two ground state hyperfine levels of the Rb atoms as $|1\rangle$ and $|2\rangle$. Before the beam splitter, let us say that all the population is in state $|1\rangle$ with momentum p_0. For convenience, we will incorporate the momentum into the state label, so that the initial state is $|1, p_0\rangle$. The initial $\pi/2$-pulse creates a superposition of two interferometer arms, one in state $|1, p_0\rangle$ and the second in state $|2, p_0 + 2\hbar k\rangle$. A π-pulse then transfers the second arm from $|2, p_0 + 2\hbar k\rangle \longrightarrow |1, p_0 + 4\hbar k\rangle$. Additional π-pulses can further accelerate the second arm by driving the sequence of transitions $|1, p_0 + 4\hbar k\rangle \longrightarrow |2, p_0 + 6\hbar k\rangle \longrightarrow |1, p_0 + 8\hbar k\rangle$ and so on.

Because Raman transitions involve two different internal states, the resonance condition for $\omega_{\rm up} - \omega_{\rm down}$ differs from that in Eq. 6.10 for the case of Bragg transitions. For pulses that deliver upward momentum kicks while changing the internal state from $|1\rangle \longrightarrow |2\rangle$ ($|2\rangle \longrightarrow |1\rangle$), the frequency difference ω_{21} between states $|2\rangle$ and $|1\rangle$ must be added to (subtracted from) $\omega_{\rm up} - \omega_{\rm down}$:

$$\omega_{\rm up} - \omega_{\rm down} = 2\frac{p_i}{\hbar}v_{\rm r} + 4\omega_{\rm r} + \omega_{21} \qquad (\text{for } |1, p_i\rangle \longrightarrow |2, p_i + 2\hbar k\rangle) \qquad (6.24)$$

$$\omega_{\rm up} - \omega_{\rm down} = 2\frac{p_i}{\hbar}v_{\rm r} + 4\omega_{\rm r} - \omega_{21} \qquad (\text{for } |2, p_i\rangle \longrightarrow |1, p_i + 2\hbar k\rangle). \qquad (6.25)$$

Assuming a configuration in which the laser beams are not retroreflected,[5] if one Raman transition $|1, p_i\rangle \longrightarrow |2, p_i + 2\hbar k\rangle$ ($|2, p_i\rangle \longrightarrow |1, p_i + 2\hbar k\rangle$) is on resonance, its neighboring Raman transitions $|2, p_i - 2\hbar k\rangle \longrightarrow |1, p_i\rangle$ and $|2, p_i + 2\hbar k\rangle \longrightarrow |1, p_i + 4\hbar k\rangle$ ($|1, p_i - 2\hbar k\rangle \longrightarrow |2, p_i\rangle$ and $|1, p_i + 2\hbar k\rangle \longrightarrow |2, p_i + 4\hbar k\rangle$) will be detuned

---

[5] In the retroreflected case, neighboring Raman transitions can be driven by the mirror image of the on-resonance laser beam pair. In this case, the neighboring transitions are detuned by an amount $\sim 4\frac{p_i}{\hbar}v_{\rm r}$, which can be large for a sufficiently high offset momentum p_i.

from resonance by a large frequency difference of magnitude $\sim 2\omega_{21}$. This implies that a Raman transition involves a very pure two level system that is not significantly polluted by neighboring transitions. We will see that Bragg transitions behave as pure two level systems under a narrower range of conditions, a feature with both advantages and disadvantages.

An important effect for Raman transitions is the optically-induced change of the frequency difference ω_{21} between the two hyperfine states from AC Stark shifts (also called light shifts). These shifts directly couple to the pulse transfer efficiency and phase, as they affect the resonance condition. AC Stark shifts change the energies of states $|1\rangle$ and $|2\rangle$ via off-resonant coupling to the excited state $|e\rangle$. We consider the AC Stark shift on state $|i\rangle$ ($i = 1$ or 2) from a spectral component with frequency ω_j and single-photon Rabi frequency Ω_{ji} coupling $|i\rangle$ and $|e\rangle$. For detuning $\Delta_{ji} \equiv \omega_j - (\omega_e - \omega_i)$ of ω_j from the $|i\rangle \longrightarrow |e\rangle$ transition, the AC Stark shift of the energy of $|i\rangle$ is (Berman, 1997)

$$\Delta E_{ji}^{AC} = \hbar \frac{\Omega_{ji}^2}{4\Delta_{ji}} \qquad (6.26)$$

for $\Delta_{ji} \gg \Omega_{ji}$, Γ. Note that the AC Stark frequency shift has the same order of magnitude as the two-photon Rabi frequency $\frac{\Omega_{ji}^2}{2\Delta_{ji}}$ associated with the spectral component ω_j we are considering. Because of the hyperfine splitting between $|1\rangle$ and $|2\rangle$ (6.8 GHz for ^{87}Rb, 3 GHz for ^{85}Rb), Δ_{j1} and Δ_{j2} are different, leading to a differential AC Stark shift between the two hyperfine levels.

The total AC Stark shift arises from summing the individual shifts ΔE_{ji}^{AC} over all the spectral components ω_j present in the laser beams. To optimize transfer efficiency and minimize residual phase shifts, it is desirable to have this total shift be zero. One technique is to adjust the power ratio between the two atom optics beams driving the Raman transitions so that the net differential AC Stark from the two beams cancels (Peters et al., 2001). For this method, there is still typically an absolute AC Stark shift of each level—this shift is just the same for the two levels. We describe this situation as having relative but not absolute AC Stark shift compensation. It is also possible to set the spectrum of each atom optics beam so that the absolute AC Stark shift from that beam of each individual hyperfine level is zero (zero absolute AC Stark shift necessarily implies zero relative shift). Having each individual beam be AC Stark shift compensated has the advantage of eliminating residual ac Stark shifts from relative intensity variations or mode mismatches of the two beams. Moreover, as discussed in Sec. 6.3.3, absolute AC Stark shift compensation is important for large area atom interferometry.

6.3.2.2 *Sequential and multiphoton Bragg transitions*

Another approach to LMT atom optics is to use Bragg transitions. One way in which this can be done is analogous to the sequential Raman method described above, but with the two-photon Raman transitions replaced by two-photon Bragg transitions. An advantage of Bragg transitions is that they do not experience unwanted detunings from AC Stark shifts, as the internal state of the atom does not change. Another effect of the fact that Bragg transitions do not change the internal state is that, in comparison to Raman transitions, they behave as pure two-level systems over a narrower set of experimental parameters. To understand this, if a Bragg transition $|p_i\rangle \longrightarrow |p_i + 2\hbar k\rangle$ is on resonance, we note that the neighboring Bragg transitions $|p_i - 2\hbar k\rangle \longrightarrow |p_i\rangle$ and $|p_i + 2\hbar k\rangle \longrightarrow |p_i + 4\hbar k\rangle$ are detuned by $\pm 8\omega_r$ (from Eq. 6.10). For Rb, $\omega_r \approx 2\pi \times$ 4 kHz, which is much smaller than the hyperfine splitting by which neighboring Raman transitions are detuned. In order to avoid strong coupling to the neighboring Bragg transitions, the two-photon Rabi frequency should be significantly below $8\omega_r$. This constraint on the Rabi frequency limits how quickly momentum can be transferred to the atoms and the Doppler width of the atoms that can be efficiently transferred. For the latter reason, Bragg transitions require colder atom sources than do Raman transitions.

The fact that neighboring Bragg transitions are detuned by only a modest amount naturally leads to the possibility of driving higher order, multiphoton Bragg transitions (Martin et al., 1988; Giltner et al., 1995; Müller et al., 2008). A $2n$ photon Bragg transition couples states $|p_0\rangle$ and $|p_0 + 2n\hbar k\rangle$, using momentum states at $2\hbar k$ intervals in between as intermediary states (see Fig. 6.3). By extension of the argument used to derive Eq. 6.10, the resonance condition for the $|p_0\rangle \longrightarrow |p_0 + 2n\hbar k\rangle$ multiphoton Bragg transition is

$$\omega_{\text{up}} - \omega_{\text{down}} = \frac{1}{n}\frac{1}{\hbar}\left[\frac{(p_0 + 2n\hbar k)^2}{2m} - \frac{p_0^2}{2m}\right] = 2\frac{p_0}{\hbar}v_r + 4n\omega_r. \tag{6.27}$$

It is possible to drive highly efficient transfer from $|p_0\rangle \longrightarrow |p_0 + 2n\hbar k\rangle$ with multiphoton Bragg transitions, but there is typically significant population in the intermediate momentum states in the middle stages of the pulse. For this reason, the dynamics of multiphoton Bragg transitions are more complex than those of a pure two level system, raising the possibility of stronger intensity dependent phase shifts.

Multiphoton Bragg transitions have the advantage of being able to transfer momentum more quickly and with fewer pulses than two-photon Bragg transitions. Since multiphoton Bragg transitions are higher-order processes, the laser intensity that it takes to drive them while keeping spontaneous emission losses acceptably low increases rapidly with n (Müller et al., 2008; Szigeti et al., 2012). With limited laser intensity, one option is to perform sequences of multiphoton Bragg transitions of intermediate order. As described in Chiow et al. 2011, we used sequential $6\hbar k$ multiphoton Bragg atom optics to demonstrate atom interferometers with momentum splittings between the arms of up to $102\hbar k$. This work was performed in a test apparatus with short interrogation time ($T \sim 10$ ms). For very large area atom interferometers in the 10 m atomic fountain, we use sequential $2\hbar k$ Bragg transitions (Kovachy et al., 2015b; Asenbaum et al., 2017).

6.3.2.3   *Adiabatic rapid passage methods*

Adiabatic rapid passage (ARP) is a commonly used technique to make the transfer between two states more robust against variations in Rabi frequency and detuning

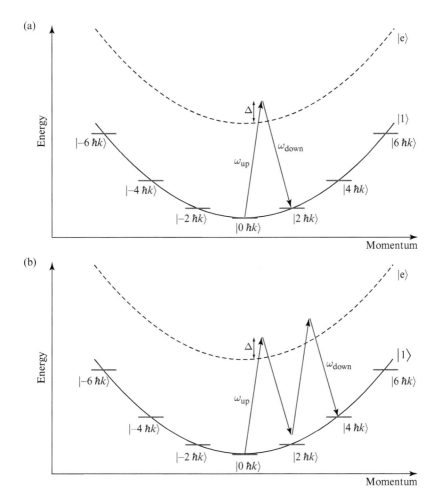

Fig. 6.3 Diagrams of Bragg transitions. (a) Bragg transition that transfers momentum $2\hbar k$. Two momentum states with the same internal state $|1\rangle$ are coupled via an intermediate excited state $|e\rangle$. The transition arises from stimulated absorption of a photon from an upward propagating laser beam with frequency ω_{up} followed by stimulated emission of a photon into a downward propagating laser beam with frequency ω_{down}. Since the internal state does not change, the resonance condition is determined by the quadratic dependence of the external kinetic energy on momentum. (b) Multiphoton Bragg transition that transfers momentum $4\hbar k$. The transition arises from stimulated absorption of two photons from the ω_{up} beam and stimulated emission of two photons into the ω_{down} beam.

(Loy, 1974; Vitanov et al., 2001; Torosov et al., 2011). A key feature of ARP is that during an ARP pulse, the detuning is swept across resonance. Variations of ARP techniques can be combined with Raman (Kotru et al., 2015) or Bragg transitions. We have demonstrated atom interferometers using multiphoton Bragg transitions with enhanced transfer efficiency via the use of ARP (Kovachy et al., 2012). Another LMT method, which we analyze in detail in Sec. 6.8, is to adiabatically accelerate atoms in an optical lattice (Hecker Denschlag et al., 2002; Cladé et al., 2009; Müller et al., 2009; McDonald et al., 2013), which can be understood as inducing sequential $2\hbar k$ ARP Bragg transitions (Peik et al., 1997).

ARP atom optics techniques are especially useful for applications in which it is difficult to make the laser intensity and transition resonance frequency constant from shot-to-shot or over the atom ensemble in a single shot. Such applications could include mobile atom-interferometry-based inertial sensors for use outside the laboratory. For example, with ARP techniques, high transfer efficiency can be achieved with a larger spread in Doppler detunings, allowing the use of hotter ensembles that contain more atoms and take less time to cool. ARP transitions require greater pulse area than their non-ARP counterparts (typically at least several times more), which leads to a proportional increase in the spontaneous emission loss, so an important tradeoff to take into account is the efficiency gain from the use of ARP versus the efficiency loss from spontaneous emission. Moreover, as discussed in Kovachy et al. 2012 and in Sec. 6.8, ARP generally leads to additional velocity- and laser-intensity-dependent phase shifts. The analysis in Sec. 6.8 gives reason for optimism that these phase shifts can be controlled at an acceptable level for macroscopic interferometers using optical lattice manipulations of the atoms. It will be interesting to experimentally investigate such interferometers in the 10 m atomic fountain in future work.

6.3.3   Challenges of large area atom interferometry

As mentioned in Sec. 6.3.1, a major technical obstacle for the realization of LMT atom interferometers has been the difficulty of preventing the interference signal from being washed out by dephasing from the LMT atom optics interactions. This task becomes dramatically more difficult for large path separation $\Delta z_{\max}$ and long pulse spacing T.

One of the key challenges for very large area atom interferometry is to prevent dephasing from laser intensity dependent phase shifts. For instance, all the LMT methods discussed in Sec. 6.3.2 are susceptible to AC Stark shifts of the atomic energies. In contrast to Raman transitions, Bragg transitions avoid the additional concern of intensity dependent detunings from the relative AC Stark shifts between the two hyperfine levels. However, both Raman and Bragg based atom optics are sensitive to the absolute AC Stark shift of each level.

In practice, the transverse laser beam profile always has intensity variations across the atom ensemble. Because of diffraction, these laser intensity variations change with height, leading to vertical intensity gradients. Through the absolute AC Stark shift, these intensity gradients cause vertical forces that are nonuniform over the atom cloud. If $\Delta z_{\max}$ is small, this is not too much of a concern, as the sensitivity of the interferometer to these forces is small. However, once $\Delta z_{\max}$ grows to tens of centimeters (Kovachy et al., 2015b; Asenbaum et al., 2017), the force sensitivity of the interferometer becomes large enough that absolute-AC-Stark-shift-induced vertical forces can lead to substantial dephasing. Moreover, for larger momentum transfer, the atom optics lasers need to be kept on longer and/or have a higher intensity, thereby increasing the overall size of any intensity dependent phase perturbations. For the work in Kovachy et al. 2015b and Asenbaum et al. 2017, we found it essential to implement a spectrum that precisely compensated the absolute AC Stark shift for each atom optics laser beam. The generation of this spectrum is discussed in detail in Kovachy et al. 2015b.

In addition to intensity dependent phase shifts, another concern is phase perturbations arising from wavefront imperfections of the atom optics laser beams. Whenever a laser beam delivers a photon recoil momentum kick to an atomic wave packet, the wavefront of the laser beam is imprinted on the wave packet's phase profile (using the terminology of Sec. 6.2.4, this can be understood as spatially dependent laser phase). Since many photon recoil kicks are delivered for LMT atom optics, these phase perturbations are imprinted many times. The amplitude of the total phase perturbation imprinted on atomic wave packet scales with the amount of momentum transferred. Since the two interferometer arms receive momentum kicks at different times spaced by T, the imprinted phase perturbations will not necessarily be common to the two interferometer arms. For example, differential imprinted phase profiles between the two arms can arise from expansion or transverse motion of the atom cloud over the pulse spacing T, effects which are magnified for longer T.

Another challenge for large area atom interferometry is the realization of high atom optics transfer efficiency. For long T, the atom cloud has more time to expand and can reach a size comparable to the width of the laser beams, leading to less efficient transfer due to Rabi frequency inhomogeneities. Moreover, atom optics that transfer more momentum are naturally more sensitive to Rabi frequency inhomogeneities. The use of wider laser beams can help to ameliorate this problem. This involves a tradeoff, however, since maintaining constant Rabi frequency while making the beams wider requires closer detuning, leading to increased spontaneous emission losses. In order to reduce the influence of large scale laser intensity and wavefront inhomogeneities in interferometers with long T, it is generally helpful to use an atom source that is transversely cold. Methods for generating such atom sources are discussed in Sec. 6.6 and in Kovachy et al. 2015a.

6.4 Ten meter atomic fountain for long duration atom interferometry

A ten meter tall atomic fountain allows us to perform interferometry while the atoms freely fall over the course of multiple seconds. Our lab is able to accommodate this fountain because the lab has an 8 m deep pit extending into the floor. A photograph of the apparatus is shown in Fig. 6.4. The ultracold atom source is located at the bottom of the pit. The atoms are cooled and trapped in a 3D magneto-optical trap (MOT) loaded by a 2D MOT and subsequently undergo evaporative cooling. Their kinetic energy spread is then further reduced by a collimating magnetic lens (Kovachy et al., 2015a). The 3D MOT vacuum chamber is attached to an upward-extending 8.8 m long vacuum tube made from aluminum, with a 10 cm diameter. An optical lattice launches the atoms upward into this vacuum tube. A solenoid is wound directly

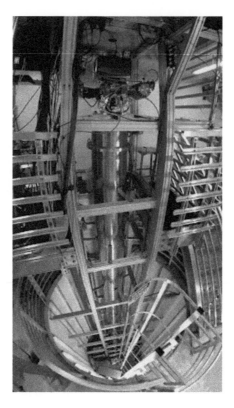

Fig. 6.4 Photograph of 10 m atomic fountain apparatus. The atom source is located at the bottom of an 8 m deep pit. An optical lattice launches the atoms upward into the fountain, which is housed inside a magnetic shield (the tall metal cylinder).

around the outer circumference of the vacuum tube. The solenoid generates a uniform bias field that establishes a quantization axis for the atoms. The vacuum tube is surrounded by a three-layer, cylindrical magnetic shield to reduce external magnetic fields (Dickerson et al., 2012).

Fig. 6.5a presents a CAD model of the apparatus. Ultra high vacuum ($< 10^{-10}$ Torr) is maintained in the 3D MOT chamber and the atomic fountain by a pair of ion pumps, one at the bottom of the tower and one at the top. When the atoms fall back to the bottom of the fountain at the end of an experimental sequence, into a detection region just above the 3D MOT, they are imaged using fluorescence detection and CCD cameras. Figure 6.5b shows a schematic of the z axis beam path for the atom optics beams used to drive Bragg transitions. The beam line is expanded to a radial waist of 2 cm by a telescope. The two atom optics beams are overlapped with orthogonal linear polarizations, which are subsequently circularized by a $\lambda/4$ waveplate. The beams are retroreflected by a mirror at the bottom of the pit. This mirror is mounted on a piezo tip-tilt stage and is rotated between interferometer pulses to compensate for Coriolis forces from the Earth's rotation (Dickerson et al., 2013). Another $\lambda/4$ waveplate is mounted in front of the retroreflection

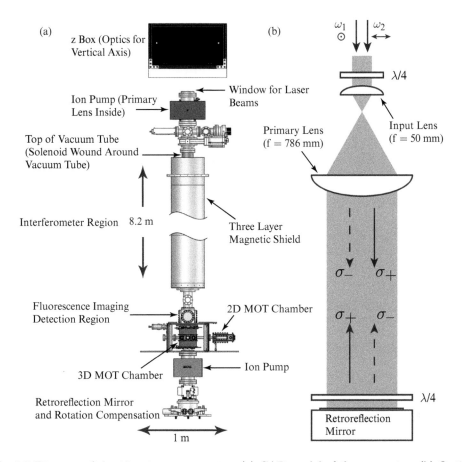

Fig. 6.5 Diagram of the 10 m tower apparatus. (a) CAD model of the apparatus. (b) Optics for the the atom optics laser beams. A telescope expands the radial beam waist to 2 cm. The two atom optics beams are labeled as ω_1 and ω_2.

mirror, so that σ^- circular polarization is rotated to σ^+ circular polarization, and vice versa. Bragg transitions are driven by counter-propagating frequency pairs with the same circular polarization. Therefore, this $\lambda/4$ waveplate ensures that one frequency component from each of the two atom optics beams drives the Bragg transitions.

After releasing the atoms from the magnetic lens, we adiabatically load them into an optical lattice potential formed by two counter-propagating laser beams. The loading occurs by ramping up the power of the laser beams while the lattice is freely falling with the atoms. The velocity of the optical lattice is determined by the frequency difference between the two beams. By ramping this frequency difference, we accelerate the lattice–and the atoms trapped in it–upward to a velocity of 13 m/s. Subsequently, we adiabatically release the atoms from the lattice by ramping down the power of the laser beams. The lattice launch and techniques used to make it highly efficient are detailed in Sec. 6.7.

6.5   Differential measurement strategies and applications of atom interferometry

6.5.1   Overview of atom interferometry applications

Atom interferometry has found a wide range of applications, ranging from the practical to the fundamental. Atom interferometers can be used as precise sensors of accelerations (Peters et al., 2001), rotations (Gustavson et al., 1997; Dickerson et al., 2013), and gravity gradients (Snadden et al., 1998). In addition, atom interferometry is a valuable tool for measuring fundamental constants (Bouchendira et al., 2011; Rosi et al., 2014) and performing fundamental physics tests, such as tests of the equivalence principle (Bonnin et al., 2013; Schlippert et al., 2014; Kuhn et al., 2014; Zhou et al., 2015; Barrett et al., 2016; Rosi et al., 2017), other aspects of general relativity (Dimopoulos et al., 2008a), quantum mechanics at macroscopic scales (Arndt and Hornberger, 2014; Müntinga et al., 2013; Bassi et al., 2017; Cotter et al., 2017; Romero-Isart, 2017), and cosmological principles through the potential implementation of atomic gravitational wave observatories (Dimopoulos et al., 2008b; Hogan et al., 2011). Here we introduce several of the applications on which our work has focused.

6.5.2   Differential measurement strategies

For many applications of atom interferometry, a crucial measurement strategy is to implement a simultaneous differential measurement. In many cases—such as gravity gradiometry (Snadden et al., 1998), tests of the equivalence principle (Bonnin et al., 2013; Schlippert et al., 2014; Kuhn et al., 2014; Zhou et al., 2015; Barrett et al., 2016; Rosi et al., 2017), gyroscopy and gyrocompassing using point source interferometry (Dickerson et al., 2013; Sugarbaker et al., 2013), and gravitational wave detection (Dimopoulos et al., 2008b; Hogan et al., 2011)—the signal of interest is differential in nature. To understand the concept behind simultaneous differential measurement, consider a collection of parallel interferometers labelled by index i,

$$|\psi_1\rangle_i + e^{i\phi_i}|\psi_2\rangle_i, \tag{6.28}$$

where $|\psi_1\rangle_i$ and $|\psi_2\rangle_i$ respectively correspond to the wavepackets associated with the two arms of interferometer i, and ϕ_i is the phase with which interferometer i interferes. These parallel interferometers, corresponding to sub-ensembles of the entire collection of atoms being used for interferometry in a given experimental shot, could represent different atom clouds separated over a long baseline (Snadden et al., 1998), clouds of different atomic species (Bonnin et al., 2013; Schlippert et al., 2014; Kuhn et al., 2014; Zhou et al., 2015; Barrett et al., 2016), or different parts of a single atom cloud (Müntinga et al., 2013; Dickerson et al., 2013). For example, in Dickerson et al. 2013 a measurement of the differential phase between parallel interferometers corresponding to the left and right side of a single atom cloud was used to make a high-precision gyroscope.

Because of the high inertial sensitivity of atom interferometers, the individual phases ϕ_i can vary by large amounts from shot-to-shot due to vibrations (Müntinga et al., 2013; Dickerson et al., 2013). However, it is possible to make this vibration-induced phase noise cancel to a high degree as a common mode in the differential phase shifts $\phi_i - \phi_j$ of the simultaneous interferometers (McGuirk et al., 2002). The simultaneity of the interferometers is critical for common mode cancellation to work. If the interferometers were not run simultaneously, different phase shifts ϕ_i and ϕ_j would not experience the noise in a common way, so that the differential signals of interest, $\phi_i - \phi_j$, would be noisy.

6.5.3 Gravity gradiometry and tidal forces across a single quantum system

As shown in Sec. 6.2.3, the interferometer phase shift is sensitive to the local gravitational acceleration g. In many applications, it is most interesting to measure gravity gradients (e.g. how quickly g changes as a function of height) (Snadden et al., 1998). An atom interferometric gravity gradiometer works by making a differential phase measurement between two displaced interferometers. A common configuration is to use a pair of interferometers separated over a vertical baseline. We applied this configuration to macroscopic scale interferometers in the 10 m fountain. The intrinsic sensitivity of these interferometers allowed us to reach a differential acceleration resolution of $10^{-10}g$ per shot and to measure a phase response of 1 rad to a local test mass (see Fig. 6.6) (Asenbaum et al., 2017).

The macroscopic spatial and temporal scales of these interferometers allowed us to observe for the first time the influence of spacetime curvature across the wavefunction of a single quantum system. In previous matter wave interferometers, the wave packet separation was sufficiently small so that the interferometer could be well-approximated as existing in a single local inertial frame. In particular, one could make a coordinate transformation into a local Lorentz frame that is freely falling with the interferometer trajectories. In this frame, the particles do not experience any gravitational influence. The interferometer phase shift arises solely from the relative motion of this frame with respect to some reference (Audretsch and Marzlin, 1994). In the case of light-pulse atom interferometry, this reference is the laser beams that drive the atom optics transitions. The situation becomes qualitatively different when the wave packet separation is large enough so that there is substantial spacetime curvature across the wavefunction. Spacetime curvature induces tidal forces across the wavefunction, so that the two interferometer arms experience different local gravitational accelerations. In the case where this tidal force is resolvable, if one boosts into a local inertial frame for one of the interferometer arms, the effect of gravity will *not* be removed from the second arm. The wavefunction probes the spacetime manifold in a nonlocal way, leaving a gravitional effect that cannot be transformed away by a particular choice of coordinates. The phase shift associated with tidal forces across the wavefunction has therefore been described as the first genuine gravitional effect in a quantum system. In order to fully characterize the influence of gravity on quantum matter, it is therefore important to explore this regime (Audretsch and Marzlin, 1994; Anandan, 1984; Bordé and Lammerzahl, 1996; Chryssomalakos and Sudarsky, 2003; Bonder and Sudarsky, 2008). We observed such a phase shift induced by tidal forces from the aforementioned local test mass (Asenbaum et al., 2017).

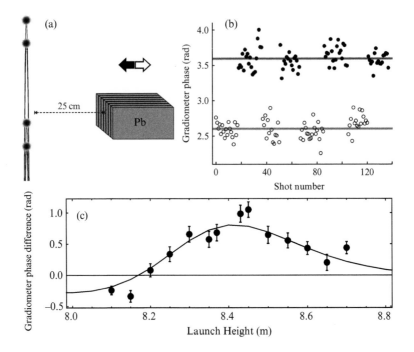

Fig. 6.6 (a) Schematic representation of the experimental setup for measuring the gravity gradient of seven lead bricks (total mass 84 kg). The differential phase is measured between a pair of interferometers separated over a vertical baseline L. (b) Measured gradiometer differential phase of a sequence with launch height $h = 8.45$ m, gradiometer baseline $L = 32$ cm, $20\hbar k$ beam splitters, and $T = 600$ ms ($\Delta z_{\max} = 7$ cm), with (solid circles) and without (open circles) the bricks present. (c) Gradiometer phase difference (with and without bricks present) as a function of launch height with $L = 10$ cm, $30\hbar k$ beam splitters, and $T = 900$ ms ($\Delta z_{\max} = 16$ cm). The black, solid curve represents a numerical phase shift calculation. Figure is from Asenbaum et al., 2017.

6.5.4   Tests of the equivalence principle

The equivalence principle is fundamental to our understanding of gravity. There are several variations of the equivalence principle, each of which makes successively stronger claims (Carroll, 2004). Many experiments focus on the weak equivalence principle (WEP), which states that all objects fall under gravity at the same rate.

Simultaneous atom interferometers that use different atomic species provide a way to test the WEP by comparing the gravitational accelerations of the different species (Bonnin et al., 2013; Schlippert et al., 2014; Kuhn et al., 2014; Zhou et al., 2015; Barrett et al., 2016). Current research in the 10 m atomic fountain is focused on a WEP test using a macroscopic scale, dual species (^{85}Rb and ^{87}Rb) interferometer. From the discussion in Sec. 6.5.2, it is essential that these dual interferometers be simultaneous so that the effects of vibration cancel as a common mode in the differential acceleration signal $g_{87} - g_{85}$ between the two species.

The macroscopic interferometer area greatly enhances the differential acceleration sensitivity. Overstreet et al. 2017 reports the achievement of a differential acceleration resolution of $\Delta g/g \approx 6 \times 10^{-11}$ per shot in the 10 m fountain, which improves on the best resolution of previous dual-species interferometers (Zhou et al., 2015) by more than three orders of magnitude. Additionally, a method proposed by Roura (Roura, 2017) to suppress the influence of gravity gradients on the differential phase shift is experimentally demonstrated (Overstreet et al., 2017). The coupling of initial kinematic offsets between the two species to gravity gradients is a leading systematic effect in atom interferometric tests of the WEP, because these offsets cause the two species to experience different local gravitational accelerations (Hogan et al., 2009). Roura's proposal involves shifting the laser frequency for the mirror pulse sequence of the interferometer by an appropriate amount so that the phase shift arising from initial-kinematics-to gravity-gradient coupling is cancelled to a high degree (Roura, 2017). We refer to this method as frequency shift gravity gradient (FSGG) compensation. In the dual species interferometer reported in Overstreet et al. 2017, this method suppresses the gravity gradient phase shift by a factor of 100 and allows gravity gradient systematics to be controlled to below one part in 10^{13}. Combined with the demonstrated differential acceleration resolution, this is an important step toward an improved atom interferometric WEP test.

In Table 6.1 we show the leading differential phase shift terms when applying FSGG compensation in the second interferometer pulse, and applying our full rotation compensation system so there are no residual couplings to earth rotations. For arithmetic purposes we assume that we have a $10\hbar k$ interferometer and the order of magnitude of $\Delta k/k = -T_{zz}T^2/2 \approx 10^{-6}$. Summing terms 1, 2, 5, and 6 gives

$$- \left(nkT_{zz}T^2 + 2n\Delta k\right)\left(\Delta z + \Delta v_z T\right), \tag{6.29}$$

Table 6.1 Differential phase shifts in a $n\hbar k$ dual-species interferometer for an arbitrary frequency shift Δk. For calculating the size of the different phase terms, it is assumed that we have a $10\hbar k$ interferometer, $\Delta k/k = -T_{zz}T^2/2$, pulse spacing $T = 900$ ms, initial position offset between ^{85}Rb and ^{87}Rb $\Delta z = 1$ mm, initial velocity offset between ^{85}Rb and ^{87}Rb $\Delta v_z = 10$ µm/s, gravity gradient $T_{zz} = -2.1 \times 10^{-6}$ s^{-1}, vertical derivative of the gravity gradient $Q_{zzz} = 10^{-7}$ m^{-1}s^{-2}, bias magnetic field $B_0 = 10$ mG, and residual magnetic field gradient $\delta B = 0.1$ mG/m.

	Phase shift	Size [rad]	Fractional size
1	$-2n\Delta k\Delta z$	-1.4×10^{-1}	2.1×10^{-10}
2	$-nkT_{zz}T^2\Delta z$	1.4×10^{-1}	2.1×10^{-10}
3	$\left(\frac{1}{m_{87}} - \frac{1}{m_{85}}\right)n^2\hbar k\Delta kT$	8.3×10^{-2}	1.3×10^{-10}
4	$\left(\frac{1}{m_{87}} - \frac{1}{m_{85}}\right)\frac{n^2}{2}\hbar k^2 T_{zz}T^3$	-8.3×10^{-2}	1.3×10^{-10}
5	$-2n\Delta k\Delta v_z T$	-1.2×10^{-2}	1.9×10^{-11}
6	$-nkT_{zz}T^3\Delta v_z$	1.2×10^{-2}	1.9×10^{-11}
7	$-B_0\delta B\left(\frac{\alpha_{87}}{m_{87}} - \frac{\alpha_{85}}{m_{85}}\right)nkT^2$	2.2×10^{-4}	3.4×10^{-13}
8	$\left(\frac{1}{m_{87}} - \frac{1}{m_{85}}\right)n^3\frac{\hbar^2 k^3}{3m}Q_{zzz}T^4$	1.4×10^{-4}	2.2×10^{-13}

and summing terms 3 and 4 gives

$$\left(\frac{1}{m_{87}} - \frac{1}{m_{85}}\right) \frac{n^2}{2} \hbar k^2 T \left(T_{zz} T^2 + 2\frac{\Delta k}{k}\right). \tag{6.30}$$

Here we recognize the gravity gradient coupling as described by Roura, whereas the second line indicates a recoil coupling to the gravity gradient due to the dual-species character of the measurement. Fortunately, both phase sensitivities can be cancelled by choosing $\Delta k = -kT_{zz} T^2/2$, which is the result as given by Roura. Overstreet et al. 2017 shows that this sensitivity can be experimentally cancelled to the 10^{-2} level. As the $\Delta z + \Delta v_z T$ term can reasonably be measured down to 1 μm, the gravity gradient coupling can be reduced within a systematic uncertainty of $\sim 2 \times 10^{-15}$. At that level the magnetic field introduces a phase shift that should (and can) be measured with 1% precision to have a comparable contribution to the systematic error budget. The final term in Table 6.1 is due to nonlinearity in the gravity gradient experienced by the spatially separated wavepacket trajectories and will have to be measured in the future. The uncertainty in this term is dominated by Q_{zzz}, which could be measured to better than 10% in the future in order to push the error budget into the 10^{-15} range.

6.5.5   Tests of quantum mechanics at macroscopic scales

The realization of interferometers with massize particles is a dramatic manifestation of the fundamentally quantum mechanical wave nature of matter and the related concepts of quantum superposition and wave-particle duality. In quantum mechanics, objects appear as particles when detected but propagate and interfere as delocalized waves (Feynman et al., 1965). The quantum mechanical features of massive particle interferometers can be illustrated with a simple thought experiment. Consider the closest classical counterpart to an interferometer for a collection of classical billiard balls. In classical mechanics, the closest possible counterpart to a beam splitter is a switch that randomly directs each billiard ball into one of two paths with equal probability. Since classical particles have definite trajectories, each billiard ball must take one path or the other. The two paths eventually overlap on a second switch that, with equal probability, randomly places each billiard ball into one of two output ports. If many billiard balls are run through this contraption, half the balls will end up in one outport port and half will end up in the other output port, up to small deviations from counting statistics. The situation in quantum mechanics is very different. Instead of following from classical probabilities, the population distribution of the two output ports can behave as if two wavelike amplitudes corresponding to propagation along the two interferometer arms are superposed at the final beam splitter, allowing the output ports to vary between full constructive and destructive interference. In this situation, each of the particles that runs through the interferometer and is ultimately detected necessarily has quantum mechanical amplitude in both interferometer paths, and the distinctly nonclassical behavior of the interferometer output ports arises from the superposition and interference of such amplitudes (Feynman et al., 1965).

In order to realize this ideal wavelike behavior, with the population varying between the two interferometer output ports with perfect contrast over a set of experimental shots, the particles must all contribute in phase to the interference signal for each experimental shot. In the notation of Eq. 6.28, the phases ϕ_i of each sub-ensemble of particles must all be the same. Additionally, to observe interference, the quantum states $|\psi_1\rangle_i$ and $|\psi_2\rangle_i$ corresponding to the two interferometer arms for each sub-ensemble i cannot become orthogonal. The ability to make all particles contribute to the interferometer in phase is directly related to the ability of the interferometer to perform differential measurements, as it implies that the differential phases $\phi_i - \phi_j$ between different sub-ensembles are well-controlled, and therefore that any differential phase perturbation that is accumulated as the particles travel over the interferometer paths will be directly reflected in the interference pattern observed by resolving different sub-ensembles (e.g. using spatially resolved detection to observe a differential interferometer phase shift between the left and right halves of an atom cloud induced by the Earth's rotation as in Dickerson et al. 2013)[6].

It can be thought of as intuitively bothersome that quantum mechanics predicts the possibility of the wavelike behavior described above for massive particles, especially when the interferometer path separation, the interferometer duration, and/or the masses of the particles involved enter the macroscopic realm. In stark contrast to classically intuitive notions, the interference is a fundamentally nonlocal effect, involving the well-correlated superposition of widely separated trajectories for each massive particle. The somewhat counter-intuitive nature of matter wave interferometry has led to speculation about whether quantum mechanics breaks down at sufficiently macroscopic scales, washing out the interference signal for matter wave interferometers with macroscopic enclosed spacetime area and/or macroscopic particle mass (Arndt and Hornberger, 2014; Bassi et al., 2017; Cotter et al., 2017; Romero-Isart, 2017). Such an effect might manifest by preventing different particles running through an interferometer in parallel from contributing in phase to an interference pattern or by destroying the phase correlation between two interferometers in a gradiometer setup. Our work with large spacetime area interferometers (Kovachy et al., 2015b; Asenbaum et al., 2017) can rule out such behavior in a macroscopic regime. In practice, we know that for macroscopic scale interferometers, it is highly challenging to prevent interference contrast loss arising from inhomogeneous interferometer phase shifts induced by environmental disturbances (e.g. see Sec. 6.3.3). Perhaps there are fundamental mechanisms of this type that intrinsically limit sufficiently macroscopic interferometers. With continued improvement in matter wave optics, matter wave interferometry can continue to probe quantum mechanics in increasingly macroscopic regimes.

---

[6] A situation in which each differential phase shift $\phi_i - \phi_j$ is nonzero but well-controlled is closely related to the situation in which all the phase shifts ϕ_i are equal. This is because in the aforementioned case, compensating differential phase shifts could, at least in principle, be applied to each $\phi_i - \phi_j$ to make all the sub-ensemble phase shifts equal, and the differential phase signals would be preserved in the knowledge of the compensations applied.

6.5.6   Gravitational wave detection

Gravitational wave observatories offer a new way to study fundamental questions in astrophysics and cosmology (Abbott et al., 2016). Atom interferometry has the potential to be a valuable tool for gravitational wave detection and to complement existing and planned gravitational wave detectors that use laser interferometry and macroscopic proof masses (Kawabe and the LIGO Collaboration, 2008; Smith and the LIGO Collaboration, 2009; Acernese et al., 2007; Larson et al., 2000). Possible advantages of atomic gravitational wave detectors include the ideal features of freely falling, neutral atoms as inertial proof masses, the ability to enhance the signal with LMT atom optics, and the ability to implement pulse sequences that are insensitive to laser phase noise (Graham et al., 2013). Proposals for atomic gravitational wave detectors involve simultaneously operated pairs of atom interferometers separated over long baselines, with the atom optics lasers shared by both interferometers (Dimopoulos et al., 2008b; Hogan et al., 2011). The gravitational wave signal enters as a differential phase shift $\Delta\phi_{\mathrm{GW}}$ between the two interferometers. For a three-pulse Mach-Zehnder interferometer sequence with pulse spacing T, a fully general relativistic calculation shows this differential phase shift to be

$$\Delta\phi_{\mathrm{GW}} = 2k_{\mathrm{eff}}hL\sin^2\left(\frac{\omega T}{2}\right)\sin\left(\theta_{\mathrm{GW}}\right), \qquad (6.31)$$

where $k_{\mathrm{eff}} = n\hbar k$ is the effective interferometer k-vector corresponding to the number n of photon momentum recoils transferred by the LMT beam splitters, h is the gravitational wave strain, L is the baseline length, ω is the gravitational wave frequency, and θ_{GW} is the gravitational wave phase. Eq. 6.31 assumes that the wavelength of the gravitational wave is much larger than L (Dimopoulos et al., 2008b).

A "back-of-the-envelope" estimate of the scaling of this phase shift can provide some (not fully rigorous) physical intuition for its origin. We will think of the gravitational wave strain as inducing a time-dependent oscillation $\sim hL\sin\left(\omega t\right)$ in the distance between the two atomic test masses separated over the baseline L. This corresponds to a relative acceleration $a_{\mathrm{GW}} \sim -hL\omega^2\sin\left(\omega t\right)$ between the two test masses. The differential phase shift between the two interferometers reflects this relative acceleration via the known interferometer acceleration response from Eq. 6.23, yielding

$$\Delta\phi_{\mathrm{GW}} \sim -k_{\mathrm{eff}}a_{\mathrm{GW}}T^2 \sim k_{\mathrm{eff}}hL\omega^2T^2\sin\left(\theta_{\mathrm{GW}}\right), \qquad (6.32)$$

where we associate ωt with the phase θ_{GW} of the gravitational wave. Equation 6.31 displays the same scaling if we Taylor expand in the limit $\omega T \ll 1$.

In Hogan et al. 2011, we performed a detailed analysis of possible experimental parameters for a satellite based atomic gravitational wave interferometric sensor (AGIS). Fig. 6.7a shows a schematic of the proposed apparatus. The two interferometers along a given baseline are respectively carried out near two satellites separated over the baseline. The atom optics pulses come from lasers on the satellites. We analyzed a broad range of possible noise backgrounds—including rotations, laser wavefront

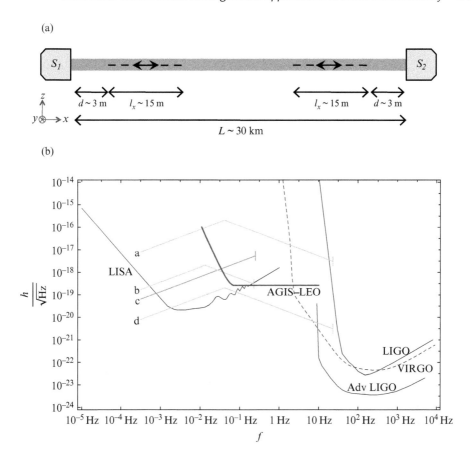

Fig. 6.7 Schematic and sensitivity curve of a proposed satellite based atomic gravitational wave detector. (a) Diagram of a baseline for a proposed atomic gravitational wave interferometric sensor in low Earth orbit (AGIS-LEO). Atom interferometers are run near a pair of satellites separated by an $L = 30$ km baseline. Each individual interferometer has a longitudinal spatial extent of approximately 15 m. (b) Strain sensitivity as a function of frequency for this detector (thick gray curve). Interferometer parameters are beam splitter momentum $\hbar k_{\mathrm{eff}} = 200\hbar k$, pulse spacing $T = 4$ s, and phase sensitivity $\delta\phi = 10^{-4}$ rad/$\sqrt{\mathrm{Hz}}$. Curves a-d show gravitational wave source strengths after integrating over the lifetime of the source or one year, whichever is shorter: (a) represents inspirals of 10^3 solar mass ($M_\odot$), $1M_\odot$ intermediate mass black hole binaries at 10 kpc, (b) inspirals of $10^5 M_\odot$, $1M_\odot$ massive black hole binaries at 10 Mpc, (c) white dwarf binaries at 10 kpc, and (d) inspirals of $10^3 M_\odot$, $1M_\odot$ intermediate mass black hole binaries at 10 Mpc. Figures are from Hogan et al. 2011.

and atom distribution fluctuations, and Newtonian gravitational backgrounds from the Earth and from satellite position jitter—and associated mitigation strategies. For example, we designed a specialized five-pulse interferometer sequence to reduce the sensitivity of the interferometer to background rotations and gravity gradients. The largest Newtonian gravitational background arises from non-spherical Earth gravity inhomogeneities, which are a significant systematic effect for frequencies below

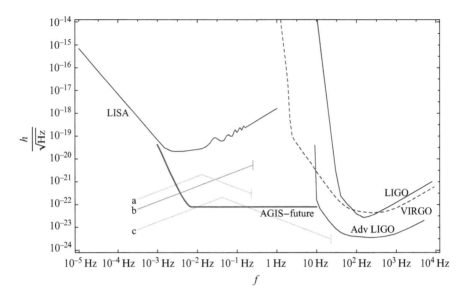

Fig. 6.8 Strain sensitivity as a function of frequency for an AGIS detector with a more ambitious set of parameters that could be realized further into the future (thick gray curve). Interferometer parameters are baseline $L = 10^4$ km, beam splitter momentum $\hbar k_{\text{eff}} = 200\hbar k$, pulse spacing $T = 40$ s, and phase sensitivity $\delta\phi = 10^{-5}$ rad/$\sqrt{\text{Hz}}$. Curves a-c show gravitational wave source strengths after integrating over the lifetime of the source or one year, whichever is shorter: (a) represents inspirals of $10^5 M_\odot$, $1 M_\odot$ massive black hole binaries at 10 Gpc, (b) white dwarf binaries at 10 Mpc, and (c) inspirals of $10^3 M_\odot$, $1 M_\odot$ intermediate mass black hole binaries at 10 Gpc. Figure is from Hogan et al. 2011.

30 mHz. The effect of these inhomogeneities on the AGIS sensor orbiting at an altitude of 1000 km was estimated using a spherical harmonic model of the near-Earth gravity field (Hogan et al., 2011). Lunar tidal forces are also a significant effect below 1 mHz. Our analysis showed that free Earth oscillations would not be a significant measurement background, nor would the effect of gravitational waves on masses within the Earth. Fig. 6.7b shows a calculated strain sensitivity curve for a proposed set of interferometer parameters that could be feasible for satellites in low Earth orbit (LEO) based on the analysis in Hogan et al. 2011. Fig. 6.8 shows a strain sensitivity curve with a more ambitious set of parameters that might be realized in the further future, with the aim of showing potential for improvements as technology continues to develop. For comparison, sensitivity curves for existing (LIGO (Kawabe and the LIGO Collaboration, 2008), Advanced LIGO (Smith and the LIGO Collaboration, 2009), and VIRGO (Acernese et al., 2007)) and proposed (LISA (Larson et al., 2000)) laser interferometry detectors with macroscopic proof masses are also included. Much of the work described in these notes has been motivated in part by developing techniques for the very large area interferometers that would be required for an atomic gravitational wave detector.

6.6   Matter wave lensing

Gaining improved control of physical systems by making them increasingly cold has led to many important advances (Osheroff, 2002; Chu, 2002; Ketterle, 2001). As discussed in Sec. 6.3.3, from the point of view of atom interferometry, transversely cold atom sources[7] are critical for very large area interferometers, especially envisioned space-based interferometers with pulse spacings of tens of seconds that would be valuable for applications such as gravitational wave detection (Dimopoulos et al., 2008b; Hogan et al., 2011). The atom source for the 10 m atomic fountain uses evaporative cooling to reach effective temperatures of 20 nK for ^{87}Rb atoms. To reach even lower effective temperatures, we make use of a technique that we refer to as lensing due to its conceptual similarity to collimating light with a lens (Chu et al., 1986).

To perform lensing on an atomic ensemble, an initially small atom cloud is allowed to freely expand for a time $t_{\text{expansion}}$. It is useful to think of the lensing process in a phase space picture. For sufficient expansion time, the atom cloud will enter the point source limit, where the position and velocity will be proportionally correlated through the relation $x \approx v t_{\text{expansion}}$. A harmonic potential $V(x) = \frac{1}{2} k x^2$ is then pulsed on for a short time δt, corresponding to a position dependent restoring force $F(x) = -kx$. Because of the position-velocity correlation, the force as a function of velocity can be expressed as $F(v) \approx -kv t_{\text{expansion}}$, leading to a velocity dependent impulse $F(v)\delta t \approx -kv t_{\text{expansion}} \delta t$. If this velocity dependent impulse is equal to $- mv$ (m is the atomic mass) for each velocity v, implying that $\delta t = \frac{m}{k t_{\text{expansion}}}$, then all the atoms are brought to rest. This process is analogous to collimating light from a point source with a lens, where the role of the harmonic potential is played by the parabolic phase profile imprinted by the lens. We refer to the example just described, with the harmonic lensing potential applied for a short time, as the thin lens limit. Lensing can also be performed in the thick lens limit, where atoms from a point source expand against a constant harmonic potential. All velocities reach a turning point at the same time, at which point the harmonic potential is suddenly turned off (Kovachy et al., 2015a). Our work uses a dual stage lensing sequence with two types of harmonic potentials–one generated by magnetic fields and one arising from the transverse intensity profile of a red-detuned Gaussian laser beam (Kovachy et al., 2015a).

There are practical limitations to how well a lens can collimate an ensemble of atoms. First, the size of the atom cloud before expansion will never be zero, so there will always be deviations from the perfect point source limit. To be as far into the point source limit as possible, it is useful to have a long expansion time. This makes our 10 m atomic fountain ideally suited to lensing. Second, just as in standard optics, it is difficult to make a perfect lens (i.e. a perfect harmonic potential). Potentials are generally harmonic only near the center, and anharmonicities in the lensing potential cause the performance of the lens to be degraded by aberrations. As shown in Fig. 6.9, high spatial frequency perturbations to the lensing potential are especially harmful.

---

[7] For the purposes of these notes, we define 'cold' from the perspective of the effective temperature that quantifies the mean kinetic energy of the atoms (Metcalf and van der Straten, 1999), as this is the relevant quantity for the applications we consider, as well as for many other applications (Kovachy et al., 2015a).

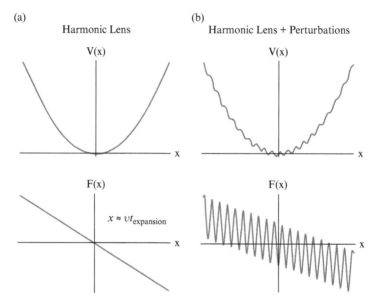

Fig. 6.9 Effect of high spatial frequency perturbations on the performance of a lens. (a) Harmonic potential V (x) and force $F(x)$ associated with an ideal lens. (b) Potential and force when high spatial frequency perturbations are added to the ideal lens. Even if the perturbations to the potential have a comparatively small amplitude, the resulting perturbations to the force can be large because they are proportional to the spatial frequency. These force perturbations can lead to significant distortion of the atom cloud.

This is because the resulting spurious forces are proportional to the spatial frequency. The work described in Kovachy et al. 2015a discusses the limitations of lensing in detail and demonstrates methods to mitigate them. These methods were used to reach effective temperatures as low as 50 pK.

6.7 Efficient optical lattice launching

6.7.1 Coherent optical lattice launch

As discussed in Sec. 6.4, we use a chirped optical lattice to launch the atoms. Conventional atomic fountains typically launch with moving optical molasses (Biedermann, 2007; Zhou et al., 2011). However, this would reheat the ultracold atoms. An optical lattice launch is an alternative, coherent process (Hecker Denschlag et al., 2002)[8]. The lattice acceleration results from repeated adiabatic rapid passage from $|p + 2n\hbar k\rangle$ to $|p + 2(n+1)\hbar k\rangle$[9]. As a result, the launch velocity is quantized in increments of

---

[8] Heating can still occur in a lattice (Pichler et al., 2010; Gerbier and Castin, 2010), but at a substantially lower level than in a moving molasses.

[9] There is also a satisfying classical picture. Consider a blue-detuned lattice with $\omega_1(t)$ the upward beam frequency and $\omega_2(t)$ the downward beam frequency. The atoms are trapped at the minima of the lattice intensity, which move in the lab frame at a velocity $v_{lat} = \frac{\omega_1 - \omega_2}{2k}$. (In a frame moving at velocity v_{lat}, the Doppler-shifted frequencies are equal, with $\omega_1 - kv_{lat} = \omega_2 + kv_{lat}$, and the

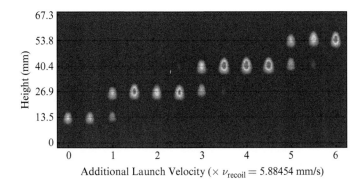

Fig. 6.10 Sequence of images demonstrating the velocity quantization of the lattice launch. For the leftmost image, 1142 photon recoil momenta are imparted to the atoms, launching them to a height of 1.25 m. Subsequent images show atoms launched with more velocity. In all cases, imaging occurs in the detection chamber at the bottom of the tower, 1.14 s after launch. As a result, atoms launched with more velocity have not fallen as far, and appear higher in the image.

$2\hbar k/m$, as can be seen in Fig. 6.10. An attempt to launch by imparting anything but an even-integer multiple of photon recoils yields the desired mean velocity, but has the additional effect of a $2\hbar k$ beamsplitter. For instance, a $(2n+1)\hbar k$ launch yields equal splitting into two clouds with momenta $(2n)\hbar k$ and $(2n+2)\hbar k$.

We therefore restrict the launch velocity to an even-integer multiple of the recoil velocity $v_L = 2n\hbar k/m$. Nevertheless, imperfections in the lattice launch (incomplete adiabaticity, Landau-Zener tunneling, etc.) cause a small fraction of the atoms to be launched into the $\pm 2\hbar k$ momentum states (Fig. 6.10). We can remove these side peaks with a Raman π-pulse with a velocity acceptance $\lesssim 2\hbar k/m$. For atoms prepared in state $|F = 2, m = 0\rangle$ prior to launch, this transfers only atoms in the central peak to $|F = 1, m = 0\rangle$, and the atoms in the side peaks can be removed with a subsequent blow-away pulse resonant with $|F = 2\rangle \rightarrow |F' = 3\rangle$.

The launch requires a deep lattice to reduce Landau-Zener tunneling while the atoms are accelerated. However, spontaneous emission losses also increase with the lattice depth. Appropriate intensities and detunings are therefore necessary to balance these effects (Sec. 6.7.3). Typical operating parameters for a full-tower launch are powers of $P_0 = 58$ mW per beam, a $1/e^2$ radial beam waist $w_0 = 1.5$ mm, and a single-photon detuning $\Delta = 2\pi \times 90$ GHz blue of the $|F = 2\rangle \rightarrow |F' = 3\rangle$ transition. These parameters yield a lattice depth of ~ 40 recoils[10].

---

lattice is at rest.) By chirping the frequency difference such that $\omega_1(t) - \omega_2(t) = 2kat$, the atoms can be accelerated with acceleration a. However, this picture doesn't explain the quantized launch velocities.

[10] For a lattice generated by retroreflection of a laser with a single-beam peak intensity $I_0 = \frac{2P_0}{\pi w_0^2}$, the spatially varying lattice intensity is $I(r, z) = 4I_0 e^{-2(r/w(z))^2}\cos^2(2\pi z/\lambda)$. This yields a potential $U(r, z) = \frac{\hbar}{4\Delta}\frac{\Gamma^2}{2}\frac{I(r,z)}{I_{sat}}$ via the Stark shift (Metcalf and van der Straten, 1999). The lattice depth in recoils is then $U(0, 0)/E_r$, where the recoil energy is $E_r = \frac{(\hbar k)^2}{2m}$.

To illustrate the lattice launching procedure, we describe a typical set of launch parameters. To adiabatically load the lattice, we ramp up the beam intensity in 8 ms while the downward laser's frequency is chirped to compensate for gravity (the lattice is at rest in the atoms' freely-falling frame as the intensity is ramped up). For a full-height (8.5 m) launch, we then chirp the upward laser's frequency at $50g$ for 29 ms to accelerate the atoms up to 13 m/s.[11] We then ramp down the lattice intensity, again over 8 ms while the downward laser compensates for gravity. The total height needed for the acceleration phase is 20 cm.

6.7.2 Lattice beam geometry

The lattice beams are arranged in a "W" configuration, as shown in Fig. 6.11b. This configuration has several benefits[12]. First, since the lattice beams need only interact with the atoms at the bottom of the tower, they can have a smaller beam waist than the atom optics beams. Introducing them off-axis facilitates focusing the lattice beams to a $1/e^2$ radial waist of $w_0 = 1.5$ mm at the retroreflection mirror. This allows for high lattice depths with modest laser powers. Introducing the lattice beams off-axis also allows them to bypass the upper quarter-wave plate (Fig. 6.11) so that their polarization can be set independently.

Another benefit to the "W" configuration is that there is no parasitic lattice. In a perfect retroreflection configuration with two beams input from above [see for example Fig. 6.11a, ignoring the circular polarization], two lattices would be formed (beam 1 downwards interfering with beam 2 upwards, and vice versa). For atoms initially at rest, one of these lattices launches the atoms upwards, but the second, parasitic lattice would launch the atoms downwards. For atoms trapped in the correct lattice moving at a finite velocity, the parasitic lattice is Doppler detuned. However, whenever the atoms transition through being at rest in the lab frame, both lattices would become resonant, adversely affecting the transfer efficiency of atoms into the desired momentum state.

Furthermore, even when Doppler-detuned, the parasitic lattice would contribute to atom loss via spontaneous emission. The "W" configuration ensures that only one downward beam and one upward beam interact with the atoms. In particular, this facilitates the full benefits of blue-detuning suppression (Sec. 6.7.3). By operating the lattice with blue single-photon detunings, the atoms accumulate at the nodes of the lattice. As a result, the effective intensity is smaller, and spontaneous emission losses are reduced.

---

[11] Just prior to acceleration by the lattice, the atoms have a downward velocity of ~ 1 m/s, so the velocity change imparted by the lattice is actually 14 m/s. The downward initial velocity results from the downward velocity of the atoms at the end of the magnetic lens

[12] One disadvantage to the "W" configuration is that there is a limited region of overlap of the downward- and upward-going beams [the diamond in Fig. 6.11b]. The opening angle of the lattice is $\theta_{lat} = \frac{2 \text{ cm}}{10 \text{ m}} = 2$ mrad. We can estimate the vertical height of the overlap region as $h_{lat} = \frac{w_0}{\theta_{lat}} = 75$ cm, which is substantially longer than the 20 cm height needed for a typical launch. Similarly, the spacing between the beams at the bottom is 3.7 mm $\approx 2.5w_0$. Thus, at a distance of w_0 from the center of the overlap region the beams on the outside of the "W" are reduced from their maximum intensity by $e^{-2\left(\frac{2.5w_0 - w_0}{w_0}\right)^2} \approx 1\%$.

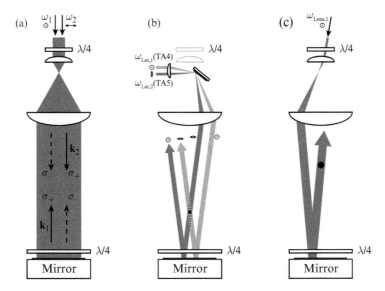

Fig. 6.11 Vertical (z) axis laser beam delivery. The laser beams are launched from an optics box above the interferometer region, and retroreflect from a mirror at the bottom. The primary telescope at the top of the tower is also shown. (a) Large-diameter beams for the atom optics, in a σ^+-σ^+ configuration. (b) Lattice launch beams. The divergence of the lattice beams is not shown for simplicity. They actually go through a focus near the retroreflection mirror. The polarizations indicated are idealized (but nearly correct prior to retroreflection). In actuality, the $\lambda/4$ at the bottom is not at $45°$ relative to the input polarization, and the retroreflected polarizations are rotated by $29°$. Each lattice beam originates from a tapered amplifier (TA). (c) Optical dipole beam path used for atom lensing (see Sec. 6.6). Angles in this figure are exaggerated.

The "W" configuration maintains many of the benefits of a retroreflection configuration. The beams originate from the same fiber with crossed polarizations and are separated by a walk-off prism. As a result, their beam profiles are nearly identical. Further, the launch angle is set by the normal vector to the retroreflection mirror and the relative angle between the two downward-going beams but is insensitive to the absolute alignment of the beam pair. The mirror angle is set by in-vacuum closed-loop precision piezo actuators, and the input beam parallelism is set by the walk-off prism. Both therefore contribute to the stability of the launch angle.

6.7.3 Optimizing lattice launch parameters

The detuning and intensity of the lattice lasers must be chosen to minimize atom losses during the launch. Again, there is a balance between Rabi frequency (the lattice depth and corresponding Landau-Zener losses) and spontaneous emission that favors large detunings and high intensities. Because the atoms are trapped in the deep lattice during launch, there is also a detuning-dependent suppression or enhancement of the spontaneous emission rate. For blue lattice detuning, the atoms are trapped at nodes of the standing wave, and spontaneous emission is suppressed. For red lattice detuning, the atoms are trapped at the regions of maximum intensity, and spontaneous emission is enhanced. While the full lattice launching process is somewhat complicated (Pichler et al., 2010; Gerbier and Castin, 2010), we can get a rough estimate of the loss rates.

The scattering rate due to spontaneous emission from a single beam with intensity I_0 and detuning Δ is given by

$$R_{\rm sc} = \frac{\Gamma}{2} \frac{(I_0/I_{\rm sat})}{(2\Delta/\Gamma)^2} \quad (\text{single beam}, \Delta \gg \Gamma). \tag{6.33}$$

For blue detuning, we can account for the suppression of spontaneous emission by finding the expectation value of the intensity $\langle I \rangle$ experienced by an atom in the lattice ground state. The suppression factor is[13]

$$\eta_{\rm lat} \equiv \frac{\langle I \rangle}{I_0} = \frac{2}{\Gamma} \frac{2\pi}{\lambda} \sqrt{\frac{\hbar\Delta}{m}} \sqrt{\frac{I_{\rm sat}}{I_0}} \quad (\text{blue-detuned pure lattice}). \tag{6.34}$$

Our lattice is not a perfect standing wave. While the upward- and downward-going beam intensities were balanced to a few percent for the data presented in this section, the orientation of the linear polarizations of the two beams was not ideal (the polarization axes differ by $\theta = 29^\circ$, see Fig. 6.11). As a result, the on-axis intensity profile has a traveling wave component in addition to the lattice:

$$I(z) = 4I_0 \sin^2\left(\frac{\theta}{2}\right) + 4I_0 \sin^2\left(\frac{2\pi}{\lambda} z\right)\left(\cos^2\left(\frac{\theta}{2}\right) - \sin^2\left(\frac{\theta}{2}\right)\right). \tag{6.35}$$

The spontaneous emission reduction factor (Eq. 6.34) is therefore slightly reduced:

$$\eta_{\rm tot} = 4\sin^2\left(\frac{\theta}{2}\right) + \eta_{\rm lat}\sqrt{\cos^2\left(\frac{\theta}{2}\right) - \sin^2\left(\frac{\theta}{2}\right)} \quad (\text{blue detuning}). \tag{6.36}$$

A similar expression can be found for the red detuning spontaneous emission enhancement factor:

$$\eta_{\rm tot} = 4\sin^2\left(\frac{\theta}{2}\right) + (4 - \eta_{\rm lat})\sqrt{\cos^2\left(\frac{\theta}{2}\right) - \sin^2\left(\frac{\theta}{2}\right)} \quad (\text{red detuning}). \tag{6.37}$$

In either case, the fraction of the atoms surviving the launch without undergoing a spontaneous emission event is

$$f_{\rm sp}^{\rm surv} = \exp\left(-\eta_{\rm tot} R_{\rm sc} \delta t_{\rm lat}\right) \tag{6.38}$$

---

[13] Near a minimum of the lattice intensity, the lattice potential can be approximated as a simple harmonic oscillator: $U \approx \frac{1}{2} m\omega_{\rm osc}^2 z^2$, where $\omega_{\rm osc}$ can be determined from the Stark shift and a Taylor expansion of the lattice intensity profile. The expectation value of the potential is then $\langle U \rangle = \frac{1}{4}\hbar\omega_{\rm osc}$. $\langle I \rangle$ can then be found from the Stark shift, $\langle U \rangle = \frac{\hbar}{4\Delta} \frac{\Gamma^2}{2I_{\rm sat}} \langle I \rangle$. This analysis assumes both that there are no spectator beams (a benefit of the "W" geometry) and that the lattice beams are single-frequency.

where δt_{lat} is the duration of the lattice launch (accounting for the intensity ramp-up and ramp-down times).

Landau-Zener tunneling also contributes to atom losses. Classically, if the lattice depth is insufficiently large during the acceleration phase, the atoms can slip out of the lattice wells. Quantum mechanical tunneling allows this loss to occur even at larger lattice depths. The Landau-Zener loss rate can be estimated by considering the launch as a sequence of Bloch oscillations. For a launch velocity of v_L, the atoms must pass through $N_L = \frac{mv_L}{2\hbar k}$ avoided crossings, each with a band gap energy of $\hbar\Omega_{\text{bg}} \sim \hbar\Omega_{\text{eff}}$.[14] The fraction of atoms surviving the launch without Landau-Zener tunneling is

$$f_{\text{LZ}}^{\text{surv}} = \left[1 - \exp\left(-\frac{\pi}{2}\frac{\Omega_{\text{bg}}^2}{\alpha}\right)\right]^{N_L} \tag{6.39}$$

where $\alpha = 2ka_L$ is the frequency chirp rate for a launch with acceleration a_L.

The total number of atoms surviving the launch without Landau-Zener tunneling or undergoing a spontaneous emission event is then

$$f_{\text{tot}}^{\text{surv}} = f_{\text{LZ}}^{\text{surv}} \times f_{\text{sp}}^{\text{surv}}. \tag{6.40}$$

For sufficiently long drift times after the launch, only $f_{\text{tot}}^{\text{surv}}$ of the atoms will remain in the final image. Figure 6.12 shows the theoretical predictions of Eq. 6.40 for three different blue detunings and various lattice intensities. The characteristic Landau-Zener cliff is apparent at low intensities, and spontaneous emission dominates at high intensities. The peak atom survival fraction $f_{\text{tot}}^{\text{surv}}$ increases with increasing detuning Δ and occurs at higher intensities I_0 for larger detunings.

Figure 6.12 also shows experimental results for launches with several intensities and blue detunings. The data support the simple theoretical model, including blue-detuning suppression. In particular, the ratio of the peak heights predicted without spontaneous emission suppression (dashed curves) is greater than that observed in the data[15]. At higher intensities, the data fall more quickly than the theory would predict. It is possible that this results from a reduced frequency purity of the lattice beams as the tapered amplifiers are driven harder, but further study is needed to confirm this.

Figure 6.13 shows data and theoretical predictions for $f_{\text{tot}}^{\text{surv}}$ for 46 GHz red detuning and 46 GHz blue detuning. While the data do not agree perfectly with the model (and the imaging duration is different between the two data sets), it is clear that the red-detuned lattice launch is less effective. Fig. 6.13b shows the detected atom distribution for red- and blue-detuned launches at optimized lattice intensities. The red-detuned launch yields a much broader distribution (broader than the camera's viewing region). As a result, peak counts, rather than total counts, are used in Fig. 6.13a. The broad profile could result in part from atoms that have undergone

---

[14] The exact band gap can be calculated numerically by solving for the lattice band structure. This has been done in Fig. 6.12 and Fig. 6.13. See also Sec. 6.8.

[15] The spontaneous emission suppression is greatest at smaller detunings where the spontaneous emission rate is higher.

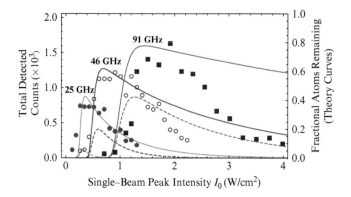

Fig. 6.12 Atom number detected as a function of the lattice intensity and detuning, for an 8.5 m launch. All indicated detunings are blue and are relative to the $|F = 2\rangle \rightarrow |F' = 3\rangle$ transition. Each point is the average of 5 separate experimental trials. A fluorescence imaging pulse 1 ms in duration yields the indicated number of total counts (the peak atom number for the 91 GHz data corresponds to $\sim 10^5$ atoms). The dashed curves are theoretical predictions accounting for Landau-Zener and spontaneous emission losses, but ignoring blue-detuning suppression. The solid curves account for blue-detuning suppression for a lattice in which the upward- and downward-going beam polarization axes differ by 29°. The relative vertical scales for the data and the theory are set by fitting the 25 GHz data to the corresponding solid curve. There are no other free parameters. The "W" configuration is used, and the upward and downward beam intensities are balanced to a few percent.

spontaneous emission and have yet to escape the imaging region. Dipole lensing from the lattice beams could also contribute to a change in the detected atom distribution. Since the magnetic lens was optimized for a blue-detuned launch, this could broaden the red-detuned distribution. In this case, the use of peak counts in Fig. 6.13a is overly optimistic, and the spontaneous emission enhancement for red-detuning would not be as dramatic as shown.

6.8   Theory of atom optics with optical lattices

We provide an analytical description of the dynamics of an atom in an optical lattice using the method of perturbative adiabatic expansion. A precise understanding of the lattice-atom interaction is essential to taking full advantage of the promising applications that optical lattices offer in the field of atom interferometry. One such application is the implementation of Large Momentum Transfer (LMT) beam splitters that can potentially provide multiple order of magnitude increases in momentum space separations over current technology. We also propose interferometer geometries where optical lattices are used as waveguides for the atoms throughout the duration of the interferometer sequence. Such a technique could simultaneously provide a multiple order of magnitude increase in sensitivity and a multiple order of magnitude decrease in interferometer size for many applications as compared to current state-of-the-art atom interferometers. This section also appeared as Kovachy et al. 2010.

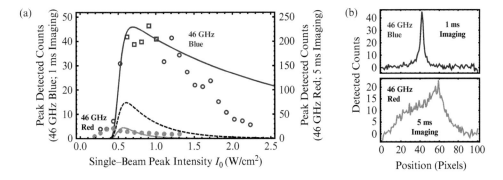

Fig. 6.13 (a) Peak counts detected as a function of the lattice intensity for red and blue detunings, for an 8.5 m launch. All indicated detunings are relative to the $|F=2\rangle \to |F'=3\rangle$ transition. The 46 GHz blue-detuned data are as in Fig. 6.12 (though peak counts are given instead of total counts). Each point is the average of 5 separate experimental trials (3 trials for the red-detuned data). The fluorescence imaging pulse duration is 1 ms for the blue-detuned data, but 5 ms for the red-detuned data (for improved signal-to-noise). As a result, the data are presented with their relative axes scaled by a factor of 5. The black dashed curve is a theoretical prediction accounting for Landau-Zener and spontaneous emission losses, but ignoring detuning suppression/enhancement. The solid curves account for detuning suppression/enhancement for a lattice in which the upward- and downward-going beam polarization axes differ by $29°$. The vertical scale for the theory curves is adapted from Fig. 6.12 to account for the switch from total to peak counts (specifically the mean ratio of peak to total counts for the five square points is used). (b) Profiles of the detected atom distribution (the pixel size is 675 μm). For the blue-detuning profile, the lattice launch intensity is 0.62 W/cm^2. For the red-detuning profile, it is 0.43 W/cm^2. The full width of the CCD is shown, so vignetting likely contributes substantially to the shape of the red-detuning profile.

6.8.1 Overview

As discussed in Sec. 6.3.2, adiabatically accelerating atoms in an optical lattice is an example of an adiabatic rapid passage method for LMT atom optics. As such, the transfer efficiency of optical lattice manipulations of the atoms is highly robust against Rabi frequency and detuning inhomogeneities. The excellent transfer efficiency of lattice-based atom optics is illustrated by the fact that the lattice launch for the 10 m atomic fountain transfers momentum $\sim 2000\hbar k$ to the atoms with $\sim 30\%$ overall efficiency (Sugarbaker, 2014).

The utility of atom interferometry for precision measurements hinges upon the ability to precisely calculate the phase accumulated along the different arms of an interferometer (Bongs et al., 2006; Dubetsky and Kasevich, 2006; Bordé, 2004), of which the phase acquired during interactions of the atoms with light is an important component. Indeed, the phase obtained by an atom during a Raman or Bragg pulse is well-understood (Hogan et al., 2009; Müller et al., 2008; Moler et al., 1992; Büchner et al., 2003). Analogously, in order to take full advantage of the potential of lattice beam splitters, we must have a detailed understanding of the phase evolution of an atom in an optical lattice. In this section, we provide a rigorous analytical treatment of this problem.

Based on this analysis, we propose atom interferometer geometries in which optical lattices are used to continuously guide the atoms, so that the atomic trajectories are precisely controlled for the duration of the interferometer sequence, with a different lattice guiding each arm of the interferometer. We point out here a distinction in terminology between a lattice waveguide and a lattice beam splitter. Here, a lattice waveguide is the use of a lattice to continuously control the trajectory of an arm of an atom interferometer. We note that two different lattice waveguides can independently control the two arms of an interferometer, or a single lattice waveguide can simultaneously control both arms. In contrast, a lattice beam splitter is an interaction of relatively short time (in comparison to a waveguide) with the primary purpose of splitting the arms of the interferometer in momentum space rather than providing continuous trajectory control.

Our analysis indicates that these lattice interferometers have the potential to offer unprecedented sensitivities for a wide variety of applications and to operate effectively over distance scales previously considered too small to be studied by precision atom interferometry. For example, one particularly interesting configuration involves using two optical lattice waveguides to continuously pull the two arms of the interferometer apart, subsequently holding the two arms a fixed distance from each other in a single lattice waveguide that is common to the two arms, and then using two lattice waveguides to recombine the arms. A pair of such interferometers could be used, for instance, as a gravity gradiometer. The sensitivity of lattice interferometers is illustrated by the fact that, given the experimental parameters stated in Hogan et al. 2009 (10^7 atoms/shot and 10^{-1} shots/s), a shot noise limited lattice gradiometer with interferometer path separation of 1 m and an interrogation time of 10 s has a differential acceleration sensitivity of $10^{-14} g / \mathrm{Hz}^{1/2}$ (this corresponds to a measurement resolution of $10^{-14} g$ after 10 s, where 10 s is the duration of a single measurement cycle). We perform phase shift calculations for these lattice interferometers using the theoretical groundwork formulated in this section, and we discuss how lattice interferometers have the potential to both exceed the performance of conventional atom interferometers in many standard applications and expand the types of measurements that can effectively be carried out using atom interferometry.

This section is organized as follows. Sec. 6.8.2 describes the Hamiltonian for an atom in an optical lattice in the different frames we use. Sec. 6.8.3 discusses the phase evolution of an atom in an optical lattice under the adiabatic approximation. Sec. 6.8.4 introduces the formalism of perturbative adiabatic expansion to calculate corrections to the adiabatic approximation, and Sec. 6.8.5 applies this formalism to calculate phase corrections to a lattice beam splitter. Sec. 6.8.6 proposes a number of interferometer geometries that make use of lattice manipulations of the atoms. Main results include Eqs. 6.65 and 6.66, which show how to obtain analytical corrections to the lowest order phase shift estimates. These corrections are surprisingly large, and understanding them is vital to realizing the full accuracy of the sensor geometries proposed in Sec. 6.8.6, as well as other geometries utilizing optical lattice manipulations of the atoms. For example, the gravitational wave detector proposed in Dimopoulos et al. 2008b may make use of lattice beam splitters and/or waveguides.

6.8.2 The Hamiltonian in different frames

An optical lattice is a periodic potential formed by the superposition of two counter-propagating laser beams. Atoms can be loaded into the ground state of the lattice by ramping up the lattice depth adiabatically, and the lattice can then be used to impart momentum to the atoms and/or to control the atoms' trajectories. Optical lattices are thus a useful tool for atom optics.

We begin our discussion of the lattice-atom interaction by finding a useful form for the Hamiltonian. As is typical for many applications of atom interferometry, we assume that we work with atomic gases dilute enough so that the effects of atom-atom interactions are negligible. We first consider the Hamiltonian in the lab frame, where for now we assume a vertical configuration with constant gravitational acceleration g so that we have a gravitational potential given by mgx. We expose the atom to a superposition of an upward-propagating beam with phase $\phi_{\text{up}}(t)$ and a downward-propagating beam with phase $\phi_{\text{down}}(t)$, which couples an internal ground state $|g\rangle$ to an internal excited state $|e\rangle$. The two-photon Rabi frequency is $\Omega(t) \equiv \frac{\Omega_{\text{up}}(t)\Omega_{\text{down}}(t)}{2\Delta}$, where we let $\Omega_{\text{up}}(t)$ denote the single-photon Rabi frequency of the upward-propagating beam, $\Omega_{\text{down}}(t)$ denote the single-photon Rabi frequency of the downward-propagating beam, and Δ denote the detuning from the excited state. We depict the physical setup in Fig. 6.14. Making the rotating wave approximation and adiabatically eliminating the excited state as is standard procedure (Berman, 1997), we obtain the following Hamiltonian, where the periodic term in the potential arises from a spatially varying AC Stark shift and where k is the magnitude of the wave vector of the laser beams (Wicht et al., 2005; Peik et al., 1997):

$$\hat{H}_{\text{Lab}} = \frac{\hat{p}^2}{2m} + 2\hbar\Omega(t)\sin^2\left[k\hat{x} - \frac{1}{2}(\phi_{\text{up}}(t) - \phi_{\text{down}}(t))\right] + mg\hat{x}. \tag{6.41}$$

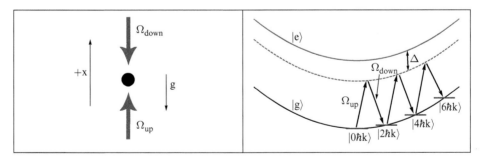

Fig. 6.14 The physical setup for applying a periodic potential. We expose atoms to counter-propagating laser beams with respective single-photon Rabi frequencies Ω_{up} and Ω_{down}. The lasers are detuned from the transition between the atom's internal ground state and excited state so that the atom's external momentum states are coupled through two-photon transitions, creating an effective lattice potential.

Note that where the difference between the frequency of the upward-propagating beam and the frequency of the downward-propagating beam is denoted by $\Delta\omega(t)$, we will have the relation $\Delta\phi(t) \equiv \phi_{\text{up}}(t) - \phi_{\text{down}}(t) = \int_0^t \Delta\omega(t')dt' + \phi_{\text{up}}(0) - \phi_{\text{down}}(0)$. For a given $\Delta\phi(t)$, the lattice standing wave will be translated by $D_{\text{Lab}}(t) \equiv \frac{\Delta\phi(t)}{2k}$ in the x direction from the origin. Thus, the velocity of the lattice in the lab frame is:

$$v_{\text{Lab}}(t) = \frac{d}{dt}D_{\text{Lab}}(t) = \frac{\Delta\omega(t)}{2k}, \tag{6.42}$$

and we rewrite the lab frame Hamiltonian as:

$$\hat{H}_{\text{Lab}} = \frac{\hat{p}^2}{2m} + 2\hbar\Omega(t)\sin^2\left[k\hat{x} - kD_{\text{Lab}}(t)\right] + mg\hat{x}. \tag{6.43}$$

In order to most readily describe the dynamics of an atom in an accelerating optical lattice, it is useful to work in momentum space. The $mg\hat{x}$ term that appears in the lab frame Hamiltonian makes such an approach difficult, especially when considering non-adiabatic corrections to the phase shift. However, we can change frames by performing a unitary transformation in order to obtain a Hamiltonian that is easier to handle analytically. In the end, we will see that approaching the problem from the point of view of dressed states provides a convenient Hamiltonian for our purposes. We consider the transformation procedure from the lab frame to the dressed state frame in Sec. 6.8.8, where we also introduce an intermediate frame that freely falls with gravity (which we call the freely falling frame). We note that the general form of the unitary transformations considered in Sec. 6.8.8 as well as the specific transformations to the different frames we consider can also be found in the Appendix of Peik et al. 1997.

It is convenient to absorb the initial velocity v_0 of the atom in the lab frame into the dressed state frame, so that velocity v_0 in the lab frame corresponds to velocity zero in the dressed state frame. The Hamiltonian in the dressed state frame is, as derived in Sec. 6.8.8:

$$\hat{H}_{\text{DS}} = \frac{\hat{p}^2}{2m} - (v_{\text{Lab}}(t) + gt - v_0)\hat{p} + 2\hbar\Omega(t)\sin^2(k\hat{x}). \tag{6.44}$$

The ability to boost to a frame in which the Hamiltonian contains no position-dependent terms outside of the lattice potential is contingent upon the assumption that the external potential in the lab frame (not including the lattice potential) is linear in x. However, real-world potentials such as the potential corresponding to Earth's gravitational field will deviate somewhat from this assumption. Any such deviations would manifest as residual position-dependent terms in the dressed state Hamiltonian, which we collectively refer to as V'. Under the semiclassical approximation, we neglect the effects of V' on the time evolution of the atomic wavepacket. This approximation is valid when the energy scale of V' over the spread of the atom's wavefunction (which is on the order of magnitude of the expectation value of V' in the atomic wavepacket) is much smaller than the energy scale of the lattice potential and is small relative to the time scale of the experiment, which is the case for a wide class of experimental parameters. For example, in the case of a rubidium atom wavepacket with a spatial spread of $\sigma \sim 100\mu m$ in the gravitational field at the Earth's surface (which has a gradient of $\lambda \sim 10^{-6} s^{-2}$), the energy scale of V' will be $\sim \frac{1}{2} m\lambda\sigma^2 \sim h \times (10^{-6} \, \text{Hz})$. This energy scale is smaller than that of a lattice of typical experimental depth ($\sim 5E_r$, where E_r is the recoil energy $\frac{\hbar^2 k^2}{2m}$) by a factor of $\sim 10^{10}$ and is small on a time scale of ~ 10 s . The effects of linear gradients and of more general potentials can be accounted for through a straightforward generalization of the results presented here, as discussed in greater depth in Kovachy 2009.

Now, we will show how working in momentum space allows us to represent $\hat{H}_{\text{DS}}$ as an infinite dimensional, discrete matrix. This matrix is discrete because the optical lattice potential term, $V_0 \sin^2(k\hat{x})$, only couples a momentum eigenstate $|p\rangle$ to the eigenstates $|p + 2\hbar k\rangle$ and $|p - 2\hbar k\rangle$ (Peik et al., 1997). For the moment, we will examine the evolution of individual eigenstates of the dressed state Hamiltonian $\hat{H}_{\text{DS}}$. These eigenstates reduce to single momentum eigenstates $|p\rangle$ when $\Omega = 0$. The knowledge of how each of these eigenstates evolves under $\hat{H}_{\text{DS}}$ will allow us to describe the dynamics of an entire wavepacket. For the moment, we will only consider momentum eigenstates corresponding to an integer multiple of $2\hbar k$, since we have boosted away the initial velocity v_0 of the atom in the lab frame. We note that it is always possible to transform to a particular dressed state frame in which a given momentum eigenstate in the lab frame corresponds to zero momentum in that dressed state frame. The results we derive here can thus be readily generalized to arbitrary momentum eigenstates in a wavepacket, as we discuss in greater detail in Sec. 6.8.9.

We consider a discrete Hilbert space spanned by the momentum eigenstates $|2n\hbar k\rangle$ for integers n, so that we can express any vector in this Hilbert space as:

$$|\Psi(t)\rangle = \sum_{n=-\infty}^{\infty} c_n(t) |2n\hbar k\rangle. \tag{6.45}$$

Since this Hilbert space is discrete, it is natural to adopt the normalization convention that $\langle 2m\hbar k|2n\hbar k\rangle = \delta_{mn}$. When considered as an operator acting on this discrete Hilbert space, $\hat{H}_{\text{DS}}$ can be written as (Müller et al., 2008; Malinovsky and Berman, 2003)

$$\hat{H}_{\text{DS}}^{\text{discrete}} = \sum_{n=-\infty}^{\infty} \frac{(2n\hbar k)^2}{2m} |2n\hbar k\rangle \langle 2n\hbar k| - \sum_{n=-\infty}^{\infty} (v_{\text{Lab}}(t) + gt - v_0)(2n\hbar k) |2n\hbar k\rangle \langle 2n\hbar k|$$

$$- \sum_{n=-\infty}^{\infty} \hbar \frac{\Omega(t)}{2} (|2n\hbar k\rangle \langle 2(n-1)\hbar k| + |2n\hbar k\rangle \langle 2(n+1)\hbar k|), \tag{6.46}$$

where we drop the common light shift. Now, it is convenient to introduce the recoil frequency $\omega_r \equiv \frac{E_r}{\hbar} = \frac{\hbar k^2}{2m}$ and the recoil velocity $v_r \equiv \frac{\hbar k}{m}$. In order to make our notation as compact as possible, we will be interested in the quantity $\alpha(t) \equiv \frac{v_{\text{Lab}}(t) + gt - v_0}{v_r}$, which is the velocity of the lattice in the dressed state frame in units of v_r. We can express the second term of $\hat{H}_{\text{DS}}^{\text{discrete}}$ in a useful way by noting that

$(v_{\text{Lab}}(t) + gt - v_0)2n\hbar k = 4n\alpha(t)E_r$. Furthermore, we define $\tilde{\Omega}(t) \equiv \frac{\Omega(t)}{8\omega_r}$. We can now write the discrete Hamiltonian in a simplified form:

$$\hat{H}_{\text{DS}}^{\text{discrete}} = 4E_r \sum_{n=-\infty}^{\infty} [n^2 |2n\hbar k\rangle \langle 2n\hbar k| - n\alpha(t)|2n\hbar k\rangle \langle 2n\hbar k|$$
$$- \tilde{\Omega}(t)\left(|2n\hbar k\rangle \langle 2(n-1)\hbar k| + |2n\hbar k\rangle \langle 2(n+1)\hbar k|\right)]. \qquad (6.47)$$

The matrix elements of this Hamiltonian are:

$$H_{mn} \equiv \langle 2m\hbar k| \hat{H}_{\text{DS}}^{\text{discrete}} |2n\hbar k\rangle = 4E_r \left[\left(n^2 - n\alpha(t)\right)\delta_{mn} - \tilde{\Omega}(t)\left(\delta_{m,n+1} + \delta_{m,n-1}\right)\right]. \qquad (6.48)$$

Having derived the above matrix elements, from now on it will be convenient to perform calculations in matrix notation. We define the following notation: we let H be the Hamiltonian matrix whose element in the mth row and nth column is given by H_{mn}, and we let $\vec{\Psi}(t)$ be the column vector whose nth entry is $c_n(t)$. Note that the vector $\vec{\Psi}(t)$ is the matrix representation of the state $|\Psi(t)\rangle$ in the basis of momentum eigenstates $|2n\hbar k\rangle$ for integers n.

6.8.3   Phase evolution under the adiabatic approximation

Now that we have determined the Hamiltonian matrix H, we have the appropriate machinery in place to describe the phase evolution of an atom in an optical lattice. We consider the process in which momentum is transferred to the atom through Bloch oscillations. Peik et al. 1997 provides a thorough and insightful description of Bloch oscillations in a number of different pictures. Given our choice of Hamiltonian, we work in the dressed state picture, which is discussed in Sec. IV.B. of Peik et al. 1997, making extensive use of Bloch's theorem, the concept of Brillouin zones, and the band structure of the lattice (Hecker Denschlag et al., 2002). As in the previous discussion, we consider the evolution of single eigenstates of the dressed state Hamiltonian, noting that we can easily generalize our results to the case of a wave packet of finite width, as we address in Sec. 6.8.9.

Initially, we consider the system to be in a momentum eigenstate. First, we adiabatically ramp up the lattice depth by increasing the laser power so that we load the system into an eigenstate of the lattice Hamiltonian. For the purposes considered here, we want the system to enter the ground eigenstate (corresponding to the zeroth band of the lattice). In order for this to be achieved, a resonance condition must be met, requiring that the velocity of the lattice must match the velocity of the atom to within v_r. The loading will be adiabatic if the adiabatic condition $\left|\langle 1| \dot{H} |0\rangle\right| \ll \frac{(\varepsilon_1 - \varepsilon_0)^2}{\hbar}$ is satisfied, where $|0\rangle$ and $|1\rangle$ respectively denote the ground state and the first excited state of the Hamiltonian (Peik et al., 1997; Hecker Denschlag et al., 2002). This condition will be easier to meet near the center of the band (where the velocity of the lattice is identical to the velocity of the atom), because the energy gap $\varepsilon_1 - \varepsilon_0$ between the zeroth band and the first band becomes smaller as the velocity difference between the lattice and the atom becomes larger. This corresponds to moving toward the border of the first Brillouin zone. The resonance condition is discussed further in Kovachy 2009.

In the lab frame, the atom accelerates under gravity, increasing the deviation between its velocity and the lattice velocity during the loading process. In the freely falling and dressed state frames, in which gravity is boosted away, this corresponds to the lattice accelerating upward while the atom remains at rest. This effect can negatively impact the loading efficiency if the loading sequence is sufficiently long so that the accrued velocity difference becomes a significant fraction of v_r. In such a scenario, the effect can be ameliorated by accelerating the lattice in the lab frame to fall with the atom, which corresponds to the lattice velocity remaining constant in the freely falling and dressed state frames.

Furthermore, we note that in the case of a lattice LMT beam splitter, the lattice should only be resonant with one arm of the interferometer, so that negligible population from the other arm is affected. Otherwise, the signal could be distorted by multi-path interference, causing a systematic error in the estimation of the interferometer phase shift. Conditions for when the negative effects of off-resonant lattices can be avoided can be estimated using the Hamiltonian matrix for an off-resonant lattice given in Eq. 6.70. We discuss off-resonant lattices quantitatively and in more detail in Sec. 6.8.6.

After the adiabatic loading of the atom into the ground state of the Hamiltonian, the frequency difference between the laser beams is swept to accelerate the lattice, periodically imparting momentum to the atom in units of $2\hbar k$. In the dressed state frame, this phenomenon can be understood in terms of avoided line crossings, which occur because the coupling between the atom and the laser beams lifts the degeneracy at the crossing points. We refer the reader to fig. 10 of Peik et al. 1997 for a clear illustration of these avoided crossings. As the frequency difference is swept so that the system passes through the avoided crossings, the system remains in the ground state of the dressed state Hamiltonian as long as the process is adiabatic. Consequently, at each of the avoided crossings, the momentum of the atom increases by $2\hbar k$, which corresponds to a Bloch oscillation. Finally, after the acceleration of the lattice, the frequency difference is held constant while the lattice depth is adiabatically ramped down, delivering the system into the momentum eigenstate $|p_i + 2N\hbar k\rangle$, where p_i is the momentum before the Bloch oscillations and N is the number of Bloch oscillations.

Fig. 6.15 depicts the lattice depth and velocity as functions of time for the process described above and shows a numerical simulation of a particular instance of this process: the adiabatic loading of the lattice from an initial state $|0\hbar k\rangle$, the transfer of $10\hbar k$ of momentum through five Bloch oscillations, and the ramping down of the lattice to deliver the system into the final state $|10\hbar k\rangle$.

Since under the adiabatic approximation we assume that the atom always stays in the ground state of the Hamiltonian, the phase $\phi(t)$ of the atom evolves as follows:

$$\phi(t) - \phi_0 = -\frac{1}{\hbar} \int_{t_0}^{t} \varepsilon_0(\alpha, \tilde{\Omega}) dt' \tag{6.49}$$

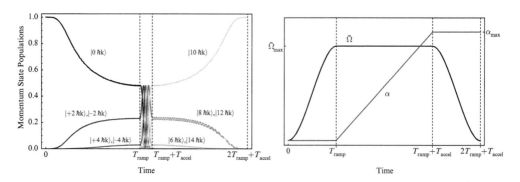

Fig. 6.15 The left panel shows a numerical simulation of a lattice acceleration in momentum space that transfers $10\hbar k$ of momentum. The lattice depth and velocity are shown as functions of time in the right panel, where in this particular case the relevant parameters are $T_{\text{ramp}} = 120(4\omega_r)^{-1}$, $T_{\text{accel}} = 16(4\omega_r)^{-1}$, $\tilde{\Omega}_{\text{max}} = \frac{19.5}{8}$, and $\alpha_{\text{max}} = 10$ (corresponding to a final lattice velocity of 10 v_r). First, we adiabatically ramp up the lattice to a depth of $19.5E_r$ so that the lattice is loaded into the ground state of the dressed state Hamiltonian. Subsequently, we accelerate and ramp down the lattice, leaving the atom in a single momentum eigenstate.

where $\varepsilon_0(\alpha,\tilde{\Omega})$ is the instantaneous ground state eigenvalue of the Hamiltonian $H(\alpha,\tilde{\Omega})$ (and it is understood that α and $\tilde{\Omega}$ are functions of time) and ϕ_0 is the initial phase of the atom. In addition to Eq. 6.49, there is also a Berry's phase term (Griffiths, 2005). However, this term is zero for a linear external potential. Therefore, there is no contribution from the Berry's phase under the semiclassical approximation, as long as the external potential is treated as linear. We discuss the validity and ramifications of this approximation in Sec. 6.8.2. Note that any such contribution would arise from the residual external potential terms of the dressed state Hamiltonian that are non-linear in x. We collectively denote these terms as V' in Sec. 6.8.2 and explain why they can often be neglected.

We now consider how the eigenvectors and eigenvalues of $H(\alpha,\tilde{\Omega})$ change with α, the dimensionless velocity of the lattice in the dressed state frame. Say that the eigenvalues of $H(\alpha,\tilde{\Omega})$ are given by $\varepsilon_n(\alpha,\tilde{\Omega})$ with corresponding eigenvectors $\vec{\Psi}_n(\alpha,\tilde{\Omega})$, where the index n runs from 0 to ∞. We choose to index the eigenvalues so that $\varepsilon_n(\alpha,\tilde{\Omega})$ denotes the the nth eigenvalue labeled in order of increasing value. Moreover, we let $c_j^{(n)}(\alpha,\tilde{\Omega})$ be the jth element of the column vector $\vec{\Psi}_n(\alpha,\tilde{\Omega})$, so that $\left|\Psi_n(\alpha,\tilde{\Omega})\right\rangle = \sum_{j=-\infty}^{\infty} c_j^{(n)}(\alpha,\tilde{\Omega})|2j\hbar k\rangle$. The transformation properties of the eigenvectors and eigenvalues under changes in α can be deduced from Bloch's theorem. It can be shown that when the lattice velocity is increased by $2v_r = \frac{2\hbar k}{m}$ (which corresponds to α being increased by two) while $\tilde{\Omega}$ is kept fixed, the new eigenvectors can be obtained through the following relation (Kovachy, 2009):

$$c_j^{(n)}(\alpha+2,\tilde{\Omega}) = c_{j-1}^{(n)}(\alpha,\tilde{\Omega}). \tag{6.50}$$

This simply represents a shift of the wavefunction in momentum space by $2\hbar k$, which is exactly what we expect, since increasing the lattice velocity by $2v_r$ corresponds to undergoing a single Bloch oscillation. The dependence of the eigenvalues on the lattice velocity can be expressed as follows:

$$\varepsilon_n(\alpha,\tilde{\Omega}) = -E_r\alpha^2 + \varepsilon_n(0,\tilde{\Omega}) + p_n(\alpha,\tilde{\Omega}) \tag{6.51}$$

where $p_n(\alpha,\tilde{\Omega})$ is periodic in α such that $p_n(\alpha+2m,\tilde{\Omega})=p_n(\alpha,\tilde{\Omega})$ for integer m holds for all α and $p_n(\alpha,\tilde{\Omega})$ vanishes when the condition $\alpha = 2m$ holds. Note that the dependence of the eigenvalues on $\tilde{\Omega}$ can be calculated using the truncated matrix approximation discussed in Sec. 6.8.4. The relevance of this dependence to the phase shift of an interferometer and how this dependence varies with momentum are discussed in Sec. 6.8.5. and Sec. 6.8.9.

The first term in Eq. 6.51 has a simple physical interpretation. Where $v_{\text{Lattice}}(t)$ is the velocity of the lattice in the dressed state frame (so that $v_{\text{Lattice}}(t) = \alpha v_r$), note that

$$E_r\alpha^2 = \frac{\hbar^2 k^2}{2m}\left(\frac{v_{\text{Lattice}}(t)}{\frac{\hbar k}{m}}\right)^2 = \frac{1}{2}mv_{\text{Lattice}}(t)^2, \tag{6.52}$$

which is simply the kinetic energy of an atom traveling along a classical trajectory defined by the motion of the lattice. The $\varepsilon_n(0,\tilde{\Omega})$ and $p_n(\alpha,\tilde{\Omega})$ terms represent the band structure of the lattice, with the $p_n(\alpha,\tilde{\Omega})$ term accounting for the bands deviating from being flat. For interferometer geometries in which lattices act as waveguides for the atoms, the net contributions to the phase shift from the $\varepsilon_0(0,\tilde{\Omega})$ and $p_0(\alpha,\tilde{\Omega})$ terms in the ground state energy and from corrections to the adiabatic approximation are often negligible, as explained in the following sections. In this case, the phase difference between the two arms of the interferometer is given by the following equation (assuming that the two arms arrive at the same endpoint). During the course of the following calculations, it is convenient to parametrize the Hamiltonian, eigenvalues, and eigenvectors using the single variable t rather than the two variables $\alpha(t)$ and $\tilde{\Omega}(t)$, and we make this switch starting here:

$$\phi_1 - \phi_2 = -\frac{1}{\hbar}\left[\int_0^T \varepsilon_0^{\text{arm1}}(t)dt - \int_0^T \varepsilon_0^{\text{arm2}}(t)dt\right]$$

$$= \frac{1}{\hbar}\left[\int_0^T \frac{1}{2}mv_{\text{Lattice}}^{\text{arm1}}(t)^2 dt - \int_0^T \frac{1}{2}mv_{\text{Lattice}}^{\text{arm2}}(t)^2 dt\right]. \tag{6.53}$$

In the above expression, T is the time elapsed during the interferometer sequence.

Observe that this expression for the phase difference can be obtained by assuming that the lattice potential acts as a constraint that forces the atoms in each arm to traverse the classical path traveled by the lattice guiding that arm. In this case the phase shift is just the difference of the respective action integrals over the two classical paths, as we would expect from the Feynman path integral formulation of quantum mechanics (Feynman and Hibbs, 1965). Since the lattice is the only potential in the freely falling and dressed state frames, the action integrals yield Eq. 6.53 (for an insightful treatment of the applications of path integrals in atom interferometry, we refer the reader to Storey and Cohen-Tannoudji 1994). The terms that we have neglected in Eq. 6.53 embody corrections to the simple picture of the lattice as a force of constraint arising from the quantum nature of the motion (e.g. a small portion of the population leaving the ground state of the lattice), which can sometimes be important. However, our simple picture provides physical intuition into the lattice phase shift and is often sufficient to derive quantitative results.

To summarize, the eigenvalues of the lattice consist of a kinetic energy term, a term that depends only on the lattice depth, and a term that is periodic in α, and the eigenvectors transform under a simple shift operation when α is changed by $2m$ for integer m. These properties are a direct result of Bloch's theorem. The symmetries that we have discussed allow us to conclude that if, for a given $\tilde{\Omega}$, we know the eigenvalues and eigenvectors of the Hamiltonian for all α within any range $[\alpha_0, \alpha_0 + 2]$, we can subsequently determine the eigenvalues and eigenvectors for arbitrary α. This result will prove to be useful from a computational standpoint, since the dynamics of the system are completely described by the solution within a finite range of α.

6.8.4 Calculating corrections to the adiabatic approximation using the method of perturbative adiabatic expansion

We now present the method of perturbative adiabatic expansion (Müller et al., 2008) to determine corrections of arbitrary order to the adiabatic approximation. We note that the particular adiabatic approximation that we correct here refers to the adiabatic evolution of the ground state of the dressed state Hamiltonian, rather than the adiabatic elimination of the excited state during the Raman process, which is treated in Müller et al. 2008. The corrections we consider will always be present to some extent, since lattice depth and velocity ramps occurring over a finite time can never be perfectly adiabatic. In addition, non-adiabatic corrections can be caused by perturbations arising from laser frequency noise and amplitude noise. Although our analytical and numerical computations indicate that the contribution of non-adiabatic corrections to the overall phase shift will be highly suppressed for many interferometer geometries, it is important to have a generalized framework with which to treat these corrections in order to determine when they are important and to precisely calculate them when necessary. Note that much of our discussion will follow a similar outline as the proof of the adiabatic theorem in Griffiths 2005. A more detailed version of the derivation presented here can be found in Kovachy 2009.

For all times t, we can express any state vector $\vec{\Psi}(t)$ in Hilbert space as a linear combination of the instantaneous eigenvectors $\vec{\Psi}_n(t)$ of the dressed state Hamiltonian matrix, where in general the coefficients of each eigenvector can vary in time. The instantaneous eigenvectors satisfy the relation $H(t)\vec{\Psi}_n(t) = \varepsilon_n(t)\vec{\Psi}_n(t)$. Note that the vectors $\vec{\Psi}_n(t)$ do not represent the momentum eigenstates $|2n\hbar k\rangle$, since the momentum eigenstates are not in general eigenstates of the dressed state Hamiltonian. Choosing coefficients with a phase $\varphi_n(t) \equiv -\frac{1}{\hbar}\int_{t_0}^{t} \varepsilon_n(t')dt'$ factored out, we can write:

$$\vec{\Psi}(t) = \sum_{n=0}^{\infty} b_n(t) e^{i\varphi_n(t)} \vec{\Psi}_n(t). \tag{6.54}$$

To simplify matters further, we choose the phase of the vectors $\vec{\Psi}_n(t)$ so that each element of $\vec{\Psi}_n(t)$ is real for all t and varies continuously with t, which we can do because the particular Hamiltonian matrix $H(t)$ we consider is a real-valued, Hermitian matrix. However, the coefficients $b_n(t)$ will in general be complex.

From here, application of the Schrodinger equation gives

$$\dot{b}_j(t) = -\sum_{n \neq j} b_n(t) \vec{\Psi}_j^\dagger(t) \frac{\partial \vec{\Psi}_n(t)}{\partial t} e^{i[\varphi_n(t) - \varphi_j(t)]}, \tag{6.55}$$

which can equivalently be written as

$$\dot{b}_j(t) = -\left(\sum_{n \in S_D(t)} b_n(t) \vec{\Psi}_j^\dagger(t) \frac{\partial \vec{\Psi}_n(t)}{\partial t} e^{i[\varphi_n(t) - \varphi_j(t)]} \right)$$
$$- \left(\sum_{n \in S_{ND}(t)} b_n(t) \frac{\vec{\Psi}_j^\dagger(t) \dot{H}(t) \vec{\Psi}_n(t)}{\varepsilon_n(t) - \varepsilon_j(t)} e^{i[\varphi_n(t) - \varphi_j(t)]} \right) \tag{6.56}$$

where $S_D(t)$ is the set of all n such that $n \neq j$ and $\varepsilon_n(t) = \varepsilon_j(t)$ and $S_{ND}(t)$ is the set of all n such that $\varepsilon_n(t) \neq \varepsilon_j(t)$ (Griffiths, 2005; Kovachy, 2009).

Eq. 6.56 illuminates the rationale behind the adiabatic approximation. Under the adiabatic approximation, we assume that $H(t)$ and hence also its eigenvectors vary slowly enough in time so that the conditions $\left| \vec{\Psi}_j^\dagger(t) \dot{H}(t) \vec{\Psi}_n(t) \right| \ll \frac{(\varepsilon_n(t) - \varepsilon_j(t))^2}{\hbar}$ (for $\varepsilon_n(t) \neq \varepsilon_j(t)$) and $\left| \vec{\Psi}_j^\dagger(t) \frac{\partial \vec{\Psi}_n(t)}{\partial t} \Delta t \right| \ll 1$ (for $\varepsilon_n(t) = \varepsilon_j(t)$ and where Δt is the time scale of the approximation) hold. The righthand side of Eq. 6.56 can therefore be approximated as zero. Then, all the coefficients $b_j(t)$ are constant in time.

To compute higher-order corrections, we employ the method of adiabatic expansion, which mathematically follows in the spirit of the Born approximation (Griffiths, 2005; Kovachy, 2009). We denote the coefficients correct to pth order as $b_j^{(p)}(t)$, where the zeroth order coefficients correspond to the solution under the adiabatic approximation. Higher-order corrections can be obtained recursively through the relation $b_j^{(p)}(t) = b_j^{(p-1)}(t_0) - \sum_{n \neq j} \int_{t_0}^{t} b_n^{(p-1)}(t') \vec{\Psi}_j^\dagger(t') \frac{\partial \vec{\Psi}_n(t')}{\partial t'} e^{i[\varphi_n(t') - \varphi_j(t')]} dt'$. In the next section, it will prove to be useful to separate the corrections at each order into terms of the following form, as illustrated for the first- and second-order cases below. For the first order case, we can write

$$b_j^{(1)}(t) = b_j^{(0)}(t_0) + \sum_{n \neq j} C_{n \to j}(t), \tag{6.57}$$

where

$$C_{n \to j}(t) \equiv - \int_{t_0}^{t} b_n^{(0)}(t_0) \vec{\Psi}_j^\dagger(t') \frac{\partial \vec{\Psi}_n(t')}{\partial t'} e^{i[\varphi_n(t') - \varphi_j(t')]} dt'. \tag{6.58}$$

The second-order solution for $b_j(t)$ will then be

$$b_j^{(2)}(t) = b_j^{(0)}(t_0) - \sum_{n \neq j} \int_{t_0}^{t} b_n^{(1)}(t') \vec{\Psi}_j^\dagger(t') \frac{\partial \vec{\Psi}_n(t')}{\partial t'} e^{i[\varphi_n(t') - \varphi_j(t')]} dt'$$

$$= b_j^{(0)}(t_0) + \sum_{n \neq j} C_{n \to j}(t) + \sum_{n \neq j} \sum_{m \neq n} C_{m \to n \to j}(t), \tag{6.59}$$

where

$$C_{m \to n \to j}(t) \equiv - \int_{t_0}^{t} C_{m \to n}(t') \vec{\Psi}_j^\dagger(t') \frac{\partial \vec{\Psi}_n(t')}{\partial t'} e^{i[\varphi_n(t') - \varphi_j(t')]} dt'. \tag{6.60}$$

The calculation of corrections of higher order is discussed in Sec. 6.8.10.

To find the eigenvectors and eigenvalues that we need to calculate the terms that make a non-negligible contribution to the expansion, we must approximate the infinite dimensional Hamiltonian matrix as a finite dimensional truncated matrix. At the end of Sec. 6.8.3, we concluded that the problem of determining the eigenvalues and eigenvectors for all α reduces to finding the eigenvalues and eigenvectors for a range $\alpha \in [\alpha_0, \alpha_0 + 2]$ for arbitrary α_0. In addition, it suffices to calculate the inner products $\vec{\Psi}_j^\dagger(t) \frac{\partial \vec{\Psi}_n(t)}{\partial t}$ just in this range of α, which follows from the symmetry $\vec{\Psi}_j^\dagger(\alpha + 2m, \tilde{\Omega}) \frac{\partial \vec{\Psi}_n(\alpha + 2m, \tilde{\Omega})}{\partial t} = \vec{\Psi}_j^\dagger(\alpha, \tilde{\Omega}) \frac{\partial \vec{\Psi}_n(\alpha, \tilde{\Omega})}{\partial t}$ for integer m (Kovachy, 2009). To make the calculation less cumbersome, we can look at the range $\alpha \in [-1, 1]$. For $\tilde{\Omega}$ not too large and α in this range, the eigenvectors with lower energies are populated almost entirely by momentum eigenstates $|2m\hbar k\rangle$ with relatively small $|m|$. This is the case because for α in this range, the diagonal elements of the Hamiltonian will be smallest for values of m close to zero. We note that for $\tilde{\Omega} = 0$, the diagonal elements are the eigenvalues. In the limit of $\tilde{\Omega} \to 0$, each eigenvector will consist of only a single momentum eigenstate, where in general eigenvectors corresponding to momentum eigenstates with m closer to zero will have lower eigenvalues. Increasing $\tilde{\Omega}$ will allow the lower eigenvectors to spread out in momentum space to a certain extent, but this will not change the fact that the lower eigenvectors will be linear combinations of momentum eigenstates corresponding to smaller values of $|m|$. The eigenvectors we care about for calculational purposes will be those with eigenvalues closer to the ground state eigenvalue. We can thus consider a truncated $(2n + 1) \times (2n + 1)$ Hamiltonian matrix centered around $m = 0$, where we choose n to be large enough so that for the particular dynamics being described, a sufficient number of eigenvectors and eigenvalues can be calculated.

6.8.5   An Example of Perturbative Adiabatic Expansion: Calculating the Non-Adiabatic Correction to the Phase Shift Evolved During a Lattice Beam Splitter

We illustrate the above method by calculating phase corrections to a lattice beam splitter. In this example, we consider the case where two optical lattices of the same depth but different accelerations are used to separate the two arms of the interferometer (after an initial momentum space splitting is achieved through Bragg diffraction). We note that the analysis here is equally applicable to the situation where two separate optical lattice waveguides are used to address the two arms of the interferometer, an example of which is illustrated in Fig. 6.18. To calculate the phase shift for applications in precision measurement, we need to determine the non-adiabatic correction to the phase difference between the two arms that accrues during the beam splitter. In practice, we do this by first calculating corrections to the ground state coefficient $b_0(t)$ and then evaluating how these corrections affect the phase difference between the arms. We note that the dominant contribution to the phase difference will come from the zeroth order term as given in Eq. 6.53.

For this example, we consider the situation shown in Fig. 6.15, where the interaction of the atoms with the lattice is divided into three distinct parts. From $t = 0$ to $t = T_{\mathrm{ramp}}$ we ramp up the lattice, from $t = T_{\mathrm{ramp}}$ to $t = T_{\mathrm{ramp}} + T_{\mathrm{accel}}$ we accelerate the lattice, and from $t = T_{\mathrm{ramp}} + T_{\mathrm{accel}}$ to $t = 2T_{\mathrm{ramp}} + T_{\mathrm{accel}} \equiv T_{\mathrm{final}}$ we ramp down the lattice. For the sake of simplicity, the ramps are chosen to be symmetric so that the lattice depth decrease ramp is the time reversed lattice depth increase ramp.

We assume that initially all of the population is in the ground state, so that $b_j^{(0)}(t) = \delta_{0j}$, and we make the lattice depth and velocity ramps adiabatic enough so that almost all of the population remains in the ground state. In order to find the non-adiabatic correction to the phase shift, we determine the non-adiabatic correction to the phase of the ground state for each arm.

Since to lowest order only the ground state is populated, the leading corrections to $b_0(t)$ will come at second order. The largest contribution comes from $C_{0\to1\to0}(t)$, and it is this term on which we focus. Because the ground state is non-degenerate, Eq. 6.60 and the fact that $\vec{\Psi}_j^\dagger(t)\frac{\partial\vec{\Psi}_n(t)}{\partial t} = \frac{\vec{\Psi}_j^\dagger(t)\dot{H}(t)\vec{\Psi}_n(t)}{\varepsilon_n(t)-\varepsilon_j(t)}$ for $\varepsilon_n(t) - \varepsilon_j(t) \neq 0$ (Griffiths, 2005) give us:

$$
C_{0\to1\to0}(t) = -\int_0^t dt_1 \left(-\int_0^{t_1} dt_2 \frac{M_{10}(t_2)}{\Delta\varepsilon_{10}(t_2)} e^{-\frac{i}{\hbar}\int_0^{t_2}\Delta\varepsilon_{10}(t_3)dt_3} \right)
$$
$$
\frac{M_{10}(t_1)}{-\Delta\varepsilon_{10}(t_1)} e^{\frac{i}{\hbar}\int_0^{t_1}\Delta\varepsilon_{10}(t_3)dt_3} . \tag{6.61}
$$

where we define $M_{10}(t) \equiv \vec{\Psi}_1^\dagger(t)\dot{H}(t)\vec{\Psi}_0(t) = \vec{\Psi}_0^\dagger(t)\dot{H}(t)\vec{\Psi}_1(t)$ and $\Delta\varepsilon_{10}(t) \equiv \varepsilon_0(t) - \varepsilon_1(t)$. Note that the two matrix elements are equal because we choose the eigenvectors to be real.

We examine the ultimate contribution of $C_{0\to1\to0}(T_{\mathrm{final}})$ to $b_0^{(2)}(T_{\mathrm{final}})$. Since during the ramp up and ramp down stages $M_{10}(t)$ depends only on $\dot{\Omega}$ but not on $\dot{\alpha}$, some portions of $C_{0\to1\to0}(T_{\text{final}})$ will be common to both arms of the interferometer, because we assume that the lattice interaction processes for the two arms differ only in the magnitude of the lattice acceleration. We denote these common terms as g_{ramp}. The remaining terms depend on the lattice acceleration and will thus differ between the arms. There will be a term g_{mixed} that depends both on $\dot{\tilde{\Omega}}$ and $\dot{\alpha}$. However, it can be shown that under the assumption that the lattice depth decrease ramp is the time reversed lattice depth increase ramp, this term is zero (Kovachy, 2009). Finally, there will be a term g_{accel} that depends quadratically on $\dot{\alpha}$ and is not explicitly dependent on $\dot{\tilde{\Omega}}$. We note that g_{accel} implicitly depends on the maximum lattice depth $\tilde{\Omega}_{\text{max}}$, which can be seen by the fact that $\tilde{\Omega}_{\text{max}}$ affects what value the energy gap $\Delta\varepsilon_{10}$ takes on during the acceleration stage (a deeper lattice leads to a larger energy gap). We can thus write

$$b_0^{(2)}(T_{\text{final}}) \approx 1 + g_{\text{ramp}} + g_{\text{accel}}. \tag{6.62}$$

In calculating g_{accel}, it is useful to note that $M_{10}(t)$ takes on a convenient form during the acceleration stage. Recalling the form of the Hamiltonian matrix from Eq. 6.48, we observe that $\dot{H}(t)$ will be a diagonal matrix with matrix elements $\dot{H}_{mn}(t) = -4E_r n \dot{\alpha}(t)\delta_{mn}$. We can thus write $M_{10}(t) = \dot{\alpha}(t)A_{10}(t)$, where $A_{10}(t) \equiv -4E_r \sum_n n \left(\vec{\Psi}_1(t)\right)_n^* \left(\vec{\Psi}_0(t)\right)_n$ is a weighted dot product.

In order to more clearly illuminate the general points we are illustrating with this example, we make the simplifying assumption that $\dot{\alpha}(t)$ is constant throughout the acceleration stage. Moreover, we assume that the lattice is deep enough so that $A_{10}(t)$ and $\Delta\varepsilon_{10}(t)$ are also constant during the acceleration stage, which is an accurate approximation for typical experimental situations. During the acceleration stage, we respectively denote these constant quantities as $\dot{\alpha}$, A_{10}, and $\Delta\varepsilon_{10_{\text{accel}}}$. Note that these assumptions, along with the assumption of mirror symmetry between the ramp up and ramp down stages, are certainly not necessary to carry out the calculation. They only serve to make the final result take a particularly simple form that provides physical insight into the process. In the absence of these assumptions, the calculation will be only slightly more complicated and can easily be performed. We note in particular that the mirror symmetry assumption is not stringent, for even when this symmetry is largely violated, the g_{mixed} term is typically an order of magnitude or more smaller than the g_{accel} term, as we verify by estimating the relevant integrals in Eq. 6.61. If needed, g_{mixed} can be calculated by evaluating these integrals. In addition, we note that the treatment given in this example can readily be generalized to the case where the lattice depth and velocity are changed simultaneously. Performing the necessary integrals, we find that

$$g_{\text{accel}} = -i\hbar \frac{\dot{\alpha}^2 A_{10}^2}{(\Delta\varepsilon_{10_{\text{accel}}})^3} T_{\text{accel}} + \hbar^2 \frac{\dot{\alpha}^2 A_{10}^2}{(\Delta\varepsilon_{10_{\text{accel}}})^4} \left[e^{\frac{i}{\hbar}\Delta\varepsilon_{10_{\text{accel}}} T_{\text{accel}}} - 1 \right]. \tag{6.63}$$

We note that for $|g_{\text{ramp}} + g_{\text{accel}}| \ll 1$, we can solve the problem by employing adiabatic expansion over a single time interval. However, there may be times when we must divide the problem into multiple parts, as discussed in Sec. 6.8.10. For typical experimental parameters, the term proportional to T_{accel} in g_{accel} will dominate both the second term in g_{accel} and the g_{ramp} term. We have verified that the g_{ramp} term (which embodies the non-adiabatic loading of the lattice) is typically much smaller than the first term in g_{accel} by estimating the integrals in Eq. 6.61 that correspond to g_{ramp} and by checking these estimates numerically. Thus, we can express the condition $|g_{\text{ramp}} + g_{\text{accel}}| \ll 1$ as

$$\left| \hbar \frac{\dot{\alpha}^2 A_{10}^2}{(\Delta \varepsilon_{10_{\text{accel}}})^3} T_{\text{accel}} \right| \ll 1. \tag{6.64}$$

This condition will often hold, since in many experimentally relevant cases the acceleration time or acceleration will be sufficiently small.

We now show how to determine the correction to the phase shift between the two arms arising from the non-adiabatic correction $g_{\text{ramp}} + g_{\text{accel}}$ in the case where this correction is small. As we recall from Eq. 6.54, the coefficient of the ground state eigenvector at time T_{final} is $b_0(T_{\text{final}}) e^{i\varphi_0(T_{\text{final}})} \approx (1 + g_{\text{ramp}} + g_{\text{accel}}) e^{i\varphi_0(T_{\text{final}})}$. From now on, we adopt the notation $g_{\text{accel}}(\dot{\alpha})$ to emphasize the dependence of this quantity on $\dot{\alpha}$, and we leave g_{ramp} without an argument to emphasize that it does not depend on $\dot{\alpha}$. The non-adiabatic correction to the phase of the coefficient of the ground state eigenvector for an arm with acceleration $\dot{\alpha}$ is

$$\phi_{\text{correction}}(\dot{\alpha}) \approx |g_{\text{ramp}}| \sin\left(\arg[g_{\text{ramp}}]\right) + |g_{\text{accel}}(\dot{\alpha})| \sin\left(\arg[g_{\text{accel}}(\dot{\alpha})]\right), \tag{6.65}$$

where we have Taylor expanded to first order in small quantities. We note that the exact expression for $\phi_{\text{correction}}(\dot{\alpha})$, which is sometimes necessary to use, is given by $\phi_{\text{correction}}(\dot{\alpha}) = \frac{1}{2i} \ln \left[\frac{1 + |g_{\text{ramp}}| e^{i \arg(g_{\text{ramp}})} + |g_{\text{accel}}(\dot{\alpha})| e^{i \arg(g_{\text{accel}}(\dot{\alpha}))}}{1 + |g_{\text{ramp}}| e^{-i \arg(g_{\text{ramp}})} + |g_{\text{accel}}(\dot{\alpha})| e^{-i \arg(g_{\text{accel}}(\dot{\alpha}))}} \right]$. To calculate the correction to the phase shift between two arms, we take the difference $\phi_{\text{correction}}(\dot{\alpha}_{\text{arm1}}) - \phi_{\text{correction}}(\dot{\alpha}_{\text{arm2}})$. The $|g_{\text{ramp}}| \sin\left(\arg[g_{\text{ramp}}]\right)$ term is common to both arms. The non-adiabatic correction to the phase difference between an arm with acceleration $\dot{\alpha}$ and an unaccelerated arm is:

$$\delta\phi(\dot{\alpha}) \equiv \phi_{\text{correction}}(\dot{\alpha}) - \phi_{\text{correction}}(0) \approx |g_{\text{accel}}(\dot{\alpha})| \sin\left(\arg[g_{\text{accel}}(\dot{\alpha})]\right). \tag{6.66}$$

Eqs. 6.65 and 6.66 provide us with a means to calculate the leading correction to the phase shift, and comparison with numerical results for a variety of experimentally conceivable lattice depth ramps shows excellent agreement. A comparison of the adiabatic expansion method with numerical calculations is illustrated in Fig. 6.16. In particular, Fig. 6.16 shows $\delta\phi(\dot{\alpha})$ as a function of $\dot{\alpha}$. The lattice depth ramps and acceleration time are identical to those of the acceleration sequence shown in Fig. 6.15, with an acceleration time of $16(4\omega_r)^{-1}$ and with $\tilde{\Omega}_{\max} = \frac{19.5}{8}$. The points shown in the figure come from a numerical simulation of the Schrodinger equation. In addition to the leading second-order term, we keep the leading fourth order correction term in $g_{\text{accel}}(\dot{\alpha})$, $-\hbar^2 \frac{\dot{\alpha}^4 A_{10}^4 T_{\text{accel}}^2}{2(\Delta \varepsilon_{10_{\text{accel}}})^6}$, calculated in Kovachy 2009, which becomes significant for larger accelerations. For example, for an acceleration of $6\left(\frac{\omega_r}{2}\right) v_r$, this term is smaller

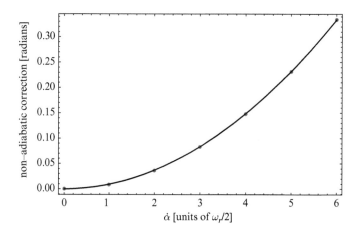

Fig. 6.16 Non-adiabatic correction to the phase difference between an arm that is accelerated by an experimentally plausible lattice beam splitter and an arm that is not accelerated (which is in our notation $\delta\phi(\dot{\alpha})$), as calculated using a simplified adiabatic expansion method in which we keep only leading terms versus numerical simulations of the Schrodinger equation. The curve represents the prediction made by the adiabatic expansion method, while the points represent the numerical results.

than the leading second-order correction by a factor of ~ 10. We neglect corrections arising from the term g_{ramp}, for these corrections are common to both arms of the interferometer to lowest order. Even the simple approximation used to obtain the corrections agrees remarkably well with the simulations (with an rms deviation of 4×10^{-4} radians), and we note that we could easily improve this approximation by including more terms in the adiabatic expansion series. As expected, we observe that the correction scales quadratically with acceleration.

The non-adiabatic correction to the phase shift between the two arms of the interferometer for arbitrary arm acceleration differences is $\phi_{\text{correction}}(\dot{\alpha}_{\text{arm1}}) - \phi_{\text{correction}}(\dot{\alpha}_{\text{arm2}}) = \delta\phi(\dot{\alpha}_{\text{arm1}}) - \delta\phi(\dot{\alpha}_{\text{arm2}})$. Note that the leading order results depicted in Fig. 6.16 will often be sufficient, but it may sometimes be necessary to calculate higher-order corrections, examples of which can be found in Kovachy 2009.

In an experimental implementation, the two arms of the beam splitter are addressed by two different lattices. Therefore, any imbalance in the depths of the two lattices will lead to a phase error between the arms. Where the intensities of the two beams forming a given lattice are I_a and I_b, we note that the lattice depth is proportional to the product $\sqrt{I_a} \times \sqrt{I_b}$, meaning that intensity imbalances lead to lattice depth imbalances. We calculate that once a lattice is ramped up, the $\varepsilon_0(0,\tilde{\Omega})$ term in the expression for the lowest eigenvalue of the dressed state Hamiltonian from Eq. 6.51 is much larger than the $p_0(\alpha,\tilde{\Omega})$ term. Thus, for an interaction lasting from time t_1 to time t_2, the dominant contribution to the phase error, which we denote as ϕ_{balance}, arising from the imbalance in the lattice depths is

$$\phi_{\text{balance}} = -\frac{1}{\hbar} \int_{t_1}^{t_2} \left[\varepsilon_0(0, \tilde{\Omega}_{\text{arm1}}) - \varepsilon_0(0, \tilde{\Omega}_{\text{arm2}}) \right] dt, \tag{6.67}$$

where we note that $\varepsilon_0(0, \tilde{\Omega})$ can be calculated using the truncated matrix approximation discussed in Sec. 6.8.4. In order to put Eq. 6.67 in a more convenient form for making order of magnitude estimates, we use the fact that for $|\tilde{\Omega}| > 0.25$, $\varepsilon_0(0, \tilde{\Omega}) \sim -8E_r|\tilde{\Omega}|$. This result is verified by direct comparison with the values of $\varepsilon_0(0, \tilde{\Omega})$ obtained with the truncated matrix approximation. Recalling that $\tilde{\Omega} \equiv \frac{\Omega}{8\omega_r}$, it is convenient to rephrase this statement in terms of Ω as follows: for $|\Omega| > 2\omega_r$, $\varepsilon_0(0, \Omega) \sim -\hbar|\Omega|$. This range of $|\Omega|$ is of considerable experimental interest and is amenable to simple approximation. However, if needed, smaller lattice depths can be treated using the general relation given in Eq. 6.67 and the truncated matrix approximation. Substituting the above result into Eq. 6.67, we obtain

$$\phi_{\text{balance}} \sim \int_{t_1}^{t_2} \left(|\Omega_{\text{arm1}}| - |\Omega_{\text{arm2}}| \right) dt = \left(\langle |\Omega_{\text{arm1}}| \rangle - \langle |\Omega_{\text{arm2}}| \rangle \right) (t_2 - t_1), \tag{6.68}$$

where $\langle |\Omega_{\text{arm1}}| \rangle$ and $\langle |\Omega_{\text{arm2}}| \rangle$ respectively denote the average values of $|\Omega_{\text{arm1}}|$ and $|\Omega_{\text{arm2}}|$ between time t_1 and time t_2. Note that fluctuations in the difference between $|\Omega_{\text{arm1}}|$ and $|\Omega_{\text{arm2}}|$ that occur at frequencies that are large with respect to the beam splitter time $t_2 - t_1$ will largely average out, suppressing their net effect on the phase shift. Furthermore, in many geometries, arm 1 will be addressed by one pair of laser beams, that we call lattice A, during the splitting of the arms and then will be addressed by a second pair of laser beams, that we call lattice B, during the recombination of the arms. Conversely, arm 2 will be addressed by lattice B during the splitting stage and by lattice A during the recombination stage. In such a geometry, the effect on the phase shift of a constant offset in depth between lattice A and lattice B will cancel, and the effect on the phase shift of fluctuations in the depth difference between lattice A and lattice B that occur at low frequencies with respect to the time scale of the interferometer sequence will be highly suppressed. Also, as we discuss in the following section, we will often be interested in the difference in the phase shifts of two interferometers in a differential configuration. If both of the interferometers remain well within the Rayleigh ranges of the laser beams so that beam divergence is a small effect, any lattice depth imbalance will be largely common to the two interferometers, suppressing its effect on the phase shift difference by orders of magnitude.

6.8.6 Applications: atom interferometers using optical lattices as waveguides

Light-pulse atom interferometer geometries have had tremendous success in performing many types of high-precision measurements. However, in many cases, we would like to be able to push the capabilities of atom interferometry by making more precise measurements using spatially compact interferometers. Atom interferometers that use optical lattices as waveguides for the atoms offer the potential to make such measurements attainable. In such a scheme, we can use an initial beam splitter composed of multiple Bragg pulses, a multiphoton Bragg pulse, or a hybrid Bragg pulse/lattice acceleration scheme as described in Hecker Denschlag et al. 2002; Cladé et al. 2009; Müller et al. 2009 to split the arms of the interferometer in momentum space. We can then control each arm independently with an optical lattice. We will once again use Bragg pulses during the π-pulse and final $\frac{\pi}{2}$-pulse stages of the interferometer sequence, with lattices acting as waveguides between these stages. The preceding analysis has developed the theoretical machinery for calculating phase shifts for these lattice interferometers. We now examine several of the most promising applications of lattice interferometers.

6.8.6.1 *Gravimetry and gravity gradiometry*

Lattice interferometers can be used to make extremely precise measurements of the local gravitational acceleration g. We proceed to calculate the phase shift for a lattice gravimeter. It is essential to note that whenever the two arms are addressed by different lattices they will be in different dressed state frames (where we recall that a dressed state frame is defined by the velocity of the corresponding lattice and by the distance that the lattice has traveled since the beginning of the interferometer sequence). Let the velocities in the lab frame of the two lattices be denoted as $v_{\text{Lab}}{}^{\text{arm1}}(t)$ and $v_{\text{Lab}}{}^{\text{arm2}}(t)$, respectively. The lattice velocities for the two arms in their respective dressed state frames will thus be $v_{\text{Lattice}}{}^{\text{arm1}}(t) = v_{\text{Lab}}{}^{\text{arm1}}(t) + gt - v_0$ and $v_{\text{Lattice}}{}^{\text{arm2}}(t) = v_{\text{Lab}}{}^{\text{arm2}}(t) + gt - v_0$. We note that v_0 is the velocity of the atom before the initial Bragg diffraction that splits the arms in momentum space. Thus, the two arms have different momenta after the Bragg diffraction. During the lattice loading period, the two lattices must be resonant (as described in Sec. 6.8.3) with the respective portions of the atomic wavefunction that they are addressing. Also, let $\Delta v(t) \equiv v_{\text{Lab}}{}^{\text{arm1}}(t) - v_{\text{Lab}}{}^{\text{arm2}}(t)$ be the velocity difference between the two arms and $\Delta d(t) \equiv \int_0^t \Delta v(t')dt'$ be the distance between the two arms. Where the interferometer sequence lasts for a time T, we can derive the phase shift for a lattice gravimeter using Eq. 6.53. We note that an additional contribution $-\frac{1}{\hbar}m(v_f - v_0 + gT)\Delta d(T)$ to the phase shift will arise if the two arms of the interferometer end up in different dressed state frames (which occurs if $\Delta d(T)$ differs from zero), where mv_f is the lab frame momentum at time T of a particular momentum eigenstate in the atomic wavepacket. After averaging, it follows from the discussion in Hogan et al. 2009 that mv_f will take on the value of the center of the momentum space wavepacket. This additional contribution is calculated by boosting both arms into the freely falling frame using the relevant transformation given in Sec. 6.8.8 and the mathematical framework discussed in Sec. 6.8.9. We can thus write the phase shift as follows, where we also include a term ϕ_{Bragg} to embody the net contribution to the phase shift arising from the Bragg pulses:

$$\Delta\phi = -\frac{1}{\hbar}\int_0^T mg\Delta d(t)dt + \frac{1}{\hbar}\int_0^T \frac{1}{2}m(v_{\text{Lab}}^{\text{arm1}}(t)^2 - v_{\text{Lab}}^{\text{arm2}}(t)^2)dt - \frac{m}{\hbar}v_f\Delta d(T) + \phi_{\text{Bragg}}.$$

$$(6.69)$$

When expressed in terms of lab frame quantities, it is apparent that the first two terms in $\Delta\phi$ correspond to the propagation phase and the third term corresponds to the separation phase, both of which typically appear in standard atom interferometer phase shift calculations (Hogan et al., 2009). We note that in the dressed state frame, ideal Bragg pulses simply yield contributions in units of $\pm\frac{\pi}{2}$ to the overall phase shift between the arms, and these contributions can easily be made to cancel so that $\phi_{Bragg} = 0$. However, we note that in some cases, corrections to the simplified picture of an ideal Bragg pulse due to such factors as gravity gradients, finite pulse and detuning effects (which can sometimes lead to a non-negligible propagation phase during the Bragg pulse), phase noise, or population loss may need to be considered. To avoid unnecessarily complicating our presentation, we will not present these corrections here. Instead, we emphasize that they are well-understood effects and refer the reader to other sources for further discussion (Hogan et al., 2009; Müller et al., 2008; Antoine, 2006). Moreover, we note that these effects gain additional suppression for interferometers in a differential configuration so that they will often be below the mrad level (Hogan et al., 2009), as in the case of the gravity gradiometer discussed below.

In the symmetric case, the velocities of the two arms in the lab frame are either opposite to each other or equal so that $v_{Lab}^{arm1}(t)^2 - v_{Lab}^{arm2}(t)^2 = 0$. Since we need the two arms of the interferometer to overlap at time T, ideally $\Delta d(T) = 0$. Thus, the first term in Eq. 6.69 will constitute the only contribution to $\Delta\phi$. However, in an experiment, the parameters in Eq. 6.69 will undergo small fluctuations around their desired values from shot to shot, so that the other terms in Eq. 6.69 act as a source of noise.

In order to cancel the effects of this noise, we can adopt a gradiometer setup in which an array of two or more gravimeters interacts with the same lattice beams. Although fluctuations in $\Delta d(t)$ will still affect phase differences between gravimeters, which take the form $-\frac{1}{\hbar}\int_0^T m(g_1 - g_2)\Delta d(t)dt$, modern phase lock techniques will typically allow us to control the phase differences between the lattice beams well enough so that these effects are smaller than shot noise (Müller et al., 2006). When measuring a gravity gradient, the value of g will vary due to the gradient over the range of a single gravimeter. For linear gradients, we can calculate the phase shift in the presence of a gravity gradient by assuming that the value of g corresponding to the gravimeter is equal to its value at the center of mass position of the atom (see Kovachy 2009 for a rigorous justification of this procedure). When higher-order derivatives of the gravitational field become sizable in comparison to the first derivative, this simple prescription may not suffice, and we can treat the problem perturbatively. If we want to measure an acceleration as well as a gravitational gradient in a noisy environment in which the fringes of the individual gravimeters are washed out, we can use dissimilar conjugate interferometers whose phase noise is strongly correlated as suggested in Chiow et al. 2009. Appropriate statistical methods can then be used to extract the desired acceleration signal (Stockton et al., 2007; Foster et al., 2002).

The effects on the phase shift of non-adiabatic corrections, lattice depth imbalances, and the finite spread of the atomic wavefunction in momentum space are considered in Sec. 6.8.4., Sec. 6.8.5, and Sec. 6.8.9. Based on the analysis in these sections, we conclude that a wide range of experimentally feasible gravimeter geometries exist that contain sufficiently adiabatic lattice depth and velocity ramps and that make use of symmetry in such a way that the net contribution of these corrections to the phase shift will be below the mrad level in a gradiometer configuration. We note that this section has developed the mathematical machinery to calculate any such corrections to arbitrary precision if necessary.

When two lattices are used to manipulate the arms of an atom interferometer, one lattice will be on resonance with a given arm, while the other will be highly detuned. As long as we keep this detuning large enough or employ geometries with sufficient symmetry between the arms, the net effect of the off-resonant lattices on the final phase shift can often be made to be smaller than the mrad level in a differential configuration (e.g. a gradiometer). For arm 1, the detuned lattice will manifest as an additional term in the discrete Hamiltonian, with matrix elements given by (using the same notation convention as in Eq. 6.48):

$$H_{mn}^{\text{detuned}} = 4E_r \left[-\tilde{\Omega}_{\text{arm2}} e^{-i\beta(t)} \delta_{m,n+1} - \tilde{\Omega}_{\text{arm2}} e^{i\beta(t)} \delta_{m,n-1} \right] \tag{6.70}$$

where $\beta(t) \equiv -\int_0^t \left[\alpha^{\text{arm1}}(t') - \alpha^{\text{arm2}}(t') \right] 4w_r dt' = -2k\Delta d(t)$ and where $\tilde{\Omega}_{\text{arm2}}$ is the lattice depth parameter corresponding to the detuned lattice that addresses arm 2 (Malinovsky and Berman, 2003). To obtain the correction to the Hamiltonian for arm 2, which comes from the detuned lattice addressing arm 1, we replace $\beta(t)$ with $-\beta(t)$ and $\tilde{\Omega}_{\text{arm2}}$ with $\tilde{\Omega}_{\text{arm1}}$. For large detunings, $\beta(t)$ will vary rapidly with time so that the contribution from the detuned Hamiltonian will be small (i.e. the rotating wave approximation). Corrections arising from the detuned Hamiltonian can be solved for perturbatively using methods such as adiabatic perturbation theory. But we emphasize again that we can often avoid situations where this will be necessary. For example, we find from perturbation theory that to avoid population loss due to the off-resonant lattice as described in Sec. 6.8.3, we should choose the off-resonant lattice to have a velocity that differs from that of the particular arm of the interferometer under consideration by an amount $\Delta v \gg |\tilde{\Omega}| v_r$ (where $\tilde{\Omega}$ is the depth parameter of the off-resonant lattice). We have verified this result with numerical simulations. However, in this regime, the off-resonant lattice can still sometimes cause a non-negligible energy shift, which we estimate with perturbation theory. The energy shift for arm 1 is $\Delta E_{\text{arm1}} \approx \frac{\hbar}{8} w_r^{-1} (v_r/\Delta v)^2 (\Omega_{\text{arm2}})^2$. The relevant quantity in determining the correction to the phase difference between the arms is the difference in energy shifts $\Delta E_{\text{arm1}} - \Delta E_{\text{arm2}}$, which is determined by the lattice depth imbalance between the arms. Where we let $\Omega_{\text{arm1}} = \Omega$ and $\Omega_{\text{arm2}} = \Omega + \Delta\Omega$:

$$\Delta E_{\text{arm1}} - \Delta E_{\text{arm2}} \approx \frac{\hbar}{4} \frac{\Omega \Delta\Omega}{w_r (\Delta v/v_r)^2}$$

$$\approx \hbar (0.4 \text{ s}^{-1}) \left(\frac{2\pi \times 3.77 \text{ kHz}}{w_r} \right) \left(\frac{20}{\Delta v/v_r} \right)^2 \left(\frac{\Omega}{5w_r} \right)^2 \left(\frac{\Delta\Omega/\Omega}{10^{-3}} \right). \tag{6.71}$$

This result agrees with numerical simulations. For ${}^{87}\text{Rb}$, the recoil frequency value $w_r = 2\pi \times 3.77 \text{ kHz}$ should be used. We can use Eq. 6.71 to estimate the size of the phase shift induced by the lattice depth imbalance for an example lattice interaction. For ^{87}Rb and assuming a lattice interaction of duration 10 ms, a lattice velocity difference of $\Delta v = 20 v_r$, a lattice depth of $\Omega = 5 w_r$, and a fractional lattice depth imbalance of $\Delta \Omega / \Omega = 10^{-3}$, the resulting phase shift is 4 mrad. For the same reasoning as in the discussion in the previous section, the net effect of lattice depth imbalances on the correction term treated here will often be further suppressed by orders of magnitude for interferometers in a differential configuration.

A lattice gravimeter can provide extraordinary levels of sensitivity. This sensitivity can be achieved over small distance scales by implementing a hold sequence in which the two arms are separated, manipulated into the same momentum eigenstate, held in place by a single lattice, and then recombined. In achieving compactness, we note that the fact that lattice interferometers are confined and can thus keep the atoms from falling under gravity during the separation and recombination stages of the interferometer as well as during the hold sequence is essential. Otherwise, for many configurations, the desired arm separation could not be reached without the atoms falling too great a distance, which would ruin the compactness of the interferometer. Hold times will be limited by spontaneous emission, which decreases contrast. Modern laser technology will allow detunings of hundreds or even thousands of GHz, making hold times on the order of 10 s within reach (Dimopoulos et al., 2008b).

Gravimeter sensitivities using the hold method greatly exceed the sensitivities of light-pulse gravimeters while simultaneously allowing for a significantly smaller interrogation region. For example, for $\sim 10^7$ atoms/shot and $\sim 10^{-1}$ shots/s, a shot noise limited conventional light-pulse interferometer with a 10 m interrogation region can achieve a sensitivity of $\sim 10^{-11} g/\mathrm{Hz}^{1/2}$. With similar experimental parameters, a shot noise limited lattice interferometer with a 10 s hold time and an interrogation region of 1 cm will have a sensitivity of $\sim 10^{-12} g/\mathrm{Hz}^{1/2}$. If we expand the interrogation region to 1 m, we obtain a sensitivity of $\sim 10^{-14} g/\mathrm{Hz}^{1/2}$. This remarkable sensitivity has a plethora of potential applications. Extremely precise gravimeters and gravity gradiometers can be constructed to perform tests of general relativity, make measurements relevant to geophysical studies, and build highly compact inertial sensors. Moreover, the fact that lattice interferometers can operate with such high sensitivities over small distance scales makes them prime candidates for exploring short distance gravity. One could set up an array of lattice gravimeters to precisely map out gravitational fields over small spatial regions, as shown in Fig. 6.17. The knowledge obtained about the local gravitational field could be useful in searching for extra dimensions (Arkani-Hamed et al., 1998) as well as in studying the composition and structure of materials. It is relevant to note the scale dependence of spatial derivatives of the gravitational field. That is, we consider a sensor a distance D_{sensor} away from a mass of radius R_{mass}. If we multiply both D_{sensor} and R_{mass} by a scale factor s, assuming fixed density, the scale dependence tells us how the size of the measured spatial derivative is affected. The mass scales as $(R_{\mathrm{mass}})^3$, while the n th spatial derivative of the gravitational field scales as the mass divided by $(D_{\mathrm{sensor}})^{n+2}$. Therefore, the scale dependence of the n th spatial derivative is $1/s^{n-1}$. Thus, the first spatial derivative (the gravity gradient) is scale invariant. Higher derivatives are proportional to powers of the inverse scale and are therefore more strongly affected by nearby masses.

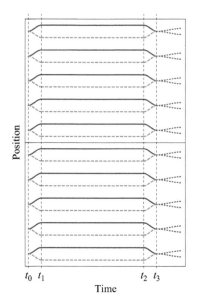

Fig. 6.17 An array of lattice gravimeters such as that shown above can be used to achieve measurements of a local gravitational field with high spatial resolution. The trajectories shown hold in the lab frame. After separating the atoms in each gravimeter by a small amount, we can implement a hold sequence to greatly increase sensitivity. Bragg pulses are used at times t_0, t_1, t_2, and t_3. Such an array could be used to study general relativistic effects, search for extra dimensions, examine local mass distributions, or measure Newton's constant. In addition, the ability of lattice interferometers with hold sequences to provide extremely precise measurements with a small interrogation region makes them ideal candidates for compact, mobile sensors.

6.8.6.2   *Tests of atom charge neutrality*

Ultra-high precision gravitational measurements are among the most promising applications of lattice interferometers, but the usefulness of lattice interferometers is not limited to the study of gravity. By exposing the two arms of a lattice interferometer to different electrostatic potentials, tests of atom charge neutrality with unprecedented accuracy could be achieved (Arvanitaki et al., 2008). The main advantage of a lattice interferometer in such a measurement is that the interrogation time can be significantly increased in comparison to the interrogation time achievable in a light-pulse geometry through the use of a hold sequence. Where t_{hold} is the duration of the hold, V is the electrostatic potential difference between the two arms of the interferometer, e is the electron charge, and ϵ is the ratio of the spurious atomic charge to the electron charge, the phase shift is given by $\frac{\epsilon e}{\hbar} V t_{\text{hold}}$ (Arvanitaki et al., 2008). Based on the results of Sec. 6.8.4., Sec. 6.8.5, and Sec. 6.8.9, and assuming an identical configuration to that described in Arvanitaki et al. 2008 except for the inclusion of a hold sequence, we estimate that the phase error induced by the undesirable effects we consider will be below the proposed shot noise limit for this experiment (1 mrad). Any systematic phase error can be characterized by the methods we have developed. For a hold time of 10 s and for integration over 10^6 shots, spurious atom charges can be probed down to the region of $\epsilon \sim 10^{-27}$. For comparison, the current best limit is $\epsilon \sim 10^{-21}$. The current limit is set by several different types of experiments, including monitoring the response of levitated objects to alternating electric fields and measuring the frequencies of electrically excited sound waves (Unnikrishnan and Gillies, 2004).

6.8.6.3 Measurements of $\frac{\hbar}{m}$ and of isotope mass ratios

The ratio $\frac{\hbar}{m}$ is of particular interest because of its direct relation to the fine structure constant. Atom interferometry has previously been used to provide precision measurements of $\frac{\hbar}{m}$ (Cadoret et al., 2008; Bouchendira et al., 2011). Eq. 6.69 indicates that if we apply different accelerations to the two arms of the interferometer, we will see a phase shift proportional to $\frac{m}{\hbar}$ that depends on the kinetic energy difference between the arms, which can be made extremely large. A differential configuration using conjugate interferometers (shown in Fig. 6.18) could reduce the net contribution of such unwanted effects as laser phase noise and cancel the gravitational phase shift up to gradients (Chiow et al., 2009). Such a geometry could provide an extremely precise measurement of $\frac{\hbar}{m}$, as illustrated by the fact that we can achieve a phase shift of $\sim 10^{11}$ radians for a 5 m interrogation region and a 0.6 s interrogation time, corresponding to a shot noise limited sensitivity of $\sim 10^{-14} \frac{\hbar}{m}/\mathrm{Hz}^{1/2}$ for the experimental parameters stated above. In this situation, the dominant unwanted effect would arise from non-adiabatic corrections, and the methods for calculating these corrections that are presented in the previous sections would need to be applied (we estimate a phase error ~ 10 rad). We note that even if the shot noise limit is not reached, the technique we have proposed could still improve the $\frac{\hbar}{m}$ limit. We emphasize again that to take full advantage of the sensitivity offered by lattice manipulations in atom interferometry, the methods we develop in this section for calculating non-adiabatic corrections are absolutely essential.

The fact that all terms in Eq. 6.69 are proportional to m (except for the ϕ_{Bragg} term, which as we have explained, will often be negligible) can be exploited to provide high-precision measurements of isotope mass ratios by using an interferometer geometry in which the two isotopes follow identical trajectories. Isotope mass ratios could be relevant to studies of advanced models of the structure of the nucleus (Lunney, 1998). Conversely, if we know the mass ratio of two isotopes sufficiently well, we can use such a geometry to precisely measure accelerations, where the two isotopes provide two dissimilar conjugate interferometers. The phase noise of these two interferometers will be extremely well correlated because they are nearly topologically identical. This also eliminates the need for the additional lattice beams required to form the second, topologically distinct interferometer that would be needed with only a single isotope. Note that this scheme to construct dissimilar, topologically identical conjugate accelerometers would not be possible for a light-pulse geometry, for the leading order phase shift of light-pulse accelerometers is independent of isotope mass.

6.8.6.4 Gyroscopes

Lattice interferometers can also be used to build compact and highly sensitive gyroscopes. There are multiple possible schemes in which optical lattices can enhance

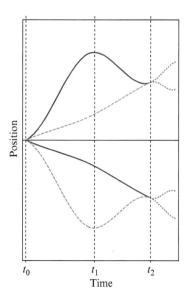

Fig. 6.18 Conjugate interferometer geometry that could be used to measure $\frac{\hbar}{m}$. The trajectories shown hold in the lab frame. Bragg pulses are used at times t_0, t_1, and t_2 (at t_0, a sequence of multiple Bragg pulses is necessary to split the system into the two arms of the two conjugate interferometers). The phase shift of the lower interferometer is subtracted from the phase shift of the upper interferometer, which suppresses the effects of laser phase noise and eliminates the gravitational phase shift up to gradients (Chiow et al., 2009). With such a scheme, a measurement of $\frac{\hbar}{m}$ to a part in 10^{14} may be possible, which could lead to the most accurate determination of the fine structure constant to date.

gyroscope sensitivity. One such scheme is to modify a typical atom-based gyroscope by replacing the Raman pulses with LMT lattice beam splitters, increasing the enclosed area of the interferometer and hence its sensitivity to rotations. The Sagnac phase shift can be written as $\frac{2m}{\hbar}\vec{\Omega}\cdot\vec{A}$, where $\vec{\Omega}$ is the rotation rate vector and $\vec{A}$ is the normal vector corresponding to the enclosed area of the interferometer (Hogan et al., 2009). The gyroscope described in Gustavson et al. 1997 achieves a sensitivity of 2×10^{-8} $\frac{\text{rad}}{\text{s}}$ Hz/$^{1/2}$. Replacing the Raman pulses in this experiment with $200\hbar k$ lattice beam splitters would increase the sensitivity by a factor of 100.

In the lattice gyroscope configuration mentioned above, we estimate that each beam splitter could introduce a non-adiabatic phase error of ~ 1 rad if the arms of the interferometer are not split symmetrically. However, beam splitter configurations that exploit symmetry between the arms can reduce this effect by orders of magnitude. Another option is to use optical lattices along multiple axes to provide complete control of the motion of the atoms in two or three dimensions (this control is only achieved in the region in which the lattices overlap, necessitating the use of wide beams). Analogously to a fiber-optic gyroscope, the atoms could be guided in repeated loop patterns, with the two arms rotating in opposite directions. Geometries in which atomic motion is controlled in multiple dimensions could also expand the possibilities for other applications of lattice interferometry (such as measurements of gravity) by allowing for the measurement of potential energy differences between arbitrary paths. For instance, a compact array of three orthogonal lattice gravity gradiometers could be used to measure the nonzero divergence of the gravitational field in free space predicted by general relativity (Dimopoulos et al., 2008a).

6.8.7   Outlook

We have presented a detailed analytical description of the interaction between an atom and an optical lattice, using the adiabatic approximation as a starting point and then proceeding to rigorously develop a method to calculate arbitrarily small corrections to this approximation using perturbative adiabatic expansion. We have applied this theoretical framework to calculate the phase accumulated during a lattice acceleration in an LMT beam splitter. Finally, we have proposed atom interferometer geometries that use optical lattices as waveguides and discussed applications of such geometries, using our theoretical methods to add rigor to this discussion. We are working toward the experimental implementation of lattice interferometers and LMT lattice beam splitters, and we hope to explore the applications we have discussed. In this experimental work, we realize that we will have to contend with a number of unwanted systematic effects, such as spatially varying magnetic fields, imperfections in the lattice beam wavefronts, and inhomogeneity of the lattice depth across the atomic cloud. We have studied these effects using both analytical and numerical methods, and we are optimistic that they can be significantly mitigated for a wide range of experimental parameters—a conclusion that we hope to verify experimentally. Many of the unwanted systematic effects that are relevant to lattice interferometers are also shared by light-pulse interferometers and can therefore be dealt with using similar methods. Therefore, we believe that it will likely be possible to realize lattice interferometers in existing apparatus originally constructed with light-pulse geometries in mind.

6.8.8   Additional calculation: boosting between different frames

For the purposes of this section, we consider unitary transformations that consist of a translation in position space, a boost in momentum space, and a time-dependent change of phase. Such a transformation has the general form

$$\hat{U}(t) = e^{\frac{i}{\hbar}d(t)\hat{p}} e^{-\frac{i}{\hbar}mv(t)\hat{x}} e^{\frac{i}{\hbar}\theta(t)}. \tag{6.72}$$

We note that the translation $d(t)$, the boost $v(t)$, and the phase $\theta(t)$ are independent parameters. We now consider the Hamiltonian in a frame that is freely falling with gravity, which takes the form

$$\hat{H}_{\text{FF}} = \frac{\hat{p}^2}{2m} + 2\hbar\Omega(t)\sin^2\left[k\hat{x} - k\left(D_{\text{Lab}}(t) + \frac{1}{2}gt^2\right)\right]. \tag{6.73}$$

We can transform from the freely falling frame to the lab frame by applying the appropriate Galilean transformation $\hat{U}_{FF}(t)$, which corresponds to specifying $d(t) = \frac{1}{2}gt^2$, $v(t) = gt$, and $\theta(t) = \frac{1}{3}mg^2t^3$, so that $\hat{H}_{Lab} = \hat{U}_{FF}(t)\hat{H}_{FF}\hat{U}_{FF}^\dagger(t) + i\hbar\left(\frac{\partial}{\partial t}\hat{U}_{FF}(t)\right)\hat{U}_{FF}^\dagger(t)$.

It is useful from a calculational standpoint to transform to a third frame, with a Hamiltonian resembling that describing the atom-light interaction from the point of view of dressed states. We note that although we could have performed a boost directly from the lab frame to the dressed state frame, it is useful to introduce the freely falling frame for pedagogical reasons, since it is the frame in which calculations for atom interferometry are often performed. In Sec. 6.8.9 , we use the transformation between the freely falling frame and the dressed state frame to highlight the parallels between a lattice beam splitter and a typical light-pulse beam splitter.

The Hamiltonian in the dressed state frame is $\hat{H}_{DS} = \frac{\hat{p}^2}{2m} - (v_{\text{Lab}}(t) + gt - v_0)\hat{p} + 2\hbar\Omega(t)\sin^2(k\hat{x})$. We absorb the initial velocity v_0 of the atom in the lab frame (and hence also in the freely falling frame) into the dressed state frame, so that velocity v_0 in the lab frame corresponds to velocity zero in the dressed state frame. The unitary transformation that transforms from the dressed state frame to the freely falling frame is $\hat{U}_{DS} = e^{\frac{i}{\hbar}mv_0\hat{x}}e^{-\frac{i}{\hbar}(D_{\text{Lab}}(t)+\frac{1}{2}gt^2)\hat{p}}e^{-\frac{i}{\hbar}(\frac{1}{2}mv_0^2 t)}$, so that $\hat{H}_{FF} = \hat{U}_{DS}(t)\hat{H}_{DS}\hat{U}_{DS}^\dagger(t) + i\hbar\left(\frac{\partial}{\partial t}\hat{U}_{DS}(t)\right)\hat{U}_{DS}^\dagger(t)$.

6.8.9   Additional calculation: generalizing to the case of a finite wavepacket

In Sec. 6.8.2, we discretized the Hamiltonian using the basis of momentum states $|2n\hbar k\rangle$ for integer n. We now generalize our results to the case of a finite wavepacket.

Throughout our analysis, we have worked mainly in the dressed state frame, since this frame is particularly convenient for describing phase evolution in a lattice. However, since different momentum states correspond to different dressed state frames, in order to calculate effects due to finite wavepacket momentum spread, we need to use a common reference frame. Phase shift calculations for light-pulse atom interferometers are often performed in the freely falling frame. Thus, we present the general results derived in this section in the freely falling frame.

In Sec. 6.8.2, we considered a particular dressed frame in which the initial velocity of the atom is boosted to zero. We now introduce an unboosted dressed state frame that is related to the freely falling frame by a translation in position space with no boost in momentum space, so that the unitary transformation $\hat{U}_{DS_0} = e^{-\frac{i}{\hbar}(D_{\text{Lab}}(t)+\frac{1}{2}gt^2)\hat{p}}$ transforms from the unboosted dressed state frame to the freely falling frame. The Hamiltonian in this frame is

$$\hat{H}_{DS_0} = \frac{\hat{p}^2}{2m} - (v_{\text{Lab}}(t) + gt)\hat{p} + 2\hbar\Omega(t)\sin^2(k\hat{x}). \tag{6.74}$$

Now, say that before the lattice acceleration, the state that we are accelerating is described by $|\Psi_{FF}(t_0)\rangle$ in the freely falling frame. We can then transform this state vector to the unboosted dressed state frame, describe its evolution to the final time t_f in this frame, and transform back to the freely falling frame. Where $\hat{T}_{\mathrm{DS_0}}(t',t)$ is the time evolution operator that takes us from time t to time t' in the unboosted dressed state frame, we can write:

$$|\Psi_{\mathrm{FF}}(t_f)\rangle = \hat{U}_{\mathrm{DS_0}}(t_f)\hat{T}_{\mathrm{DS_0}}(t_f,t_0)\hat{U}^{\dagger}_{\mathrm{DS_0}}(t_0)|\Psi_{\mathrm{FF}}(t_0)\rangle. \tag{6.75}$$

Denoting the initial momentum space wavefunction in the freely falling frame as $\Psi_{\mathrm{FF}}(p,t_0) \equiv \langle p|\Psi_{\mathrm{FF}}(t_0)\rangle$, we can express Eq. 6.75 as:

$$|\Psi_{\mathrm{FF}}(t_f)\rangle = \int dp\, \hat{U}_{\mathrm{DS_0}}(t_f)\hat{T}_{\mathrm{DS_0}}(t_f,t_0)|p\rangle\, e^{\frac{i}{\hbar}(D_{\mathrm{Lab}}(t_0)+\frac{1}{2}gt_0^2)p}\Psi_{\mathrm{FF}}(p,t_0). \tag{6.76}$$

We now consider how each momentum eigenstate $|p\rangle$ evolves in the unboosted dressed state frame. That is, we must calculate $\hat{T}_{\mathrm{DS_0}}(t_f,t_0)|p\rangle$ for each p. In order to do so, we introduce a class of boosted dressed state frames DS_p parameterized by p, so that momentum p in the unboosted dressed state frame (which is just the frame DS_0) corresponds to momentum zero in the frame DS_p. In essence, where $v \equiv \frac{p}{m}$, frame DS_p travels with velocity v with respect to the unboosted dressed state frame. In frame DS_p, the Hamiltonian takes the form

$$\hat{H}_{\mathrm{DS}_p} = \frac{\hat{p}^2}{2m} - (v_{\mathrm{Lab}}(t)+gt-v)\hat{p} + 2\hbar\Omega(t)\sin^2(k\hat{x}), \tag{6.77}$$

where we note that the unitary transformation that transforms from frame DS_p to the unboosted dressed state frame is

$$\hat{U}_p(t) = e^{\frac{i}{\hbar}p\hat{x}}e^{-\frac{i}{\hbar}\left[-(D_{\mathrm{Lab}}(t)+\frac{1}{2}gt^2)mv+\frac{1}{2}mv^2t\right]}. \tag{6.78}$$

Where $\hat{T}_{\mathrm{DS}_p}(t_f,t_0)$ is the time evolution operator in frame DS_p, we can write:

$$\hat{T}_{\mathrm{DS_0}}(t_f,t_0)|p\rangle = \hat{U}_p(t_f)\hat{T}_{\mathrm{DS}_p}(t_f,t_0)\hat{U}^{\dagger}_p(t_0)|p\rangle$$
$$= e^{\frac{i}{\hbar}\left[-(D_{\mathrm{Lab}}(t_0)+\frac{1}{2}gt_0^2)mv+\frac{1}{2}mv^2t_0\right]}\hat{U}_p(t_f)\hat{T}_{\mathrm{DS}_p}(t_f,t_0)|0\rangle. \tag{6.79}$$

The problem of determining the evolution of the momentum eigenstate $|p\rangle$ in the unboosted dressed state frame thus reduces to evolving the momentum eigenstate $|0\rangle$ in the frame DS_p, which we know how to do from the preceding sections. We assume that the momentum space wavefunction is narrow enough so that the momentum eigenstates we consider are resonant with the lattice (that is, as we recall from Sec. 6.8.3, the magnitudes of their velocities with respect to the lattice are less than v_r). Where the lattice acceleration is chosen so as to transfer a momentum of $2n\hbar k$ to the atom, the result from Sec. 6.8.3 tells us that the time evolution in Eq. 6.79 yields

$$\hat{T}_{\mathrm{DS}_p}(t_f,t_0)|0\rangle = e^{i\phi_p}|2n\hbar k\rangle \tag{6.80}$$

for the following phase factor relevant to the phase shift of the interferometer:

$$\phi_p = \frac{m}{2\hbar} \int_{t_0}^{t_f} v_{\text{Lattice}}^p(t)^2 dt \tag{6.81}$$

where $v_{\text{Lattice}}^p(t) \equiv v_{\text{Lab}}(t) + gt - v$ is the lattice velocity in frame DS_p. For the sake of pedagogy, at the moment we neglect the phase arising from the lattice depth, the phase arising from the small periodic variations in the lowest eigenvalue, and the phase arising from non-adiabatic corrections, where we note that we could calculate these contributions if needed. Evaluating the integral in Eq. 6.81, substituting this result into Eq. 6.79, and applying the transformation $\hat{U}_p(t_f)$, we obtain

$$\hat{T}_{\text{DS}_0}(t_f, t_0)\,|p\rangle = e^{i\frac{m}{2\hbar}\int_{t_0}^{t_f}(v_{\text{Lab}}(t)+gt)^2 dt}\,|p + 2n\hbar k\rangle. \tag{6.82}$$

We can now evaluate Eq. 6.76, which gives us the final result

$$|\Psi_{\text{FF}}(t_f)\rangle\rangle = \int dp\,|p + 2n\hbar k\rangle\,e^{i\phi_{\text{FF}}(p)}\,\Psi_{\text{FF}}(p, t_0), \tag{6.83}$$

where

$$\phi_{\text{FF}}(p) = \frac{m}{2\hbar}\int_{t_0}^{t_f}(v_{\text{Lab}}(t)+gt)^2 dt - 2nk\left(D_{\text{Lab}}(t_f) + \frac{1}{2}gt_f^2\right) - \frac{p}{\hbar}\Delta D \tag{6.84}$$

for $\Delta D \equiv (D_{\text{Lab}}(t_f) + \frac{1}{2}gt_f^2) - (D_{\text{Lab}}(t_0) + \frac{1}{2}gt_0^2)$.

Observe that the second term in Eq. 6.84 is what would typically be called the laser phase in a light-pulse atom interferometer for an n th-order beam splitter. We note that as long as the resonance condition is met for the momentum eigenstates we are considering so that they are accelerated, the phase evolved during a lattice acceleration in the DS_0 dressed state frame is independent of p, as shown in Eq. 6.82. (This is true up to non-adiabatic corrections and the small periodic variations in the ground state eigenvalue.) This makes dressed state frames particularly convenient for performing calculations involving wavepackets. In contrast, in the freely falling frame, the accumulated phase ϕ_{FF} is dependent on p, but this dependence cancels in the final expression for the phase shift between the two arms of an interferometer as long as the distance travelled by the atom while locked into a lattice is the same for both arms. Note that momentum dependent contributions to the total phase shift must arise when we treat the problem purely in terms of dressed states, since the total phase shift is an observable quantity and must therefore be independent of the frame in which it is calculated. The key point to realize is that if the total distance traveled in the lattice is not the same for both arms, then the two arms will end up in two different dressed state frames.

We now consider the effect on phase evolution of the periodic term $p_0(\alpha, \tilde{\Omega})$ in the expression for the ground state eigenvalue of the dressed state Hamiltonian given in Eq. 6.51. This term leads to a momentum dependent correction $\delta\phi_p$ to the evolved phase ϕ_p described in Eq. 6.81, where

$$\delta\phi_p = -\frac{1}{\hbar}\int_{t_0}^{t_f} p_0\left(\frac{v_{\text{Lattice}}^p(t)}{v_r}, \tilde{\Omega}(t)\right) dt. \tag{6.85}$$

We remind the reader here that $p_0\left(\frac{v_{\text{Lattice}}^p(t)}{v_r}, \tilde{\Omega}(t)\right)$ is an energy, so the right side of the above equation indeed has units of phase. We note that $p_0\left(\frac{v_{\text{Lattice}}^p(t)}{v_r}, \tilde{\Omega}(t)\right)$ can be calculated using the truncated matrix approximation described in Sec. 6.8.4. Since $p_0\left(\frac{v_{\text{Lattice}}^p(t)}{v_r}, \tilde{\Omega}(t)\right)$ is periodic in $v_{\text{Lattice}}{}^P(t)$ with period $2v_r$, the contribution $\delta\phi_p^{\text{arm1}} - \delta\phi_p^{\text{arm2}}$ of this correction term to the phase difference between the two arms of an interferometer will be highly suppressed if both arms load the atom into the lattice near the center of the zeroth band (as discussed in Sec. 6.8.3) and undergo a nearly integral number of Bloch oscillations so that the effects of this periodic variation in the ground state eigenvalue will be largely common to the two arms, as verified by estimation of the integral in Eq. 6.85 and by numerically solving the Schrödinger equation.

6.8.10   Additional calculation: perturbative adiabatic expansion at higher orders

In order to calculate non-adiabatic corrections at arbitrary order, we define $C_{a\to b\to\cdots\to m\to n\to j}(t)$ recursively in the natural way based on the notation of Sec. 6.8.4:

$$C_{a\to b\to\cdots\to m\to n\to j}(t) \equiv -\int_{t_0}^{t} C_{a\to b\to\cdots\to m\to n}(t')\vec{\Psi}_j^\dagger(t')\frac{\partial\vec{\Psi}_n(t')}{\partial t'}e^{i[\varphi_n(t')-\varphi_j(t')]}dt'. \tag{6.86}$$

For arbitrary order p, we can then write:

$$b_j^{(p)}(t) = C_j(t) + \sum_{n\neq j}C_{n\to j}(t) + \sum_{n\neq j}\sum_{m\neq n}C_{m\to n\to j}(t) + \cdots$$
$$+ \sum_{n\neq j}\sum_{m\neq n}\cdots\sum_{a\neq b}C_{a\to b\to\cdots\to m\to n\to j}(t) \tag{6.87}$$

where the last term includes p sums and $C_j(t) \equiv b_j^{(0)}(t_0)$.

The convergence of the perturbative series depends on the time interval of the solution. As long as the Hamiltonian does not vary at an infinitely fast rate, we can always work on a small enough time scale so that the series converges rapidly. To solve the problem on time scales for which the series does not converge quickly, we can simply break the problem into multiple parts. This method provides us with a means to describe the system for a Hamiltonian that changes arbitrarily fast in time. Having a slowly varying Hamiltonian just serves to allow us to solve the problem without dividing it into as many parts (much of the time we will not have to divide the problem at all, and in Sec. 6.8.5 we derive conditions for when this will be the case), thus making the calculation significantly easier.

6.9 Conclusion and outlook

The work described in these notes offers promising prospects for future experiments. Ongoing work in the 10 m fountain is aimed at continuing the WEP test. In the future, macroscopic scale atom interferometers could be used for additional tests of general relativity (Dimopoulos et al., 2008a), a measurement of Newton's gravitational constant (Rosi et al., 2014), or a measurement of the gravitational Aharanov-Bohm effect (Hohensee et al., 2012). In addition, many of the techniques we have discussed are relevant for proposed space-based experiments, in which even longer interferometer durations could be realized (Altschul et al., 2015; Williams et al., 2016). Finally, the incorporation of squeezed states (Hosten et al., 2016; Cox et al., 2016) into macroscopic scale atom interferometers could further enhance the interferometer resolution.

References

Abbott, B. P. et al. (2016, Feb.). *Phys. Rev. Lett.*, **116**(6), 061102.

Acernese, F. et al. (2007). *AIP Conference Proceedings*, **924**, 187.

Altschul, B. et al. (2015). *Advances in Space Research*, **55**(1), 501–24.

Anandan, J. (1984). *Phys. Rev. D*, **30**(Oct.), 1615–24.

Antoine, C. (2006, Jul.). *Applied Physics B*, **84**(4), 585–97.

Arkani-Hamed, N., Dimopoulos, S., and Dvali, G. (1998). *Physics Letters B*, **429**, 263–72.

Arndt, M., and Hornberger, K. (2014). *Nat. Phys.*, **10**(Apr.), 271–7.

Arvanitaki, A. et al. (2008, Mar.). *Phys. Rev. Lett.*, **100**(12), 120407.

Asenbaum, P. et al. (2017, May). *Phys. Rev. Lett.*, **118**, 183602.

Audretsch, J., and Marzlin, K.-P. (1994, Sep.). *Phys. Rev. A*, **50**(3), 2080–95.

Barrett, B. et al. (2016). *Nat. Comm.*, **7**, 13786.

Bassi, A., Groardt, A., and Ulbricht, H. (2017). *Classical and Quantum Gravity*, **34**(19), 193002.

Berman, P. R. (ed.) (1997). *Atom Interferometry*. Academic Press, New York.

Biedermann, G. (2007). Doctoral thesis, Stanford University.

Bonder, Y., and Sudarsky, D. (2008). *Class. Quantum Gravity*, **25**(10), 105017.

Bongs, K., Launay, R., and Kasevich, M. A. (2006, Aug.). *Applied Physics B*, **84**(4), 599–602.

Bonnin, A. et al. (2013, Oct.). *Phys. Rev. A*, **88**(4), 043615.

Bordé, Christian J. (2004). *General Relativity and Gravitation*, **36**(3), 475–502.

Bordé, Ch. J., and Lammerzahl, C. (1996). *Phys. Bl.*, **52**, 238–40.

Bouchendira, R. et al. (2011, Feb.). *Phys. Rev. Lett.*, **106**(8), 080801.

Büchner, M. et al. (2003, Jul.). *Physical Review A*, **68**(1), 013607.

Cadoret, M. et al. (2008, December). *Phys. Rev. Lett.*, **101**(23), 230801.

Carroll, S. M. (2004). *Spacetime and Geometry: An Introduction to General Relativity*. Addison-Wesley, San Francisco.

Chiow, S.-W. et al. (2009, Jul.). *Phys. Rev. Lett.*, **103**(5), 050402.

Chiow, S.-W. et al. (2011, September). *Phys. Rev. Lett.*, **107**(13), 130403.

Chiow, S.-W. et al. (2012, Sep.). *Opt. Lett.*, **37**(18), 3861–3.

Chryssomalakos, C., and Sudarsky, D. (2003). *Gen. Relativ. Gravit.*, **35**(4), 605–17.

Chu, S. (2002). In *Nobel Lectures, Physics 1996–2000* (ed. G. Ekspong). World Scientific Publishing Co.

Chu, S. et al. (1986, Feb.). *Optics Letters*, **11**(2), 73.

Cladé, P. et al. (2009, Jun.). *Phys. Rev. Lett.*, **102**(24), 240402.

Cotter, J. P. et al. (2017). *Science Advances*, **3**(8).

Cox, K. C. et al. (2016, Mar.). *Phys. Rev. Lett.*, **116**, 093602.

Cronin, A. D., Pritchard, D. E., and Schmiedmayer, J. (2009, Jul.). *Reviews of Modern Physics*, **81**(3), 1051–1129.

Dickerson, S. (2014). Doctoral thesis, Stanford University.

Dickerson, S. et al. (2012, Jun.). *The Review of Scientific Instruments*, **83**(6), 065108.

Dickerson, S. M. et al. (2013, Aug.). *Phys. Rev. Lett.*, **111**(8), 083001.

Dimopoulos, S. et al. (2008*a*, Aug.). *Phys. Rev. D*, **78**(4), 042003.

Dimopoulos, S. et al. (2008*b*, Dec.). *Phys. Rev. D*, **78**(12), 122002.

Dubetsky, B., and Kasevich, M. (2006, Aug.). *Phys. Rev. A*, **74**(2), 023615.

Feynman, R. P., and Hibbs, A. R. (1965). *Quantum Mechanics and Path Integrals*. McGraw-Hill, New York.

Feynman, R. P., Leighton, R. B., and Sands, M. (1965). *The Feynman Lectures on Physics Volume III*. Addison-Wesley, Reading, MA.

Foster, G. T. et al. (2002). *Optics Letters*, **27**(11), 951–3.

Gerbier, F., and Castin, Y. (2010, Jul.). *Phys. Rev. A*, **82**, 013615.

Giltner, D. M., McGowan, R. W., and Lee, S. A. (1995). *Phys. Rev. Lett.*, **75**(14), 2638–41.

Graham, P. W. et al. (2013, Apr.). *Phys. Rev. Lett.*, **110**(17), 171102.

Griffiths, D. J. (2005). *Introduction to Quantum Mechanics*. Pearson Education, Upper Saddle River, NJ.

Gustavson, T. L., Bouyer, P., and Kasevich, M. A. (1997). *Phys. Rev. Lett.*, **78**(11), 2046–9.

Hecker Denschlag, J. et al. (2002). *J. Phys. B.: At. Mol. Opt. Phys.*, **35**, 3095–110.

Hogan, J. M. et al. (2011, May). *General Relativity and Gravitation*, **43**(7), 1953–2009.

Hogan, J. M., Johnson, D. M. S., and Kasevich, M. A. (2009, Jun.). In *Proc. Int. School Phys. Enrico Fermi, Course CLXVIII* (ed. E. Arimondo, W. Ertmer, and W. P. Schleich), pp. 411–47. IOS Press.

Hohensee, M. A. et al. (2012, Jun.). *Phys. Rev. Lett.*, **108**(23), 230404.

Hosten, O. et al. (2016). *Nature*, **529**(7587), 505–8.

Kasevich, M., and Chu, S. (1991, Jul.). *Phys. Rev. Lett.*, **67**(2), 181–4.

Kawabe, K., and the LIGO Collaboration (2008). *Journal of Physics: Conference Series*, **120**(3), 032003.

Ketterle, W. (2001). In *The Nobel Prizes 2001* (ed. T. Frängsmere), Volume 1995, pp. 118–54. Nobel Foundation.

Kotru, K. et al. (2015, September). *Phys. Rev. Lett.*, **115**(10), 103001.

Kovachy, T. (2009). Senior thesis, Harvard University.

Kovachy, T. et al. (2015*a*, Apr.). *Phys. Rev. Lett.*, **114**(14), 143004.

Kovachy, T. et al. (2015*b*, Dec.). *Nature*, **528**(7583), 530–3.

Kovachy, T., Chiow, S.-W., and Kasevich, M. A. (2012, Jul.). *Phys. Rev. A*, **86**(1), 011606.

Kovachy, T. et al. (2010, Jul.). *Physical Review A*, **82**(1), 013638.

Kozuma, M. et al. (1999). *Phys. Rev. Lett.*, **82**(5), 871–5.

Kuhn, C. C. N. et al. (2014). *New J. Phys.*, **16**(7), 073035.

Larson, S. L., Hiscock, W. A., and Hellings, R. W. (2000, Aug.). *Phys. Rev. D*, **62**(6), 062001.

Loy, M. M. T. (1974). *Phys. Rev. Lett.*, **32**(15), 814–17.

Lunney, D. (1998). In *Nuclei in the Cosmos V* (ed. N. Prantzos and S. Harissopulos), Paris, pp. 296–302. Editions Frontieres.

Malinovsky, V. S., and Berman, P. R. (2003, Aug.). *Phys. Rev. A*, **68**(2), 023610.

Martin, P. J. et al. (1988). *Phys. Rev. Lett.*, **60**(6), 515–18.

McDonald, G. D. et al. (2013, Nov.). *Phys. Rev. A*, **88**(5), 053620.

McGuirk, J. M. et al. (2002, Feb.). *Phys. Rev. A*, **65**(3), 033608.

McGuirk, J. M., Snadden, M. J., and Kasevich, M. A. (2000, Nov.). *Phys. Rev. Lett.*, **85**(21), 4498–501.

Metcalf, H. J., and van der Straten, P. (1999). *Laser Cooling and Trapping*. Springer-Verlag, New York.

Moler, K. et al. (1992). *Physical Review A*, **45**(1), 342–8.

Müller, H., Chiow, S.-W., and Chu, S. (2008, Feb.). *Phys. Rev. A*, **77**, 023609.

Müller, H. et al. (2009, Jun.). *Phys. Rev. Lett.*, **102**(24), 240403.

Müller, H. et al. (2008, May). *Phys. Rev. Lett.*, **100**(18), 180405.

Müller, H. et al. (2006, Jul.). *Applied Physics B*, **84**(4), 633–42.

Müntinga, H. et al. (2013, Feb.). *Phys. Rev. Lett.*, **110**(9), 093602.

Osheroff, D. D. (2002). In *Nobel Lectures, Physics 1996-2000* (ed. G. Ekspong). World Scientific Publishing Co.

Overstreet, C. et al. (2017). *arXiv preprint arXiv:1711.09986*.

Peik, E. et al. (1997). *Phys. Rev. A*, **55**(4), 2989–3001.

Peters, A., Chung, K. Y., and Chu, S. (2001). *Metrologia*, **38**, 25–61.

Pichler, H., Daley, A. J., and Zoller, P. (2010, Dec.). *Phys. Rev. A*, **82**, 063605.

Romero-Isart, Oriol (2017). *New Journal of Physics*, **19**(12), 123029.

Rosi, G. et al. (2017). *Nat. Comm.*, **8**, 15529.

Rosi, G. et al. (2014). *Nature*, **510**(7506), 518–21.

Roura, A. (2017, Apr.). *Phys. Rev. Lett.*, **118**, 160401.

Schlippert, D. et al. (2014, May). *Phys. Rev. Lett.*, **112**(20), 203002.

Smith, J. R. and the LIGO Collaboration (2009). *Classical and Quantum Gravity*, **26**(11), 114013.

Snadden, M. J. et al. (1998). *Phys. Rev. Lett.*, **81**(5), 971–4.

Stockton, J. K., Wu, X., and Kasevich, M. A. (2007, Sep.). *Phys. Rev. A*, **76**(3), 033613.

Storey, P., and Cohen-Tannoudji, C. (1994). *J. Phys. II (France)*, **4**, 1999.

Sugarbaker, Alex (2014). Doctoral thesis, Stanford University.

Sugarbaker, A. et al. (2013, Sep.). *Phys. Rev. Lett.*, **111**(11), 113002.

Szigeti, S. S. et al. (2012). *New J. Phys.*, **14**(2), 023009.

Torosov, B. T., Guérin, S., and Vitanov, N. V. (2011, Jun.). *Phys. Rev. Lett.*, **106**(23), 233001.

Unnikrishnan, C. S., and Gillies, G. T. (2004). *Metrologia*, **41**(5), S125.

Vitanov, N. V. et al. (2001). *Annu. Rev. Phys. Chem.*, **52**, 763–809.

Wicht, A., Sarajlic, E., Hensley, J. M., and Chu, S. (2005, Aug.). *Phys. Rev. A*, **72**(2), 023602.

Williams, J. et al. (2016). *New J. Phys.*, **18**(2), 025018.

Zhou, L. et al. (2015, Jul.). *Phys. Rev. Lett.*, **115**(1), 013004.

Zhou, L. et al. (2011, Jul.). *General Relativity and Gravitation*, **43**(7), 1931–42.

7

Quantum jumps, Born's rule, and objective classical reality via quantum Darwinism

WOJCIECH HUBERT ZUREK

Theory Division, MS B213, LANL
Los Alamos, NM, 87545, USA

Zurek, W. H. "Quantum jumps, Born's rule, and objective classical reality via quantum Darwinism." In *Current Trends in Atomic Physics*. Edited by Antoine Browaeys, Thierry Lahaye, Trey Porto, Charles S. Adams, Matthias Weidemüller, and Leticia F. Cugliandolo. Oxford University Press (2019). © Oxford University Press. DOI: 10.1093/oso/9780198837190.003.0007

Chapter Contents

Emergence of the classical world from the quantum substrate of our Universe is a long-standing conundrum. I describe three insights into the transition from quantum to classical that are based on the recognition of the role of the environment. I begin with derivation of preferred sets of states that help define what exists—our everyday classical reality. They emerge as a result of breaking of the unitary symmetry of the Hilbert space which happens when the unitarity of quantum evolutions encounters nonlinearities inherent in the process of amplification—of replicating information. This derivation is accomplished without the usual tools of decoherence, and accounts for the appearance of quantum jumps and emergence of preferred *pointer states* consistent with those obtained via environment-induced superselection, or *einselection*. Pointer states obtained this way determine what can happen—define events—without appealing to Born's rule for probabilities. Therefore, $p_k = |\psi_k|^2$ can be now deduced from the entanglement-assisted invariance, or *envariance*—a symmetry of entangled quantum states. With probabilities at hand one also gains new insights into foundations of quantum statistical physics. Moreover, one can now analyze information flows responsible for decoherence. These information flows explain how perception of objective classical reality arises from the quantum substrate: effective amplification they represent accounts for the objective existence of the einselected states of macroscopic quantum systems through the redundancy of pointer state records in their environment—through *quantum Darwinism*.

7.1 Introduction and preview

This essay is not a comprehensive review. It is nevertheless a brief review of several interrelated developments that can be collectively described as "Quantum Theory of Classical Reality". Its predecessor[1] was intended to be a brief annotated guide to some of the (then) recent results.

Present lecture notes evolved away from the annotated guide to literature in the direction of a review in two ways: they are less complete as a guide (as quite a few relevant papers have appeared since 2008, and while I mention some of them, I am sure there are significant omissions). On the other hand, I went beyond the annotated guide canon, and I review some of the key advances in more depth. Still, this is no substitute for the original papers, or for a fully fledged review.

Two mini-reviews in *Nature Physics* (Zurek, 2009) and *Physics Today* (Zurek, 2014) are also available. A more detailed review (Zurek, 2007a) is by now somewhat out-of-date, as several relevant results were obtained since 2007 when it was written. Moreover, a book that will cover this same ground as the present lectures, as well as theory of decoherence and other related subjects is (slowly) being written (Zurek, in preparation). Nevertheless, it is hoped that readers may appreciate, in the interim, an update as well as the more informal presentation style of this overview, including the "frequently asked questions" in Sec. 7.6.

---

[1] A concise predecessor of these notes has appeared as Chapter 13 in *Many Worlds? Everett, Quantum Theory, and Reality*, S. Saunders et al., eds (Oxford University Press, 2010). It was prepared in 2008, and is now somewhat out-of-date. Nevertheless, the framing of the discussion in the "Relative States" (but not "Many Worlds"!) reading of Everett's interpretation has survived the very significant update that led to the present manuscript.

The "Relative State Interpretation" set out fifty years ago by Hugh Everett III (Everett, 1957a; 1957b) is a convenient starting point for our discussion. Within its context, one can reevaluate basic axioms of quantum theory as extracted, for example, from the classic textbook (Dirac, 1958). Everettian view of the Universe is a good way to motivate exploring the effect of the environment on the state of the system. (Of course, a complementary motivation based on a non-dogmatic reading of (Bohr, 1928) is also possible.)

The basic idea we shall pursue here is to accept a relative state explanation of the "collapse of the wavepacket" by recognizing, with Everett, that observers perceive the state of the "rest of the Universe" *relative* to their own state, or—to be more precise—relative to the state of their records. This allows quantum theory to be universally valid. (This does *not* mean that one has to accept a "many worlds" ontology; see (Zurek, 2007a) for discussion.)

Much of the heat in various debates on the foundations of quantum theory seems to be generated by the expectation that a *single* idea should provide a complete solution. When this does not happen—when there is progress, but there are still unresolved issues—the possibility that an idea responsible for this progress may be a step in the right direction—but that more than one idea, one step, is needed—is often dismissed. As we shall see, developing quantum theory of our classical everyday reality requires the solution of *several* problems and calls for several ideas. In order to avoid circularities, they need to be introduced in the right order.

7.1.1 Preferred pointer states from einselection

Everett explains perception of the collapse. However, his relative state approach raises three questions absent in Bohr's Copenhagen Interpretation (Bohr, 1928) that relied on the independent existence of an *ab initio* classical domain. Thus, in a completely quantum Universe one is forced to seek sets of preferred, effectively classical but ultimately quantum states that can define what exists—branches of the universal state vector—and that allow observers keep reliable records. Without such **preferred basis** relative states are just "too relative", and the relative state approach suffers from *basis ambiguity* (Zurek, 1981).

Decoherence selects preferred *pointer states* (Zurek, 1981; 1982; 1991), so this issue was in fact resolved some time ago. The principal consequence of the environment-induced decoherence is that, in open quantum systems—systems interacting with their environments—only certain quantum states retain stability in spite of the immersion of the system in the environment: Superpositions are unstable, and quickly decay into mixtures of the einselected, stable pointer states (Zurek, 1981; 1982; 1991; 2003b; 2007a; 2014; Zeh, 1990; Paz and Zurek, 2001; Joos et al., 2003; Schlosshauer, 2004; 2007). This is einselection—a nickname for *environment*—*induced* super*selection*. Thus, while the significance of the environment in suppressing quantum behavior was pointed out by Dieter Zeh already in 1970 (Zeh, 1970; 2006), the role of decoherence in einselecting preferred pointer states and, hence its role in inducing the transition from quantum to classical became clear only later (Zurek, 1981; 1982; 1991; Schlosshauer, 2007).

7.1.2   Born's rule from envariance

Einselection can account for preferred sets of states, and, hence, for Everettian "branches". But this is achieved at a very high price—the usual practice of decoherence is based on averaging (as it involves reduced density matrices defined by a partial trace). This means that one is using Born's rule to relate amplitudes to probabilities. However, as emphasized by Everett, Born's rule should not be postulated in an approach that is based on purely unitary quantum dynamics. The assumption of the universal validity of quantum theory raises the issue of **the origin of Born's rule**, $p_k = |\psi_k|^2$, which—following the original conjecture (Born, 1926)—is simply postulated in textbook discussions.

Here we shall see that Born's rule can be derived from entanglement—assisted invariance, or envariance—from the symmetry of entangled quantum states. Envariance is a purely quantum symmetry, as it is critically dependent on the telltale quantum feature—entanglement. Envariance sheds new light on the origin of probabilities relevant for the everyday world we live in, e.g. for statistical physics and thermodynamics. Moreover, fundamental derivation of objective probabilities allows one to discuss information flows in our quantum Universe, and, hence, understand how perception of classical reality emerges from quantum substrate.

7.1.3   Classical reality via quantum Darwinism

Even preferred quantum states defined by einselection are still ultimately quantum. Therefore, they cannot be found out by initially ignorant observers through direct measurement without getting disrupted (reprepared). Yet, states of macroscopic systems in our everyday world seem to exist objectively—they can be found out by anyone without getting disrupted. This ability to find out an unknown state is in fact an operational definition of "objective existence". So, if we are to explain the emergence of everyday objective classical reality, we need to identify **quantum origin of objective existence**.

We shall do that by dramatically upgrading the role of the environment: in decoherence theory the environment is the collection of degrees of freedom where quantum coherence (and, hence, phase information) is lost. However, in "real life" the role of the environment is in effect that of a witness (see e.g. (Zurek, 2000; 2003b)) to the state of the system of interest, and of a communication channel through which the information reaches us, the observers. This mechanism for the emergence of the classical objective reality is the subject of the theory of quantum Darwinism.

7.2   Quantum postulates and relative states

We start from a well-defined solid ground—the list of quantum postulates that are explicit in Dirac (1928), and at least implicit in most quantum textbooks.

The first two deal with the mathematics of quantum theory:

(i) *The state of a quantum system is represented by a vector in its Hilbert space $\mathcal{H}_S$.*
(ii) *Evolutions are unitary (e.g. generated by the Schrödinger equation).*

These two postulates provide an essentially complete summary of the mathematical structure of quantum physics. They are often (DeWitt, 1971; DeWitt and Graham, 1973) supplemented by a composition postulate:

(o) *States of composite quantum systems are represented by a vector in the tensor product of the Hilbert spaces of its components.*

Physicists sometimes differ in assessing how much of postulate (o) follows from (i). We shall not be distracted by this issue, and move on to where the real problems are. Readers can follow their personal taste in supplementing (i) and (ii) with whatever portion of (o) they deem necessary. It is nevertheless useful to list (o) explicitly to emphasize the role of the tensor structure it posits: it is crucial for entanglement, quantum phenomenon we will depend on.

Using (o), (i) and (ii), suitable Hamiltonians, etc., one can calculate. Yet, such quantum calculations are only a mathematical exercise—without additional postulates one can predict nothing of experimental consequence from their results. What is so far missing is physics—a way to establish correspondence between abstract state vectors in $\mathcal{H}_S$ and laboratory experiments (and/or everyday experience) is needed to relate quantum mathematics to our world.

Establishing this correspondence starts with the next postulate:

(iii) *Immediate repetition of a measurement yields the same outcome.*

Immediate repeatability is an idealization (it is hard to devise such non-demolition measurements, but it can be done). Yet postulate (iii) is uncontroversial. The notion of a "state" is based on predictability, and the most rudimentary prediction is that the state is what it is known to be. This key ingredient of quantum physics goes beyond the mathematics of postulates (o)–(ii). It enters through the repeatability postulate (iii). Moreover, a classical equivalent of (iii) is taken for granted (unknown classical state can be discovered without getting disrupted), so repeatability does not clash with our classical intuition.

Postulate (iii) is the last uncontroversial postulate on the textbook list. This collection comprises our *quantum core* postulates—our *credo*, the foundation of the quantum theory of the classical.

In contrast to classical physics (where unknown states can be found out by an initially ignorant observer) the very next quantum axiom limits the predictive attributes of the state compared to what they were in the classical domain:

(iv) *Measurement outcomes are limited to an orthonormal set of states (eigenstates of the measured observable). In any given run of a measurement an outcome is just one such state.*

This *collapse postulate* is controversial. To begin with, in a completely quantum Universe it is inconsistent with the first two postulates: starting from a general pure state $|\psi_S\rangle$ of the system (postulate (i)), and an initial state $|A_0\rangle$ of the apparatus $\mathcal{A}$, and assuming unitary evolution (postulate (ii)) one is led to a superposition of outcomes:

$$|\psi_S\rangle|A_0\rangle = \left(\sum_k a_k|s_k\rangle\right)|A_0\rangle \Rightarrow \sum_k a_k|s_k\rangle|A_k\rangle \, , \qquad (7.1)$$

which is in contradiction with, at least, a literal interpretation of the "collapse" anticipated by axiom (iv). This conclusion follows for an apparatus that works as intended in tests (i.e. $|s_k\rangle|A_0\rangle \Rightarrow |s_k\rangle|A_k\rangle$) from linearity of quantum evolutions that is in turn implied by unitarity of postulate (ii).

Everett settled (or at least bypassed) the "collapse" part of the problem with (iv)—observer perceives the state of the rest of the Universe relative to his / her records. This is the essence of the Relative State Interpretation.

However, from the standpoint of our quest for classical reality perhaps the most significant and disturbing implication of (iv) is that quantum states do not exist—at least not in the objective sense to which we are used to in the classical world. The outcome of the measurement is typically *not* the preexisting state of the system, but one of the eigenstates of the measured observable.

Thus, whatever quantum state is, "objective existence" independent of what is known about it is clearly not one of its attributes. This malleability of quantum states clashes with the classical idea of what the state should be. Some even go as far as to claim that quantum states are simply a description of the information that an observer has, and have essentially nothing to do with "existence".

I believe this denial of existence under any circumstances is going too far—after all, there are situations when a state can be found out, and the repeatability postulated by (iii) recognizes that its existence can be confirmed. But, clearly, (iv) limits the "quantum existence" of states to situations that are "under the jurisdiction" of postulate (iii) (or slightly more general situations where the preexisting density matrix of the system commutes with the measured observable).

Collapse postulate (iv) complicates interpreting quantum formalism, as has been appreciated since Bohr (1928) and von Neumann (1932). Therefore, at least before Everett, it was often cited as an indication of the ultimate insolubility of the "quantum measurement problem". Yet, (iv) is hard to argue with—it captures what happens in the laboratory measurements.

To resolve the clash between the mathematical structure of quantum theory and our perception of what happens in the laboratory, in the real world measurements, one can accept—with Bohr—the primacy of our experience. The inconsistency of (iv) with the mathematical core of the quantum formalism—superpositions of (i) and unitarity of (ii)—can then be blamed on the nature of the apparatus. According to the Copenhagen Interpretation the apparatus is classical, and, therefore, not a subject to the quantum principle of superposition (which follows from (i)). Measurements straddle the quantum-classical border, so they need not abide by the unitarity of (ii). Therefore, collapse can happen on the "lawless" quantum-classical border.

This quantum-classical duality posited by Bohr challenges the unification instinct of physicists. One way of viewing decoherence is to regard einselection as a mechanism that accounts for effective classicality by suspending the validity of the quantum principle of superposition in a subsystem while upholding it for the composite system that includes the environment (Zurek, 1991; 2003b).

Everett's alternative to Bohr's approach was to abandon the literal collapse and recognize that, once the observer is included in the wavefunction, one can consistently interpret the consequences of such correlations. The right hand side of Eq. 7.1 contains all the possible outcomes, so the observer who records outcome #17 perceives the branch of Universe that is consistent with that event reflected in his records. This view of the collapse is also consistent with repeatability of postulate (iii); re-measurement by the same observer using the same (non-demolition) device yields the same outcome.

Nevertheless, this relative state view of the quantum Universe suffers from a basic problem: the principle of superposition (the consequence of axiom (i)) implies that the state of the system or of the apparatus after the measurement can be written in infinitely many unitarily equivalent basis sets in the Hilbert spaces of the apparatus (or of the observer's memory);

$$\sum_k a_k |s_k\rangle |A_k\rangle = \sum_k a'_k |s'_k\rangle |A'_k\rangle = \sum_k a''_k |s''_k\rangle |A''_k\rangle = \dots \qquad (7.2)$$

This is the *basis ambiguity* (Zurek, 1981). It appears as soon as—with Everett—one eliminates axiom (iv). The bases employed above are typically non-orthogonal, but in the Everettian relative state setting there is nothing that would preclude them, or that would favor, e.g., the Schmidt basis of $\mathcal{S}$ and $\mathcal{A}$ (the orthonormal basis that is unique, provided that the absolute values of the *Schmidt coefficients* in such a *Schmidt decomposition* of an entangled bipartite state differ).

In our everyday reality we do not seem to be plagued by such basis ambiguity problems. So, in our Universe there is something that (in spite of (i) and the egalitarian superposition principle it implies) picks out preferred states, and makes them effectively classical. Axiom (iv) anticipates this.

Consequently, before there is an (apparent) collapse in the sense of Everett, a set of preferred states—one of which is selected by (or at the very least, consistent with) observer's records—must be chosen. There is nothing in the writings of Everett that would even hint that he was aware of basis ambiguity and questions it leads to.

The next question concerns probabilities: how likely is it that, after I measure, my state will be, say, $|\mathcal{I}_{17}\rangle$? Everett was keenly aware of this issue, and even believed that he solved it by deriving Born's rule. In retrospect, it is clear that the argument he proposed—as well as the arguments proposed by his followers, including DeWitt (1970; 1971), DeWitt and Graham (1973) and Geroch (1984) who noted the failure of Everett's original approach, and attempted to fix the problem—did not accomplish as much as was hoped for, and did not amount to a derivation of Born's rule (see (Squires, 1990; Stein, 1984; Kent, 1990) for influential critical assessments).

In textbook versions of the quantum postulates probabilities are assigned by another (Born's rule) axiom:

(v) *The probability p_k of an outcome $|s_k\rangle$ in a measurement of a quantum system that was previously prepared in the state $|\psi\rangle$ is given by $|\langle s_k|\psi\rangle|^2$.*

Born's rule fits very well with Bohr's approach to quantum-classical transition (e.g. with postulate (iv)). However, Born's rule is at odds with the spirit of the relative state approach, or any approach that attempts (as we do) to deduce perception of the classical everyday reality starting from the quantum laws that govern our Universe. This does not mean that there is a mathematical inconsistency here: one can certainly use Born's rule (as the formula $p_k = |\langle s_k|\psi\rangle|^2$ is known) along with the relative state approach in averaging to get expectation values and the reduced density matrix.

Indeed, until the derivation of Born's rule in a framework of decoherence was proposed, decoherence practice relied on probabilities given by $p_k = |\langle s_k|\psi\rangle|^2$. They enter whenever one assigns physical interpretation to reduced density matrices, a key tool of the decoherence theory. Everett's point was not that Born's rule is wrong, but, rather, that it should be *derived* from the other quantum postulates, and we shall show how to do that.

7.3   Quantum origin of quantum jumps

To restate briefly the three problems identified above, we need to derive the essence of the collapse postulates (iv) and Born's rule (v) from our *credo*—the core quantum postulates (o)–(iii). Moreover, even when we accept relative state origin of "single outcomes" and "collapse", we still need to justify emergence of the preferred basis that is the essence of (iv).

This issue (which in our summary of textbook axiomatics of quantum theory is a part of the collapse postulate) is so important that it is often captured by a separate postulate which declares that "Observables are Hermitian". This, in effect, means that the outcomes of measurements should correspond to orthogonal states in the Hilbert space. Furthermore, we should do it without appealing to Born's rule—without decoherence, or at least without its usual tools such as reduced density matrices that rely on Born's rule. Once we have preferred states, we will also have a set of candidate *events*. Once we have events we shall be able to pose questions about their probabilities.

The *preferred basis problem* was settled by the environment-induced superselection (*einselection*), usually regarded as a principal consequence of decoherence. This is discussed elsewhere (Zurek, 1981; 1982). Preferred pointer states and einselection are usually justified by appealing to decoherence. Therefore, they come at a price that would have been unacceptable to Everett: decoherence and einselection employ reduced density matrices and trace, and so their predictions are based on averaging, and thus, on probabilities—on Born's rule.

Here we present an alternative strategy for arriving at preferred states that—while not at odds with decoherence—does not rely on the Born's rule-dependent tools of decoherence. Our overview of the origin of quantum jumps is brief. However, we direct the reader to references where different steps of that strategy are discussed in more detail. In short, we describe how one should go about doing the necessary physics, but we only sketch what needs to be done, and we do not explain all the details—the requisite steps are carried out in the references we provide: our discussion is meant as a guide to the literature, and not a substitute.

Decoherence done "in the usual way" (which, by the way, is a step in the right direction, in understanding the practical and even many of the fundamental aspects of the quantum-classical transition!) is not a good starting point in addressing the more fundamental aspects of the origins of the classical.

In particular, decoherence is not a good starting point for the derivation of Born's rule. We have already noted the problem with this strategy: it courts circularity. It employs Born's rule to arrive at the pointer states by using reduced density matrix which is obtained through trace—i.e., averaging, which is where Born's rule is implicitly invoked (see e.g. (Nielsen and Chuang, 2000)). So, using decoherence to derive Born's rule is at best a consistency check.

While I am most familiar with my own transgressions in this matter (Zurek, 1998b), this circularity also afflicts other approaches, including the proposal based on decision theory (Deutsch, 1999; Wallace, 2003; Saunders, 2004), as noted also by others (see e.g. Forrester, 2007; Dawid and Thebault, 2015). Therefore, one has to start the task from a different end.

To get anywhere—e.g., to define "events" essential in the introduction of probabilities—we need to show how the mathematical structure of quantum theory (postulates (o), (i) and (ii)—Hilbert space and unitarity) supplemented by the uncontroversial postulate (iii) (immediate repeatability, hence predictability) leads to preferred sets of states.

7.3.1 Quantum jumps from quantum core postulates

Surprisingly enough, deducing preferred states from "quantum credo" turns out to be simple. The possibility of repeated confirmation of an outcome is all that is needed to establish an effectively classical domain within the quantum Universe, and to define events such as measurement outcomes.

One can accomplish this with minimal assumptions ("quantum core" postulates (o)—(iii) on the above list) as described in (Zurek, 2007b; 2013). Here we review the basic steps. We assume that $|v\rangle$ and $|w\rangle$ are among the possible repeatably accessible outcome states of $\mathcal{S}$.

$$|v\rangle|A_0\rangle \Longrightarrow |v\rangle|A_v\rangle, \tag{7.3a}$$

$$|w\rangle|A_0\rangle \Longrightarrow |w\rangle|A_w\rangle. \tag{7.3b}$$

So far, we have employed postulates (i) and (iii). The measurement, when repeated, would yield the same outcome, as the pre-measurement states have not changed. Thus, postulate (iii) is indeed satisfied.

We now assume the process described by Eq. 7.3 is fully quantum, so postulate (ii)—unitarity of evolutions—must also apply. Unitarity implies that the overlap of the states before and after must be the same. Hence:

$$\langle v|w\rangle(1 - \langle A_v|A_w\rangle) = 0. \tag{7.4}$$

Our conclusions follow from this simple equation. There are two possibilities that depend on the overlap $\langle v|w\rangle$.

Suppose first that $\langle v|w\rangle \neq 0$. One is then forced to conclude that the measurement was unsuccessful since the state of $\mathcal{A}$ was unaffected by the process above. That is, the transfer of information from $\mathcal{S}$ to $\mathcal{A}$ must have failed completely as in this case $\langle A_v|A_w\rangle = 1$ must hold. In particular, the apparatus can bear no imprint that distinguishes between states $|v\rangle$ and $|w\rangle$ that aren't orthogonal.

The other possibility, $\langle v|w\rangle = 0$, allows for an arbitrary $\langle A_v|A_w\rangle$, including a perfect record, $\langle A_v|A_w\rangle = 0$. Thus, outcome states must be orthogonal if—in accord with postulate (iii)—they are to survive intact a successful information transfer in general or a quantum measurement in particular, so that immediate re-measurement can yield the same result.

The same derivation can be carried out for $\mathcal{S}$ with a Hilbert space of dimension $\mathcal{N}$ starting with a system state vector $|\psi_{\mathcal{S}}\rangle = \sum_{k=1}^{\mathcal{N}} \alpha_k|s_k\rangle$, where (as before)—a priori $\{|s_k\rangle\}$ need to be only linearly independent.

The simple reasoning above leads to a surprisingly decisive conclusion: orthogonality of the outcome states of the system is absolutely essential for them to imprint even a minute difference on the state of any other system while retaining their identity. The overlap $\langle v|w\rangle$ must be 0 *exactly* for $\langle A_v|A_w\rangle$ to differ from unity.

Imperfect or accidental information transfers (e.g. to the environment in course of decoherence) can also define preferred sets of states providing that the crucial non-demolition demand of postulate (iii) is imposed on the unitary evolution responsible for the information flow.

A straightforward extension of the above derivation to where it can be applied not just to measured quantum systems (where nondemlition is a tall order), but to the measuring devices (where repeatability is essential) is possible (Zurek, 2007b; 2013). The derivation is somewhat more demanding technically, as one needs to allow for mixed states and for decoherence in a model of a presumably macroscopic apparatus, but the conclusion is the same: records maintained by the apparatus or repeatably accessible states of macroscopic but ultimately quantum systems must correspond to orthogonal subspaces of their Hilbert space.

It is important to emphasize that we are not asking for clearly distinguishable records (i.e. we are not demanding orthogonality of the states of the apparatus, $\langle A_v|A_w\rangle = 0$). Indeed, in the macroscopic case (Zurek, 2013) one does not even ask for the state of the system to remain unchanged, but only for the outcomes of the consecutive measurements to be identical (i.e. the evidence of repeatability is in the outcomes). Still, even under these rather weak assumptions one is forced to conclude that *quantum states can exert distinguishable influences and remain unperturbed only when they are orthogonal*. To arrive at this conclusion we have relied on postulate (i), on the fact that when two vectors in the Hilbert space are identical then physical states they correspond to must also be identical.

7.3.2   Discussion

Emergence of orthogonal outcome states is established above on the foundation of very basic (and very quantum) assumptions. It leads one to conclude that observables are indeed associated with Hermitan operators.

Hermitian observables are usually introduced in a very different manner—they are the (postulated!) quantum versions of the familiar classical quantities. This emphasizes physical significance of their spectra (especially when they correspond to conserved quantities). Their orthogonal eigenstates emerge from the mathematics, once their Hermitian nature is assumed. Here we have deduced their Hermiticity by proving orthogonality of their eigenstates—possible outcomes—from the quantum core postulates by focusing on the effect of information transfer on the measured system.

The restriction to an orthogonal set of outcomes yields **preferred basis**: the essence of the collapse axiom (iv) need not be postulated! It follows from the uncontroversial quantum core postulates (o)-(iii).

We note that the preferred basis arrived at in this manner essentially coincides with the basis obtained long time ago via einselection, (Zurek, 1981; 1982). It is just that here we have arrived at this familiar result without implicit appeal to Born's rule, which is essential if we want to take the next step, and derive postulate (v).

We have relied on unitarity, so we did not derive the actual *collapse* of the wavepacket to a single outcome—single event. Collapse is nonunitary, so one cannot deduce it starting from the quantum core that includes postulate (ii). However, we have accounted for one of the key collapse attributes: the necessity of a symmetry breaking—of the choice of a single *orthonormal* set of states from amongst various possible basis sets each of which can equally well span the Hilbert space of the system—follows from the core quantum postulates. This sets the stage for collapse—for quantum jumps.

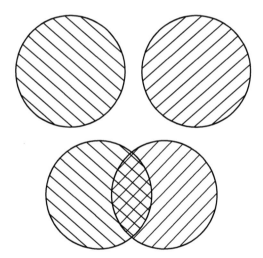

Fig. 7.1 The fundamental (pre-quantum) connection between distinguishability and repeatability of measurements. The two circles represent two states of the measured system. They correspond to two outcomes—e.g., two properties of the underlying states (represented by two cross-hatchings). A measurement that can result in either outcome—that can produce a record correlated with these two properties—can be repeatable only when the two corresponding states (the two circles) do not overlap (case illustrated at the top). Repeatability is impossible without distinguishability: when two states overlap (case illustrated in the bottom), repetition of the measurement can always result in a system switching the state (and, thus, defying repeatability). In the quantum setting this pre-quantum connection between repeatability and distinguishability leads to the derivation of orthogonality of repeatable measurement outcomes (and the two cross-hatchings can be thought of as two linear polarizations of a photon—orthogonal on the top, but not below), but the basic intuition demanding distinguishability as a prerequisite for repeatability does not rely on quantum formalism.

As we have already briefly noted, this reasoning can be extended (Zurek, 2013) to when repeatedly copied states belong to a macroscopic, decohering system (e.g., an apparatus pointer). In that case microstate *can* be perturbed by copying (or by the environment). What matters then is not the "nondemolition" of the microstate of the pointer, but persistence of the record its macrostate (corresponding to a whole collection of microstates) represents. To formulate this demand precisely one can rely on repeatability of copies: for instance, even though microstates of the pointer change upon readout due to the interaction with the environment, its macrostate should still represent the same measurement outcome—it should still contain the same "actionable information" (Zurek, 2013). This more general discussion addresses also other issues (e.g. connection between repeatability, distinguishability, and POVM's, raised in **FAQ #4**) that arise in realistic settings (see Fig. 7.1 for the illustration of the key idea).

7.4   Probabilities from entanglement

Derivation of events allows and even forces one to enquire about their probabilities or—more specifically—about the relation between probabilities of measurement outcomes and the initial pre-measurement state. As noted earlier, several past attempts at the derivation of Born's rule turned out to be circular. Here we present key ideas behind a circularity-free approach.

We emphasize that our derivation of events did not rely on Born's rule. In particular, we have not attached any physical interpretation to the values of scalar products, and the key to our conclusions rested on whether the scalar product is (or is not) 0 or 1, or neither.

We now briefly review envariant derivation of Born's rule based on the symmetry of entangled quantum states—on entanglement-assisted **invariance** or **envariance**. The study of envariance as a physical basis of Born's rule started with (Zurek, 2003a; 2005; 2003b), and is now the focus of several other papers (Schlosshauer and Fine, 2005; Barnum, 2003; Herbut, 2007). The key idea is illustrated in Fig. 7.2.

As we shall see, the eventual loss of coherence between pointer states can be also regarded as a consequence of quantum symmetries of the states of systems entangled with their environment. Thus, the essence of decoherence arises from symmetries of entangled states. Indeed, some of the consequences of einselection (including emergence of preferred states, as we have seen it in the previous section) can be studied without employing the usual tools of decoherence theory (reduced density matrices and trace) that, for their physical significance, rely on Born's rule.

Decoherence that follows from envariance also allows one to justify additivity of probabilities, while the derivation of Born's rule by Gleason (1957) assumed it (along with the other Kolmogorov's axioms of the measure-theoretic formulation of the foundations of probability theory, and with the Copenhagen-like setting). Appeal to symmetries leads to additivity also in the classical setting (as was noted already by Laplace (1820); see discussion in Gnedenko (1968). Moreover, Gleason's theorem (with its rather complicated proof based on "frame functions" introduced especially for this purpose) provides no motivation why the measure he obtains should have any physical significance—i.e. why should it be regarded as probability. As illustrated in Fig. 7.2 and discussed below, envariant derivation of Born's rule has a transparent physical motivation.

Additivity of probabilities is a highly nontrivial point. In quantum theory the overarching additivity principle is the quantum principle of superposition. Anyone familiar with the double slit experiment knows that probabilities of quantum states (such as the states corresponding to passing through one of the two slits) do *not* add, which in turn leads to interference patterns.

The presence of entanglement eliminates local phases (thus suppressing quantum superpositions, i.e. doing the job of decoherence). This leads to additivity of probabilities of events associated with preferred pointer states.

7.4.1 Decoherence, phases, and entanglement

Decoherence is the loss of phase coherence between preferred states. It occurs when $\mathcal{S}$ starts in a superposition of pointer states singled out by the interaction (represented below by the Hamiltonian $\mathbf{H}_{\mathcal{S}\mathcal{E}}$). As in Eq. 7.3, states of the system leave imprints—become 'copied'—but now $\mathcal{S}$ is 'measured' by $\mathcal{E}$, its environment:

$$(\alpha|\uparrow\rangle + \beta|\downarrow\rangle)|\varepsilon_0\rangle \overset{\mathbf{H}_{\mathcal{S}\mathcal{E}}}{\Longrightarrow} \alpha|\uparrow\rangle|\varepsilon_\uparrow\rangle + \beta|\downarrow\rangle|\varepsilon_\downarrow\rangle = |\psi_{\mathcal{S}\mathcal{E}}\rangle. \tag{7.5}$$

Equation 7.4 implied that the untouched states are orthogonal, $\langle\uparrow|\downarrow\rangle = 0$. Their superposition, $\alpha|\uparrow\rangle + \beta|\downarrow\rangle$ turns into an entangled $|\psi_{\mathcal{S}\mathcal{E}}\rangle$. Thus, neither $\mathcal{S}$ nor $\mathcal{E}$ alone have a pure state. This loss of purity signifies decoherence. One can still assign a mixed state that represents surviving information about $\mathcal{S}$ to the system.

Phase changes can be detected: In a spin $\frac{1}{2}$–like system state $|\rightarrow\rangle = \frac{|\uparrow\rangle+|\downarrow\rangle}{\sqrt{2}}$ is orthogonal to $|\leftarrow\rangle = \frac{|\uparrow\rangle-|\downarrow\rangle}{\sqrt{2}}$. Phase shift operator $\mathbf{u}_{\mathcal{S}}^\varphi = |\uparrow\rangle\langle\uparrow| + e^{i\varphi}|\downarrow\rangle\langle\downarrow|$ alters phase that distinguishes them: for instance, when $\varphi = \pi$, it converts $|\rightarrow\rangle$ to $|\leftarrow\rangle$. In experiments $\mathbf{u}_{\mathcal{S}}^\varphi$ would shift the interference pattern.

We assume perfect decoherence, $\langle\varepsilon_\uparrow|\varepsilon_\downarrow\rangle = 0$: $\mathcal{E}$ has a perfect record of pointer states. What information survives decoherence, and what is lost?

Consider someone who knows the initial pre-decoherence state, $\alpha|\uparrow\rangle + \beta|\downarrow\rangle$, and would like to make predictions about the decohered $\mathcal{S}$. We now show that when $\langle\varepsilon_\uparrow|\varepsilon_\downarrow\rangle = 0$ phases of α and β no longer matter for $\mathcal{S}$—phase φ has no effect on *local* state of $\mathcal{S}$, so measurements on $\mathcal{S}$ cannot detect phase shift, as there is no interference pattern to shift.

Phase shift $\mathbf{u}_{\mathcal{S}}^\varphi \otimes \mathbf{1}_{\mathcal{E}}$ (acting on an entangled $|\psi_{\mathcal{S}\mathcal{E}}\rangle$) cannot have any effect on its local state because it can be undone by $\mathbf{u}_{\mathcal{E}}^{-\varphi} = |\varepsilon_\uparrow\rangle\langle\varepsilon_\uparrow| + e^{-i\varphi}|\varepsilon_\downarrow\rangle\langle\varepsilon_\downarrow|$, a 'countershift' acting on a distant $\mathcal{E}$ decoupled from the system:

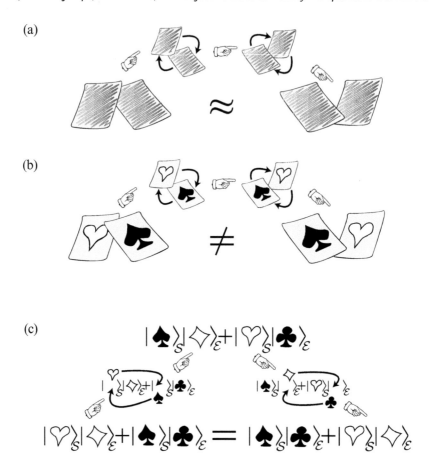

Fig. 7.2 Envariance is a symmetry of entangled states. It allows one to demonstrate Born's rule (Zurek, 2003a; 2003b; 2005) using a combination of (i) an old intuition of Laplace (Laplace, 1820) about invariance and the origins of probability and (ii) quantum symmetries of entanglement. **(a)** Laplace's *principle of indifference* (illustrated with playing cards) aims to establish symmetry using invariance under swaps. A player who doesn't know face values of cards is indifferent—does not care—if they are swapped before he gets the one on the left. For Laplace, this indifference was the evidence of a (subjective) symmetry: It implied *equal likelihood*—equal probabilities of the invariantly swappable alternatives. For the two cards above, subjective probability $p_\spadesuit = \frac{1}{2}$ would be inferred by someone who does not know their face value, but knows that one and only one of the two cards is a spade. When probabilities of a set of elementary events are provably equal, one can compute probabilities of composite events and thus develop a theory of probability. Even the additivity of probabilities can be *established,* see, e.g., (Gnedenko, 1968). This is in contrast to Kolmogorov's measure-theoretic axioms (which *include* additivity of probabilities). Above all, Kolmogorov's theory does not assign probabilities to elementary events (physical or otherwise), while our approach yields probabilities when symmetries of elementary events under swaps are known. **(b)** The problem with Laplace's principle of indifference is its subjectivity. The actual physical state of the system (the two cards) is altered by the swap. A related problem is that the assessment of indifference is based on ignorance: it as was argued, e.g., by supporters of the relative frequency approach (regarded by many as more "objective" foundation of probability) that it is impossible to deduce anything (including probabilities) from ignorance. This (along with subjectivity) was a reason why equal likelihood was regarded with suspicion as a basis of probability in classical physics.

Fig. 7.2 Continued (c) In quantum physics symmetries of entanglement can be used to deduce objective probabilities starting with a known state. The relevant symmetry is the *entanglement*—assisted in*variance* or *envariance*. When a pure entangled state of a system S and another system we call "an environment $\mathcal{E}$" (anticipating connections with decoherence) $|\psi_{S\mathcal{E}}\rangle = \sum_{k=1}^{N} a_k |s_k\rangle |\varepsilon_k\rangle$ can be transformed by $U_S = u_S \otimes 1_{\mathcal{E}}$ acting solely on S, but the effect of U_S can be undone by acting solely on $\mathcal{E}$ with an appropriately chosen $U_{\mathcal{E}} = 1_S \otimes u_{\mathcal{E}}$, it is envariant under u_S. For such composite states one can rigorously establish that the local state of S remains unaffected by u_S. Thus, for example, the phases of the coefficients in the Schmidt expansion $|\psi_{S\mathcal{E}}\rangle = \sum_{k=1}^{N} a_k |s_k\rangle |\varepsilon_k\rangle$ are envariant, as the effect of $u_S = \sum_{k=1}^{N} \exp(i\phi_k) |s_k\rangle\langle s_k|$ can be undone by a countertransformation $u_{\mathcal{E}} = \sum_{k=1}^{N} \exp(-i\phi_k) |\varepsilon_k\rangle\langle \varepsilon_k|$ acting solely on the environment. This envariance of phases implies their irrelevance for the local states—in effect, it implies decoherence. Moreover, when the absolute values of the Schmidt coefficients are equal (as in (c) above), a swap $|\spadesuit\rangle\langle\heartsuit| + |\heartsuit\rangle\langle\spadesuit|$ in S can be undone by a 'counterswap' $|\clubsuit\rangle\langle\diamondsuit| + |\diamondsuit\rangle\langle\clubsuit|$ in $\mathcal{E}$. So, as can be established more carefully (Zurek, 2005), $p_\spadesuit = p_\heartsuit = \frac{1}{2}$ follows from the objective symmetry of such an entangled state. This proof of equal probabilities is based not on ignorance (as in Laplace's indifference) but on a perfect knowledge of the "wrong thing"—of the global observable that rules out (via quantum indeterminacy) any information about complementary local observables. When supplemented by simple counting, envariance leads to Born's rule (Zurek, 2003a; 2003b; 2005; 2011).

$$\mathbf{u}_{\mathcal{E}}^{-\varphi}(\mathbf{u}_{S}^{\varphi}|\psi_{S\mathcal{E}}\rangle) = \mathbf{u}_{\mathcal{E}}^{-\varphi}(\alpha|\uparrow\rangle|\varepsilon_\uparrow\rangle + e^{i\varphi}\beta|\downarrow\rangle|\varepsilon_\downarrow\rangle) = |\psi_{S\mathcal{E}}\rangle. \tag{7.6}$$

Phases in $|\psi_{S\mathcal{E}}\rangle$ can be changed in a faraway $\mathcal{E}$ decoupled from but entangled with S. Therefore, they can no longer influence local state of S. (This follows from quantum theory alone, but is essential for causality—if they could, measuring S would reveal this, enabling superluminal communication!)

Decoherence is caused by the loss of phase coherence. Superpositions decohere as $|\uparrow\rangle, |\downarrow\rangle$ are recorded by $\mathcal{E}$. This is not because phases become "randomized" by interactions with $\mathcal{E}$, as is sometimes said (Dirac, 1958). Rather, they become delocalized: they lose significance for S alone. They are a global property of the composite state—they no longer belong to S, so measurements on S cannot distinguish states that started as superpositions with different phases for α, β. Consequently, information about S is lost—it is displaced into correlations between S and $\mathcal{E}$, and local phases of S become a global property—global phases of the composite entangled state of $S\mathcal{E}$.

We have considered this information loss here without reduced density matrices, the usual decoherence tool. Our view of decoherence appeals to symmetry, invariance of S—entanglement-assisted in*variance* or *envariance* under phase shifts of pointer state coefficients, Eq. 7.6. As S entangles with $\mathcal{E}$, its local state becomes invariant under transformations that could have affected it before.

Rigorous proof of coherence loss uses quantum core postulates (o)-(iii) and relies on quantum *facts 1—3*:

1. *Locality: A unitary must act on a system to change its state.* State of S that is not acted upon doesn't change even as other systems evolve (so $1_S \otimes (|\varepsilon_\uparrow\rangle\langle\varepsilon_\uparrow| + e^{-i\varphi}|\varepsilon_\downarrow\rangle\langle\varepsilon_\downarrow|)$ does not affect S even when $S\mathcal{E}$ are entangled, in $|\psi_{S\mathcal{E}}\rangle$);
2. *State of a system is all there is to predict measurement outcomes;*

3. *A composite state determines states of subsystems* (so local state of $\mathcal{S}$ is restored when the state of the whole $\mathcal{SE}$ is restored).

Facts help characterize local states of entangled systems without using reduced density matrices. They follow from quantum theory: locality is a property of interactions. The other two facts define the role and the relation of the quantum states of individual and composite systems in a way that does not invoke density matrices (to which we are not entitled in absence of Born's rule). Thus, phase shift $\mathbf{u}_{\mathcal{S}}^{\varphi} \otimes \mathbf{1}_{\mathcal{E}} = (|\uparrow\rangle\langle\uparrow| + e^{i\varphi}|\downarrow\rangle\langle\downarrow|) \otimes \mathbf{1}_{\mathcal{E}}$ acting on pure pre-decoherence state matters: measurement can reveal φ. In accord with facts 1 and 2, $\mathbf{u}_{\mathcal{S}}^{\varphi}$ changes $\alpha|\uparrow\rangle + \beta|\downarrow\rangle$ into $\alpha|\uparrow\rangle + e^{i\varphi}\beta|\downarrow\rangle$. However, the same $\mathbf{u}_{\mathcal{S}}^{\varphi}$ acting on $\mathcal{S}$ in an entangled state $|\psi_{\mathcal{SE}}\rangle$ does not matter for $\mathcal{S}$ alone, as it can be undone by $\mathbf{1}_{\mathcal{S}} \otimes (|\varepsilon_{\uparrow}\rangle\langle\varepsilon_{\uparrow}| + e^{-i\varphi}|\varepsilon_{\downarrow}\rangle\langle\varepsilon_{\downarrow}|)$, a countershift acting on a faraway, decoupled $\mathcal{E}$. As the global $|\psi_{\mathcal{SE}}\rangle$ is restored, by fact 3 the local state of $\mathcal{S}$ is also restored even if $\mathcal{S}$ is not acted upon (so that, by fact 1, it remains unchanged). Hence, local state of decohered $\mathcal{S}$ that obtains from $|\psi_{\mathcal{SE}}\rangle$ could not have changed to begin with, and so it cannot depend on phases of α, β.

The only pure states invariant under such phase shifts (unaffected by decoherence) are pointer states. Resilience we saw, Eqs. 7.1, 7.3, lets them preserve correlations. For instance, entangled state of the measured system $\mathcal{S}$ and the apparatus, $|\psi_{\mathcal{SA}}\rangle$, decoheres as $\mathcal{A}$ interacts with $\mathcal{E}$:

$$(\alpha|\uparrow\rangle|A_{\uparrow}\rangle + \beta|\downarrow\rangle|A_{\downarrow}\rangle)|\varepsilon_0\rangle \overset{\mathbf{H}_{\mathcal{AE}}}{\Longrightarrow} \alpha|\uparrow\rangle|A_{\uparrow}\rangle|\varepsilon_{\uparrow}\rangle + \beta|\downarrow\rangle|A_{\downarrow}\rangle|\varepsilon_{\downarrow}\rangle = |\Psi_{\mathcal{SAE}}\rangle \qquad (7.7)$$

Pointer states $|A_{\uparrow}\rangle$, $|A_{\downarrow}\rangle$ of $\mathcal{A}$ survive decoherence by $\mathcal{E}$. They retain perfect correlation with $\mathcal{S}$ (or an observer, or other systems) in spite of $\mathcal{E}$, independently of the value of $\langle\varepsilon_{\uparrow}|\varepsilon_{\downarrow}\rangle$. Stability under decoherence is—in our quantum Universe—a prerequisite for effective classicality: familiar states of macroscopic objects also have to survive monitoring by $\mathcal{E}$ and, hence, retain correlations.

Decohered $\mathcal{SA}$ is described by a *reduced density matrix*,

$$\rho_{\mathcal{SA}} = \mathrm{Tr}_{\mathcal{E}}|\Psi_{\mathcal{SAE}}\rangle\langle\Psi_{\mathcal{SAE}}| . \qquad (7.8a)$$

When $\langle\varepsilon_{\uparrow}|\varepsilon_{\downarrow}\rangle = 0$, pointer states of $\mathcal{A}$ retain correlations with the outcomes:

$$\rho_{\mathcal{SA}} = |\alpha|^2|\uparrow\rangle\langle\uparrow||A_{\uparrow}\rangle\langle A_{\uparrow}| + |\beta|^2|\downarrow\rangle\langle\downarrow||A_{\downarrow}\rangle\langle A_{\downarrow}| \qquad (7.8b)$$

Both $\uparrow$ and $\downarrow$ are present: There is no 'literal collapse'. We will use $\rho_{\mathcal{SA}}$ to examine information flows. Thus, we will need probabilities of the outcomes.

Trace is a mathematical operation. However, regarding the reduced density matrix $\rho_{\mathcal{SA}}$ as statistical mixture of its eigenstates—states $\uparrow$ and $\downarrow$ and $A_{\uparrow}$, $A_{\downarrow}$ (pointer state) records—relies on Born's rule, that allows one to view tracing as averaging. This is no longer just mathematics—this is where a mathematical operation leads to reduced density matrix and to physical consequences. We didn't use it till Eq. 7.8 to avoid circularity. Now we derive $p_k = |\psi_k|^2$, Born's rule as we shall need it: we need to prove that the probabilities are indeed given by the eigenvalues $|\alpha|^2$, $|\beta|^2$ of $\rho_{\mathcal{SA}}$. This is the postulate (v), obviously crucial for relating quantum formalism to experiments. We want to deduce Born's rule from the quantum core postulates (o)−(iii).

We note that this brief and somewhat biased discussion of the envariant origin of decoherence is not a substitute for more complete presentations that do not (as we did, for good reasons in the present context) shy away from employing usual tools of decoherence theory, including in particular reduced density matrices (Zurek, 1991; 2003b; Joos et al., 2003).

7.4.2 Probabilities from symmetries of entanglement

In quantum physics one seeks probability of measurement outcome starting from a known state of $\mathcal{S}$ and ready-to-measure state of the apparatus pointer $\mathcal{A}$. Entangled state of the whole is pure, so (at least prior to the decoherence by the environment) there is no ignorance in the usual sense.

However, *envariance* in a guise slightly different than before (when it accounted for decoherence) implies when mutually exclusive outcomes have certifiably equal probabilities: Suppose $\mathcal{S}$ starts as $|\rightarrow\rangle = |\uparrow\rangle + |\downarrow\rangle$, so interaction with $\mathcal{A}$ yields $|\uparrow\rangle|A_\uparrow\rangle + |\downarrow\rangle|A_\downarrow\rangle$, an *even* (equal coefficient) state. (Here and below we skip normalization to save on notation).

Unitary *swap* $|\uparrow\rangle\langle\downarrow| + |\downarrow\rangle\langle\uparrow|$ permutes states in $\mathcal{S}$:

$$|\uparrow\rangle\,|A_\uparrow\rangle + |\downarrow\rangle\,|A_\downarrow\rangle \quad \longrightarrow \quad |\downarrow\rangle|A_\uparrow\rangle + |\uparrow\rangle|A_\downarrow\rangle. \tag{7.9a}$$

After the swap $|\downarrow\rangle$ is as probable as $|A_\uparrow\rangle$ was (and still is), and $|\uparrow\rangle$ as $|A_\downarrow\rangle$. Probabilities in $\mathcal{A}$ are unchanged (as $\mathcal{A}$ is untouched) so $p_\uparrow$ and $p_\downarrow$ must have been swapped. To prove equiprobability we now swap records in $\mathcal{A}$:

$$|\downarrow\rangle\,|A_\uparrow\rangle + |\uparrow\rangle\,|A_\downarrow\rangle \quad \longrightarrow \quad |\downarrow\rangle|A_\downarrow\rangle| + |\uparrow\rangle|A_\uparrow\rangle. \tag{7.9b}$$

Swap in $\mathcal{A}$ restores pre-swap $|\uparrow\rangle|A_\uparrow\rangle + |\downarrow\rangle|A_\downarrow\rangle$ without touching $\mathcal{S}$, so (by fact 3) the local state of $\mathcal{S}$ is also restored (even though, by fact 1, it could not have been affected by the swap of Eq. 7.9b). Hence (by fact 2), all predictions about $\mathcal{S}$, *including probabilities*, must be the same! Probability of $|\uparrow\rangle$ and $|\downarrow\rangle$, (as well as of $|A_\uparrow\rangle$ and $|A_\downarrow\rangle$) are exchanged yet unchanged. Therefore, they must be equal. Thus, in our two state case $p_\uparrow = p_\downarrow = \frac{1}{2}$. For N envariantly equivalent alternatives, $p_k = \frac{1}{N}\ \forall k$.

Getting rid of phases beforehand was crucial: swaps in an isolated pure state will, in general, change the phases, and, hence, change the state. For instance, in the notation of Fig. 7.2, $|\spadesuit\rangle + i|\heartsuit\rangle$, after a swap $|\spadesuit\rangle\langle\heartsuit| + |\heartsuit\rangle\langle\spadesuit|$, becomes $i|\spadesuit\rangle + |\heartsuit\rangle$, i.e., is orthogonal to the pre-swap state.

The crux of the proof of equal probabilities was that the swap does not change anything *locally*. This can be established for entangled states with equal coefficients but—as we have just seen—is simply not true for a pure unentangled state of just one system.

In the real world the environment will become entangled (in course of decoherence) with the preferred states of the system of interest (or with the preferred states of the apparatus pointer). We have already seen how postulates (i)–(iii) lead to preferred sets of states. We have also pointed out that—at least in idealized situations—these states coincide with the familiar pointer states that remain stable in spite of decoherence. So, in effect, we are using the familiar framework of decoherence to derive Born's rule. Fortunately our conclusions about decoherence can be reached without employing the usual (Born's rule-dependent) tools of decoherence (reduced density matrix and trace).

So far we have only explained how one can establish equality of probabilities for the outcomes that correspond to Schmidt states associated with coefficients that differ at most by a phase. This is not yet Born's rule. However, it turns out that this is the hard part of the proof: once such equality is established, a simple counting argument (a version of that employed in (Zurek, 1998b; Deutsch, 1999; Wallace, 2003; Saunders, 2004) leads to the relation between probabilities and unequal coefficients (Zurek, 2003a; 2003b; 2005).

Thus, for an uneven state $|\phi_{SA}\rangle = \alpha|\uparrow\rangle|A_\uparrow\rangle + \beta|\downarrow\rangle|A_\downarrow\rangle$ swaps on S and A yield $\beta|\uparrow\rangle|A_\uparrow\rangle + \alpha|\downarrow\rangle|A_\downarrow\rangle$, and not the pre-swap state, so $p_\uparrow$ and $p_\downarrow$ are not equal. However, uneven case reduces to equiprobability via *finegraining*, so envariance, Eq. 7.9, yields Born's rule, $p_{s|\psi} = |\langle s|\psi\rangle|^2$, in general.

To see how, we take $\alpha \propto \sqrt{\mu}$, $\beta \propto \sqrt{\nu}$, where μ, ν are natural numbers (so the squares of α and β are commensurate). To finegrain, we change the basis; $|A_\uparrow\rangle = \sum_{k=1}^{\mu} |a_k\rangle/\sqrt{\mu}$, and $|A_\downarrow\rangle = \sum_{k=\mu+1}^{\mu+\nu} |a_k\rangle/\sqrt{\nu}$, in the Hilbert space of A:

$$|\phi_{SA}\rangle \propto \sqrt{\mu}\,|\uparrow\rangle|A_\uparrow\rangle + \sqrt{\nu}\,|\downarrow\rangle|A_\downarrow\rangle =$$

$$= \sqrt{\mu}\,|\uparrow\rangle \sum_{k=1}^{\mu} |a_k\rangle/\sqrt{\mu} + \sqrt{\nu}\,|\downarrow\rangle \sum_{k=\mu+1}^{\mu+\nu} |a_k\rangle/\sqrt{\nu}. \qquad (7.10a)$$

We simplify, and imagine an environment decohering A in a new orthonormal basis. That is, $|a_k\rangle$ correlate with $|e_k\rangle$ so that;

$$|\Phi_{SAE}\rangle \propto \sum_{k=1}^{\mu} |\uparrow a_k\rangle|e_k\rangle + \sum_{k=\mu+1}^{\mu+\nu} |\downarrow a_k\rangle|e_k\rangle \qquad (7.10b)$$

as if $|a_k\rangle$ were the preferred pointer states decohered by the environment so that $\langle e_k|e_l\rangle = \delta_{kl}$.

Now swaps of $|\uparrow a_k\rangle$ with $|\downarrow a_k\rangle$ can be undone by counterswaps of the corresponding $|e_k\rangle$'s. Counts of the finegrained equiprobable ($p_k = \frac{1}{\mu+\nu}$) alternatives labelled with $\uparrow$ or $\downarrow$ lead to Born's rule:

$$p_\uparrow = \frac{\mu}{\mu+\nu} = |\alpha|^2, \quad p_\downarrow = \frac{\nu}{\mu+\nu} = |\beta|^2. \qquad (7.11)$$

Amplitudes 'got squared' as a result of Pythagoras' theorem (Euclidean nature of Hilbert spaces). The case of incommensurate $|\alpha|^2$ and $|\beta|^2$ can be settled by an appeal to continuity of probabilities as functions of state vectors.

7.4.3 Discussion

In physics textbooks Born's rule is a postulate. Using entanglement we derived it here from the quantum core axioms. Our reasoning was purely quantum: knowing a state of the composite classical system means knowing state of each part. There are no entangled classical states, and no objective symmetry to deduce classical equiprobability, the crux of our derivation. Entanglement—made possible by the tensor structure of composite Hilbert spaces, introduced by the composition postulate (o)—was key. Appeal to symmetry—subjective and suspect in the classical case—becomes rigorous thanks to objective envariance in the quantum case. Born's rule, introduced by textbooks as postulate (v), follows.

Relative frequency approach (found in many probability texts) starts with "events". It has not led to successful derivation of Born's rule. We used entanglement symmetries to identify equiprobable alternatives. However, by employing envariance one can also deduce frequencies of events by considering M repetitions, i.e.,

$$(\alpha|\uparrow\rangle|A_\uparrow\rangle + \beta|\downarrow\rangle|A_\downarrow\rangle)^{\otimes M},$$

of an experiment, and deduce departures from the average frequencies that are also expected when M is finite. Moreover, one can even show the inverse of Born's rule. That is, one can demonstrate that the amplitude should be proportional to the square root of frequency (Zurek, 2011).

As the probabilities are now in place, one can think of quantum statistical physics. One could establish its foundations using probabilities we have just deduced. But there is an even simpler and more radical approach (Deffner and Zurek, 2016; Zurek, 2018) that arrives at the microcanonical state without the need to invoke ensembles and probabilities. Its detailed explanation is beyond the scope of this section, but the basic idea is to regard an even state of the system entangled with its environment as the microcanonical state. This is a major conceptual simplification of the foundations of statistical physics: one can get rid of the artifice of invoking infinite collections of similar systems to represent a state of a single system in a manner that allows one to deduce relevant thermodynamic properties.[2]

7.5 Quantum Darwinism, classical reality, and objective existence

Quantum Darwinism recognizes that observers use the environment as a communication channel to acquire information about pointer states indirectly, leaving the system of interest untouched and its state unperturbed. Observers can find out the state of the system without endangering its existence (which would be inevitable in direct measurements). Indeed, the reader of this text is—at this very moment—intercepting a tiny fraction of the photon environment scattered by this page with his eyes to gather all of the information he needs.

---

[2] We note that envariance has been successfully tested in several recent experiments (Vermeyden et al., 2015; Harris et al., 2016; Deffner, 2017; Ferrari and Amoretti, 2018).

This is how virtually all of our information is acquired. A direct measurement is not what we do. Rather, we count on redundancy, and settle for information that exists in many copies. This is how objective existence—cornerstone of classical reality—arises in the quantum world.

7.5.1   Mutual information in quantum correlations

To develop theory of quantum Darwinism we need to quantify information between fragments of the environment and the system. Mutual information is a convenient tool that we shall use for this purpose.

The mutual information between the system $\mathcal{S}$ and a fragment $\mathcal{F}$ (that will play the role of the apparatus $\mathcal{A}$ of Eq. 7.8 in the discussion above) can be computed using the density matrices of the systems of interest to obtain their von Neumann entropies $H_X = -\text{Tr}\rho_X \lg \rho_X$:

$$I(\mathcal{S}:\mathcal{F}) = H_{\mathcal{S}} + H_{\mathcal{F}} - H_{\mathcal{S},\mathcal{F}} = -(|\alpha|^2 \lg |\alpha|^2 + |\beta|^2 \lg |\beta|^2) \tag{7.12}$$

We have used the density matrices of the $\mathcal{S}$ and $\mathcal{A}$ (as a "stand-in" for $\mathcal{F}$) from Eq. 7.8 to obtain the specific value of mutual information above.

We already noted the special role of the pointer observable. It is stable and, hence, it leaves behind information-theoretic progeny—multiple imprints, copies of the pointer states—in the environment. By contrast, complementary observables are destroyed by the interaction with a single subsystem of $\mathcal{E}$. They can in principle still be accessed, but only when *all* of the environment is measured. Indeed, because we are dealing with a quantum system, things are much worse than that: the environment must be measured in precisely the right (typically global) basis to allow for such a reconstruction. Otherwise, the accumulation of errors over multiple measurements will lead to an incorrect conclusion and re-prepare the state and environment, so that it is no longer a record of the state of $\mathcal{S}$, and phase information is irretrievably lost.

7.5.2   Objective reality form redundant information

Quantum Darwinism was introduced relatively recently. Previous studies of the records "kept" by the environment were focused on its effect on the state of the system, and not on their utility. Decoherence is a case in point, as are some of the studies of the decoherent histories approach (Gell-Mann and Hartle, 1998; Halliwell, 1999). The exploration of quantum Darwinism in specific models has started at the beginning of this millenium (Blume-Kohout and Zurek, 2005; 2006; 2008; Ollivier et al., 2004; 2005). We do not intend to review all of the results obtained to date in detail. The basic conclusion of these studies is, however, that the dynamics responsible for decoherence is also capable of imprinting multiple copies of the pointer basis on the environment. Moreover, while decoherence is always implied by quantum Darwinism, the reverse need not be true. One can easily imagine situations where the environment is completely mixed, and, thus, cannot be used as a communication channel, but would still suppress quantum coherence in the system.

For many subsystems, $\mathcal{E} = \bigotimes_k \mathcal{E}^{(k)}$, the initial state $(\alpha|\uparrow\rangle + \beta|\downarrow\rangle)|\varepsilon_0^{(1)}\varepsilon_0^{(2)}\varepsilon_0^{(3)}...\rangle$ evolves into a "branching state":

$$|\Upsilon_{\mathcal{SE}}\rangle = \alpha|\uparrow\rangle|\varepsilon_\uparrow^{(1)}\varepsilon_\uparrow^{(2)}\varepsilon_\uparrow^{(3)}...\rangle + \beta|\downarrow\rangle|\varepsilon_\downarrow^{(1)}\varepsilon_\downarrow^{(2)}\varepsilon_\downarrow^{(3)}...\rangle \qquad (7.13)$$

Linearity assures all branches persist: collapse to one outcome is not in the cards. However, large $\mathcal{E}$ can disseminate information about the system. The state $|\Upsilon_{\mathcal{SE}}\rangle$ represents many records inscribed in its fragments, collections of subsystems of $\mathcal{E}$ (Fig. 7.3). This means that the state of $\mathcal{S}$ can be found out by many, independently, and indirectly—hence, without disturbing $\mathcal{S}$. This is how symptoms of objective existence arise in our quantum world.

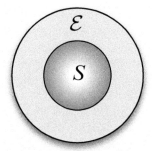

(a) <u>Decoherence Paradigm:</u>
Universe is divided into
System & Environment

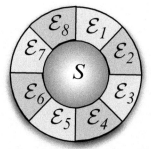

(b) <u>Redundancy Paradigm:</u>
Environment is divided
into *Subenvironments*

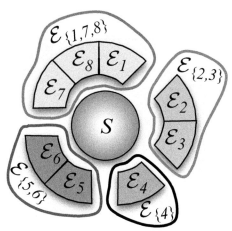

(c) *Subenvironments* are combined
into *Fragments* that each have
nearly complete information.

Fig. 7.3 Quantum Darwinism recognizes that environments consist of many subsystems, and that observers acquire information about system of interest $\mathcal{S}$ by intercepting copies of its pointer states deposited in $\mathcal{E}$ as a result of decoherence.

An environment fragment $\mathcal{F}$ can act as apparatus with a (possibly incomplete) record of $\mathcal{S}$. When $\mathcal{E}\backslash\mathcal{F}$ ('the rest of the $\mathcal{E}$') is traced out, $\mathcal{SF}$ decoheres, and the reduced density matrix describing joint state of $\mathcal{S}$ and $\mathcal{F}$ is:

$$\rho_{\mathcal{SF}} = \mathrm{Tr}_{\mathcal{E}\backslash\mathcal{F}}|\Psi_{\mathcal{SE}}\rangle\langle\Psi_{\mathcal{SE}}| = |\alpha|^2|\uparrow\rangle\langle\uparrow||F_\uparrow\rangle\langle F_\uparrow| + |\beta|^2|\downarrow\rangle\langle\downarrow||F_\downarrow\rangle\langle F_\downarrow|. \tag{7.14}$$

When $\langle F_\uparrow|F_\downarrow\rangle = 0$, $\mathcal{F}$ contains perfect record of the preferred states of the system. In principle, each subsystem of $\mathcal{E}$ may be enough to reveal its state, but this is unlikely. Typically, one must collect many subsystems of $\mathcal{E}$ into $\mathcal{F}$ to find out about $\mathcal{S}$.

The redundancy of the data about pointer states in $\mathcal{E}$ determines how many times the same information can be independently extracted—it is a measure of objectivity. The key question of quantum Darwinism is then: *How many subsystems of $\mathcal{E}$— what fraction of $\mathcal{E}$—does one need to find out about $\mathcal{S}$?*. The answer is provided by the mutual information $I(\mathcal{S}:\mathcal{F}_f) = H_\mathcal{S} + H_{\mathcal{F}_f} - H_{\mathcal{SF}_f}$, information about $\mathcal{S}$ available from $\mathcal{F}_f$, fraction $f = \frac{\#\mathcal{F}}{\#\mathcal{E}}$ of $\mathcal{E}$ (where $\#\mathcal{F}$ and $\#\mathcal{E}$ are the numbers of subsystems).

In case of perfect correlation a single subsystem of $\mathcal{E}$ would suffice, as $I(\mathcal{S}:\mathcal{F}_f)$ jumps to $H_\mathcal{S}$ at $f = \frac{1}{\#\mathcal{E}}$. The data in additional subsystems of $\mathcal{E}$ are then redundant. Usually, however, larger fragments of $\mathcal{E}$ are needed to find out enough about $\mathcal{S}$. Black plot in Fig. 7.4 illustrates this: $I(\mathcal{S}:\mathcal{F}_f)$ still approaches $H_\mathcal{S}$, but only gradually. The length of this plateau can be measured in units of f_δ, the initial rising portion of $I(\mathcal{S}:\mathcal{F}_f)$. It is defined with the help of the *information deficit* δ observers tolerate:

$$I(\mathcal{S}:\mathcal{F}_{f_\delta}) \geq (1-\delta)H_\mathcal{S} \tag{7.15}$$

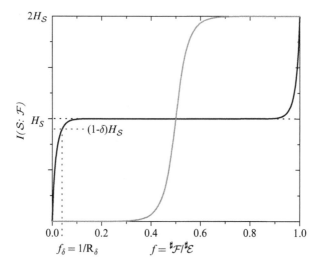

Fig. 7.4 Information about the system contained in a fraction f of the environment. Black plot shows a typical $I(\mathcal{S}:\mathcal{F}_f)$ established by decoherence. Rapid rise means that nearly all classically accessible information is revealed by a small fraction of $\mathcal{E}$. It is followed by a plateau: additional fragments only confirm what is already known. Redundancy $R_\delta = 1/f_\delta$ is the number of such independent fractions. Grey plot shows $I(\mathcal{S}:\mathcal{F}_f)$ for a random state in the composite system $\mathcal{SE}$.

Redundancy is the number of such records of $\mathcal{S}$ in $\mathcal{E}$:

$$\mathcal{R}_\delta = 1/f_\delta \qquad (7.16)$$

$\mathcal{R}_\delta$ sets the upper limit on how many observers can find out the state of $\mathcal{S}$ from $\mathcal{E}$ independently and indirectly. In models (Ollivier et al., 2004; 2005; Blume-Kohout and Zurek, 2005; 2006; 2008; Riedel and Zurek, 2010; 2012; Zwolak et al., 2014), including especially photon scattering analyzed extending the decoherence model of Joos and Zeh (1985), $\mathcal{R}_\delta$ is huge (Riedel and Zurek, 2010; 2012; Zwolak et al., 2014) and depends on δ only weakly (logarithmically).

This is 'quantum spam': $\mathcal{R}_\delta$ imprints of pointer states are broadcast through the environment. Many observers can access them independently and indirectly, assuring objectivity of pointer states of $\mathcal{S}$. Repeatability is key: states must survive copying to produce many imprints.

7.5.3   Discussion

Our discussion of quantum jumps shows when, in spite of the no-cloning theorem (Wootters and Zurek, 1982; Dieks, 1982), repeatable copying is possible. Discrete preferred states set the stage for quantum jumps. Copying yields branches of records inscribed in subsystems of $\mathcal{E}$. Initial superposition yields superposition of branches, Eq. 7.13, so there is no literal collapse. However, fragments of $\mathcal{E}$ can reveal only one branch (and not their superposition). Such evidence will suggest 'quantum jump' from superposition to a single outcome, in accord with (iv).

Not all environments are good in this role of a witness. Photons excel: they do not interact with the air or with each other, faithfully passing on information. Small fraction of photon environment usually reveals all we need to know. Scattering of sunlight quickly builds up redundancy: a 1μ dielectric sphere in a superposition of 1μ size increases $\mathcal{R}_{\delta=0.1}$ by $\sim 10^8$ every microsecond (Riedel and Zurek, 2010; 2012). Mutual information plot illustrating this case is shown in Fig. 7.5.

Air is also good in decohering, but its molecules interact, scrambling acquired data. Objects of interest scatter both air and photons, so both acquire information about position, and favor similar localized pointer states.

Quantum Darwinism shows why it is so hard to undo decoherence (Zwolak and Zurek, 2013). Plots of mutual information $I(\mathcal{S}:\mathcal{F}_f)$ for initially pure $\mathcal{S}$ and $\mathcal{E}$ are antisymmetric (see Fig. 7.4) around $f = \frac{1}{2}$ and $H_\mathcal{S}$ (Blume-Kohout and Zurek, 2005) when typical fragments of the environment are assumed. Hence, a counterpoint of the initial quick rise at $f \leq f_\delta$ is a quick rise at $f \geq 1 - f_\delta$, as last few subsystems of $\mathcal{E}$ are included in the fragment $\mathcal{F}$ that by now contains nearly all $\mathcal{E}$. This is because an initially pure $\mathcal{SE}$ remains pure under unitary evolution, so $H_{\mathcal{SE}} = 0$, and $I(\mathcal{S}:\mathcal{F}_f)|_{f=1}$ must reach $2H_\mathcal{S}$. Thus, a measurement on *all* of $\mathcal{SE}$ could confirm its purity in spite of decoherence caused by $\mathcal{E}\backslash\mathcal{F}$ for all $f \leq 1 - f_\delta$. However, to verify this one has to intercept and measure all of $\mathcal{SE}$ in a way that reveals pure state $|\Upsilon_{\mathcal{SE}}\rangle$, Eq. 7.13. Other measurements destroy phase information. So, undoing decoherence is in principle possible, but the required resources and foresight preclude it.

In quantum Darwinism decohering environment acts as an amplifier, inducing branch structure of $|\Upsilon_{\mathcal{SE}}\rangle$ distinct from typical states in the Hilbert space of $\mathcal{SE}$:

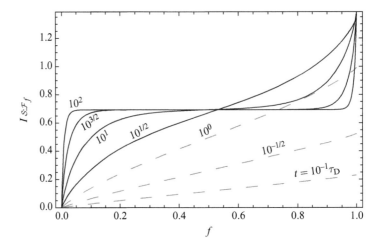

Fig. 7.5 The quantum mutual information $I(\mathcal{S}:\mathcal{F}_f)$ vs. fragment size f at different elapsed times for an object illuminated by a point-source black-body radiation. Individual curves are labeled by the time t in units of the decoherence time τ_D. For $t \leq \tau_D$ (dashed lines), the information about the system available in the environment is low. The linearity in f means each piece of the environment contains new, independent information. For $t > \tau_D$ (solid lines), the shape of the partial information plot indicates redundancy; the first few pieces of the environment increase the information, but additional pieces only confirm what is already known.

$I(\mathcal{S}:\mathcal{F}_f)$ of a random state is given by the gray plot in Fig. 7.4, with no plateau or redundancy. Antisymmetry means that $I(\mathcal{S}:\mathcal{F}_f)$ 'jumps' at $f = \frac{1}{2}$ to $2H_{\mathcal{S}}$.

Environments that decohere $\mathcal{S}$, but scramble information because of interactions between subsystems (e.g. air) eventually approach such random states. Quantum Darwinism is possible only when information about $\mathcal{S}$ is preserved in fragments of $\mathcal{E}$, so that it can be recovered by observers. There is no need for perfection. Partially mixed environments or imperfect measurements correspond to noisy communication channels: their capacity is depleted, but we can still get the message (Zwolak et al., 2009; 2010).

Quantum Darwinism settles the issue of the origin of classical reality by accounting for all of the operational symptoms of objective existence in a quantum Universe: a single quantum state cannot be found out through a direct measurement. However, pointer states usually leave multiple records in the environment. Observers can use these records to find out the (pointer) state of the system of interest. Observers can afford to destroy photons while reading the evidence—the existence of multiple copies implies that other observers can access the information about the system indirectly and independently, and that they will all agree about the outcome. This is how objective existence arises in our quantum world.[3]

---

[3] There has been significant progress in the study of the acquisition and dissemination of the information by the environments (Paz and Roncaglia, 2009; Brandao et al., 2015; Riedel et al., 2016; Riedel, 2017; Pleasance and Garraway, 2017; Knott et al., 2018). More detailed discussion of the results obtained in these papers is, unfortunately, beyond the scope of our brief review.

7.6   Discussion: frequently asked questions

The subject of this paper has a long history. As a result, there are different ways of talking, thinking, and writing about it. It is almost as if different points of view have developed different languages. As a result, one can find it difficult to understand the ideas, as one often has to learn "the other language" used to discuss the same problem. This is further complicated by the fact that all of these languages use essentially the same words, but charged with a very different meanings. Concepts like "existence", "reality", or "state" are good examples.

The aim of this section is to acknowledge this problem and to deal with it to the extent possible within the framework of a brief guide. We shall do that in a way inspired by modern approach to languages (and to travel guides): rather than study vocabulary and grammar, we shall use "conversations" based on a few "frequently asked questions". The hope is that this exercise will provide the reader with some useful hints of what is meant by certain phrases. This is very much in the spirit of the "travel guide", where a collection of frequently used expressions is often included.

FAQ #1: *What is the difference between "decoherence" and "einselection"?*

Decoherence is the process of the *loss of phase coherence* caused by the interaction between the system and the environment. Einselection is an abbreviation of "environment-induced superselection", which designates *selection of preferred set of pointer states* that are immune to decoherence. Decoherence will often (but not always) result in einselection. For instance, interaction that commutes with a certain observable of a system will preserve eigenstates of that *pointer observable, pointer states* that are einselected, and do *not* decohere. By contrast, superpositions of such pointer states will decohere. This picture can be (and generally will be) complicated by the evolution induced by the Hamiltonian of the system, so that perfect pointer states will not exist, but approximate pointer states will be still favored—will be much more stable then their superpositions. There are also cases when there is decoherence, but it treats all the states equally badly, so that there is no einselection, and there are no pointer state. Perfect depolarizing channel (Nielsen and Chuang, 2000) is an example of such decoherence that does not lead to einselection. Section 7.3 emphasizes the connection between predictability and einselection, and leads to a derivation of preferred states that does not rely on Born's rule.

FAQ #2: *Why does axiom (iv) conflict with "objective existence" of quantum states?*

The criterion for objective existence used here is pragmatic and operational: Finding out a state without prior knowledge is a necessary condition for a state to objectively exist (Zurek, 2003b; Ollivier et al., 2004; 2005; Blume-Kohout and Zurek, 2005; 2006; 2008). Classical states are thought to exist in this sense. Quantum states do not: quantum measurement yields an outcome—but, according to axiom (iv), this is one of the eigenstates of the measured observable, and not a preexisting state of the system. Moreover, according to axiom (iii) (or collapse part of (iv)) measurement re-prepares the system in one of the eigenstates of the measured observable. A sufficient condition for objective existence is the ability of many observers to independently find out the state of the system without prior knowledge, and to agree about it. Quantum Darwinism makes this possible.

FAQ #3: *What is the relation between the preferred states derived using their predictability (axiom (iii)) in Section 7.3 and the familiar "pointer states" that obtain from einselection?*

In the idealized case (e.g. when perfect pointer states exist) the two sets of states are necessarily the same. This is because the key requirement (stability in spite of the monitoring / copying by the environment or an apparatus) that was used in the original definition of pointer states in (Zurek, 1981) is essentially identical to "repeatability"—key ingredient of axiom (iii). It follows that when interactions commute with certain observables (e.g. because they depend on them), these observables are constants of motion under such an interaction Hamiltonian, and they will be left intact. For example, interactions that depend on position will favor (einselect) localized states, and destroy (decohere) non-local superpositions. Using *predictability sieve* to implement einselection (Zurek, 1993; 2003b; Paz and Zurek, 2001; Schlosshauer, 2007) is a good way to appreciate this.

FAQ #4: *Repeatability of measurements, axiom (iii), seems to be a very strong assumption. Can it be relaxed (e.g. to include POVM's)?*

Nondemolition measurements are very idealized (and hard to implement). In the interest of brevity we have imposed a literal reading of axiom (iii). This is very much in the spirit of Dirac's textbook, but it is also more restrictive than necessary (Zurek, 2007b), and does not cover situations that arise most often in the context of laboratory measurements. All that is needed in practice is that the *record* made in the apparatus (e.g. the position of its pointer) must be "repeatably accessible". Frequently, one does not care about repeated measurements of the quantum system (which may be even destroyed in the measurement process). Axiom (iii) captures in fact the whole idea of a record—it has to persist in spite of being read, copied, etc. So one can impose the requirement of repeatability at the macroscopic level of an apparatus pointer with a much better physical justification than Dirac did for the microscopic measured system. The proof of Section 7.3 then goes through essentially as before (Zurek, 2013), but details (and how far can one take the argument) depend on specific settings. This "transfer of the responsibility for repeatability" from the quantum system to a (still quantum, but possibly macroscopic) apparatus allows one to incorporate non-orthogonal measurement outcomes (such as POVM's) very naturally: the apparatus entangles with the system, and then acts as an ancilla in the usual projective measurement implementation of POVM's (see e.g. Nielsen and Chuang, 2000).

FAQ #5: *Probabilities—why do they enter? One may even say that in Everettian setting "everything happens", so why are they needed, and what do they refer to?*

Axiom (iii) interpreted in relative state sense "does the job" of the collapse part of axiom (iv). That is, when observer makes a measurement of an observable he will record an outcome. Repetition of that measurement will confirm his previous record. That leads to the symmetry breaking derived in Section 7.3 and captures the essence of the "collapse" in the relative state setting (Zurek, 2003b; Zurek, 2007b). So, when an observer is about to measure a state (e.g. prepared previously by another measurement) he knows that there are as many possible outcomes as there are eigenvalues of the measured observable, but that he will end up recording just one of them. Thus, even if "everything happens", a specific observer would remember a specific sequence of past events that happened to him. The question about the probability of an outcome—a future event that is about to happen—is then natural, and it is most naturally posed in this "just before the measurement" setting. The concept of probability does not (need not!) concern alternatives that already exist (as in classical discussions of probability, or some "Many Worlds" discussions). Rather, (see Zurek (2005), Sebens and Caroll (2018)) it concerns future potential events one of which will become a reality upon a measurement.

FAQ #6: *Derivation of Born's rule here and in (Zurek, 2003a; 2003b; 2005; 2007a), and even derivation of the orthogonality of outcome states use scalar products. But scalar product appears in Born's rule. Isn't that circular?*

Scalar product is an essential part of *mathematics* of quantum theory. Derivation of Born's rule relates probabilities of various outcomes to amplitudes of the corresponding states using symmetries of entanglement. So it provides a connection between mathematics of quantum theory and experiments—physics. Hilbert space (with the scalar product) is certainly an essential part of the input. And so are entangled states and entangling interactions. They appear whenever information is transferred between systems (e.g. in measurements, but also as a result of decoherence). All derivations proceed in such a way that only two values of the scalar product—0 and 1—are used as input. Both correspond to certainty.

FAQ #7: *How can one infer probability from certainty?*

Symmetry is they key idea. When there are several (say, n) mutually exclusive events that are a part of a state invariant under their swaps, their probabilities must be equal. When these events exhaust all the possibilities, probability of any one of them must be $\frac{1}{n}$. In contrast to the classical case discussed by Laplace, tensor nature of states of composite quantum systems allows one to exhibit *objective* symmetries (Zurek, 2003a; 2003b; 2005; 2007a). Thus, one can dispense with Laplace's *subjective* ignorance (his "principle of indifference"), and work with objective symmetries of entangled states. The key to the derivation of probabilities are the proofs; (i) that phases of Schmidt coefficients do not matter (this amounts to decoherence, but is established without the reduced density matrix and partial trace, the usual Born's rule—dependent tools of decoherence theory) and; (ii) that equal amplitudes imply equal probabilities. Both proofs are based on *entanglement-assisted invariance* (or *envariance*). This symmetry allows one to show that certain (Bell state-like) entangled states of the whole imply equal probabilities for local states. This is done using symmetry and certainty as basic ingredients. In particular, one relies on the ability to undo the effect of local transformations (such as a "swap") by acting on another part of the composite system, so that the preexisting state of the whole is recovered with certainty. Using envariance one can even show that the amplitude of 0 necessarily implies probability of 0 (i.e. impossibility) of the corresponding outcome.[4] One can also prove additivity of probabilities (Zurek, 2005) using modest assumption—the fact that probabilities of an event and its complement sum up to 1. Additivity is a consequence of the loss of phase coherence implied by envariance.

---

[4] This is because in a Schmidt decomposition that contains n such states with zero coefficients one can always combine two of them to form a new state, which then appears with the other $n - 2$ states, still with the amplitude of 0. This purely mathematical step should have no implications for the probabilities of the $n - 2$ states that were not involved. Yet, there are now only $n - 1$ states with equal coefficients. So the probability w of any state with zero amplitude has to satisfy $nw = (n - 1)w$, which holds only for $w = 0$ (Zurek, 2007a).

FAQ #8: *Why are the probabilities of two local states in a Bell-like entangled state equal? Is the invariance under re-labeling of the states the key to the proof?*

Envariance is needed precisely because re-labeling is not be enough. For instance, states can have intrinsic properties they "carry" with them even when they get re-labeled. Thus, a superposition of a ground and excited states $|g\rangle + |e\rangle$ is invariant under re-labeling, but this does not change the fact that the energy of the ground state $|g\rangle$ is less than the energy of the excited $|e\rangle$. So there may be intrinsic properties of quantum states (such as energy) that "trump" relabeling, and it is a priori possible that probability is like energy in this respect. This is where envariance saves the day. To see this, consider a Schmidt decomposition of an entagled state $|\heartsuit\rangle|\diamondsuit\rangle + |\spadesuit\rangle|\clubsuit\rangle$ where the first ket belongs to $\mathcal{S}$ and the second to $\mathcal{E}$. Probabilities of Schmidt partners must be equal, $p_\heartsuit = p_\diamondsuit$ and $p_\spadesuit = p_\clubsuit$. (This "makes sense", but can be established rigorously, e.g. by showing that the amplitude of $|\clubsuit\rangle$ vanishes in the state left after a projective measurement that yields $\heartsuit$ on $\mathcal{S}$.) Moreover, after a swap $|\spadesuit\rangle\langle\heartsuit| + |\heartsuit\rangle\langle\spadesuit|$, in the resulting state $|\spadesuit\rangle|\diamondsuit\rangle + |\heartsuit\rangle|\clubsuit\rangle$, one has $p_\spadesuit = p_\diamondsuit$ and $p_\heartsuit = p_\clubsuit$. But probabilities in the environment $\mathcal{E}$ (that was not acted upon by the swap) could not have changed. It therefore follows that $p_\heartsuit = p_\spadesuit = \frac{1}{2}$, where the last equality assumes (the usual) normalization of probabilities with $p(\texttt{certain event}) = 1$.

FAQ #9: *Probabilities are often justified by counting, as in the relative frequency approach. Is counting involved in the envariant approach?*

There is a sense in which envariant approach is based on counting, but one does not count the actual events (as is done is statistics) or members of an imaginary ensemble (as is done in relative frequency approach) but, rather, the number of potential invariantly swappable (and, hence, equiprobable) mutually exclusive events. Relative frequency statistics can be recovered (very much in the spirit of Everett) by considering branches in which certain number of events of interest (e.g. detections of $|\heartsuit\rangle$, $|1\rangle$, "spin up", etc.) has occured. This allows one to quantify probabilities in the resulting fragment of the "multiverse", with *all* of the branches, including the "maverick" branches that proved so difficult to handle in the past (DeWitt, 1970; 1971; DeWitt and Graham, 1973; Geroch, 1984; Squires, 1990; Stein, 1984; Kent, 1990). They are still there (as they certainly have every right to be!) but appear with probabilities that are very small, as can be established using envariance (Zurek, 2005). These branches need not be "real" to do the counting—as before, it is quite natural to ask about probabilities before finding out (measuring) what actually happened.

FAQ #10: *What is the "existential interpretation"? How does it relate to "Many Worlds Interpretation"?*

Existential interpretation is an attempt to let quantum theory tell us how to interpret it by focusing on how effectively classical states can emerge from within our Universe that is "quantum to the core". Decoherence was a major step in solving this problem: it demonstrated that in open quantum systems only certain states (selected with the help of the environment that monitors such systems) are stable. They can persist, and, therefore—in that very operational and "down to earth" sense—exist. Results of decoherence theory (such as einselection and pointer states) are interpretation independent. But decoherence was not fundamental enough—it rested on assumptions (e.g. Born's rule) that were unnatural for a theory that aims to provide a fundamental view of the origin of the classical realm starting with unitary quantum dynamics. Moreover, it did not go far enough: einselection focused on the stability of states in presence of environment, but it did not address the question of what states can survive measurement by the observer, and why. Developments described briefly in this "guide" go in both directions. Axiom (iii) that is central in Section 7.3 focuses on repeatability (which is another symptom of persistence, and, hence, existence). Events it defines provide a motivation (and a part of the input) for the derivation of Born's rule sketched in Section 7.4. These two sections shore up "foundations". Quantum Darwinism explains why states einselected by decoherence are detected by the observers. Thus, it reaffirms the role of einselection by showing that pointer states are usually reproduced in many copies in the environment, and that observers find out the state of the system indirectly, by intercepting fragments of the environment (which now plays a role of the communication channel). These advances rely on unitary evolutions and Everett's "relative state" view of the collapse. However, none of these advances depends on adopting orthodox "Many Worlds" point of view, where each of the branches is "equally real".

7.7 Conclusions

The advances discussed in this paper include derivation of preferred pointer states (key to postulate (iv)) that does not rely on the usual tools of decoherence, the envariant derivation of probabilities (postulate (v)), and quantum Darwinism. Taken together, and in the right order, they illuminate the relation of quantum theory with the classical domain of our experience. They complete what I term the *existential interpretation* based on the operational definition of objective existence, and justify confidence in quantum mechanics as the ultimate theory that needs no modifications to account for the emergence of the classical.

Of the three advances mentioned above, we have summed up the main idea of the first (the quantum origin of quantum jumps), provided an illustration of the second (the envariant origin of Born's rule), and briefly explained quantum Darwinism. As noted earlier, this is, at best, a mini-review, and an incomplete guide to the literature.

Everett's insight—the realization that relative states settle the problem of collapse—was the key to these developments (and to progress in understanding fundamental aspects of decoherence). But it is important to be careful in specifying what exactly we need from Everett and his followers, and what can be left behind. There is no doubt that the concept of relative states is crucial. Even more important is the idea that one can apply quantum theory to anything—that there is nothing *ab initio* classical. But the combination of these two ideas does not yet force one to adopt a "Many Worlds Interpretation" in which all of the branches are equally real.

Quantum states combine ontic and epistemic attributes. They cannot be "found out", so they do not exist as classical states did. But once they are known, their existence can be confirmed. This interdependence of existence and information brings to mind two contributions of John Wheeler: his early assessment of relative states interpretation, which he saw as an extension of Bohr's ideas (Wheeler, 1957), and also his "It from Bit" program (Wheeler, 1990), where information was the source of existence.

This interdependence of existence and information was very much in evidence in this paper. Stability in spite of the (deliberate or accidental) information transfer led to preferred pointer states, and is the essence of einselection. Entanglement deprives local states of information (which is transferred to correlations) and forces one to describe these local states in probabilistic terms, leading to Born's rule. Robust existence emerges ("It from Many Bits", to paraphrase Wheeler) through quantum Darwinism. The selective proliferation of information makes amplified, redundant records immune to measurements, and allows einselected states to be found out indirectly—without endangering their existence.

I would like to thank C. Jess Riedel and Michael Zwolak for stimulating discussions. This research was funded by DoE through LDRD grant at Los Alamos, and, in part, by FQXi.

References

Barnum, H. (2003). No-signalling-based version of Zurek's derivation of quantum probabilities: a note on "Environment-assisted invariance, entanglement, and probabilities in quantum physics", quant-ph/0312150.

Blume-Kohout, R., and Zurek, W. H. (2005). A simple example of "quantum Darwinism": redundant information storage in many-spin environments. *Found. Phys.* 35, 1857–6.

Blume-Kohout, R., and Zurek, W. H. (2006). Quantum Darwinism: entanglement, branches, and the emergent classicality of redundantly stored quantum information. *Phys. Rev.* A 73, 062310.

Blume-Kohout, R., and Zurek, W. H. (2008). Quantum Darwinism in quantum Brownian motion. *Phys. Rev. Lett.* 101, 240405.

Bohr, N. (1928). The quantum postulate and the recent development of atomic theory. *Nature* 121, 580–90.

Born, M. (1926). Zur Quantenmechanik der Stoßvorgänge. *Z. Phys.* 37, 863–7.

Brandao, F. G. S. L., Piani, M., and Horodecki, P. (2015). Generic emergence of classical features in quantum Darwinism. *Nat. Commun.* 6, 7908.

Dawid, R., and Thebault, K. P. Y. (2015). Many worlds: decoherent or incoherent? *Synthese* 192, 1559–80.

Deffner, S., and W. H. Zurek (2016). Foundations of statistical mechanics from symmetries of entanglement. *New J. Phys.* 18, 063013.

Deffner, S. (2017). Demonstration of entanglement assisted invariance on IBM's quantum experience. *Heliyon* 3, e00444.

Deutsch, D. (1999). Quantum theory of probability and decisions. *Proc. R. Soc. Lond.* A 455, 3129–37.

DeWitt, B. S. (1970). Quantum mechanics and reality. *Phys. Today* 23, 30–5.

DeWitt, B. S. (1971). The many-universes interpretation of quantum mechanics. In *Foundations of Quantum Mechanics* (ed. B. d'Espagnat), pp. 211-62. (New York, NY: Academic Press); reprinted in Ref. DeWitt and Graham (1973).

DeWitt, B. S., and Graham, N., eds (1973). *The Many-Worlds Interpretation of Quantum Mechanics* (Princeton University Press, Princeton).

Dieks, D. (1982). Communication by EPR devices. *Phys. Lett. A* 92, 271–2.

Dirac, P. A. M. (1958). *Quantum Mechanics* (Clarendon Press, Oxford).

Everett III, H. (1957a). 'Relative state' formulation of quantum mechanics. *Rev. Mod. Phys.* 29, 454–62.

Everett III, H. (1957b). The theory of the universal wave function. PhD dissertation, Princeton University. (Reprinted in Ref. DeWitt and Graham (1973)).

Ferrari, D, and Amoretti, M. (2018). Demonstration of envariance and parity learning on the IBM 16 qubit processor. arXiv:1801.02363.

Forrester, A. (2007). Decision theory and information propagation in quantum physics. *Stud. Hist. Philos. Mod. Phys.* 38, 815–31.

Gell-Mann, M., and Hartle, J. B. (1998). Strong decoherence. In *Proc. 4th Drexel Conf. on Quantum Non-Integrability: The Quantum-Classical Correspondence* (ed. D.-H. Feng, B.-L. Hu), pp. 3–35 (Hong Kong: International Press of Boston). arXiv/gr-qc/9509054.

Geroch, R. (1984). The Everett interpretation. *Noûs* 18, 617–33.

Gleason, A. M. (1957). Measures on the closed subspaces of a Hilbert space. *J. Math. Mech.* 6, 885–93.

Gnedenko, B. V. (1968). *The Theory of Probability* (Chelsea, New York).

Halliwell, J. J. (1999). Somewhere in the universe: where is the information stored when histories decohere? *Phys. Rev. D* 60, 105031.

Harris, J., et al. (2016). Quantum probabilities from quantum entanglement: experimentally unpacking the Born rule. *New J. Phys.* 18, 053013.

Herbut, F. (2007). Quantum probability law from 'environment-assisted invariance' in terms of pure-state twin unitaries. *J. Phys. A* 40, 5949–71.

Joos, E., and Zeh, H. D. (1985). The emergence of classical properties through interaction with the environment. *Z. Phys.* 59, 223–43.

Joos, E., et al. (2003). *Decoherence and the Appearance of a Classical World in Quantum Theory* (Springer, Berlin).

Kent, A. (1990). Against many-worlds interpretations. *Int. J. Mod. Phys.* A5, 1745–62.

Knott, P. A., et al. (2018). Generic emergence of objectivity of observables in infinite dimensions, arxiv:1802.05719.

Laplace, P. S. (1820). *A Philosophical Essay on Probabilities*, English translation of the French original by F. W. Truscott and F. L. Emory (Dover, New York 1951).

Nielsen, M. A., and Chuang, I. L. (2000). *Quantum Computation and Quantum Information* (Cambridge University Press).

Ollivier H., Poulin, D., and Zurek, W. H. (2004). Objective properties from subjective quantum states: environment as a witness. *Phys. Rev. Lett.* 93, 220401.

Ollivier, H., Poulin, D., and Zurek, W. H. (2005). Environment as a witness: selective proliferation of information and emergence of objectivity in a quantum universe. *Phys. Rev. A* 72, 042113.

Paz, J.-P., and Zurek, W. H. (2001). Environment-induced decoherence and the transition from quantum to classical. In *Coherent Atomic Matter Waves,* Les Houches lectures (ed. R. Kaiser, C. Westbrook, and F. David), pp. 533–614. (Springer, Berlin).

Paz, J.-P. and Roncaglia, A. J. (2009). Redundancy of classical and quantum correlations during decoherence. *Phys. Rev. A* 80, 042111.

Pleasance, G. and Garraway, B. M. (2017). An application of quantum Darwinism to a structured environment. *Phys. Rev. A* 96, 062105.

Riedel, C. J., and Zurek, W. H. (2010). Quantum Darwinism in an everyday environment: huge redundancy in scattered photons. *Phys. Rev. Lett.* 105, 020404.

Riedel, C. J., and Zurek, W. H. (2011). Redundant information from thermal illumination: quantum Darwinism in scattered photons. *New J. Phys.* 13, 073038.

Riedel, C. J., Zurek, W. H., and Zwolak, M. (2016). The objective past of a quantum universe: redundant records of consistent histories. *Phys. Rev. A* 93, 032126.

Riedel, C. J. (2017). Classical branch structure from spatial redundancy in a many-body wavefunction. *Phys. Rev. Lett.* 118, 120402.

Saunders, S. (2004). Derivation of the Born rule from operational assumptions. *Proc. R. Soc. Lond. A* 460, 1771–88.

Schlosshauer, M. (2004). Decoherence, the measurement problem, and interpretations of quantum mechanics. *Rev. Mod. Phys.* 76, 1267–1305.

Schlosshauer, M. (2007). *Decoherence and the Quantum to Classical Transition* (Springer, Berlin).

Schlosshauer, M., and Fine, A. (2005). On Zurek's derivation of the Born rule. *Found. Phys.* 35, 197–213.

Sebens, C. T. and Carroll, S. M. (2018). Self-locating uncertainty and the origin of probability in Everettian quantum mechanics. *Br. J. Phil. Sci.* 69, 25–74.

Squires, E. J. (1990). On an alleged 'proof' of the quantum probability law. *Phys. Lett. A* 145, 67–8.

Stein, H. (1984). The Everett interpretation of quantum mechanics: many worlds or none? *Noûs* 8, 635–52.

Vermeyden, L. et al. (2015). Experimental test of envariance. *Phys. Rev. A* 91, 012120.

von Neumann, J. (1932). *Mathematical Foundations of Quantum Theory*, translated from German original by R. T. Beyer (Princeton University Press, Princeton, 1955).

Wallace, D. (2003). Everettian rationality: defending Deutsch's approach to probability in the Everett interpretation. *Stud. Hist. Philos. Mod. Phys.* 34, 415–39.

Wheeler, J. A. (1957). Assessment of Everett's 'relative state' formulation of quantum theory. *Rev. Mod. Phys.* 29, 463–65.

Wheeler, J. A. (1990). Information, physics, quantum: the search for links. In *Complexity, Entropy, and the Physics of Information* (ed. W. H. Zurek), pp. 3–28. (Redwood City, CA: Addison Wesley).

Wootters, W. K., and Zurek, W. H. (1982). A single quantum cannot be cloned. *Nature* 299, 802–3.

Zeh, H. D. (1970). On the interpretation of measurement in quantum theory. *Found. Phys.* 1, 69–76.

Zeh, H. D. (1990). Quantum mechanics and algorithmic complexity. In *Complexity, Entropy, and the Physics of Information* (ed. W. H. Zurek), pp. 405–22. (Redwood City, CA: Addison Wesley).

Zeh, H. D. (2006). Roots and fruits of decoherence. In *Quantum Decoherence* (ed. B. Duplantier, J.-M. Raimond, and V. Rivasseau), pp. 151–75. (Basel, Switzerland: Birkhäuser).

Zurek, W. H. (1981). Pointer basis of quantum apparatus: into what mixture does the wave packet collapse? *Phys. Rev. D* 24, 1516.

Zurek, W. H. (1982). Environment-induced superselection rules. *Phys. Rev. D* 26, 1862.

Zurek, W. H. (1991). Decoherence and the transition from quantum to classical. *Phys. Today* 44, 36; see also an 'update', quant-ph/0306072.

Zurek, W. H. (1993). Preferred states, predictability, classicality and the environment-induced decoherence. *Prog. Theor. Phys.* 89, 281–312.

Zurek, W. H. (1998a). Decoherence, chaos, quantum-classical correspondence, and the algorithmic arrow of time. *Phys. Scr.* T76, 186–98.

Zurek, W. H. (1998b). Decoherence, einselection, and the existential interpretation (the rough guide). *Phil. Trans. R. Soc. Lond. A* 356, 1793–1821.

Zurek, W. H. (2000). Einselection and decoherence from an information theory perspective. *Ann. Physik* (Leipzig) 9, 855–64.

Zurek, W. H. (2003a). Environment-assisted invariance, entanglement, and probabilities in quantum physics. *Phys. Rev. Lett.* 90, 120404.

Zurek, W. H. (2003b). Decoherence, einselection, and the quantum origins of the classical. *Rev. Mod. Phys.* 75, 715–75.

Zurek, W. H. (2005). Probabilities from entanglement, Born's rule $p_k = |\psi_k|^2$ from envariance. *Phys. Rev. A* 71, 052105.

Zurek, W. H. (2007a). Relative states and the environment: einselection, envariance, quantum Darwinism, and the existential interpretation. (https://arxiv.org/abs/0707.2832)

Zurek, W. H. (2007b). Quantum origin of quantum jumps: breaking of unitary symmetry induced by information transfer in the transition from quantum to classical. *Phys. Rev. A* 76, 052110.

Zurek, W. H. (2009). Quantum Darwinism. *Nat. Phys.* 5, 181–88.

Zurek, W. H. (2011). Entanglement symmetry, amplitudes, and probabilities: inverting Born's rule. *Phys. Rev. Lett.* 106, 250402.

Zurek, W. H. (2013). Wave-packet collapse and the core quantum postulates: discreteness of quantum jumps from unitarity, repeatability, and actionable information. *Phys. Rev. A* 87, 052111.

Zurek, W. H. (2014). Quantum Darwinism, classical reality, and the randomness of quantum jumps. *Phys. Today* 67, 44–50.

Zurek, W. H. (2018). Eliminating ensembles from equilibrium statistical physics: Maxwell's demon, Szilard's engine, and thermodynamics via entanglement. *Physics Reports,* 755, 1–21.

Zurek, W. H. *Decoherence, Quantum Darwinism, and the Quantum Theory of the Classical,* in preparation.

Zwolak, M., and Zurek, W. H. (2013). Complementarity of quantum discord and classically accessible information. *Sci. Rep.* 3, 1729.

Zwolak, M., Quan, H.-T., and Zurek, W. H. (2009). Quantum Darwinism in a mixed environment. *Phys. Rev. Lett.* 103, 110402.

Zwolak, M., Quan, H.-T., and Zurek, W. H. (2010). Redundant imprinting of information in nonideal environments: objective reality via a noisy channel. *Phys. Rev. A* 81, 062110.

Zwolak, M., Riedel, C. J., and Zurek, W. H. (2014). Amplification, redundancy, and the quantum Chernoff information. *Phys. Rev. Lett.* 112, 140406.

8

Generation of high-order harmonics and attosecond pulses

ANNE L'HUILLIER

Department of Physics, Lund University
Lund, Sweden

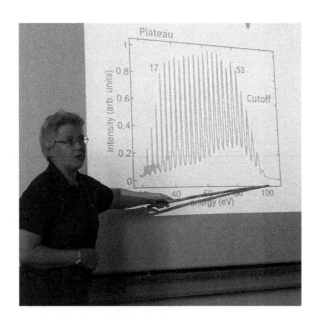

L'Huillier, A. "Generation of high-order harmonics and attosecond pulses." In *Current Trends in Atomic Physics*. Edited by Antoine Browaeys, Thierry Lahaye, Trey Porto, Charles S. Adams, Matthias Weidemüller, and Leticia F. Cugliandolo. Oxford University Press (2019).
© Oxford University Press. DOI: 10.1093/oso/9780198837190.003.0008

Chapter Contents

These notes have been written following a series of lectures given at the Les Houches summer school in July 2016. There are three lectures. The first one (Sec. 8.1) provides an introduction to high-order harmonic generation in gases in strong laser fields. A semi-classical model is described. The second one (Sec. 8.2) discusses the propagation of harmonics in the nonlinear medium and in particular the phase matching condition. The third (Sec. 8.3) presents an interferometric technique used for both the measurement of attosecond pulses and for the study of photoionization dynamics. None of the students were specialized in this field of research so that the lectures are intended for non-specialists.

8.1   High-order harmonic generation in strong laser fields

8.1.1   Introduction

High-order harmonics are generated when an intense laser field is focused into a gas cell. The radiation, which is in the extreme ultraviolet (XUV) range of the electromagnetic spectrum, is emitted along the propagation axis. Fig. 8.1 shows a typical spectrum obtained in a gas of neon. The laser used in this experiment is a titanium sapphire laser, with fundamental wavelength of 800 nm ($\hbar\omega = 1.55\,\text{eV}$) and pulse duration of 40 fs. The laser polarization is linear. Only harmonics of odd order are generated, because of the dipole selection rules (see also Exercise 8.4 below). The rather spectacular "plateau" consisting of high-order harmonics with the same intensity indicates that the light-matter interaction leading to high-order harmonic generation (HHG) is highly non-linear and non-perturbative. The reduced harmonic intensity observed at low energy is simply due to the decreased efficiency of the grating used to analyse the emitted radiation. One can also see small peaks corresponding to the second diffraction order of the grating.

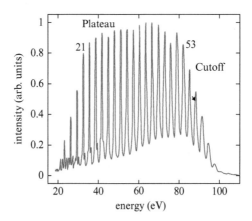

Fig. 8.1 HHG spectrum, with the characteristic plateau and cutoff.

Exercise 8.1   Consider harmonics between 21 and 53 as in Fig. 8.1 and assume that they have the same phase at a given time, for example, at $t = 0$ and the same amplitude. How does the temporal structure of the combined field look like? What is the pulse duration of the individual pulses?

Proof

$$\sum_{q=21,\text{odd}}^{53} e^{iq\omega t} = e^{i21\omega t} \sum_{n=0}^{16} e^{2in\omega t} = e^{i21\omega t} \frac{1 - e^{i34\omega t}}{1 - e^{i2\omega t}},$$

$$\left| \sum_{q=21,\text{odd}}^{53} e^{iq\omega t} \right|^2 = \left| \frac{\sin(17\omega t)}{\sin(\omega t)} \right|^2. \tag{8.1}$$

The resulting temporal structure is that of a train of pulses separated by half a laser cycle $T/2 = \pi/\omega$. The pulse duration is approximately such that $17\omega\tau = \pi$, i.e. $\tau = \pi/(17\omega) = T/34$, the laser period divided by twice the number of harmonics which add in phase. For 800 nm radiation, $T = 2.6$ fs and $\tau = 76$ as. These equations are similar to those encountered when describing diffraction through multiple slits, with space replaced by time. ∎

Directly after the first experiments showing HHG at the end of the 80s (McPherson et al. 1987, Ferray et al. 1988), the question whether harmonics were phase-locked was raised (Farkas and Toth 1992, Harris et al. 1993). At the same time, the behavior of an atom in a strong laser field was simulated by solving the time-dependent Schrödinger equation within the single active electron approximation,

$$i\hbar \frac{\partial \Psi}{\partial t} = -\frac{\hbar^2}{2m} \nabla^2 \Psi + \left[V(r) + e\vec{E}.\vec{r} \right] \Psi. \tag{8.2}$$

Here $V(r)$ is the atomic potential, equal to $-e^2/4\pi\epsilon_0 r$ for hydrogen, and usually described within a single active electron approximation; e, m are the charge and mass of the electron and ϵ_0 is the vacuum permittivity. The laser-atom interaction is described within the dipole approximation. The electromagnetic field $\vec{E}$ can be written as $E_0 \sin(\omega t)\hat{x}$, where $\hat{x}$ is the polarization direction and E_0 is a function slowly varying with time (variation which we neglect in the following). From the time-dependent wavefunction $\Psi(t)$, the dipole moment,

$$d(t) = < \Psi(t)|ex|\Psi(t) >, \tag{8.3}$$

is calculated, and the harmonic spectrum emitted by a single atom is equal to the square of the Fourier transform of the acceleration, i.e. the second derivative of the dipole moment.

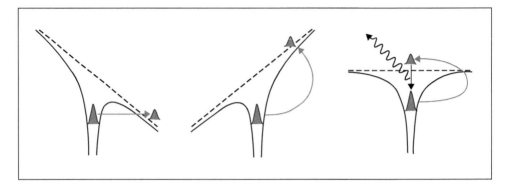

Fig. 8.2 Illustration of the three-step model: tunnel ionization; propagation of the electron in the continuum in presence of the laser field; recombination and emission of an XUV photon.

The success of these calculations to reproduce the plateau and cutoff of HHG spectra (Kulander and Shore 1989, Krause et al. 1992) led to the elaboration of a very intuitive semiclassical model (Schafer et al. 1993, Corkum 1993), that we describe in the next section, partly as an exercise.

8.1.2   Three-step model

The semi-classical model (see Fig. 8.2) describes the generation of attosecond pulses as follows: (1) the laser field induces a distortion of the atomic potential so that an electron can be ionized by tunnel through the distorted barrier; (2) the electron is first driven away by the laser field. When the laser field changes sign, it is driven back towards the ionic core; (3) it may return with a high kinetic energy and be captured by its parent ion, emitting an XUV photon with energy $E_p = I_p + E_{\text{kin}}$, where I_p is the ionization energy and E_{kin} the kinetic energy of the electron when it returns.

Exercise 8.2   Derive the equation for the electronic trajectories within classical mechanics, assuming that the electron is born at $t = t_i$, $x = 0$ with zero velocity ($\dot{x} = 0$). Give an expression for E_{kin} at the time of return as a function of the ponderomotive energy $U_p = e^2 E_0^2/(4m\omega^2)$, the return time t_r and the ionization time t_i.

Proof   We solve the Newton's equation:

$$\ddot{x}(t) = -\frac{eE_0}{m}\sin(\omega t) \tag{8.4}$$

with the initial conditions $x(t_i) = 0$ and $\dot{x}(t_i) = 0$. We obtain

$$\dot{x}(t) = \frac{eE_0}{m\omega}[\cos(\omega t) - \cos(\omega t_i)],$$

$$x(t) = \frac{eE_0}{m\omega^2}[\sin(\omega t) - \sin(\omega t_i) - \omega(t - t_i)\cos(\omega t_i)]. \tag{8.5}$$

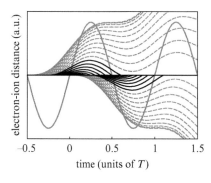
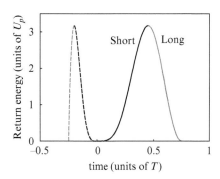

Fig. 8.3 Electron trajectories (left) and kinetic energy distribution (right). The gray solid line indicates the laser field. The trajectories in black are those returning to the ionic core and thus contributing to HHG. Those in dashed gray contribute to ionization. The kinetic energy at return (right) is proportional to the square of the slopes of the trajectories when they cross the horizontal axis. The solid (dashed) line indicates the return energy as a function of return (ionization) time.

The kinetic energy can be obtained from Eq. 8.5 as,

$$E_{\text{kin}} = \frac{1}{2}m\dot{x}^2(t_r) = \frac{e^2E_0^2}{2m\omega^2}\left[\cos(\omega t_r) - \cos(\omega t_i)\right]^2 = 2U_p\left[\cos(\omega t_r) - \cos(\omega t_i)\right]^2. \quad (8.6)$$

∎

Examples of electron trajectories are shown in Fig. 8.3, together with a plot of the kinetic energy, which can be obtained from Eq. 8.6 by calculating the return times which are solutions of

$$\sin(\omega t_r) - \sin(\omega t_i) - \omega(t_r - t_i)\cos(\omega t_i) = 0. \quad (8.7)$$

The kinetic energy reaches a maximum (a cutoff) equal to $3.17\,U_p$. All energies (except the cutoff) can be reached by two trajectories, which we call short and long respectively. Finally, even when considering one type of trajectories, the attosecond emission is slightly "chirped", i.e. the different frequency components are emitted at different times. The chirp is positive for the short trajectories and negative for the long trajectories. Considering the whole bandwidth emitted for the short trajectories, the dispersion of the return times is $\simeq 0.4\,T$, i.e. of the order of one femtosecond at 800 nm.

Exercise 8.3 Here, we calculate some orders of magnitude. The laser intensity is 5×10^{14} W cm^{-2}. What is the corresponding field amplitude? Compare to the Coulomb field for a hydrogen atom at the Bohr radius. What is the ponderomotive energy at the above intensity and at a laser wavelength of 800 nm? What is the order of magnitude of the electron excursion in the field?

Proof   The relation between intensity and amplitude is $I = c\epsilon_0 E_0^2/2$; $\omega = 2\pi c/\lambda$.

$$E_0 = \sqrt{\frac{2I}{c\epsilon_0}} = 6.1 \times 10^{10} \text{ V m}^{-1}.$$

$$E_c = \frac{e}{4\pi\epsilon_0 a_0^2} = 5.1 \times 10^{11} \text{ V m}^{-1}. \qquad (8.8)$$

$$U_p = \frac{e^2 E_0^2}{4m\omega^2} = \frac{e^2\lambda^2 I}{8\pi^2 c^3 \epsilon_0 m} = 3 \text{ eV}.$$

$$\bar{x} = \frac{eE_0}{m\omega^2} = 1.9 \text{ nm}.$$

The Coulomb field and that induced by the laser are comparable. The excursion by the electron is quite large compared to, e.g., the Bohr radius.   ∎

The discussion so far concerns what happens due to the interaction of the atom and the laser field during half a laser cycle. The three steps are repeated every half laser cycle with a sign change of the electromagnetic field, and therefore of the emitted radiation. This leads to the emission of a train of attosecond pulses, with a sign flip between consecutive pulses. The interferences between consecutive attosecond pulses result in a frequency comb of high-order harmonics.

Exercise 8.4   Consider two identical attosecond pulses separated by half a laser cycle ($T/2$), with a sign flip. Calculate the interference pattern.

Proof   Let $\mathcal{E}(t)$ denote the total electric field and $\mathcal{E}_1(t)$ that corresponding to the first attosecond pulse. We have

$$\mathcal{E}(t) = \mathcal{E}_1(t) - \mathcal{E}_1\left(t - \frac{T}{2}\right). \qquad (8.9)$$

Taking the Fourier transform of Eq. 8.9 and introducing the spectral amplitudes $\mathcal{A}(\Omega)$ and $\mathcal{A}_1(\Omega)$,

$$\mathcal{A}(\Omega) = \mathcal{A}_1(\Omega) - \int \mathcal{E}_1\left(t - \frac{T}{2}\right) e^{-i\Omega t} dt = \mathcal{A}_1(\Omega) - \int \mathcal{E}_1(t) e^{-i\Omega\left(t + \frac{T}{2}\right)} dt,$$

$$\mathcal{A}(\Omega) = \mathcal{A}_1(\Omega)\left(1 - e^{-i\frac{\Omega T}{2}}\right). \qquad (8.10)$$

Consequently

$$|\mathcal{A}(\Omega)|^2 = 4|\mathcal{A}_1(\Omega)|^2 \sin^2\left(\frac{\Omega T}{4}\right). \qquad (8.11)$$

The spectrum presents maxima when $\Omega T/4 = (2q + 1)\pi/2$, q integer, i.e. when $\Omega = 2\pi(2q + 1)/T = (2q + 1)\omega$ (odd harmonics), and zeros at the even harmonics of the laser field. ∎

This derivation can be generalised to $2N$ pulses. Eq. 8.9 becomes

$$\mathcal{E}(t) = \sum_{n=1}^{2N}(-1)^{n-1}\mathcal{E}_1\left[t - (n - 1)\frac{T}{2}\right], \tag{8.12}$$

and Eq. 8.10

$$\mathcal{A}(\Omega) = \sum_{n=0}^{N-1}\mathcal{A}_1(\Omega)\left[1 - e^{-i\frac{\Omega T}{2}}\right]e^{-in\Omega T}. \tag{8.13}$$

After some mathematical manipulation, we obtain

$$|\mathcal{A}(\Omega)|^2 = 4|\mathcal{A}_1(\Omega)|^2\frac{\sin^2\left(\frac{\Omega T}{4}\right)\sin^2\left(\frac{N\Omega T}{2}\right)}{\sin^2\left(\frac{\Omega T}{2}\right)} \tag{8.14}$$

which features a spectrum of odd high-order harmonics modulating a broad spectrum $[\mathcal{A}_1(\Omega)]$ generated by a half laser cycle. In this simple approximation where all of the pulses are assumed to be identical (except for an overall sign that changes from pulse to pulse), the width of the harmonic peaks is given approximately by $N T\delta\Omega = 2\pi$, i.e. $\delta\Omega = \omega/N$.

In conclusion, we have shown how attosecond pulses arise from phase-locked odd-order harmonics and vice versa how high-order harmonics are due to the interferences of attosecond pulses. We have also presented the model explaining the emission of an attosecond pulse twice per laser cycle. In the next section, we describe the emission by a macroscopic medium.

8.2 Macroscopic aspects

This chapter presents a general introduction to nonlinear optics, followed by a description of the particularities of high-order harmonic generation.

8.2.1 Propagation equations

We start by the (well known) Maxwell equations, with $\mathcal{E}$, $\mathcal{B}$ the electric and magnetic fields respectively, $\mathcal{D}$ the electric displacement, and $\mathcal{J}$ the current density.

$$\nabla \times \mathcal{E} = -\frac{\partial \mathcal{B}}{\partial t},$$
$$\nabla \times \mathcal{B} = \mu_0\left(\mathcal{J} + \frac{\partial \mathcal{D}}{\partial t}\right). \tag{8.15}$$

which, after Fourier transform, introducing the spectral components $\tilde{\mathcal{E}} = \int \mathcal{E}e^{-i\Omega t}dt$, etc., become

$$\nabla \times \tilde{\mathcal{E}} = -i\Omega\tilde{B},$$
$$\nabla \times \tilde{B} = \mu_0 \left(\tilde{J} + i\Omega\tilde{D}\right). \tag{8.16}$$

We separate $\tilde{D}$ into a linear and nonlinear part as $\tilde{D} = \epsilon_0\varepsilon_\Omega\tilde{\mathcal{E}} + \tilde{P}$, where ε_Ω is the relative permittivity describing the dispersion in the medium at frequency Ω and $\tilde{P}$ the polarization. We use the transformation

$$\nabla \times \nabla \times \tilde{\mathcal{E}} = \nabla(\nabla\tilde{\mathcal{E}}) - \nabla^2\tilde{\mathcal{E}}. \tag{8.17}$$

The first term on the right hand side is zero since $\nabla\tilde{\mathcal{E}} = 0$. We incorporate the current density $\tilde{J}$ in the nonlinear polarization $\tilde{P}$ and we obtain the propagation equation:

$$\nabla^2\tilde{\mathcal{E}} + \frac{\Omega^2}{c^2}\varepsilon_\Omega\tilde{\mathcal{E}} = -\mu_0\Omega^2\tilde{P}. \tag{8.18}$$

Furthermore, we define the wave vector at frequency Ω, $k_\Omega = \sqrt{\varepsilon_\Omega}\Omega/c = n_\Omega\Omega/c$, where n_Ω is the refractive index at frequency Ω. Eq. 8.18 becomes

$$\nabla^2\tilde{\mathcal{E}} + k_\Omega^2\,\tilde{\mathcal{E}} = -\mu_0\Omega^2\tilde{P}. \tag{8.19}$$

8.2.2   Propagation equations for high-order harmonics

The conversion efficiency for high-order harmonics remains low (at most 10^{-5}) so that the polarization field can be assumed to be predominantly created by the fundamental field. We here consider separately the propagation of each harmonic field, since the wave equations are decoupled from each other. The propagation of the qth harmonic field $\tilde{\mathcal{E}}_q$ with frequency $\Omega = q\omega$ is decribed by

$$\nabla^2\tilde{\mathcal{E}}_q + k_q^2\,\tilde{\mathcal{E}}_q = -\mu_0 q^2\omega^2\tilde{P}_q. \tag{8.20}$$

We now introduce the envelope $E_q = \tilde{\mathcal{E}}_q\exp(ik_q z)$, z denoting the propagation axis and make the slowly-varying approximation:

$$\frac{\partial^2\left(E_q e^{-ik_q z}\right)}{\partial z^2} \approx -2ik_q\frac{\partial E_q}{\partial z}e^{-ik_q z} - k_q^2 E_q e^{-ik_q z}. \tag{8.21}$$

Eq. 8.20 becomes

$$\nabla_\perp^2 E_q - 2ik_q\frac{\partial E_q}{\partial z} = -\mu_0 q^2\omega^2 P_q e^{-i(qk_1-k_q)z}, \tag{8.22}$$

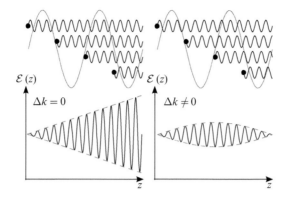

Fig. 8.4 Build up of the harmonic field when phase matching is realized (left) and not realized (right).

where the symbol $\perp$ refers to derivation with respect to the transverse directions x and y and where we have introduced $\tilde{\mathcal{P}}_q = P_q \exp(iqk_1 z)$, where k_1 is the fundamental wave vector. The quantity $\Delta k = qk_1 - k_q$ is also called phase mismatch between the polarization field at frequency $q\omega$, which follows the fundamental field (phase velocity c/n_1) and the generated harmonic field, which propagates in the nonlinear medium with the wave vector k_q (phase velocity c/n_q).

Figure 8.4 illustrates the generation of high-order harmonics in the nonlinear medium in a phase matched (left) and a non-phase matched (right) case. In the left figure, the field builds up continuously since the phase velocities of the fundamental and harmonic fields are identical, while in the right figure, at a certain distance called coherence length $L_c = \pi/\Delta k$, the field created at $z > L_c$ becomes opposite in phase with that created earlier in the medium, leading to destructive interference and decrease of the total field.

Exercise 8.5   Solve Eq. 8.22 in one dimension, for a uniform medium of length L. We assume a loose focusing geometry so that P_q can be considered as constant over the medium length. Discuss the dependence of $|E_q|^2$ as a function of L and Δk.

Proof   The propagation equation takes a simple form:

$$\frac{\partial E_q}{\partial z} = -\frac{ik_q}{2\epsilon_0} P_q\, e^{-i\Delta kz}, \tag{8.23}$$

so that

$$E_q = -\frac{ik_q}{2\epsilon_0} P_q \int_{-L/2}^{L/2} e^{-i\Delta kz}\,dz \approx -\frac{ik_q P_q}{\epsilon_0} \frac{\sin(\Delta kL/2)}{\Delta k}, \tag{8.24}$$

$$|E_q|^2 \propto \left| \frac{\sin(\Delta k L / 2)}{\Delta k} \right|^2. \tag{8.25}$$

$|E_q|^2$ varies as a function of L as a $\sin^2$ function with maximum at $\Delta k L = \pi$, i.e. $L = L_c$. If $\Delta k = 0$, $|E_q|^2 \propto L^2$. $|E_q|^2$ varies with Δk as a sinc function. The width at half maximum is approximately such that $\Delta k L = \pi$. For an infinite medium, Δk must be equal to zero to get a nonzero harmonic field.   ∎

High-order harmonics are often above the ionization threshold, and absorption of the harmonic field by the medium occurs, thus limiting the buildup of the field. Eq. 8.25 can be generalized to (Constant et al. 1999, Ruchon et al. 2008)

$$|E_q|^2 \propto e^{-\kappa_q L} \frac{\cosh(\kappa_q L) - \cos(\Delta k L)}{\Delta k^2 + \kappa_q^2}, \tag{8.26}$$

where κ_q denotes the absorption coefficient at frequency $q\omega$. Fig. 8.5 shows how the harmonic yield varies as a function of the medium length for different absorption lengths L_{abs}. The dashed gray line line is obtained with $L_{abs} = \infty$ and $L_c = 0$.

In the last section, we discuss the different contributions to Δk for high-order harmonic generation.

8.2.3   Phase mismatch in high-order harmonic generation

We can identify four contributions:

- The dispersion of the neutral medium $\Delta k_n = (n_1 - n_q) q\omega/c$. At the fundamental frequency, $n_1 > 1$ while at the harmonic frequencies above the ionization threshold $n_q < 1$, so that this contribution is positive. The refractive index is related

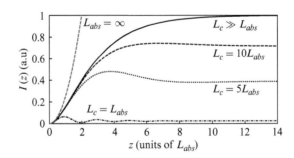

Fig. 8.5 Variation of the harmonic intensity in the nonlinear medium for different ratios between the coherence and absorption lengths.

to the dynamic susceptibility χ and the atomic polarizability α through the formula:

$$n = \sqrt{1+\chi} \approx 1 + \frac{\chi}{2} \approx 1 + \frac{\mathcal{N}_n \alpha}{2\epsilon_0}, \tag{8.27}$$

where $\mathcal{N}_n$ is the density of neutral atoms, so that

$$\Delta k_n = \frac{q\omega}{c} \frac{\mathcal{N}_n}{2\epsilon_0} [\alpha(\omega) - \alpha(q\omega)]. \tag{8.28}$$

- The dispersion due to free electrons

$$\Delta k_e = \frac{q\omega}{c} \frac{\mathcal{N}_e}{2\epsilon_0} \left(-\frac{e^2}{m\omega^2} + \frac{e^2}{mq^2\omega^2} \right), \tag{8.29}$$

where $\mathcal{N}_e$ is the density of free electrons. Note that the second term in the parenthesis above can be neglected, which means that the free electrons act mainly on the fundamental field, making it propagating faster than in the neutral medium. Δk_e is negative and opposite in sign to Δk_n. In absence of all other contributions, $\Delta k_e = \Delta k_n$ means that the phase velocities of the fundamental and q th harmonic field are the same, which is achieved when a few percent of the atoms in the nonlinear medium are ionized.

- The dispersion due to focusing. When a laser beam goes through a focus, the phase varies (in addition to the usual $-kz$). For a Gaussian beam the phase variation is the Gouy phase shift $\zeta(z) = \tan^{-1}(z/z_0)$, where z_0 is the Rayleigh length. When $z \ll z_0$, $\zeta(z) \approx z/z_0$ so that the corresponding contribution to the phase mismatch is negative and approximately equal to

$$\Delta k_f \approx -\frac{q}{z_0}. \tag{8.30}$$

- The dispersion induced by the electron trajectory in the continuum. The phase accumulated by the electron wavepacket (also called dipole phase) is approximately proportional to the laser intensity leading to:

$$\Delta k_d = -\beta \frac{\partial I}{\partial z}, \tag{8.31}$$

where β is positive. This term changes sign across the focus, being negative when $z \leq 0$ and positive when $z \geq 0$ (Salières et al. 1995).

We note that since both Δk_e and Δk_d are time-dependent, phase matching of high-oder harmonics is also time-dependent.

8.3   An introduction to attosecond physics with attosecond pulse trains

In this last section, we describe the interferometric technique called RABBIT (reconstruction of attosecond beating by interference of two-photon transitions) which has been used for the first characterization of attosecond pulses (Paul et al. 2001) and which is the basis of one of the most fascinating applications of attosecond pulses, namely, the determination of time delays in photoionization (Schultze et al. 2010, Klünder et al. 2011).

8.3.1   Principle of the RABBIT technique

To understand the principle of this technique, we start from Fermi's golden rule for photoionization. Time-dependent perturbation theory, combined with the rotating wave approximation, allows us to calculate the ionization rate for an atom subject to a perturbation $V \exp(-i\omega t)$,

$$R_{fi} = \frac{\pi}{2\hbar^2} |<f|V|i>|^2 \delta(\omega_{fi} - \omega), \tag{8.32}$$

where $|i>$ is the initial state, $|f>$ the final state [Fig. 8.6a] and $\omega_{fi} = (E_f - E_i)/\hbar$ (E_f, E_i are the energies of the final and initial states respectively). This result is called Fermi's golden rule. It can be generalized to two-photon ionization [Fig. 8.6b]:

$$R_{fi} = \frac{\pi}{8\hbar^4} \left| \sum_m \frac{<f|V|m><m|V|i>}{\omega_{mi} - \omega} \right|^2 \delta(\omega_{fi} - 2\omega), \tag{8.33}$$

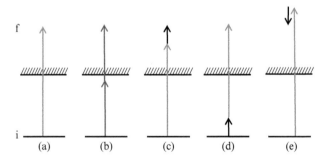

f				
(a)	(b)	(c)	(d)	(e)

i

Fig. 8.6 Different ionization processes. (a) Single photoionization; (b) two-photon ionization, with the same photon energy; (c) two-photon ionization with an XUV and an IR photon; it is also an "above-threshold-ionization" process, since the energy of the XUV photon is above the ionization threshold; (d) same as (c), except the the IR photon is absorbed first. This process is usually negligible compared to (c); (e) photoionization with XUV-photon absorption and IR-photon stimulated emission. The RABBIT technique combines (c) and (e).

where the summation is over all possible intermediate states $|m>$ both in the discrete and continuum spectrum. In the next step, we replace V by $-ex\tilde{\mathcal{E}}$ and consider two-photon ionization with the q th harmonic field and a weak fraction of the fundamental field. Two processes can be considered within the rotating wave approximation, as indicated in Fig. 8.6c, d. The process indicated in (d) is usually negligible compared to (c), since the energy of the first photon is far from resonance, while in (c) it coincides with the energy of a continuum state. Eq. 8.33 becomes

$$R_{fi} = \frac{\pi}{8\hbar^4}|M_{\text{abs}}|^2\delta(\omega_{fi} - (q+1)\omega), \tag{8.34}$$

where

$$M_{\text{abs}} = \lim_{\varepsilon \to 0} \sum_m \frac{<f|ex|m><m|ex|i>}{\omega_{mi} - q\omega + i\varepsilon}\tilde{\mathcal{E}}_1\tilde{\mathcal{E}}_q. \tag{8.35}$$

The infinitesimal quantity ε is here introduced to avoid a discontinuity (since the first photon absorption leads to a continuum state and is therefore resonant). In the RABBIT technique, atoms are ionized by a frequency comb of high-order harmonics. Absorption of the $(q+2)^{\text{th}}$ harmonic followed by emission of a laser photon [Fig. 8.6e] leads to the same final state as (c) and therefore interferes with it. The amplitude of this process can be written as

$$M_{\text{em}} = \lim_{\varepsilon \to 0} \sum_m \frac{<f|ex|m><m|ex|i>}{\omega_{mi} - (q+2)\omega + i\varepsilon}\tilde{\mathcal{E}}_1^*\tilde{\mathcal{E}}_{q+2}, \tag{8.36}$$

and the total ionization rate is:

$$R_{fi} = \frac{\pi}{8\hbar^4}|M_{\text{abs}} + M_{\text{em}}|^2\delta(\omega_{fi} - (q+1)\omega). \tag{8.37}$$

with

$$|M_{\text{abs}} + M_{\text{em}}|^2 = |M_{\text{abs}}|^2 + |M_{\text{em}}|^2 + 2|M_{\text{abs}}||M_{\text{em}}|\cos\left[\arg(M_{\text{abs}}) - \arg(M_{\text{em}})\right]. \tag{8.38}$$

8.3.2 Phase contributions

We now examine the different contributions to the phase of M_{abs}.

$$\arg(M_{\text{abs}}) = \arg\tilde{\mathcal{E}}_1 + \arg\tilde{\mathcal{E}}_q + \arg\left[\lim_{\varepsilon \to 0} \sum_m \frac{<f|ex|m><m|ex|i>}{\omega_{mi} - q\omega + i\varepsilon}\right]. \tag{8.39}$$

We assume the fundamental field to be Fourier transform limited. It is delayed compared to the XUV field by a delay τ so that $\arg \tilde{\mathcal{E}}_1 = \omega\tau$. For the qth harmonic field, $\arg \tilde{\mathcal{E}}_q = \phi_q$. The third phase, φ_q, arises from the presence of a pole in the denominator which leads to the following separation between a real part (called Cauchy principal part) and an imaginary part:

$$\lim_{\varepsilon \to 0} \sum_m \frac{<f|ex|m><m|ex|i>}{\omega_{mi} - q\omega + i\varepsilon} = \sum_{PP} \frac{<f|ex|m><m|ex|i>}{\omega_{mi} - q\omega} + i\pi <f|ex|n><n|ex|i>,$$

$$(8.40)$$

$|n>$ denoting the state at resonance. We therefore obtain

$$\arg(M_{\text{abs}}) = \omega\tau + \phi_q + \varphi_q \qquad (8.41)$$

and similarly

$$\arg(M_{\text{em}}) = -\omega\tau + \phi_{q+2} + \varphi_{q+2} \qquad (8.42)$$

so that

$$\arg(M_{\text{abs}}) - \arg(M_{\text{em}}) = 2\omega\tau + \phi_q - \phi_{q+2} + \varphi_q - \varphi_{q+2}. \qquad (8.43)$$

These phase differences can be interpreted as follows:

$$\frac{\phi_{q+2} - \phi_q}{2\omega} \approx \frac{\partial\phi}{\partial\Omega}, \qquad (8.44)$$

is the derivative of the spectral phase of the attosecnd pulses (also called group delay) sampled at frequency $(q + 1)\omega$, while

$$\frac{\varphi_{q+2} - \varphi_q}{2\omega} \approx \frac{\partial\varphi}{\partial\Omega}, \qquad (8.45)$$

is a delay induced by the two-photon ionization.

8.3.3 Applications of the RABBIT technique. Photoionization time delays

The RABBIT technique was invented to characterize attosecond pulses in a train of pulses (Paul et al. 2001). Fig. 8.7 shows an experimental photoelectron spectrum as a function of τ. The sideband peaks (corresponding to the frequencies $(q + 1)\omega$, where q is odd) strongly oscillate with delay. If we neglect the effect of ionization on the delay (Eq. 8.45), the phase of the sideband oscillations represents the spectral phase of the attosecond pulses. Knowing $|\tilde{\mathcal{E}}_q|^2$ and $\partial\phi/\partial\Omega|_{(q+1)\omega}$, we can reconstruct, up to a constant phase factor, $\mathcal{E}(t)$ for an average attosecond pulse in the train.

Recently, a lot of research has been devoted to the ionization delay $\tau_{2\mathrm{ph}}$, which, in some cases, can be decomposed into two components

$$\tau_{2\mathrm{ph}} = \tau_{1\mathrm{ph}} + \tau_{\mathrm{cc}}, \tag{8.46}$$

with the interpretation that the first term represents the photoionization time delay, i.e. the time it takes for the electron to propagate in the Coulomb potential after photoionization, while the second term is a correction due to the probe (infrared) field (Klünder et al. 2011, Dahlström et al. 2012). Experiments consist in measuring a difference in delay, e.g. between 3s and 3p ionization in Ar (Klünder et al. 2011) or a variation of delay across an autoionization resonance (Kotur et al. 2016). In this way, the influence of the attosecond pulse group delay can be eliminated and the electron wavepacket fully characterized.

In conclusion, the interferometric technique presented in this section has been essential to the first experimental evidence of attosecond pulses, and its first characterization. More importantly, it has led to new physics, which consists in characterizing electron wavepackets both in phase and amplitude, in the spectral or in the temporal domain.

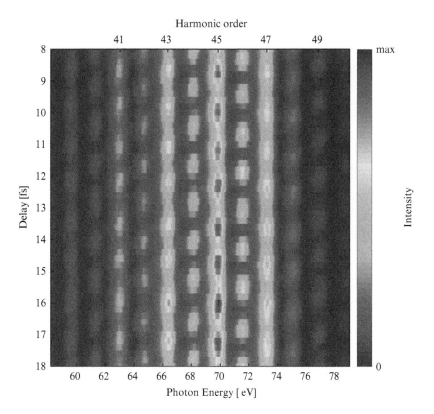

Fig. 8.7 Experimental RABBIT scan in Ne. The photoelectron spectrum is shown as a function of delay.

Acknowledgments

The author acknowledges the support of the European Research Council (PALP) as well as the Swedish Research Council.

References

Constant, E. (1999) *Phys. Rev. Lett.* **82**, 1668.
Corkum, P. B. (1993) *Phys. Rev. Lett.* **71**, 1994.
Dahlström, J. M. *J. Phys. B* **45**, 183001.
Farkas, G. and Toth, C. (1992) *Phys. Lett. A.* **168**, 447.
Ferray, M. et al. (1988) *J. Phys. B*, **21**, L31.
Harris, S. E. et al. (1993) *Opt. Commun.* **100**, 487.
Klünder, K. (2011) *Phys. Rev. Lett.* **106**, 143002.
Krause, J. L. et al. (1992) *Phys. Rev. Lett.* **68**, 3535.
Kulander, K. C. and Shore, B. W. (1989) *Phys. Rev. Lett.* **62**, 524.
McPherson, A. et al. (1987) *J. Opt. Soc. Am. B* **4** 595.
Paul, P. M. et al. (2001) *Science* **292**, 1689.
Ruchon, T. et al. (2008) *New J. Phys.* **10**, 025027.
Salières, P. et al. (1995) *Phys. Rev. Lett.* **74**, 3776.
Schafer, K. J. (1993) *Phys. Rev. Lett.* **70**, 1599.
Schultze, M. et al. (2010) *Science* **328**, 1658.

9

Ultrafast electron dynamics as a route to explore chemical processes

ALEXANDER I. KULEFF

Theoretical Chemistry, Physical Chemistry Institute
Heidelberg University, Germany

Kuleff, A. I. "Ultrafast electron dynamics as a route to explore chemical processes." In *Current Trends in Atomic Physics*. Edited by Antoine Browaeys, Thierry Lahaye, Trey Porto, Charles S. Adams, Matthias Weidemüller, and Leticia F. Cugliandolo. Oxford University Press (2019). © Oxford University Press. DOI: 10.1093/oso/9780198837190.003.0009

Chapter Contents

9.1 Problem overview

9.1.1 Chemistry as dynamics of quantum particles

The purpose of these lecture notes is to give a concise overview of the problems one faces in the attempt to describe quantum mechanically a chemical reaction, to draw some conclusions from those problems, and, by providing the basics for describing electronic coherences, to give some interesting perspectives for designing schemes to influence the chemical reactivity of a molecule by manipulating the pure electron dynamics.

On a molecular level, the chemical reactions represent processes in which bonds between atoms are formed or broken, or the molecule undergoes structural changes like isomerization. Therefore, every chemical reaction involves some motion of the nuclei, as well as some reorganization of the electronic cloud, i.e., involves dynamics of both types of particles building up the molecule—the nuclei and the electrons. It is also intuitively clear that the motion of the electrons and nuclei throughout a chemical reaction is strongly correlated. To better understand and quantify this mutual influence, let us see how one can describe the evolution of a molecule.

As we want to describe this evolution on a quantum mechanical level, we need to solve the time-dependent Schrödinger equation

$$i\hbar \frac{\partial}{\partial t} \Psi(\underline{x}, t) = \hat{H}(\underline{x}) \Psi(\underline{x}, t), \tag{9.1}$$

where $\underline{x}$ denotes the set of all dynamic variables of the particles constituting the system, and $\hat{H}(\underline{x})$ is the Hamiltonian of the system, being the sum of its kinetic and potential energy.

We can try to solve this equation using the method of separation of variables, namely to look for a solution of Eq. 9.1 which can be written in the product form $\Psi(\underline{x}, t) = \varphi(\underline{x})\vartheta(t)$. Substituting this expression for the wave function in Eq. 9.1 and dividing both sides by $\varphi(\underline{x})\vartheta(t)$, we get

$$i\hbar \frac{\dot{\vartheta}(t)}{\vartheta(t)} = \frac{\hat{H}(\underline{x})\varphi(\underline{x})}{\varphi(\underline{x})}, \tag{9.2}$$

where we have used the standard dot-notation for the time derivative $\dot{\vartheta}(t)$. Since the two sides of Eq. 9.2 depend on different variables, the only possibility to satisfy this equation is that both sides are equal to a constant, which we can denote as E. We, thus, obtain two equations for each of the functions in the product. The equation for $\vartheta(t)$ can be directly integrated yielding

$$\vartheta(t) = \vartheta_0 e^{-\frac{i}{\hbar}Et},$$

with ϑ_0 being the initial condition of $\vartheta(t)$. The equation for $\varphi(\underline{x})$ is the time-independent Schrödinger equation

$$\hat{H}(\underline{x})\varphi(\underline{x}) = E\varphi(\underline{x}), \tag{9.3}$$

which depends on the particular form of the Hamiltonian of the system $\hat{H}(\underline{x})$.

Let us assume for a moment that we can solve Eq. 9.3, i.e., we can find the eigenvalue E and the corresponding eigenfunction $\varphi_E(\underline{x})$. We can, therefore, write the solution of the time-dependent Schrödinger (9.1) as

$$\Psi(\underline{x},t) = \varphi_E(\underline{x})e^{-\frac{i}{\hbar}Et}, \tag{9.4}$$

where we have absorbed the constant ϑ_0 into $\varphi_E(\underline{x})$. Although containing all the information about the evolution of the system, the wave function $\Psi(\underline{x},t)$ itself is not an observable. According to the traditional interpretation of quantum mechanics, the quantity that can be directly related to the outcome of a measurement performed at time t is the probability density, given by the absolute square of the wave function. It is not difficult to see that the wave function in Eq. 9.4 has a rather trivial time dependence in the form of a phase factor $e^{-\frac{i}{\hbar}Et}$ and thus

$$|\Psi(\underline{x},t)|^2 = |\varphi_E(\underline{x})|^2,$$

meaning that the evolution of the system determined by the wave function (9.4) cannot be observed!

This "paradox" can be resolved by realizing that the wave function in Eq. 9.4 is a *particular* solution of the Eq. 9.1 involving only a single eigenstate of the system φ_E and that every linear combination of particular solutions is also a solution of Eq. 9.1. Therefore, the *general* solution of the time-dependent Schrödinger equation is given by the linear combination of all possible particular solutions

$$\Psi(\underline{x},t) = \sum_{i=1}^{\infty} \alpha_i \varphi_i(\underline{x})e^{-\frac{i}{\hbar}E_i t}. \tag{9.5}$$

In the case of a continuous spectrum of the system Hamiltonian, the sum in the above expression has to be replaced by an integration over the energy.

It is not difficult to check that if we take a linear combination of even just two particular solutions, i.e.,

$$\Psi(\underline{x},t) = \alpha\varphi_1(\underline{x})e^{-\frac{i}{\hbar}E_1 t} + \beta\varphi_2(\underline{x})e^{-\frac{i}{\hbar}E_2 t},$$

the probability density of the system will show a time dependence, arising from the interference between the two solutions

$$|\Psi(\underline{x},t)|^2 = |\alpha|^2|\varphi_1(\underline{x})|^2 + |\beta|^2|\varphi_2(\underline{x})|^2 + 2\mathrm{Re}\left\{\alpha^*\beta\varphi_1^*(\underline{x})\varphi_2(\underline{x})e^{-\frac{i}{\hbar}(E_1-E_2)t}\right\}.$$

It is easy to show that the above structure is reproduced also for the expectation value of any Hermitian operator, i.e., for any physical observable. Therefore, in order to get a non-trivial time dependence of an observable, a *wave packet*, i.e., a linear combination of eigenstates corresponding to different eigenenergies, has to be created. Eq. 9.5 also tells us that if we are able to solve the time-*independent* Schrödinger

Eq. 9.3, i.e., to find all relevant eigenstates $\varphi_i(\underline{x})$ and eigenenergies E_i of the system, we will be able to reconstruct its time evolution. It is also worth noting that the time scale of the triggered dynamics is determined by the the energy difference between the populated eigenstates—large energy differences will describe fast dynamics, while the population of close lying states will trigger slow dynamics.

9.1.2 Molecular states and Born-Oppenheimer approximation

Let us now see what kind of eigenstates we have to deal with when we study a molecule. For this purpose, we need to construct a realistic molecular Hamiltonian. To a very good approximation, the electrons and nuclei in a molecule can be taken as point charges and point masses interacting only via the two-body electrostatic (Coulomb) interaction. Although other types of interactions between the particles in a molecule exist (e.g. spin-orbit coupling, spin-spin coupling, finite-nuclear-size effects, etc.), those are much weaker than the Coulomb interaction and rarely play a role in the chemical reactivity.

The molecular Hamiltonian, describing N electrons and M nuclei, has thus the form

$$\hat{H}_{mol} = \underbrace{\sum_{i=1}^{N} -\frac{\hbar^2}{2m_e}\nabla_i^2}_{\hat{T}_e} + \underbrace{\sum_{i=1}^{N}\sum_{j>i}^{N}\frac{e^2}{|\vec{r}_i - \vec{r}_j|}}_{V_{ee}} + \underbrace{\sum_{i=1}^{N}\sum_{\alpha=1}^{M}\frac{-e^2 Z_\alpha}{|\vec{r}_i - \vec{R}_\alpha|}}_{V_{en}}$$

$$+ \underbrace{\sum_{\alpha=1}^{M} -\frac{\hbar^2}{2M_\alpha}\nabla_\alpha^2}_{\hat{T}_n} + \underbrace{\sum_{\alpha=1}^{M}\sum_{\beta>\alpha}^{M}\frac{e^2 Z_\alpha Z_\beta}{|\vec{R}_\alpha - \vec{R}_\beta|}}_{V_{nn}} \tag{9.6}$$

where $\vec{r}$ refers to the electronic coordinate, while $\vec{R}$ to the nuclear one. Correspondingly, ∇_i acts on the electronic coordinates and ∇_α on the nuclear ones. We have used also the following standard notations: e is the electronic (elementary) charge, m_e is the electronic mass, and Z_α and M_α are the atomic number and mass, respectively, of nucleus α.

With the above Hamiltonian we need to solve the time-independent Schrödinger equation

$$\hat{H}_{mol}\Psi(\underline{r},\underline{R}) = \mathcal{E}\Psi(\underline{r},\underline{R}), \tag{9.7}$$

where with $\underline{r}$ and $\underline{R}$ we have denoted the full set of electronic $\{\vec{r}_1,\ldots,\vec{r}_N\}$ and nuclear $\{\vec{R}_1,\ldots,\vec{R}_M\}$ coordinates, respectively. Note that the eigenfunction $\Psi(\underline{r},\underline{R})$ is defined in the full coordinate space. Unfortunately, the above equation can be solved in an analytically closed form only for systems containing two particles, e.g., the hydrogen atom, containing a single electron and a single nucleus. Thus, in principle, even the treatment of the simplest molecule (H_2), relies on some approximations.

The most important approximation is based on the fact that the electrons are much lighter than the nuclei (the proton itself is about 1836 times heavier than the electron) and thus in order to have comparable momenta the electrons have to move much faster. Therefore, in a dynamical sense, the electrons can be regarded as particles that follow the nuclear motion *adiabatically*, meaning that they adapt to the new nuclear positions instantaneously, without requiring a finite relaxation time. This allows us to decouple the electronic and the nuclear motion. Thus, we may consider the time-independent Schrödinger equation for the electrons only, at a fixed nuclear geometry $\underline{R}$

$$\hat{H}_e \Phi(\underline{r}; \{\underline{R}\}) \equiv (\hat{T}_e + V_{ee} + V_{en}) \Phi(\underline{r}; \{\underline{R}\}) = E(\underline{R}) \Phi(\underline{r}; \{\underline{R}\}). \tag{9.8}$$

Obviously, $\hat{H}_e$ is an operator in the electronic space that depends parametrically on the nuclear coordinates $\underline{R}$ (see, Eq. 9.6). This permits us also to write the full molecular wave function as a product of an electronic $\Phi(\underline{r}; \{\underline{R}\})$ and a nuclear $\chi(\underline{R})$ wave functions

$$\Psi(\underline{r}, \underline{R}) = \Phi(\underline{r}; \{\underline{R}\}) \chi(\underline{R}).$$

If we now insert this product form of the wave function into the Schrödinger equation for the full molecular Hamiltonian (9.7) and use Eq. 9.8, we obtain

$$\hat{H}_{mol}(\Phi\chi) = [E(\underline{R}) + V_{nn}](\Phi\chi) + \sum_\alpha -\frac{\hbar^2}{2M_\alpha} \left[\Phi\nabla_\alpha^2 \chi + 2(\nabla_\alpha \Phi) \cdot (\nabla_\alpha \chi) + \chi \nabla_\alpha^2 \Phi \right]. \tag{9.9}$$

The last two terms in Eq. 9.9 involve derivatives of the electronic wave function with respect to the nuclear coordinates. These terms are proportional to powers of the ratio of the electron to nuclear mass and are typically much smaller than the remaining terms. The neglect of these terms is known as the *Born-Oppenheimer approximation*. It brings Eq. 9.9 into a Schrödinger equation for the nuclear wave function χ (after dividing both sides by the electronic wave function Φ)

$$[\hat{T}_n + \underbrace{E(\underline{R}) + V_{nn}}_{V(\underline{R})}]\chi(\underline{R}) = \mathcal{E}\chi(\underline{R}). \tag{9.10}$$

The nuclear wave function is thus an eigenfunction of an adiabatic Hamiltonian, whose effective potential $V(\underline{R})$ is composed by the averaged electron-electron and electron-nuclear forces, contained in $E(\underline{R})$, and the instantaneous nuclear-nuclear repulsion V_{nn}. This effective potential is called electronic potential-energy surface (PES).

Since its introduction in 1927, the Born-Oppenheimer approximation has become the cornerstone in the theory of molecules [for a comprehensive review, see (Cederbaum, 2004)]. One should be aware that even the notion of molecular *electronic states*, so widely used in electron spectroscopy, is a consequence of the approximate separability of the electronic and nuclear motion, i.e., of the Born-Oppenheimer approximation. Moreover, in many cases it is possible to separate the vibrational and the rotational motion of the nuclei, thus bringing the molecular wave function into a product of electronic, vibrational, and rotational states. The adequacy of this approximation has been confirmed by numerous molecular spectroscopy data. The molecular states of even the simplest diatomic molecule reveals the structure, schematically shown in Fig. 9.1. The transitions between the valence electronic states

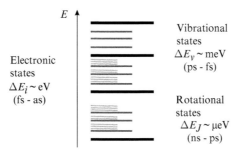

Fig. 9.1 Schematic representation of a typical structure of molecular eigenstates.

lie typically in the optical to ultraviolet (UV) spectral range. Vibrational transitions occur between different vibrational levels of the same electronic state and lie usually in the infrared (IR) spectrum. Rotational transitions, respectively, occur mostly between rotational levels of the same vibrational state and belong to the microwave regime.

As we saw already in Sec. 9.1.1, there is an intimate relation between the energy spacing of the eigenstates and the time scale of the dynamics triggered by their coherent population. The spectral ranges of the different transitions, therefore, tell us the respective time scales. The electron dynamics take place typically in the femtosecond (10^{-15} s) to attosecond (10^{-18} s) regime. The nuclear vibrations are in the pico- (10^{-12} s) to femtosecond time scale, while the nuclear rotations are usually a few orders of magnitude slower.

One has to remember, however, that this decoupling of the electronic and nuclear degrees of freedom is only an approximation. Although accurate in many cases, the Born-Oppenheimer approximation may fail when the electronic PESs become very close to each other and even break down completely when the PESs become degenerate forming the so-called conical intersections (Domcke et al., 2004). It has to be noted that for polyatomics with their dense electronic spectrum and a large number of nuclear degrees of freedom, such conical intersections of electronic states, where the electronic and the nuclear motion is strongly coupled, appear very often. However, even when it fails, the Born-Oppenheimer approximation remains a useful concept and a reference on which we try to systematically improve.

9.1.3 Describing correlated electrons

In addition to its transparent physical picture, the Born-Oppenheimer approximation greatly simplifies the computational treatment of molecules, separating the electronic and the nuclear problem. By solving the electronic problem at a series of different nuclear configurations, we can build the effective potential $V(\underline{R})$ seen by the nuclei. These PESs can then be used to integrate the nuclear equation and study the chemical reactivity of the molecule. Many of the molecular properties, however, can be deduced by solving only the electronic problem at the most stable nuclear configuration (the equilibrium geometry), which is the objective of the vast field called electronic-structure theory. Let us now briefly discuss the electronic problem, or how one approaches Eq. 9.8, also known as clamped-nuclei Schrödinger equation.

The electronic-structure theory is a huge and fast developing field. A large number of methods for solving the molecular electronic problem have been proposed over the past few decades. Here we are not aiming at giving an overview of the existing methods, but rather to give some flavor about the problems, and introduce some concepts and notations that will be used in the next section.

Even at fixed nuclear geometry, the electronic problem cannot be solved analytically if we have to deal with more than one electron. The "problematic" part in the many-electron Hamiltonian is the non-local V_{ee} term which introduces correlation in the electronic motion. The simplest way to deal with this difficulty, suggested already in the early days of quantum theory, is the so-called orbital approximation, or the Hartree-Fock (HF) model [see, e.g., (Szabo and Ostlund, 1989; McWeeny, 1989; Helgaker, Jørgensen and Olsen, 2000)], where each electron is assumed to move *independently* in the mean field of the remaining electrons. In the HF approximation, the Hamiltonian of the system becomes a sum of one-electron Hamiltonians, and the many-electron wave function—a product of one-electron functions $\varphi(\vec{r})$ called *orbitals*. To incorporate the indistinguishability of the electrons and the requirement of the Pauli exclusion principle, this product has to be antisymmetrized leading to the following approximate form of the electronic ground-state of the molecule, known as Slater determinant

$$|\Phi_0\rangle \equiv \Phi_0(\vec{r}_1, \ldots, \vec{r}_N) = \frac{1}{\sqrt{N!}} \begin{vmatrix} \varphi_1(\vec{r}_1) & \cdots & \varphi_N(\vec{r}_1) \\ \vdots & \ddots & \vdots \\ \varphi_N(\vec{r}_1) & \cdots & \varphi_N(\vec{r}_N) \end{vmatrix}, \tag{9.11}$$

where we have introduced the standard ket notation $|\cdot\rangle$ for states.

The incorporation of the Pauli exclusion principle actually represents the simplest form of electron correlation and is usually referred to as Fermi correlation. Although this partially correlates the electrons, the Hartree-Fock method does not take into account their correlated motion. That is why, the Hartree-Fock wave function is traditionally regarded as a *fully uncorrelated* one and the electron correlation is correspondingly defined as the difference between the exact and the Hartree-Fock solution of the many-electron problem. It is worth noting that although the Hartree-Fock approximation substantially simplifies the electronic problem, it leads to non-linear integro-differential equations that are difficult to solve for polyatomic systems. Thus, in practice, the orbitals are expanded in finite basis sets and the equations are solved iteratively—the so-called self-consistent field (SCF) procedure.

The Hartree-Fock approach yields a very transparent and intuitive picture of the molecule. The electrons occupy orbitals[1] that are grouped in shells. This concept is so vastly used in the chemical literature and so helpful in extracting and predicting some molecular properties that, as in the case with the Born-Oppenheimer approach, one often forgets that it is only an approximation based on a rather "crude" independent-particle model. The way the Hartree-Fock method addresses the many-electron problem leads to an over-estimation of the electron-electron repulsion and in many

---

[1] In principle, each orbital can accommodate two electrons with opposite spin.

cases to completely wrong results. It is, therefore, desirable to go beyond the HF picture and improve the way we treat the electron-electron interaction.

There is a large number of approaches to this problem developed over the years. An important class of such approaches, the so-called post-Hartree-Fock or *ab initio* methods, is based on the fact that the many-electron wave function can be improved by linearly combining determinants constructed from the eigenfunctions of the Hartree-Fock Hamiltonian. The most prominent example of this concept is the so-called configuration interaction (CI) method. In the CI approach [see, e.g., (Szabo and Ostlund, 1989; Helgaker, Jørgensen and Olsen, 2000)], the many-electron wave function is expanded in excited determinants, or configurations

$$|\Psi\rangle = c_0|\Phi_0\rangle + \sum_{a,i} c_i^a|\Phi_i^a\rangle + \sum_{a<b,i<j} c_{ij}^{ab}|\Phi_{ij}^{ab}\rangle + \cdots . \tag{9.12}$$

In what follows, we will use $|\Psi\rangle$ as the exact many-electron wave function.

The first term in Eq. 9.12, or the ground-state configuration $|\Phi_0\rangle$, is the solution of the HF problem (9.11), while the remaining terms are generated by performing single, double, etc., excitations from the ground-state configuration. These excitation determinants are constructed in the following way. The HF eigenvalue problem yields a set of orbitals $\{\varphi_p(\vec{r})\}$ and orbital energies $\{\varepsilon_p\}$. In principle, there are infinitely many solutions of the HF equations, but in practice, since the equations are solved using a finite set of basis functions, we obtain a finite number of orbitals and orbital energies, typically substantially larger than the number of electrons in the molecule N. The orbitals corresponding to the lowest N orbital energies are called occupied orbitals (we will use indices i, j, k, ... to denote them), while the remaining are called unoccupied, or virtual orbitals (we will use indices a, b, c, ... to denote them[2]). The ground-state configuration $|\Phi_0\rangle$ is, therefore, the Slater determinant composed of all occupied orbitals, while the singly-excited determinant $|\Phi_i^a\rangle$ is constructed by removing the occupied orbital φ_i from $|\Phi_0\rangle$ and adding the virtual orbital φ_a, which can be regarded as an approximation to the excited sate of the system in which an electron from orbital φ_i is promoted to orbital φ_a. The higher excitation classes are constructed analogously.

The CI approach consists of finding the expansion coefficients c in Eq. 9.12 which is done variationally. Being variational, CI provides an exact solution of the many-electron problem within the one-particle basis set used. Unfortunately, even for small molecules and moderate size basis sets, the number of configurations and the respective size of the CI matrix is enormous, making the so-called full-CI (when all possible configurations within the basis set used are taken in expansion (9.12)) accessible only for few-electron systems. One usually has to truncate the expansion (9.12) after a certain excitation class which, unfortunately, leads to the so-called size-consistency problem, making the quality of the description to decrease with the increase of the system size.

---

[2] We will reserve p, q, r, ... for general indices.

Fortunately, there are other approaches that do not have this drawback, like for example the very successful coupled-cluster (CC) methods [see, e.g., (Bartlett and Musiał, 2007)], or the methods based on perturbation theory, like the Møller-Plesset (MP) methods [see, e.g., (Cremer, 1989)]. Although based on different approaches to the electron-correlation problem, the approximations for the many-electron wave function obtained by these methods can be represented in the form of Eq. 9.12 and analyzed in terms of determinants representing different electronic configurations. In this respect the CI expansion introduces an important concept in which the electron correlation manifests in the electronic states. We can still keep the orbital picture for the electronic structure, but the many-body or the correlation electronic effects appear as additional excitations of the electrons in the one-body or orbital space.

In the next section we will concentrate on describing the effects and manifestations of the pure electron dynamics in molecules. As we discussed above, for observing non-trivial dynamics a wave packet or a linear combination of eigenstates has to be created. These can be the electronically excited states of the neutral molecule or of the cation, if the external perturbation triggering the dynamics has ionized the system. The CI and the methods briefly mentioned above are all applicable not only to systems in their neutral and charged ground states, but can also be used to obtain excited states. However, in practice, only the lowest few exited states can be obtained with a reasonable computational cost. A very successful alternative for obtaining properties of an excited or ionized system are the Green's function (GF) or propagator methods. An important advantage of the GF based methods, compared to the mentioned above wave function based ones, is that they allow one to compute the entire spectrum of ionic or excited states of the system in a single run. Several GF approaches to the electronic problem are known in the literature (Linderberg and Öhrn, 2004). Here we would like to mention the class of GF methods, known as algebraic diagrammatic construction (ADC) schemes (Cederbaum, 1998), which are very suitable for the problem at hand. Computationally, the ADC methods are relatively inexpensive and can be used to treat system with a few tens of atoms.

9.2 Correlated electron dynamics following ionization

We saw until know that, due to the usually large difference in the electron and nuclear dynamics time scales, we can approximately decouple the electronic and nuclear motion and treat them separately. Being much faster, the electron motion can thus be described assuming that the nuclei are not moving. It is therefore interesting to examine the situation of pure electron dynamics which take place before the nuclei have time to move substantially and thus may be considered as fixed.

In this section we will present the physics behind the electron dynamics triggered by ionization, as well as the formalism for its description. We note, however, that similar effects appear also after a broadband excitation of a molecule and one can use similar formalism to study them. As we will see, very often due to the electron correlation the removal of an electron from a molecular system will lead to the creation of an electronic wave packet, or, in other words, to pure electron dynamics. The hole charge created upon the ionization may migrate thoughout the system due to the quantum coherence between the populated cationic states. To distinguish this process from the usual charge transfer, which is driven by the slower nuclear dynamics, it was termed *charge migration* (Cederbaum and Zobeley, 1999). Let us see now how we can describe such dynamics.

9.2.1 The hole density

A possible way to trace and analyze the electron dynamics following ionization of a molecular system is to compute the evolution of the positive charge left after the removal of the electron. This can be done by evaluating the density of the initially created hole, or the hole density, which is given by (Cederbaum and Zobeley, 1999)

$$Q(\vec{r},t) := \langle \Psi_0 | \hat{\rho}(\vec{r},t) | \Psi_0 \rangle - \langle \Phi_i | \hat{\rho}(\vec{r},t) | \Phi_i \rangle = \rho_0(\vec{r}) - \rho_i(\vec{r},t). \tag{9.13}$$

In the above definition $|\Psi_0\rangle$ is the ground state of the neutral molecule, $\hat{\rho}$ is the local density operator, and $|\Phi_i\rangle$ is the initially prepared cationic state. The first term in Eq. 9.13, ρ_0, is the time-independent ground-state density of the neutral system and the second one, ρ_i, is time-dependent, since we assume that $|\Phi_i\rangle$ is not an eigenstate of the cation. In Heisenberg picture, the time-dependent part reads

$$\rho_i(\vec{r},t) = \langle \Phi_i | e^{i\hat{H}t} \hat{\rho}(\vec{r}) e^{-i\hat{H}t} | \Phi_i \rangle, \tag{9.14}$$

where $\hat{H}$ is the cationic Hamiltonian, or the clamped-nuclei Hamiltonian of the system with one electron less. In what follows, we will use atomic units, $e = m_e = \hbar = 1$. Using the resolution of the identity within the complete set of cationic eigenstates $\{|I\rangle\}$, i.e., $1 = \sum_I |I\rangle\langle I|$, and inserting it in Eq. 9.14, one gets

$$\rho_i(\vec{r},t) = \sum_{I,J} \langle \Phi_i | I \rangle \langle I | e^{i\hat{H}t} \hat{\rho}(\vec{r}) e^{-i\hat{H}t} | J \rangle \langle J | \Phi_i \rangle$$

$$= \sum_{I,J} x_I^* \rho_{IJ}(\vec{r}) x_J e^{-i(E_J - E_I)t},$$

where E_I is the ionization energy corresponding to the state $|I\rangle$, $x_I = \langle \Phi_i | I \rangle$ is the transition amplitude with respect to the initial cationic state $|\Phi_i\rangle$, and $\rho_{IJ} = \langle I | \hat{\rho}(\vec{r}) | J \rangle$ is the charge density matrix between the states $|I\rangle$ and $|J\rangle$. Without loss of generality, one can assume the quantities x_I and ρ_{IJ} to be real, bringing the expression for $\rho_i(\vec{r},t)$ in the following form

$$\rho_i(\vec{r},t) = \sum_{I,J} x_I \rho_{IJ}(\vec{r}) x_J \cos\left[(E_J - E_I)t\right].$$

To evaluate the hole density we can use the second-quantization representation of the density operator within a one-particle basis (orbitals)

$$\hat{\rho}(\vec{r}) = \sum_{p,q} \varphi_p^*(\vec{r}) \varphi_q(\vec{r}) \hat{a}_p^\dagger \hat{a}_q,$$

with $\hat{a}_p^\dagger$ and $\hat{a}_p$ being the corresponding creation and annihilation operators, respectively. The operator $\hat{a}_p^\dagger$ creates an electron in orbital φ_p and the operator $\hat{a}_p$ destroys an electron (creates a hole) in orbital φ_p.

The operators $\hat{a}_p^\dagger$ and $\hat{a}_p$ anticommute, i.e., $\hat{a}_p^\dagger \hat{a}_q + \hat{a}_q \hat{a}_p^\dagger = \delta_{pq}$, and have the following properties when acting on the N-electron HF ground state $|\Phi_0\rangle$

$$\hat{a}_p^\dagger|\Phi_0\rangle = \begin{cases} |\Phi^p\rangle, & p \in \text{virt.} \\ 0, & p \in \text{occ.} \end{cases}, \qquad \hat{a}_p|\Phi_0\rangle = \begin{cases} 0, & p \in \text{virt.} \\ |\Phi_p\rangle, & p \in \text{occ.} \end{cases},$$

where $|\Phi^p\rangle$ is an $(N+1)$-electron determinant with an additional electron in the virtual orbital φ_p, and $|\Phi_p\rangle$ is an $(N-1)$-electron determinant with a hole (an electron missing) in the occupied orbital φ_p.

Within this representation, the hole density, Eq. 9.13, takes the following form (Breidbach and Cederbaum, 2003)

$$Q(\vec{r},t) = \sum_{p,q} \varphi_p^*(\vec{r})\varphi_q(\vec{r})N_{pq}(t), \tag{9.15}$$

where the matrix $\mathbf{N}(t)$, referred to as hole-density matrix, is given by

$$N_{pq}(t) = \langle\Psi_0|\hat{a}_p^\dagger\hat{a}_q|\Psi_0\rangle - \langle\Phi_i|\hat{a}_p^\dagger\hat{a}_q|\Phi_i\rangle + \sum_{I,J} x_I\langle I|\hat{a}_p^\dagger\hat{a}_q|J\rangle x_J\left[1 - \cos\left((E_I - E_J)t\right)\right]. \tag{9.16}$$

Diagonalizing the hole-density matrix $\mathbf{N}(t)$ at each time point t we get

$$Q(\vec{r},t) = \sum_p |\tilde{\varphi}_p(\vec{r},t)|^2 \tilde{n}_p(t), \tag{9.17}$$

where $\tilde{\varphi}_p(\vec{r},t)$ are called *natural charge orbitals*, and $\tilde{n}_p(t)$ are their *hole-occupation numbers*. The hole-occupation number, $\tilde{n}_p(t)$, contains the information what part of the initially created hole charge is in the natural charge orbital $\tilde{\varphi}_p(\vec{r},t)$ at time t. At each instant of time, the natural charge orbitals are different linear combinations of the initial orbital basis $\{\varphi_p(\vec{r})\}$.

From the definition of the hole density, Eq. 9.13, it is clear that $Q(\vec{r},t)$ is normalized at all times, namely

$$\int d\vec{r}\, Q(\vec{r},t) = 1,$$

which leads to the following relation for the hole-occupation numbers

$$\sum_p \tilde{n}_p(t) = 1.$$

The hole-occupation numbers, together with the natural charge orbitals, are central quantities in the analysis and interpretation of the multielectron dynamics taking place after the removal of an electron.

The description of the electron dynamics following ionization via the above defined hole density implies that the process of ionization and the rearrangement of the electronic cloud during this process are not taken into account. In other words, we suppose that the ionized electron is removed from the system on a much shorter time scale than that of the triggered electron dynamics and, therefore, the initially created ionic state can be described by a separable manyelectron wave function, neglecting the interaction between the ionized electron and the remaining ionic core. This is the so-called sudden approximation. Such a situation can be realized when the ionization is performed, for example, by a high-energy photon (well above the corresponding ionization threshold) such that the removed electron has a high kinetic energy and thus leaves rapidly the interaction volume.

9.2.2   Choice of cationic basis and initial state

If we assume a weakly correlated ground state, i.e., $|\Psi_0\rangle \approx |\Phi_0\rangle$, the first term in Eq. 9.16 will be simply $n_p \delta_{pq}$, with n_p being the occupation number of orbital $\varphi_p(\vec{r})$[3]. For the ease of presentation and for keeping the discussion as transparent as possible, we will assume that the system is ionized out of a weakly correlated ground state. We note, however, that the methodology presented is not restricted to such cases.

Eq. 9.16 is further simplified if we suppose that the initial cationic state is created via the removal of an electron from a single orbital, i.e., $|\Phi_i\rangle = \hat{a}_i|\Phi_0\rangle$. This particular choice of initial state will reduce the second term in Eq. 9.16 to $n_p \delta_{pq} - \delta_{pi}\delta_{qi}$, bringing the hole-density matrix to the following simpler form

$$N_{pq}(t) = \delta_{pi}\delta_{qi} + \sum_{I,J} x_I \langle I|\hat{a}_p^\dagger \hat{a}_q|J\rangle x_J \left[1 - \cos\left((E_I - E_J)t\right)\right]. \qquad (9.18)$$

The initial state can, of course, be constructed such that it corresponds to the removal of an electron from a linear combination of HF orbitals. This allows one to reproduce the hole density of practically every particular initial vacancy. In this case the first term in Eq. 9.18 will be $\sum_r w_r \delta_{pr}\delta_{qr}$ with w_r being the weight with which each orbital contributes to the initial hole density and satisfying the normalization condition $\sum_r w_r = 1$. However, to avoid investigating many linear combinations of HF orbitals of interest in what follows we will assume that the initial hole is created by removal of an electron from a specific HF orbital. In this way we can unambiguously identify the basic mechanisms leading to charge migration (see Sec. 9.2.3). Since the time-dependent Schrödinger equation which governs the electron dynamics is a linear equation, these mechanisms are also operative when other choices of initial state are used.

As we mentioned in the preceding section, we suppose that the ionization is performed on an ultrashort time scale. Irrespective of the particular way of removing the electron, in the sudden-ionization limit the remaining electrons will not have time to relax and therefore one may assume that the electron is removed from a single molecular orbital, being a result of an independent-particle model. Since the Hartree-Fock approximation provides the best independent-particle theory to describe

---

[3] As the name suggests, $n_p = 1$ if orbital φ_p is occupied, and $n_p = 0$ if φ_p is unoccupied.

ionization[4], we may assume that the initial hole created upon the sudden ionization is described favorably by a HF-orbital. We have to note that the time needed for the other electrons to respond to a sudden creation of a hole is about 50 asec (Breidbach and Cederbaum, 2005). It was shown (Breidbach and Cederbaum, 2005) that this time is universal, i.e., it does not depend on the particular system, and as such can be interpreted as the characteristic time scale of electron correlation. Thus, in practice, sudden ionization is equivalent to ionization performed faster than the electron correlation.

Let us now discuss the question of the choice of cationic states $|I\rangle$ needed for constructing the hole-density matrix. Since the exact cationic states and energies are not at our disposal, appropriate approximations for these quantities have to be used.

The simplest approximation for obtaining the hole density is to use the Hartree-Fock method. For the ground state of the system one can use the single-determinant HF ground state $|\Phi_0\rangle$, while for the cationic states the one-hole (1h) configurations with respect to $|\Phi_0\rangle$. In this case, the cationic states are constructed by removing an electron from an occupied orbital, i.e., $|I\rangle = \hat{a}_k|\Phi_0\rangle$. The ionization energies of these states are simply the HF orbital energies ε_k [this is known as Koopmans' theorem (Szabo and Ostlund, 1989)]. Therefore, the hole-density matrix will get the form

$$N_{pq}(t) = \delta_{pi}\delta_{qi} + x_p x_q \left[1 - \cos\left((\varepsilon_p - \varepsilon_q)t\right)\right] n_p n_q. \tag{9.19}$$

We see that because of the factor $n_p n_q$, only occupied orbitals participate in possible charge dynamics. However, the hole density will be time-dependent only if the involved orbitals have non-vanishing transition amplitudes x_p between the initially prepared cationic state and the cationic basis used. Within the Hartree-Fock approach and Koopmans' approximation, the transition amplitudes reduce to Kronecker delta symbols $(x_p = \langle\Phi_i|I\rangle = \langle\Phi_0|\hat{a}_i^\dagger \hat{a}_p|\Phi_0\rangle = \delta_{ip})$ and, therefore, the time-dependent term in Eq. 9.19 vanishes. As the combined Hartree-Fock and Koopmans' approximation defines the situation in which the electron correlation effects are exactly absent, this result leads to the important statement that *in the absence of electron correlation and relaxation the hole charge will stay in the orbital in which it is originally created and no charge migration will take place* (Cederbaum and Zobeley, 1999). In other words, the charge-migration phenomenon that we will discuss in the following sections is *solely* due to the many-body effects, namely the electron correlation and electron relaxation.

One possibility to include many-body effects in the construction of cationic states $|I\rangle$ is to expand the wave function in electronic configurations as it is traditionally done in the configuration-interaction formalism discussed in Sec. 9.1.3. Using the HF- based configurations, the CI expansion of the cationic state reads

$$|I\rangle = \sum_j c_j^{(I)} \hat{a}_j |\Phi_0\rangle + \sum_{a,k<l} c_{akl}^{(I)} \hat{a}_a^\dagger \hat{a}_k \hat{a}_l |\Phi_0\rangle + \cdots, \tag{9.20}$$

---

[4] With the Hartree-Fock theory the many-body corrections to the ionization energies begin to contribute at a higher order of perturbation theory than with other choices of independent-particle approximation (Szabo and Ostlund, 1989).

where $c^{(I)}$'s are expansion coefficients. The terms $\hat{a}_j|\Phi_0\rangle$ are the one-hole config-urations mentioned above, since one electron has been removed from the corre-sponding occupied orbital, the terms $\hat{a}_a^{\dagger}\hat{a}_k\hat{a}_l|\Phi_0\rangle$ are referred to as two-hole-one-particle (2h1p) configurations, indicating that in addition to the removal of one electron another one is excited to a virtual orbital, and so forth.

Since the states $|I\rangle$ are eigenstates of the cationic Hamiltonian, they can be chosen orthonormal, $\langle I|J\rangle = \delta_{IJ}$. Using the orthonormality condition and the standard second-quantization formalism of creation and annihilation operators, the term $\langle I|\hat{a}_p^{\dagger}\hat{a}_q|J\rangle$ of Eq. 9.18 can easily be evaluated. For every pair of $|I\rangle$ and $|J\rangle$ it represents a matrix whose elements are functions of the expansion coefficients $c^{(I)}$ and $c^{(J)}$. The explicit form of this matrix can be found in (Breidbach and Cederbaum, 2003).

As we mentioned in Sec. 9.1.3, the CI method is only one of the approaches that can be used for obtaining the cationic eigenstates. In fact, it is not a very practical one since, as a rule, the CI calculations are very demanding, due to the extended configuration space involved, and the lack of size-consistency represents an additional obstacle for obtaining meaningful results in applications to large systems. As noted, the GF methods like ADC are especially suitable here. The CI form of the cationic states, Eq. 9.20, is, however, very transparent and helpful for analysis, as we will see in the next section.

A final comment is in order. Until now we have assumed that one can diagonalize the cationic Hamiltonian matrix and using the eigenstates $|I\rangle$ and eigenenergies E_I to compute the hole density via Eqs. 9.15–9.17. However, depending on the size of the system and the one-particle basis set used, the size of the Hamiltonian secular matrix can become extremely large making the full-diagonalization approach prohibitively expensive. An alternative approach is to use a direct-propagation technique, a method known as *multielectron wave-packet propagation* (Kuleff, Breidbach, and Cederbaum, 2005). The method relies on constructing the many-body cationic Hamiltonian using a Green's function approach and performing the direct propagation of the initial state chosen with the help of the short-iterative Lanczos algorithm. With this approach, systems with more than 100 correlated electrons can be studied.

9.2.3 Basic mechanisms

The general case of electron dynamics solely driven by electron correlation is difficult to understand due to the many effects that can contribute. This makes sometimes the numerical calculations based on the methodology described above difficult to analyze. For that reason, it is useful to examine cases in which the dynamics are governed by a single mechanism. It appeared that there is an intimate relationship between the ionization spectrum of the system and electron dynamics that can be expected after removing of an electron from a particular orbital (Breidbach and Cederbaum, 2003). That is why it is illuminating to take a closer look at a typical ionization spectrum.

A typical ionization spectrum of a molecule computed at fixed nuclear geometry is given in Fig. 9.2. It consists of vertical lines, where each line represents a cationic eigenstate $|I\rangle$, see Eq. 9.20, positioned at the corresponding ionization energy E_I. The intensity, or the height, of each line is given by the square of the transition amplitude,

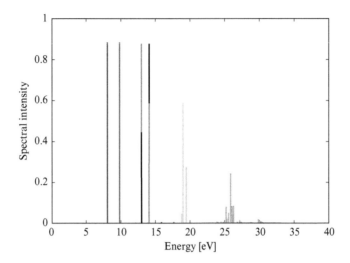

Fig. 9.2 Typical ionization spectrum of a molecule. Each vertical line is related to a cationic eigenstate and is located at the corresponding ionization energy. The height of the lines, or the spectral intensity, is given by the square of the transition amplitude. The different shades in the lines correspond to the contributions of different 1h configurations, describing an electron missing from a particular orbital.

$x_I = \langle I|\hat{a}_i|\Psi_0\rangle$, a quantity closely related to the partial-channel ionization cross section (Cederbaum et al., 1986). Within the assumption of weakly correlated ground state, i.e., $|\Psi_0\rangle \approx |\Phi_0\rangle$, it is clear that only the 1h configurations $\hat{a}_k|\Phi_0\rangle$ will contribute to the transition amplitude x_I.

Without correlation effects in the ion, i.e., within an independent-particle model, the spectrum will consist of a finite series of lines, one for every occupied in the ground state orbital φ_i, with intensities equal to 1, and positioned at the corresponding orbital energies ε_i (Koopmans' theorem). If the correlation effects are weak—and this is typically the case in the outer-valence shells of the system—the ionization spectrum will consist of *main lines* which have large overlap with the 1h configurations (see the 4 lowest in energy lines in Fig. 9.2). In this case the molecular orbital picture is still valid.

If the correlation effects are stronger, beside the main line *satellite lines* may appear (see the group of lines just below 20 eV in Fig. 9.2) representing excitations on top of the ionization. The intensities of the satellite lines are weaker than those of the main lines, since they correspond to cationic states that are dominated by 2h1p configurations and have only small or moderate overlap with the 1h configurations. Two types of satellites can be distinguished (Cederbaum et al., 1986): *relaxation satellites*, where at least one of the two holes in the 2h1p configuration is identical to the 1h orbital of the main line, and *correlation satellites*, where both holes differ from the 1h orbital of the main line. For completeness, we mention that there is a third type of satellites, the *ground-state-correlation satellites*, stemming from the correlation effects present in the ground state of the neutral, but we will not discuss them further here, because if the ground-state correlation effects are weak such kind of states will have very small intensities, i.e., will be very weakly populated in the ionization process.

In the inner valence, where the correlation effects are strong, the distinction between the main lines and the satellites ceases to exist and the spectrum becomes a quasi-continuum of lines with small to moderate intensities (see the group of lines above 25 eV in Fig. 9.2). This phenomenon is known as *breakdown of the molecular orbital picture of ionization* (Cederbaum et al., 1986).

Depending on the structure of the ionic states involved, three basic mechanisms of charge migration have been identified (Breidbach and Cederbaum, 2003): the hole-mixing case, the dominant-satellite case, and the breakdown-of-the-molecular-orbital-picture case. In what follows we will briefly describe them.

(*i*) **Hole-mixing case.** For simplicity, we will consider the two-hole mixing, i.e., the situation when two lines in the spectrum correspond to ionic states which are linear combinations of two 1h configurations. In Fig. 9.2 these are the third and the fourth line where the contributions of the different 1h configurations are shown with different shades of gray—dark gray for, say, a hole in orbital φ_i, and light gray for a hole in φ_j. These two states can be written as

$$|I\rangle = \alpha\,\hat{a}_i|\Phi_0\rangle + \beta\,\hat{a}_j|\Phi_0\rangle,$$

$$|J\rangle = \beta\,\hat{a}_i|\Phi_0\rangle - \alpha\,\hat{a}_j|\Phi_0\rangle,$$

where, due to the orthonormality of the states, the two coefficients α and β satisfy the equation $\alpha^2 + \beta^2 = 1$. In this idealized case, if we create the initial hole in one of the orbitals, say, φ_i, it can be shown easily (Breidbach and Cederbaum, 2003) that the hole-occupation numbers will take the following form:

$$\tilde{n}_{i/j}(t) = \frac{1}{2} \pm \frac{1}{2}\sqrt{1 - 4(\alpha\beta)^2\sin^2(\omega t)},$$

where ω is the difference between the energies of the two lines in the spectrum, i.e., $\omega = (E_I - E_J)$. the created hole will, therefore, oscillate between the two natural charge orbitals $\tilde{\varphi}_i(\vec{r},t)$ and $\tilde{\varphi}_j(\vec{r},t)$ with a period $T = 2\pi/\omega$. It has to be mentioned again that the hole-occupation numbers give the occupation of the natural charge orbitals, which at each time point are different linear combinations of the HF orbitals φ_i and φ_j. The time-evolution of the hole-occupation numbers for this case is shown in Fig. 9.3, where the contributions of the HF orbitals to the natural charge orbitals are color-coded. We see that at each odd quarter cycle an avoided crossing of the hole occupations $\tilde{n}_i(t)$ and $\tilde{n}_j(t)$ appears and the two natural charge orbitals interchange their character. Therefore, the charge dynamics represent an oscillation between the two HF orbitals φ_i and φ_j involved in the hole mixing. If these two orbitals are localized on two different sites of the molecule, the hole-mixing mechanism will lead to an oscillation of the initially created positive charge between these two sites. A schematic representation of such a situation is shown in Fig. 9.4, where the hole density $Q(z,\,t)$, Eq. 9.17, is plotted, supposing that the two mixed orbitals φ_i and φ_j are localized on the two sides of a long-chain molecule. The quantity $Q(z,\,t)$ has

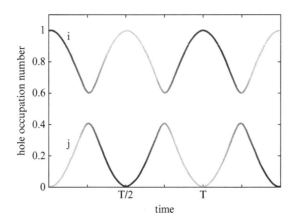

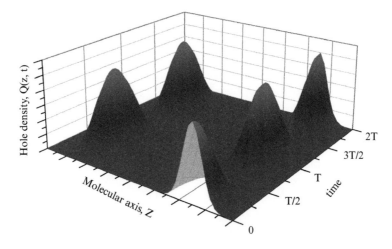

Fig. 9.3 Time-evolution of the hole-occupation numbers in the case of two-hole mixing. The contributions of the two Hartree-Fock orbitals (φ_i and φ_j) to the natural charge orbitals ($\tilde{\varphi}_i$ and $\tilde{\varphi}_j$) are color-coded. As seen, at each odd quarter period an avoided crossing takes place and the character of the two natural charge orbitals is swapped. Within the first half period the initially ionized orbital (φ_i) "loses" its hole charge, which is transferred to the other orbital involved in the hole mixing (φ_j).

Fig. 9.4 Schematic representation of the evolution of the hole density along the chain axes of a long-chain molecule for the case of hole mixing in which the two orbitals involved are localized on the two ends of the molecule.

been obtained from $Q(\vec{r},t)$ by integrating over the remaining axes, perpendicular to the molecular-chain axes z.

(ii) **Dominant-satellite case.** Let us consider the situation when we have two ionic states, a main state and a satellite, both having overlap with the original 1h configuration. In CI language these states can be written as

$$|I_m\rangle = \alpha\,\hat{a}_i|\Phi_0\rangle + \beta\,\hat{a}_a^\dagger\hat{a}_k\hat{a}_l|\Phi_0\rangle, \tag{9.21a}$$

$$|I_s\rangle = \beta \hat{a}_i |\Phi_0\rangle - \alpha \hat{a}_a^\dagger \hat{a}_k \hat{a}_l |\Phi_0\rangle, \tag{9.21b}$$

where again the two coefficients α and β satisfy the equation $\alpha^2 + \beta^2 = 1$.

As was mentioned above, two situations can be distinguished. When all involved occupied orbitals φ_i, φ_k, and φ_l are different, i.e., the case of a correlation satellite, and when one of the holes in the 2h1p configuration is identical to the that of the 1h configuration ($k = i$ or $l = i$), i.e., the case of a relaxation satellite. In the correlation-satellite case, one gets the following expressions for the hole-occupation numbers (Breidbach and Cederbaum, 2003):

$$\tilde{n}_i(t) = 1 - 2(\alpha\beta)^2 (1 - \cos\omega t),$$
$$\tilde{n}_k(t) = \tilde{n}_l(t) = -\tilde{n}_a(t) = 2(\alpha\beta)^2 (1 - \cos\omega t),$$

where $\omega = E_{I_m} - E_{I_s}$ and the initial ionization is performed from orbital φ_i. As seen, the dynamics are again oscillatory with a period determined by the energy differences between the states $|I_m\rangle$ and $|I_s\rangle$. The hole initially localized on orbital φ_i will migrate to the orbital φ_k (or φ_l) accompanied by an excitation from orbital φ_l (or φ_k) to the virtual orbital φ_a. The time-evolution of the hole-occupation numbers in this case is shown in the left panel of Fig. 9.5. One can imagine several different situations in which, depending on the spatial localizations of the four orbitals involved, the net positive charge can migrate between different sites of the system.

In the case of a relaxation satellite, when for example $l = i$ in Eq. 9.21, one observes very different dynamics (Lünnemann, Kuleff, and Cederbaum, 2009). The hole-occupation numbers in this case will be (Lünnemann, Kuleff, and Cederbaum, 2009)

$$\tilde{n}_i(t) = 1,$$
$$\tilde{n}_k(t) = -\tilde{n}_a(t) = 2\alpha\beta \sin^2(\omega t/2).$$

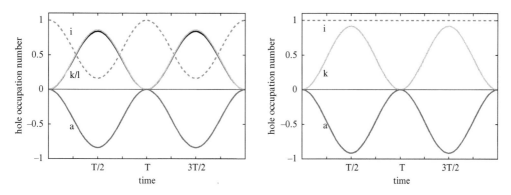

Fig. 9.5 Time-evolution of the hole-occupation numbers in the case of dominant correlation satellite (left graph) and dominant relaxation satellite (right graph). In both cases the coefficients α and β are chosen to be $\sqrt{0.7}$ and $\sqrt{0.3}$, respectively. For convenience, in the dominant-correlation-satellite case the $\tilde{n}_k(t)$ and $\tilde{n}_l(t)$ curves are artificially made to differ slightly from each other.

Therefore, the hole initially localized on orbital φ_i will stay put, while an excitation-deexcitation alternations between orbitals φ_k and φ_a will take place with an oscillation period determined by the energy difference between the main state and the relaxation satellite. The time-evolution of the hole-occupation numbers in this case is shown in the right panel Fig. 9.5.

Although the relaxation satellite does not affect the dynamics of the initially created charge (the initial hole remains stationary), depending on the spatial distribution of the orbitals involved, the dominant relaxation satellite mechanism can still lead to a charge migration via hole screening. If the electron is ejected from a localized orbital i and the unoccupied orbital a is localized in the same region of space then promoting an electron to the orbital a will "cancel out" in physical space the initially created hole in orbital i, i.e., the initial hole will be screened by the excitation $k \to a$. Thus, in the idealized case when the orbitals i and a are localized on one site of the system, while the orbital k is localized on a different site, the dominant relaxation satellite mechanism will lead to oscillations in the real space of the total hole charge between these two parts of the system. This is, for example, the situation observed after core ionization of the molecule nitrosobenzene (Kuleff et al., 2016). We note here that although identified also in valence ionization spectra (Lünnemann, Kuleff, and Cederbaum, 2009), strong relaxation satellites appear quite often after removing a core electron.

(*iii*) **Breakdown of the molecular orbital picture case.** In the inner-valence region of the spectrum, where the quasi-continuum of lines appears, one can distinguish two general cases depending on whether the states are below or above the double ionization threshold of the system. Supposing that the quasi-continuum of states has a Lorentzian shape, in both cases the initially ionized orbital will "lose" its positive charge exponentially with time. This behavior is depicted in Fig. 9.6. If the states are below the double-ionization threshold, the charge will be typically shared among

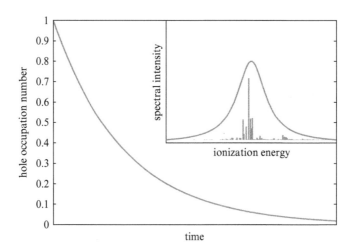

Fig. 9.6 Time-evolution of the hole-occupation number of the initially ionized orbital in the case of a breakdown of the molecular orbital picture. For a continuous Lorentzian line shape in the ionization spectrum, the hole occupation of the initially ionized orbital will decay exponentially.

many other orbitals and at the end of the process will be spread more-or-less uniformly over the whole cation. If the states are above the double ionization threshold, i.e., an electronic decay channel is open, this mechanism will describe the process of emission of a secondary electron. In this case an electron from a higher orbital will fill the initial vacancy and a secondary electron will be ejected from another orbital into the continuum.

Finally, we would like to stress again that if the removal of the electron from a particular orbital populates only a single state, like the pure 1h states shown in Fig. 9.2 (the first two lowest in energy states), the charge will stay put where it is originally created and no electron dynamics can be expected. Such a situation is usually realized after ionization out of the outermost shells of a molecule. We note, however, there are many systems in which the electron correlation is strong enough already in the outermost shell leading to rich dynamical effects.

9.3   Attochemistry

We saw that the ultrafast removal of an electron from a molecule will very often lead to ultrafast dynamics of the created hole. Depending on the number and the type of cationic eigenstates populated, this hole may migrate throughout the system on an ultrafast time scale. There are by now ample *ab initio* computed examples which show that charge migration is a rich phenomenon with many facets that are rather characteristic of the molecule studied [see, e.g., (Kuleff and Cederbaum, 2014)]. A typical example of charge migration is shown in Fig. 9.7, where the evolution of the hole density $Q(\vec{r},t)$, Eq. 9.17, during the first 6 fs after an inner-valence ionization of the oligopeptide metylamidated diglycine (CH_3-NH-Gly-Gly) is depicted (Kuleff, Lünnemann and Cederbaum, 2013). The mechanism underlying this particular process is 5-hole mixing.

In computing and analyzing the charge migration we supposed that the nuclei of the system are fixed, allowing us to study only the evolution of the purely electronic wave packet created upon ionization. Although in many cases this might be quite a good approximation, especially during the first few femtoseconds after the ionization, it is clear that after some time the slower nuclear dynamics will enter the picture

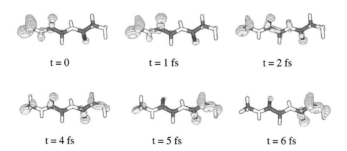

$t = 0$ $t = 1$ fs $t = 2$ fs

$t = 4$ fs $t = 5$ fs $t = 6$ fs

Fig. 9.7 A typical example of charge migration. Snapshots of the evolution of the density of the hole-charge created after ultrafast inner-valence ionization of the oligopeptide CH_3-NH-Gly-Gly. We see that starting from the methylamine site, the charge almost completely migrates to the remote glycine in about 6 fs.

and will inevitably influence the electronic motion. There are many scenarios what the impact of the nuclear motion could be. Particularly attractive is the situation in which the hole, initially created at one end of the molecule, migrates very fast, before the nuclei move, to another remote site where the bonding is weak. In the presence of the hole this weak bond will start to break. Due to its fast migration, the hole may still escape and oscillate back to the initial site. The oscillations may proceed for some time and become slower until the bond eventually breaks. Dissociation of the molecule is a clear possibility to dispose of the excess electronic energy. This could explain the experimental observations of Weinkauf and Schlag [see, e.g., (Schlag et al., 2007)], in which, after a localized ionization on a specific site of a peptide chain (typically the chromophore), a bond breaking on a remote site of the chain occurs. This process was observed in a number of peptide chains with different length and composition. A typical example is shown in Fig. 9.8 for the oligopeptide Ala-Ala-Ala-Tyr (Weinkauf et al., 1997). Due to the large number of nuclear degrees of freedom, the observed bond-breaking cannot be explained by the usual intramolecular vibrational energy redistribution (IVR) process. Being purely statistical, the IVR mechanism will deposit much less energy on the bond in question than the energy needed for its rupture. It was, therefore, suggested that the energy is transferred together with the positive charge, coining the term *charge-directed reactivity*.

An important step towards elucidating the mechanism for the charge-directed reactivity was the time-resolved experiments on the prototypical molecule 2-Phenylethyl-N, N-dimethylamine (PENNA) (Lehr et al., 2005). The molecule and the observed process is schematically presented in Fig. 9.9. Again, after the localized ionization of the chromophore, it was observed that the positive charge is transferred to the remote amine site and the molecule dissociates, breaking the C_1-C_2 bridging bond. The whole process was measured to take 80 ± 28 fs (Lehr et al., 2005). Detailed *ab initio* calculations (Lünnemann, Kuleff and Cederbaum, 2008) have suggested the following explanation of the process in terms of concerted electron-nuclear dynamics. The

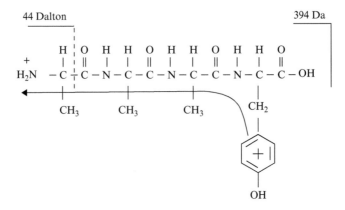

Fig. 9.8 Schematic representation of the observed charge-directed reactivity in the oligopeptide Ala-Ala-Ala-Tyr. After localized ionization of the tyrosine chromophore, the positive charge is transferred to the N-terminal, accompanied by the breaking of the nearest C-C bond. The latter results in the creation of ionic fragment with mass of 44 Da.

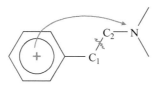

Fig. 9.9 Schematic representation of the observed charge-directed reactivity in the PENNA molecule. After a localized ionization of the chromophore, the positive charge is transferred to the remote amine site and the molecule dissociates, breaking the C_1-C_2 bridging bond. The whole process was measured to take 80 ± 28 fs.

localized initial ionization on the chromophore triggers an ultrafast charge migration and some part of the positive charge starts to bounce between the chromophore and the N-terminal. The period of the pure electronic oscillations was calculated to be less than 8 fs. As time proceeds, the nuclear dynamics enter the picture and with the elongation of the C_1-C_2 bond more and more charge starts to oscillate between the two sites. Eventually, the bond breaks and the charge is trapped at the energetically more favorable N-terminal fragment.

The pure electronic charge migration, therefore, can be regarded as a first step of an effective transfer of the charge from one moiety of the system to another. In the same time, since the electronic motion governs the effective potential seen by the nuclei, it is clear that the charge migration strongly influences the nuclear motion. The molecular site on which the hole-charge is finally trapped will strongly influence the follow-up nuclear rearrangement (for example, the molecule may dissociate or not). Therefore, if we are able to control the charge-migration process, i.e., to modulate this initial step, we might be able to indirectly steer the succeeding charge-transfer process, or the entire charge-directed reactivity. This shows that apart from its fundamental importance, the study of the charge-migration phenomenon may have far reaching practical consequences. If successful, one will be able to speak about "attochemistry" in analogy with the research area known nowadays as femtochemistry (Zewail, 1994). Contrary to the femtochemistry, where with femtosecond pulses one tries to directly influence the movement of the nuclei and guide the evolution of the system into the desired direction, the dream here is that by influencing the electron dynamics via ultrashort (attosecond) laser pulses, one would be able to preselect and put the system on a particular reaction pathway at a very early stage of its quantum evolution.

We would like to note that in the last two decades the attosecond pulse generation techniques, as well as the ultrafast time-resolved spectroscopies, have been developed tremendously (see the lecture notes of A. L'Huillier in the present volume). With pulses as short as few tens of attoseconds available nowadays, one is able to initiate and probe processes that take place before the nuclear dynamics come into play, i.e., to study electron dynamics on their natural time scale. Observing and controlling the dynamics of correlated electrons in larger molecules is, however, a very challenging task. First experiments have been reported only recently (Calegari et al., 2014; Kraus et al., 2015). The development of the necessary cutting-edge experimental techniques can be done only hand in hand with the development of theoretical methods able to describe such processes. Not only that. In many cases, the interpretation of the experimental results cannot be done without detailed calculations. The deep understanding of the underlying concepts is indispensable for designing successful experiments. We hope that the present lecture notes will contribute to this and will inspire theorists and experimentalits to enter the fascinating field of ultrafast electron dynamics.

References

Bartlett R. J., and Musial, M. (2007). *Rev. Mod. Phys.* **79**, 291.

Breidbach J., and Cederbaum, L. S. (2003). *J. Chem. Phys.* **118**, 3983.

Breidbach J., and Cederbaum, L. S. (2005). *Phys. Rev. Lett.* **94**, 033901.

Calegari, F. et al. (2014). *Science* **346**, 336.

Cederbaum, L. S. et al. (1986). *Adv. Chem. Phys.* **65**, 115.

Cederbaum, L. S. (1998). in *Encyclopedia of Computational Chemistry*, P. von Ragué Schleyer et al. (eds), Wiley, Chichester.

Cederbaum L. S., and Zobeley, J. (1999). *Chem. Phys. Lett.* **307**, 205.

Cederbaum, L. S. (2004). Born-Oppenheimer approximation and beyond. *Conical Intersections: Electronic structure, Dynamics and Spectroscopy*. World Scientific, Singapore.

Cremer, D. (1998). in *Encyclopedia of Computational Chemistry*, P. von Ragué Schleyer et al. (eds), Wiley, Chichester.

Domcke, W., Yarkoni, D., and Köppel, H. (eds) (2004). *Conical Intersections: Electronic structure, Dynamics and Spectroscopy*. World Scientific, Singapore.

Helgaker, T., Jørgensen, P., and Olsen J. (2000). *Molecular Electronic-Structure Theory*. Wiley, Chichester.

Kraus, P. M. et al. (2015). *Science* **350**, 790.

Kuleff, A. I., Breidbach, J., and Cederbaum, L. S. (2005). *J. Chem. Phys.* **123**, 044111.

Kuleff, A. I., Lünnemann, S., and Cederbaum, L. S. (2013). *Chem. Phys.* **414**, 100.

Kuleff, A. I. and Cederbaum, L. S. (2014). *J. Phys. B* **47**, 124002.

Kuleff, A. I. et al. (2016). *Phys. Rev. Lett.* **117**, 093002.

Lehr, L. et al. (2005). *J. Phys. Chem. A* **109**, 8074.

Linderberg, J. and Öhrn, Y. (2004). *Propagators in Quantum Chemistry* (2nd edn) Wiley, New York.

Lünnemann, S., Kuleff, A. I., and Cederbaum, L. S. (2008). *Chem. Phys. Lett.* **450**, 232.

Lünnemann, S., Kuleff, A. I., and Cederbaum, L. S. (2009). *J. Chem. Phys.* **130**, 154305.

McWeeny, R. (1989). *Methods of Molecular Quantum Mechanics* (2nd edn). Academic Press, London.

Schlag, E. et al. (2007). *Angew. Chem. Int. Ed.* **46**, 3196.

Szabo, A., and Ostlund, N. S. (1989). *Modern Quantum Chemistry*. McGraw-Hill, New York.

Weinkauf, R. et al. (1997). *J. Phys. Chem. A* **101**, 7702.

Zewail, A. H. (1994) *Femtochemistry* (Vol. I and II). World Scientific, Singapore.

10

Matter-wave physics with nanoparticles and biomolecules

CHRISTIAN BRAND[1],
SANDRA EIBENBERGER[1,2], UGUR SEZER[1],
AND MARKUS ARNDT[1]

1) University of Vienna, Faculty of Physics, Vienna, Austria
2) Fritz-Haber-Institut der Max-Planck-Gesellschaft,
Department of Molecular Physics, Berlin, Germany

Brand, C., et al. "Matter-wave physics with nanoparticles and biomolecules." *In Current Trends in Atomic Physics.* Edited by Antoine Browaeys, Thierry Lahaye, Trey Porto, Charles S. Adams, Matthias Weidemüller, and Leticia F. Cugliandolo. Oxford University Press (2019).
© Oxford University Press. DOI: 10.1093/oso/9780198837190.003.0010

Chapter Contents

10.1   Introduction

The Les Houches summer school was witness of how far quantum physics with atoms has already been advanced. While the non-classical aspects of quantum physics— such as mesoscopic superpositions or entanglement—require us to rethink our notions of reality, locality, logic, or space-time, they have also led to emerging technologies around computation, simulation, metrology, and sensing.

Our present contribution adds to both lines of research: We discuss how to visualize quantum mechanics in experiments with bodies of increasing mass and complexity, and we demonstrate how quantum interference can enhance molecule metrology and provide a tool for physical chemistry and biomolecular physics.

Several quantum dualities can be illustrated in such experiments, including the duality of *determinism and randomness*, of *wave and particle*, and the *mutual uncertainty* of conjugate variables. Richard Feynman once pointed out that all these characteristics can be found in a single demonstration (Feynman et al., 1965), the double slit experiment with massive particles. The discrete, local particle nature appears in the detection process, while the indistinguishability of alternative paths through the experiment—the delocalized wave nature—explains the observed interference pattern. This is why Feynman considered the double slit experiment to have "in it the heart of quantum mechanics", containing its "only mystery".

Today, we might interject that an important aspect of quantum physics was missing, namely *entanglement* as the fundamental non-separability of two or more quantum systems. However, entanglement is not foreign to matter-wave physics. It appears in the form of mode entanglement and more relevantly in the form of quantum decoherence–which is even useful as a resource for molecule metrology.

Double- and multi-slit diffraction experiments with massive matter have been realized with electrons (Jönsson, 1961), neutrons (Zeilinger et al., 1988), atoms (Keith et al., 1988; Carnal et al., 1991) and their clusters (Schöllkopf and Toennies, 1994), as well as small (Chapman, M. et al., 1995, Schöllkopf et al., 2004) and large molecules (Arndt et al., 1999). The combination of several diffraction elements into full matter-wave interferometers allows accessing states of increasing macroscopicity: Nowadays, it is possible to delocalize individual atoms on the half-meter scale (Kovachy et al., 2015a) and to demonstrate spatial superposition states of systems ranging from single electrons (Hasselbach, 2010) up to organic molecules exceeding 10^4 atomic mass units (amu) (Eibenberger et al., 2013). All studies together already span a factor of 10^7 in mass and are still fully consistent with Schrödinger's quantum mechanics, as developed ninety years ago (Schrödinger, 1926).

Here we report on explorations of quantum physics with strongly bound, warm objects of high internal complexity. We study matter-wave interference of organic nanomatter that may bind dozens or more than a thousand atoms into one single quantum object (Hornberger et al., 2012; Arndt and Hornberger, 2014). Such experiments aim at pushing quantum physics towards two complementary frontiers: high mass and high complexity; gravity and life. Both frontiers offer research questions and opportunities for several decades to come. A number of reviews have already been written about the expected or speculative modifications of matter-wave physics at high mass (Bassi et al., 2013b; Arndt and Hornberger, 2014). Here, we will focus on the second aspect as a guideline to discuss and develop new techniques for quantum optics with molecules of biological interest. This is driven by five complementary goals:

Curiosity In our daily lives one often tends to make qualitative distinctions between inorganic and organic matter, the inanimate and the animate world or even non-conscious objects and conscious beings. Is there any fundamental limit or is it a distinction of complexity only? Can we still delocalize peptides, proteins, DNA, viruses, or cells? This question has triggered a long-term journey, on a mass and time scale comparable with testing the gravitational limits of quantum physics.

Nanotechnology It has traditionally been difficult to prepare neutral beams of size-selected massive particles in the mass range of $10^3 - 10^6$ amu. Cluster aggregation sources deliver intense beams of particles, but often the signal is distributed over many mass peaks. In contrast to that, life defines and selects the size and shape of biomolecules by their functionality. Can we exploit this 'nature-made nanotechnology' for quantum experiments, even if we do not ask any biological question? Biomolecules are fragile and not easy to volatilize, but progress is being made and some of it is reported here.

Complexity Throughout recent years, we have witnessed enormous progress in the handling of few-particle quantum systems and in obtaining control over millions of quantum degenerate ultra-cold atoms, which may be as cold as a few picokelvin. Complementary to that we accept the challenge of exploring quantum physics in a complexity range that is closer to our daily lives, with molecules composed of many hundreds of *covalently* bound atoms. Because of their complexity, one can assign an internal temperature to macromolecules, even when they are prepared in spatial Schrödinger cat states. We find that the de Broglie wavelength of macromolecules can stay coherent over many milliseconds even when the internal temperatures exceed several hundred Kelvins.

Decoherence With increasing complexity decoherence becomes an important agent in the game. The rich internal arrangement, in particular of biomolecules like proteins with their different structure levels—from primary to quarternary structure— opens interaction channels with their environment that do not exist for electrons, neutrons, atoms, or dielectric nanoparticles.

Metrology The high sensitivity of macromolecules to external perturbations is both a curse and a blessing: it complicates attempts to prepare pure quantum states but it enables new ways of sensitively measuring molecular properties—by monitoring their matter-wave fringe shifts in the presence of external forces. Such measurements build on the same principles as modern atom gravimeters and rotation sensors. While molecule interferometers will not compete with atom interferometers as inertial sensors, their unique potential is rooted in their capability to act universally on a variety of particle classes and to expose them to external fields—providing a test bed for physical chemistry and biomolecular physics, with the potential of becoming a viable technology.

In these lecture notes, we will present a tutorial introduction to molecular quantum physics, focusing on the coherent state preparation, on the diversity of diffraction techniques for complex molecules, on methods how to exploit quantum interference for obtaining molecular spectra and examples of challenges in preparing and detecting neutral biomolecular beams for quantum experiments.

10.2 Delocalization and diffraction

10.2.1 General source and coherence requirements

Throughout this text we are focusing on quantum aspects of the center-of-mass motion of complex molecules, i.e. on the coherent splitting of their de Broglie wave fronts, their recombination and interference. In our experiments, de Broglie wavelengths range between $\lambda_{\mathrm{dB}} = h/mv = 0.3 - 5 \times 10^{-12}$ m, which is typically $10^3 - 10^4$ times smaller than the size of the molecule itself and comparable to the shortest wavelengths in high-resolution transmission electron microscopy. Here h is Planck's constant, m is the mass of the molecule, and v its velocity. For particles lighter than about 1000 amu, it is often possible to sublimate or evaporate them in thermal Knudsen cells. This even holds for a sizable set of biomolecules such as nucleobases, vitamins, antibiotics, vanillin, caffeine and various other compounds. Thermal beams have therefore proven useful for many of our quantum interference studies so far.

In molecular beams of small divergence—even in near-field interferometers the molecular beam is typically constrained to a divergence angle of 1 mrad—we can distinguish longitudinal and transverse coherence, which provide a measure for the interference capability of the emerging molecules. It is determined by the spectral and geometric properties of the source.

Longitudinal coherence, as in light optics, is determined by the spectral purity of the source. A practical value is $L_c = \lambda_{\mathrm{dB}}^2 / \Delta\lambda_{\mathrm{dB}}$, defined by the relative spread $\Delta\lambda_{\mathrm{dB}}$ of de Broglie wavelengths λ_{dB}. It remains constant with distance from the source but can be improved by spectral filtering, i.e. velocity selection. In thermal sources, this value can be as short as $L_c \simeq 2\lambda_{\mathrm{dB}}$. However, this does not prevent successful interference since we are only dealing with the description of the center-of-mass motion of a single particle.

Transverse coherence is at the origin of the indistinguishability of different molecular paths through space. It can be estimated both from Heisenberg's uncertainty principle in quantum physics and the van Cittert-Zernike theorem in optics. The transverse coherence $X_c \simeq 2L\lambda_{\mathrm{dB}}/D$ grows linearly with the distance from the source L and inversely proportional with the source size D. Being based on a diffraction phenomenon, it increases with increasing de Broglie wavelength.

While path-indistinguishability is a key condition for matter-wave interference, we may also ask whether there is any chance to establish two-particle interference, as in Hanbury-Brown-Twiss experiments, which have already been realized for atoms (Westbrook et al., 1998). This would require identity in all degrees of freedom. However, any N-atom system has $3N$ mechanical degrees of freedom, in addition to electronic, fine, hyperfine and Zeeman levels in a complex configuration space. Moreover, large molecules can adopt a number of conformations, have vibrationally-induced dipole moments, and may undergo conformational changes upon photo-absorption (photo-isomerization). Hence, the chances that two molecules are identical in all of these degrees of freedom, is extremely slim.

This is why macromolecule interferometry deals only with the superposition of single-particle wave functions. Molecule lasers, in the spirit of successfully realized atom lasers (Bloch et al., 1999; Andrews et al., 1997; Hagley et al., 2001), would require sub-microkelvin temperatures in all degrees of freedom and even then a mechanism to ensure that all molecules end up in the same absolute energy minimum of the conformational landscape. While this feat has been achieved for diatomic molecules— it remains an outstanding challenge for macromolecules.

In thermal sources the mechanical degrees of freedom are equilibrated with the source temperature in the range of 500–1000 K by collisions. However, as soon as the molecules leave the cell and enter high- or ultra-high vacuum, the degrees of freedom are decoupled on the time scale of the interference experiments. In the absence of thermal emission processes, the micro-canonical temperature then determines the rotational and vibrational motion of the molecule—and the exchange of energy between the different vibrational modes. The important point is that the internal dynamics remains decoupled from the center-of-mass-motion. Since matter-wave physics requires only a well-defined center-of-mass state of an ensemble of molecules whose internal states are sufficiently similar, de Broglie interference can actually be observed for surprisingly large composite objects.

10.2.1.1   *Beam splitting of molecular matter-waves*

Coherent beam splitters are the foundation of every interferometer. They divide, redirect, and recombine the incident waves. In analogy to classical optics, we distinguish two species also in matter-wave interferometry.

Wave front beam splitters modulate the incident wave front using a spatially periodic structure, i.e. a diffraction grating, which may either modulate the matter-wave phase or amplitude. *Phase gratings* for atoms or molecules can be realized by exploiting the dipole potential emerging from the interaction between a particle's polarizability and the electric field amplitude of a standing light-wave grating. *Absorptive gratings* may be built using nanomechanical masks or *photo-depletion* mechanisms such as single- and two-photon-ionization, or photo-fragmentation in the anti-node of a standing light field.

Amplitude beam splitters, such as Ramsey-Bordé (Bordé, 1989) or Raman beam splitters (Kasevich et al., 1991) do not realize stationary gratings in real space, but create entanglement between the internal atomic state and the external center-of-mass motion during the coherent interaction between the atom and one or two laser beams. We will see further below that even in the absence of any coherent cycling as in atoms, single-photon absorption can contribute to the beam splitting process of hot molecules (Cotter et al., 2015).

Here we review the concepts of beam splitters for complex molecules and discuss the dephasing mechanisms that emerge in particular in experiments with biomolecules, which are typically polar.

10.2.1.2   *Far-field diffraction at a single grating*

First demonstrations of matter-wave physics with electrons (Davisson and Germer, 1927; Thomson, G. P. 1927), atoms, and diatomic molecules (Estermann and Stern, 1930) were done using reflective diffraction at clean crystal surfaces. With the advent of nanotechnologies, free-standing nanomechanical transmission gratings were realized for electrons (Jönsson, 1961), atoms (Keith et al., 1988), and molecules (Chapman, M. et al. (1995) Schöllkopf et. al., 2004; Arndt et al., 1999). Such beam splitters are open masks that remove portions of the incident wave front by absorption or scattering. The spatially periodic confinement of the wave function in the slit openings leads to a spread of the transverse momentum behind the mask, in accordance with Heisenberg's uncertainty principle and diffraction rules as known from wave optics. In contrast to light optics, the matter-wave transmission function $\tilde{t}(x)$ is not described by a binary filter $t(x)$ alone, but also by the attractive interaction between the particles and the grating walls.

In the simplest case, we can treat the local interactions by considering particles that pass the grating wall in a straight line at distance x. In the Eikonal approximation we assume that this trajectory receives a transverse momentum kick which is soft enough that the related position shift will only become noticeable a long distance behind the grating. The interaction time is determined by the grating thickness b and the molecular velocity v_z along the forward direction. For gratings with a thickness of hundreds of nanometers and openings as narrow as 50 nm it is fair to assume that we can neglect retardation effects and treat the Casimir-Polder (CP) interaction in the near-field approximation, i.e. using a van der Waals scaling (Grisenti et al., 1999; Brand et al., 2015a): The attraction of only one of two walls in each slit then affects the molecular wave transmission function in dependence of the distance x of the particle to this wall:

$$\tilde{t}(x) = t(x) \exp\left(\frac{i}{\hbar v_z} \int_0^b \frac{C_3}{x^3} dz\right) \tag{10.1}$$

The acquired phase is determined by the molecular polarizability $\alpha(\omega)$ at all frequencies ω as well as by the dielectric function $\varepsilon(\omega)$ of the grating material, which is encoded in the surface reflection coefficient $r(\omega) = (\varepsilon(\omega) - 1)/(\varepsilon(\omega) + 1)$.

$$C_3 = \frac{\hbar}{16\pi^2\epsilon_0} \int_0^\omega \alpha(\omega)r(\omega)d\omega \tag{10.2}$$

In this way, the analysis of atomic diffraction patterns can be used to extract detailed information about atomic polarizabilities (Holmgren et al., 2010) and atom-surface interactions (Perreault et al., 2005; Grisenti et al., 1999).

For molecules the description gets more demanding due to their complex internal structure and dynamics. They may have a permanent or a vibration-induced electric dipole moment and they may appear in a plethora of different conformational arrangements. In principle, the attractive interaction to the wall depends on all these factors. However, at a source temperature of 500–1000 K the molecules are highly excited and conformations interconvert on the time scale of 1–10 picoseconds. This has to be compared to the molecular transit time through the grating, which amounts to 500 ps for $v_z = 200$ m/s and $b = 100$ nm. Hence, some of the molecule-wall interactions can be described by an average value over molecular starting conditions and orientations. On the other hand, additional effects become relevant. Vibrationally induced dipole moments, for instance, will increase the effective static polarizability (Gring et al., 2010) and the molecule may exchange excitations with the grating during the transit, leading to additional attractive or repulsive contributions to the potential (Brand et al., 2015a).

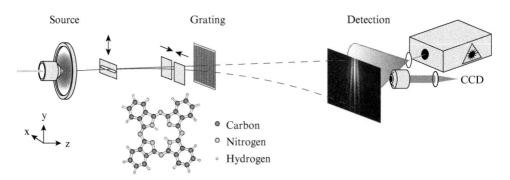

Fig. 10.1 Molecules like phthalocyanine shown here are thermally evaporated by a continuous laser beam (421 nm) focused down to a spot size of ~1.5 μm. The resulting molecular beam is shaped in x- and y-direction before it reaches the grating after 1.55 m of free flight. Here, the molecules illuminate 50–90 slits coherently. After additional 0.59 m they arrive at the quartz window of the detector. The quantum nature of the diffraction is revealed by analyzing the position distribution of the arriving particles, which we image using laser-induced fluorescence microscopy.

The current setup of the far-field diffraction apparatus is shown in Fig. 10.1 (Juffmann et al., 2012). It is housed in a 2.14 m long vacuum chamber which is evacuated to a pressure below 10^{-7} mbar. This is necessary to guarantee a collision-free flight of the molecules through the whole apparatus. The journey of the molecules begins in the source section where they are deposited as a thin film on the inner surface of a vacuum window. When a laser is focused through the glass onto this film, the molecules absorb the light and are thermally evaporated. The size of the laser spot defines the source size D and is chosen to be as small as possible. For a spot size of 1−2 μm the transverse coherence of a particle with a wavelength of $\lambda_{\mathrm{dB}} = 3$ pm amounts to $2L\lambda_{\mathrm{dB}}/D = 5$–9 μm at a distance $L = 1.55$ m behind the source. For common gratings with a period of 100 nm this corresponds to a coherent illumination of 50 to 90 slits. After diffraction the molecules propagate for another 59 cm before they impinge on a thin quartz plate where they stick. In order to visualize the molecular pattern we need a method to read out the position of each particle with high accuracy. This is done by exciting the molecules electronically with laser light of suitable wavelength. Once elevated to the excited state the molecules release the excess energy by the emission of a photon. For systems with a high fluorescence quantum yield this cycle can be repeated many times before the molecules undergo a photochemical reaction and are lost to the detection process. Collecting a large number of photons allows for a very precise localization of every molecule down to 10 nm (Juffmann et al., 2012).

A typical diffraction pattern is shown in Fig. 10.2. It shows the signal of the organic dye phthalocyanine (PcH$_2$) diffracted at a 45 nm thick membrane made of silicon nitride (Brand et al., 2015b). Into this membrane our collaborators around Ori Cheshnovsky at Tel Aviv University milled a mask using a focused gallium ion beam. From the diffaction pattern we can extract various information. First, we see high-contrast diffraction, up to the 9th order. This proves the de Broglie wave nature of PcH$_2$ molecules and corroborates the assumption that quantum mechanics in the translational degrees of freedom can persist, even if the internal degrees of freedom

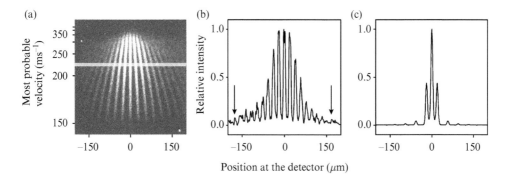

Fig. 10.2 (a) Molecular diffraction of phthalocyanine at a 45 nm thick silicon nitride grating. The different molecular velocities are sorted by gravity and range between 350 and 150 m/s. The trace in (b) corresponds to the highlighted area in (a) and the arrows point at the $\pm 9^{th}$ diffraction order. The population of high diffraction orders can be traced back to the influence of Casimir-Polder interactions. The huge impact of these can be estimated when we compare the experimentally observed trace to a simulated one neglecting any attractive interactions (c).

are highly excited. It also underlines that the rotational symmetry, which we assume for atoms and could also take for granted in C_{60} diffraction experiments (Arndt et al., 1999) is no prerequisite for the preparation of a clean translational quantum state. The spacing of the diffraction orders increases from the top to the bottom of the image. Since slower molecules fall further down in the gravitational potential of the earth before they reach the detector, their height encodes the molecular speed and thus the de Broglie wavelength. The highest diffraction orders on both sides of the image are separated by 18 grating momenta. For comparison, we calculate how many momenta of light would be necessary to realize the same separation for rubidium in a light grating at 780 nm. This yields

$$\Delta p = 18 \cdot \frac{h}{d} = \frac{18}{100\,\text{nm}} \cdot \frac{h}{2\pi} \cdot \frac{2\pi}{780\,\text{nm}} \cdot 780\,\text{nm} = 140\,\hbar k_{Rb}, \qquad (10.3)$$

which compares very favorably with large momentum transfer beam splitters in state-of-the-art atom interferometry (Müller et al., 2008; Kovachy et al., 2015).

The enhanced polarizability of complex molecules may thus actually be useful for achieving a wide arm separation in nanomechanical beam splitters, since the population of high order diffraction orders is enhanced by the van der Waals interaction. Future efforts will still have to focus on improvements of the grating's coherence, i.e. the regularity of the grating period to about 10 ppm over a few millimeters in a single mask as well as between different masks of the same fabrication batch. For 160 nm thick gratings written photo-lithographically this is already proven state of the art for more than two decades (Savas et al., 1995). Modern focused ion beam (FIB) machines with interferometrically controlled translation stages are approaching a similar precision. A technological exploitation of such large momentum beam splitters requires either blazed gratings or more intense molecular sources, since the signal is still distributed over many diffraction orders.

A proper description of diffraction patterns also includes details of the grating bars, such as their geometry (Savas et al., 1995; Grisenti et al., 2000) and material inhomogeneities. The theoretical description can be based on virtual photons exchanged between the grating and the matter-wave. For dielectric gratings it is often sufficient to include only the scattering of a virtual photon from the molecule to the grating and back. However, if neccessary, scattering inside the material can be corrected for (Brand et al., 2015a). Any remaining discrepancy between theory and experiment provides a hint on additional interactions, such as charges inside the thin dielectric membrane. Our experiments with PcH$_2$ diffracted at a FIB written grating showed an attractive potential 5-8 times stronger than expected based on the CP-interaction alone (Brand et al., 2015a). This effect was greatly reduced for masks that were generated in photolithography and reactive ion etching at low energies.

For non-polar molecules the near-field CP-force and the attraction between a neutral particle and a residual charge have the same distance scaling and can be hardly distinguished. However, for polar molecules the interaction with the grating depends on the orientation of the molecular dipole moment and the rotation axis to the wall. This has clear consequences and is of particular importance for the diffraction of biomolecules at nanomechanical masks. All biomolecules are equipped with functional groups such as OH, NH$_2$, or COOH, which determine the interaction with their surroundings and are necessary for their biological function. An example is hypericin, which acts as a neurotransmitter, antibiotic, and antiviral agent and is, naturally occurring in the medicinal herb Saint John's wort (Knobloch et al., 2016). Fig. 10.3 shows its diffraction pattern at a carbon grating of 20 nm thickness. The diffraction peaks are less sharp than those of phthalocyanine in Fig. 10.2, which we attribute to the interaction of the permanent electric dipole moment of hypericin of about 4 Debye with the grating. To elucidate the role of the molecular dipole moment further, we have compared the diffraction pattern of the non-polar molecule tetraphenylporphyrin (TPP) and its polar derivative (MeO)TPP. Both substances are close in mass and polarizability, but differ in their dipole moment because of an

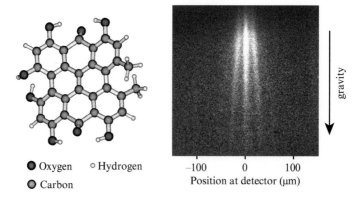

Fig. 10.3 Even the polar neurotransmitter and antibiotic hypericin still shows interference, when diffracted at a nanomechanical carbon grating of 20 nm thickness and 100 nm period.

OCOCH$_3$ group attached to one phenyl ring. In Fig. 10.4 their diffraction patterns are compared for two different gratings (Knobloch et al., 2016), both milled by a focused ion beam, but into materials with greatly differing conductivity. While the carbon membrane is a weak electric conductor the silicon dioxide grating is an insulator. Hence, charges buried inside the gratings are more likely to be neutralized for carbon than for silicon dioxide.

Non-polar TPP shows high contrast interference at both materials. The polar molecule (MeO)TPP, however, exhibits already a slightly blurry pattern behind the carbon grating which is entirely washed out behind the silicon dioxide mask. Not a single diffraction order can be resolved behind the insulating grating and slow molecules even seem to be deflected to beyond the detector area.

This observation is attributed to the dynamics of the permanent electric dipole moment in an inhomogeneous electric field. In contrast to non-polar molecules, where the interaction is always attractive, polar molecules may experience attractive or repulsive forces depending on their orientation with respect to local charges. These may be implanted during focused ion beam writing and reach a surface density of up to $10^{12}/\text{cm}^2$ in SiO$_2$ (Yogev et al., 2008).

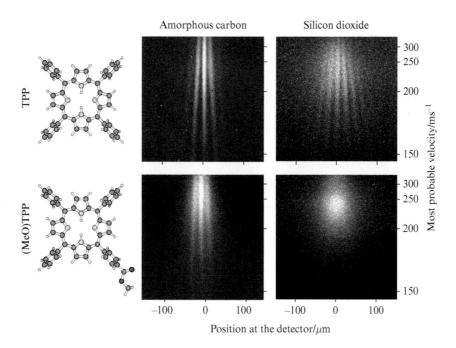

Fig. 10.4 Molecular diffraction patterns of the non-polar TPP (C$_{44}$H$_{30}$N$_4$, $m = 615$ amu) and the polar (MeO)TPP (C$_{46}$H$_{32}$N$_4$O$_2$, $m = 673$ amu, dipole moment 1.8 Debye) at mechanical gratings written in amorphous carbon (period $d = 100$ nm; left) and silicon dioxide ($d = 160$ nm; right). The resistivity of amorphous carbon (1.55×10^{-5} Ωm) is 20 orders of magnitude smaller than of silicon dioxide (10^{15} Ωm). This readily explains why a carbon grating can be substantially less or more homogeneously charged and better suited for diffracting of polar molecules.

In thermal beams, neither the initial orientation nor the rotational state are well controlled. Furthermore, the electric field inside the dielectric can be stochastically distributed in amplitude and direction. In consequence, each molecule will experience a different local force and we accumulate the incoherent sum of many coherent but shifted single-molecule diffraction patterns on the screen. Even though there is no proper decoherence, path entanglement, or which-path information shared with the environment, the matter-wave contrast disappears because of this random phase averaging due to the molecule-wall interaction. This is of particular importance for large polar biomolecules, such as for instance insulin which is known to have an electric dipole moment of 72 Debye in solution (Laogun et al., 1984).

However, even for non-polar molecules the CP-interaction influences the molecular diffraction pattern substantially, as shown in Fig. 10.2. The simplest approach to modeling this pattern is to encode the attractive interaction in an 'effective width' of the single-slit diffraction pattern, which defines the fringe envelope (Grisenti et al., 1999). Fitting this width allows extracting a qualitative measure for the path-integrated interaction strength. During the diffraction of PcH_2 at a 45 nm thick SiN_x grating, for instance, the geometrical slit width of 50 nm appears reduced to an effective width of only 15 nm (Brand et al., 2015b). Larger molecules with a higher polarizability will see a further reduced slit width and may eventually not even pass the grating any more.

For quantum experiments with massive particles it therefore appears essential to reduce the grating thickness—and thus the relevance of CP-potentials—to the smallest conceivable value. In recent years, 2D-membranes have become available, that range in thickness between a few nanometers and a single atomic layer. Our collaboration partners at Tel Aviv University made it possible to manufacture gratings with 100 nm period into such membranes with high quality (Brand et al., 2015b). In Fig. 10.5 we show the diffraction patterns of PcH_2 behind several of these gratings. They all are about 1 nm thick and we observe a clear reduction of the CP-force compared to thicker masks. The smallest interaction is indeed observed for single-layer graphene, which is only one carbon atom thin. However, even this ultimate nanomechanical element for molecular quantum optics exhibits an effective slit width of 35 nm, still 40 percent smaller than the geometrical opening. Although the molecule spends only a few picoseconds 'inside' the slit—the molecule is actually larger than the grating thickness (!)—it is still affected by the CP-interaction. These experiments also allude to a discussion between Nils Bohr and Albert Einstein, on whether information about the path of a particle through a double slit can be retrieved by monitoring the recoil of the mask. If this were possible, it would destroy the diffraction pattern (Bohr, 1949). Single-layer graphene is about the lightest and thinnest durable mechanical mask one can think of, making it a valid candidate for testing this question. According to Bohr the diffraction remains unnoticed and coherence maintained as long as the momentum transfer between the molecule and the mask is smaller than the intrinsic momentum uncertainty of the grating. The following back-of-the-envelope estimate may elucidate this relation: The momentum uncertainty for the matter-wave diffracted at a rectangular slit is (Nairz et al., 2003): $\Delta x_{mol} \cdot \Delta p_{mol} \geq 0.89\, h$. This relation looks different from the one typically shown in textbooks, since we define Δp_{mol} as the experimentally observable full width at half maximum (FWHM) of the diffraction

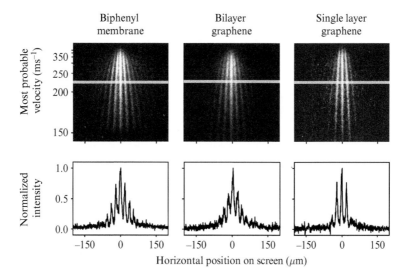

Fig. 10.5 Molecular diffraction at gratings with a membrane thickness of 1 nm. While the biphenyl membrane (left) is an insulator, bilayer (middle) and single-layer (right) graphene are conducting. The reduced population of higher diffraction orders is due to the reduction of the molecule-wall interaction.

envelope and Δx_{mol} as geometrical width of a single diffraction slit. For the grating we take $\Delta x_{\mathrm{grat}} \cdot \Delta p_{\mathrm{grat}} \geq \hbar$, where the uncertainties are usually taken to be the $1/\sqrt{e}$ values of a Gaussian distribution in position and momentum. The threshold for coherent diffraction is $\Delta p_{\mathrm{grat}} = \Delta p_{\mathrm{mol}}$. As soon as Δp_{mol} gets larger than Δp_{grat} coherent diffraction vanishes. When we take the minimal uncertainty in momentum of the grating $\Delta p_{\mathrm{grat}} = \hbar/\Delta x_{\mathrm{grat}}$ and insert this into the other equation, we get

$$\Delta x_{\mathrm{mol}} \geq \frac{0.89\,h}{\Delta p_{\mathrm{mol}}} \Rightarrow \Delta x_{\mathrm{mol}} \geq \frac{0.89\,h \cdot \Delta x_{\mathrm{grat}}}{\hbar} \Rightarrow \Delta x_{\mathrm{mol}} \geq 1.78\pi \Delta x_{\mathrm{grat}}. \tag{10.4}$$

This is the lower boundary for the geometrical opening of the grating at which diffraction is still coherent. For graphene nanoribbons the measured position uncertainty amounts to $\Delta x_{\mathrm{grat}} = 0.1 - 0.5$ nm (Sapmaz et al., 2003; Garcia-Sanchez et al., 2008). Hence, each slit would have to be smaller than about 3 nm. In this limit, van der Waals forces would extract the molecules from the beam, anyhow.

10.2.2 Optical gratings

10.2.2.1 Phase gratings

In contrast to material gratings, which inevitably introduce an orientation-dependent phase shift, optical beam splitters can act on a large variety of particles, even when they are charged or exhibit an electric or magnetic dipole moment. The internal complexity of large molecules renders coherent cycling between the ground and the excited states difficult and excludes Ramsey-Bordé or Raman interferometers. We can, however, exploit all optical processes that induce a spatially periodic phase shift or entail a spatially periodic depletion of the molecular beam.

The diffraction at a standing light wave has entered the literature already in 1933, when Piotr Kapitza and Paul Dirac suggested electron reflection at an optical Bragg grating (Kapitza and Dirac, 1933). The original Kapitza-Dirac (KD) effect is based on the ponderomotive potential that the electron experiences in the presence of a light beam (Batelaan, 2007). The electron itself has no known internal structure but charge: It follows the electric field and in motion it sees the Lorentz force of the magnetic field component, too. The first demonstration of KD-diffraction with electrons was achieved in 2001 (Freimund et al., 2001).

Phase gratings for atoms were already realized fifteen years earlier than for electrons (Gould et al., 1986) and all-optical Mach-Zehnder atom interferometers were demonstrated with thin (Rasel et al., 1995; Delhuille et al., 2001) and thick phase masks (Giltner et al., 1995). Here, the spatially periodic phase imprint is based on the dipole interaction between the electric field of the standing laser wave and the particle's optical polarizability α_{opt}. This interaction induces an electric dipole moment oscillating at the laser frequency ω_{L}, far-off any molecular resonance. The coupling of the dipole moment to the electric field E in the standing wave results in the potential $V(x, z, t) = -\alpha_{\text{opt}}|E(x, z, t)|^2/4$.

The spatially periodic transmission function $t(x) = \exp(i\phi(x))$ can be calculated from the phase $\phi(x)$ accumulated during the molecular transit across the standing laser light wave of power P_{L} and period $d = \lambda_{\text{L}}/2$ along the x-axis (transverse to the molecular beam) with a Gaussian beam waists w_y along the y-axis and w_z along the z-axis (longitudinal) (Arndt et al., 2014). For a pure phase grating, the accumulated phase $\phi(x)$ is given by

$$\phi(x) = \phi_0 \cdot \cos^2(k_{\text{L}}x) = \frac{4\sqrt{2\pi}\alpha_{\text{opt}}P_{\text{L}}}{hc\varepsilon_0 w_y v_z} \cdot \cos^2(k_{\text{L}}x), \tag{10.5}$$

For Eq. 10.5 we have assumed that the molecular absorption cross section is sufficiently small that we can neglect any uptake of a real photon by the molecule.

The far-field diffraction pattern at the sinusoidal phase grating is determined by the Fourier transform of the transmission function and proportional to a sum over Bessel functions J_n,

$$\psi(p) \propto \sum_{n=-\infty}^{\infty} J_n(\phi_0) \exp(-in2k_{\text{L}}x), \text{with } n \in \mathbb{N}. \tag{10.6}$$

We learn from Eq. 10.6 that the momentum exchange between the molecule and the light wave always involves a multiple of two photon momenta, $2\hbar k_{\text{L}}$. This can be interpreted as the absorption and stimulated emission of virtual photons, but the derivation does not require the photon picture at all. It suffices to consider the spatial geometry of the potential landscape. This is also the reasons why the de Broglie wavelength does not enter in this consideration. For a structure of period d the momentum transfer always amounts to multiples of h/d.

Far-field diffraction experiments with fullerenes (Nairz et al., 2001) showed that this idea can also be used for large molecules, in spite of their internal complexity, high thermal excitation and transition lines as wide as 20–40 nm (Dresselhaus et al., 1998). Even though textbooks always display grating diffraction with an interference maximum at the center of the pattern, Eq. 10.6 explains why this is no longer the case for phase gratings: only at low laser power or small molecular polarizability the sum over the Bessel functions will mimic the far-field diffraction pattern of an absorptive grating, like the graphene nanomasks in Sec. 10.2.1. At high laser power, the central peak can be suppressed.

Kapitza-Dirac Talbot-Lau interferometry in the near-field Beam splitters based on wave front division in a phase grating are an interesting option for composite particles, but the observation of wave interference in the far-field requires the collimation angle to be smaller than the diffraction angle. With decreasing de Broglie wavelength, we therefore need to collimate the molecular beam increasingly well. This becomes demanding for particles the size of a protein.

A solution to this problem is near-field interferometry which allows working with wide and spatially incoherent particle beams. This results in a huge gain in signal. However, it comes at the price that we now need two or more diffraction elements to prepare and probe the molecular de Broglie wave nature. The idea of using Talbot-Lau interferometry for that purpose goes back to John Clauser (Clauser, 1997) who also demonstrated it for thermal potassium beams (Clauser and Li, 1994).

This approach was successfully implemented at the University of Vienna for native (Brezger et al., 2002) and fluorinated fullerenes (Hackermüller et al., 2003) as well as for porphyrins using large micromechanical gold gratings with 500 nm thickness and grating periods of $d = 990$ nm. However, it soon became clear that the dispersive forces of massive organic molecules near mechanical walls would render the source requirements exigent, in particular with regard to velocity selectivity (Brezger et al., 2003). The solution to this problem was to combine an absorptive grating with a phase grating and again an absorptive grating. If the molecular beam was spatially coherent right from the source, the first grating could also be a phase grating with 100% transmission. This has been demonstrated for cold atoms (Cahn et al., 1997) and Bose-Einstein condensates (Denschlag, 2000). However, the preparation of coherent macromolecular beams has remained a challenge throughout the years and even though there is good progress in the development of various source types, coherence still needs to be enhanced by virtue of spatial selection.

In Fig. 10.6 we show a sketch of the near-field Kapitza-Dirac Talbot-Lau interferometer (KDTLI) (Gerlich et al., 2007). Its mechanical gratings G1 and G3 are etched into a 160 nm thick nanomechanical SiN_x membrane with a period of $d = 266$ nm and open slits as narrow as $s = 110$ nm. While the width of each individual opening is only defined by the etching process to within 2–5 nm accuracy, the average grating period is defined to better than 50 ppm.

The first grating G1 represents an array of 'nano-illuminators' (Jahns and Lohmann, 1979). Molecules impinging onto G1 under any angle of incidence will have their transverse wave function restricted and therefore their momentum coherently

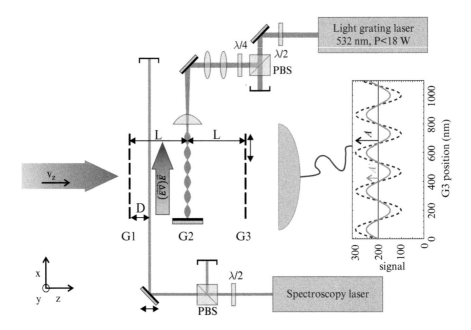

Fig. 10.6 Kapitza-Dirac Talbot-Lau interferometer (KDTLI): Molecular coherence is pre-pared by diffraction at each slit of the first grating G1. The coherence function spreads out and covers more than two anti-nodes of the optical grating G2. The spatially periodic phase imprinted by the standing light wave and subsequent interference lead to the formation of a molecular density pattern at the third grating, G3. It resolves the molecular fringe pattern before the molecules are ionized and counted in the quadrupole mass spectrometer. An intense spectroscopy laser and high-voltage deflection electrodes can interact with the molecules to shift their interference pattern and to retrieve information about their internal molecular properties.

spread, in accordance with Heisenberg's uncertainty relation. We can also see it as the preparation of an approximate cylindrical Huygens wavelet, the wave front of which is well defined because the point of origin is so small. It was E. Lau (Lau, 1948) who realized that spatially coherent illumination of the second grating, at least over two neighboring slits, can be achieved if the distance L between G1 and G2 is chosen close to the Talbot length $L_T = d^2/\lambda_{dB}$. Diffraction at the central optical grating G2 then leads to the lens-less formation of a grating self-image, which can only be explained by near-field wave interference. The transmitted molecules thus form a particle density pattern in free space, with a characteristic length scale of $L \simeq L_T$ behind G2, which can be captured and imaged on a screen (Juffmann et al., 2009).

It is, however, often favorable to use a third mask, G3, whose period equals that of the emerging molecular fringes. The interference pattern is then encoded in the number of molecules transmitted through G3 as a function of the transverse position of G3. Recording the interference pattern by measuring the molecular count rate as a function of transverse position of G3 has become the standard technique in Kapitza-Dirac Talbot-Lau interferometry and has allowed to demonstrate matter-wave interference of the most massive particles to date (see Fig. 10.7). This achievement became possible because of the highly developed capabilities in chemical synthesis in the group of Marcel Mayor at the University of Basel. They provided porphyrin derivatives where numerous peripheral hydrogen atoms were substituted by massive perfluoroalkyl chains. This trick allowed increasing the particle mass, maintaining a low polarizability-to-mass ratio and therefore achieving a good vapor pressure in a sublimation or vaporization process (Tüxen et al., 2011). We measured high-contrast matter-wave interference with these particles, even though they are fragile and ~ 500 K hot (Eibenberger et al., 2013).

These functionalized molecules are already in the complexity class of proteins and we compare them pictorially to the structure of the proteins insulin and cytochrome C in Fig. 10.7. TPPF20 ($C_{284}H_{190}F_{320}N_4S_{12}$) is made from 810 covalently bound atoms with a total molecular mass of 10123 amu and is the most massive object for which matter-wave interference has been observed so far (Eibenberger et al., 2013). Insulin, for example, is composed of 777 atoms in 51 amino acids ($C_{254}H_{377}N_{65}O_{75}S_6$) with a total mass of about 5743 amu, and cytochrome C is composed of 104 amino acids, 1776 atoms and has a total mass of 12430 amu ($C_{541}H_{910}FeN_{145}O_{175}S_4$). Interference with such proteins is still work in preparation, since even for the same interferometer concept and scale as before, the source preparation and molecule detection are substantially more demanding for large biomolecules.

The role of absorption during the interaction with the optical grating

So far we have described an idealized phase grating in G2 and neglected absorption. However, when the optical power or the molecular absorption cross section are high, photons can be absorbed in the standing light wave. The probability of photon absorption is highest in the anti-nodes and lowest in the nodes of the standing light wave. The mean number of absorbed photons n_0 during the transit through a light grating formed by a Gaussian laser beam depends on its waist w_y, the molecular velocity through the grating v_z, the molecular absorption cross section σ_{abs}, and the wavelength of the laser λ_L,

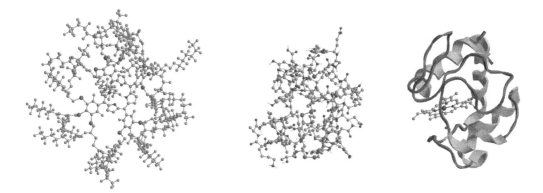

Fig. 10.7 The functionalized porphyrin TPPF20 (left) is the largest object for which matter-wave interference has been observed so far. It is comparable in complexity and mass with insulin (middle) or cytochrome C (right). The extension of TPPF20 can reach up to 50 Å.

$$n(x) = n_0 \cdot \cos^2(k_{\mathrm{L}}x) = \frac{8\sigma_{\mathrm{abs}}\lambda_{\mathrm{L}}P_{\mathrm{L}}}{\sqrt{2\pi}hcw_yv_z} \cdot \cos^2(k_{\mathrm{L}}x). \qquad (10.7)$$

The absorption of a photon is accompanied by a momentum transfer of $\Delta p = \hbar k_{\mathrm{L}}$ to the molecule. If the molecules were exposed to incoherent thermal light they would 'know' whether the photon came from the left or the right and receive a corresponding momentum kick. Inside the coherent standing laser light field, however, the photon itself is in a superposition of 'coming from the left' and 'coming from the right'. Upon absorption, the molecular matter-wave thus also ends up in a superposition of moving to the left *and* to the right.

In far-field diffraction we cannot distinguish between those two possibilities since both statistical and coherent absorption populate the same momentum states. Inside the KDTL interferometer the situation is different: The wave function of a single molecule may pass two neighboring anti-nodes of G2 in momentum states that would not overlap at G_3. The additional recoil may, however, refocus them and thus enable their interference.

The observed process is also intriguing with regard to measurement and the possibility of entanglement. Absorption labels a molecule by its internal energy: Every 532 nm photon raises the internal temperature of a C_{70} molecule by about 139 K (Hornberger et al., 2005). This creates a correlation between the internal and the center-of-mass state—very much like in a Ramsey-Bordé beam splitter (Bordé, 1989). It differs, however, in that there is no simple Rabi cycling in macromolecules to reverse the absorption process coherently.

In order to interfere with itself the molecule needs to remain in the same internal state along all paths. Upon absorption the molecule loses its ability to interfere with all parts of the superposition state which have not changed their internal energy. Absorption thus transfers each delocalized molecule into a mixture of temperature states, depending on the number of photons it absorbs, and on the instant it does. However, in spite of all complexity, the internal dynamics remains unitary within each temperature class. The quantum random walk in momentum space is thus accompanied by a division of the molecular coherence into different temperature classes. Within each internal energy class, de Broglie coherence is maintained. Absorptive heating in the light wave realizes a special case of a photo-depletion grating (see below), which 'depletes' a given molecular temperature class with a spatially periodic grating structure. Similar to diffraction at a mechanical grating, where molecules 'know' about the nearby grating bars even when they are transmitted through the open slits, the unheated molecules 'know' about the possibility of being heated by the photons in the standing light wave. Interestingly, and different from the situation at nanomechanical gratings, the 'depleted' states can form their own interference class.

A proper model has to take into account all aspects of the molecule-grating interaction: the phase shift imparted by virtue of the dipole force, the effects of coherent absorption and the measurement-induced grating. These contributions were analyzed in detail by our theory partners around Klaus Hornberger at the University of Duisburg-Essen.

In Fig. 10.8 we compare their model with our time-of-flight resolved experiments with C_{70} molecules. Different molecular velocities correspond to different Talbot lengths L_T. In the KDTLI, with a distance L between consecutive gratings, the experimentally accessible velocities for C_{70} span over several Talbot orders L/L_T, allowing us to study the interference visibility as a function of Talbot order. The dashed-dotted line in Fig. 10.8b compares our data with the assumption of a pure phase grating, the dashed line represents a model assuming incoherent stochastic absorption, and the solid line represents a full quantum model including coherent absorption in an optical grating (Cotter et al., 2015). The agreement with the experiment is clearly best for the complete model.

One may wonder why decoherence by photon emission, as observed for atoms (Chapman et al., 1995; Kokorowski et al., 2001) and fullerenes (Hackermüller et al., 2004; Hornberger et al., 2005), does not destroy the interferogram. In many complex molecules the emission process can be suppressed, because of vibrational relaxation on the picosecond scale and non-radiative transitions to meta-stable states on the nanosecond scale (Dresselhaus et al., 1998). This eliminates the stochastic recoil of the spontaneous emission process and the distribution of which-path information to the environment. In our experiments fluorescence and phosphorescence are not entirely excluded and might localize the molecule to half of the wavelength of the emitted radiation. This has been taken into account in a recent theoretical model, too (Walter et al., 2016).

Universal photo-depletion gratings Measurement-induced gratings are not only implicitly present in KDTL interferometry, but have been also explicitly used in

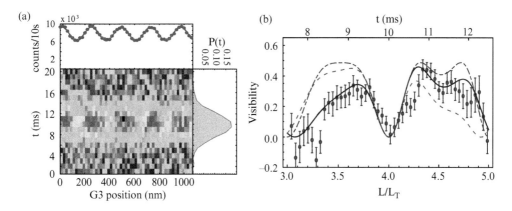

Fig. 10.8 Coherent absorption in KDTLI: (a) The time-of-flight resolved molecular interference pattern has an almost Gaussian velocity distribution (right) and a near-sinusoidal integrated molecular density distribution in real space, as also expected in theory. (b) Interference fringe visibility as a function of the scaled interferometer length $L/L_T = Lmv/hd^2$. Both a pure phase grating model (dash-dotted) and a model assuming a random walk in momentum space (dashed) fail to reproduce the experiment. The full quantum model (continuous line) accounts for coherent single-photon absorption, periodic phase shifts, as well as a measurement-induced grating (see text).

atom optics, before. Diffraction at depopulation gratings was for instance observed with argon atoms that were optically pumped from a detectable metastable state to the undetected ground state (Abfalterer et al., 1997). Pulsed optical gratings in the time-domain were also essential in realizing a complete atom interferometer (Fray et al., 2004).

The biggest difference between quantum optics of atoms and many-body systems consists in the way we can address the latter. Resonant interactions are typically precluded, but photon absorption can still deplete particle beams in various ways.

Single-photon ionization gratings were suggested to enable new tests of high-mass quantum physics in metal cluster interferometry (Reiger et al., 2006). The idea was theoretically corroborated (Nimmrichter et al., 2011), and experimentally demonstrated (Haslinger et al., 2013) in OTIMA interferometry, a matter-wave interferometer with three optical photo-depletion gratings in the time-domain. Vacuum ultraviolet nanosecond laser pulses allowed establishing the probably shortest conceivable optical grating period (with $d = 78.5$ nm) and to provide a photon energy high enough to ionize all metal clusters, many dielectric particles, and tryptophan-rich peptides. Single-photon ionization is independent of any specific internal resonance and therefore counts as an ingredient for a 'universal' diffraction element in molecular quantum optics.

Even particles with ionization energies exceeding the energy of a single photon can often be treated using a similar concept. Weakly bound van der Waals clusters may absorb a photon and rather dissociate than ionize—and thus also be removed from the particle beam. Diffraction gratings based on that effect have recently been demonstrated for hexafluorobenzene and vanillin (Dörre et al., 2014).

10.3   Quantum enhanced measurements

The sensitivity of quantum phases to external perturbations is friend and foe at the same time. While it increases the demand for ever better experimental skills, it also creates new opportunities for refined measurements on single particles in ultra-high vacuum. Chemistry, biology, pharmacy, and medicine depend on detailed knowledge of molecular properties, many of which may be obtained through interactions with static or dynamic electrical or magnetic fields, with radiation at all wavelengths or collisions with controlled probe particles.

Matter-wave interferometry can contribute to the field by generating free-flying nanoscale structured molecular density patterns that will shift in response to the external forces. The physics behind this idea is similar to that of atom interferometric sensors, which are typically geared towards measuring inertial forces, caused by gravity and rotations (Cronin et al., 2009; Tino and Kasevich, 2014).

Since the spatial resolution of beam shifts in matter-wave experiments can exceed that of classical deflectometers by far, the quantum advantage bears great potential for molecule metrology, too. This is best exploited if interferometers can handle a large range of diverse atoms, molecules, clusters, covering several orders of magnitude in mass, different internal configurations and excitations. This is what KDTL and OTIMA interferometry can offer.

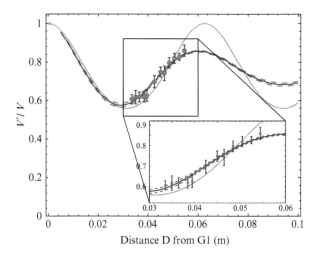

Fig. 10.9 The visibility reduction factor $R = V'/V$ depends on the location where the laser hits the molecule relative to the inter-grating separation. The perturbation is strongest close to G2. A fit of the data to the reduced visibility allows extracting the absolute absorption cross section with high resolution.

Over the past years many different particle properties have been studied in Vienna. This comprises measurements of the static (Berninger et al., 2007) and the optical (Hackermüller et al., 2007; Hornberger et al., 2009) polarizability, as well as permanent (Eibenberger et al., 2011), and vibration-induced electric dipole moments (Gring et al., 2010). Molecule interferometry was shown to complement mass spectrometry in discriminating constitutional isomers—which only differ in their atomic arrangement, but not their mass (Tüxen et al., 2010)—and to attribute the origin of molecular fragmentation (Gerlich et al., 2008). Here we focus on a recent experiment demonstrating the potential of matter-wave assisted spectroscopy on molecules (Nimmrichter et al., 2008; Eibenberger et al., 2014). The idea finds again a close analogy in atom optics, where a photon recoil was used to determine h/m_{Cs} and from this the fine structure constant with high precision (Bouchendira et al., 2011).

Our interest is directed towards the absolute absorption cross section σ_{abs} which has been notoriously difficult to assess for large organic molecules in dilute beams. There are multiple challenges in classical molecular beam physics: firstly, many large organic species have a low vapor pressure, typically smaller than 0.1 mbar, even close to their decomposition temperature. If the absorption cross section is as small as $\sigma_{\mathrm{abs}} = 10^{-17}$ cm^{-2}, which is a typical value for aromatic amino acids, a thermal vapor would need to be extended over 40 cm to reduce the incident light to a fraction of $1/e$. But already in a unperturbed molecular beam, extracted from this vapor and after free-flight over about 2 m distance, the density has dropped to about 10^{10} cm^{-3} where direct absorption would require a prohibitively long vapor cell. Very dilute molecular beams can often still be observed in fluorescence or using an 'action mechanism' such as the evaporation of an adsorbed noble gas atom (Fielicke et al., 2004). But this does not apply to fragile cluster aggregates, non-fluorescent biomolecules, or many molecules of chemical or astrophysical interest. Starting from the experiment, sketched in Fig. 10.6, we realize that the wave function associated with every individual molecules is first delocalized, diffracted and then recombined to obtain the interferogram that is probed by scanning grating G3. In KDTLI, the molecular nanostructure has a periodicity of 266 nm, which can be resolved with a position accuracy of about 10 nm for the signal-to-noise that is typical for molecules below 1000 amu. We now add a *running* laser beam at distance D from G1 along the grating k-vector which is sufficiently intense to ensure the absorption of 0.1–0.3 photons per molecule on average. A delocalized molecule that absorbs a photon will experience a recoil by the photon momentum $\Delta p = \hbar k_K$ and be excited to a higher-lying electronic state. It exhibits an interferogram displaced by the distance $s = \lambda_{dB} D / \lambda_K$, where λ_K is the wavelength of the running laser beam. In our particular example the photon wavelength of 532 nm was comparable to the wave packet separation close to the second grating. And yet, high-contrast de Broglie interference can be maintained since the absorption process is phase coherent. The total pattern is the sum of all shifted and unshifted ones, but there are only two discrete options. Thus, the overall contrast is reduced, but by an amount which can be unambiguously correlated with the number of absorbed photons. When the laser beam parameters are well known, one can extract the absolute absorption cross section with few percent precision. Photon absorption diminishes the unperturbed interference fringe visibility V and leads to a reduced visibility $V' = \langle R \rangle_v V$, with the velocity averaged reduction factor $\langle R \rangle_v$ (Eibenberger et al., 2014)

$$\langle R \rangle_v = \left| \int_0^\infty dv_z P(v_z) \exp\left(-n_0 \left[1 - \exp(2\pi i s / d)\right]\right) \right|. \tag{10.8}$$

The realization of this idea allowed characterizing the absolute cross section for the fullerenes C_{70} to within 2%. For this proof-of-principle experiment a high-power continuous solid-state laser was used. Future upgrades can be based on widely tunable radiation as provided by a dye laser, titanium sapphire laser and their higher harmonics. The scheme is also open to a number of multiphoton combinations (Rodewald et al., 2016).

10.4   Molecular beam sources for nanoscale organic matter and biomolecules

If we want to pursue quantum experiments with high-mass objects in analogy to advanced atom interferometry, it is necessary to prepare beams of isolated neutral particles that are sufficiently slow and cold, selected in mass, and to some extent also geometry. The neutrality requirement is not fundamental—as a matter of fact electron interferometry is an integral part of electron microscopy and holography (Tonomura, 1999)—however it is of practical importance, since charged particles are easily perturbed by external fields. At present, the helium cation He^+ is still the most massive species in fundamental matter-wave physics with charged particles (Hasselbach, 2010). Other than that, we require the particle momentum to be smaller than 10^7 amu $\times$ m/s, corresponding to a minimum de Broglie wavelength of $\lambda_{dB} = 40$ fm. This is the target value for the next-generation of macromolecule interferometers in Vienna.

In single-grating far-field diffraction, mass selection is important, since the scattering angle scales inversely with particle mass. In interferometric near-field self-imaging, however, the period of the fringe pattern is always determined by the interferometer geometry. The particle mass then determines the fringe contrast, via the match or mismatch between the Talbot length $L_T = d^2/\lambda_{dB}$ or Talbot time $L_T = d^2 m/h$ and multiples of the grating distance. In practice, advanced near-field interferometers can tolerate a span of $\Delta\lambda_{dB}/\lambda_{dB} \simeq 10\%$. Deviations from that rule of thumb emerge for $m > 10^5$ amu, because of dispersive phase shifts.

A key difference between molecule and atom interferometry is the internal particle structure. In our current experiments, we are mostly interested in de Broglie interference, that is the evolution of the particle's center-of-mass motion. However, we recall that an N −body system can store energy in 3 translational, 3 rotational and $3N − 6$ vibrational modes, most of which will be (highly) excited at room temperature.

Vibrational modes, with an energy of $0.01-0.3$ eV, may be frozen out in the presence of cryogenically cold buffer gas. However, even then a large molecule may be trapped in a local energy minimum of a vast conformational landscape.

It is even more challenging to approach the rotational ground state. An insulin molecule, for example, with a mass of 5808 amu, is composed of 51 amino acids with a total moment of inertia of about 9×10^{-42} kg m^2. Its rotational energy levels therefore scale like $E_{rot} \simeq 40~\mu K \times J(J + 1)$ and are excited to an average rotational quantum number around a few thousand at room temperature and still around $J \simeq 100$ in a superfluid helium droplet at $T = 380$ mK (Toennies et al., 2001). Human hemoglobin ($m = 68600$ amu), containing about 4800 atoms in a toroidal diameter of about 60 Å has a moment of inertia of 4×10^{-40} kg m^2, pushing the temperature requirements for rotational ground state cooling by another factor of ten.

When we further consider all electronic, fine, hyperfine, and Zeeman sublevels, it is safe to say that all molecules will populate different statistical combinations of their internal states and every particle will only interfere with itself, never with another one. High-contrast de Broglie interferences is still possible as long as the internal states do not provide information about the molecular position.

10.4.1 Thermal beams and thermal molecules in supersonic beams

Thermal sublimation from the solid phase or evaporation from the liquid phase works well for many molecules up to about 1000 amu. This includes all nucleobases, various porphyrins (600–800 amu), the fullerenes C_{60} (720 amu) and C_{70} (840 amu), but also vitamins, such as beta-carotene (pre-vitamin A) or vitamin E (α-tocopherol). Most of them have been studied in KDTL interferometry (Mairhofer et al., 2017).

Vanillin and caffeine have sufficient thermal vapor pressure to form a thermal molecular beam that can be clustered into complex aggregates when injected into an adiabatically expanding supersonic noble gas jet. This has been successfully used in OTIMA interferometry (Dörre et al., 2014). Since they are bound by van der Waals forces, these cluster are prone to dissociation after photon absorption. This makes them susceptible for photo-depletion gratings and photo-depletion spectroscopy, in general.

10.4.2   Tailoring organic materials for improved volatility

The required sublimation or evaporation temperature can be reduced by chemical functionalization. Perfluoroalkyl chains such as $S(CH_2)_2C_8F_{17}$ have been found to substantially reduce the intermolecular binding (Tüxen et al., 2011), since the high electro-negativity of the fluorine atoms binds the charge and reduces the electrical polarizability. Such functionalization has allowed to volatilize porphyrin derivatives for successful quantum interference experiments beyond 10^4 amu (Eibenberger et al., 2013). The same idea has enabled thermal beams of oligopeptides, too (Schätti et al., 2017). Even though one might think that such modifications are artificial, it turns out that fluorinated compounds are commonly used in medicine and pharmacy (Purser et al., 2008). This suggests that also quantum-interference enhanced measurements on *tailored* biomolecules can provide relevant data.

10.4.3   Laser injection of large peptides into (humid) expanding noble gases

Many biomolecules will denature or fragment abundantly, when they are heated for extended periods of time. This can be prevented by reducing the interaction time: Focusing a nanosecond laser beam onto an area of 1 mm^2 with mJ pulse energy can heat a surface at a rate of $\sim 10^{11}$ K/s. Interactions binding the molecules to the surface are thus rapidly broken and even fragile molecules may leave the hot surface before too much energy is absorbed. In order to prevent post-desorption fragmentation, the particles are entrained in an adiabatically expanding noble gas, released by a nearby short-pulse nozzle. At stagnation pressures of several bars and gas pulse times around 20–30 μs the number of collisions is high enough to cool even large peptides. While the current limits of this technique are still being explored, it has been shown that even polypeptides composed of more than a dozen amino acids could successfully be brought into the gas phase using this technique (Geyer et al., 2016). We can also proceed one step further and ask for the influence of microhydration, i.e. the controlled addition of water molecules to the free biomolecule. Such studies shall identify the role of the native environment on the molecular properties and the possible role of evaporation as an enabling agent in optical diffraction gratings or a cause of evaporative decoherence (Geyer et al., 2016).

The implementation of such a source is shown in Fig. 10.10a: Biomolecular powder is picked up by a felt wheel and pressed onto a glassy carbon wheel (Gahlmann et al., 2008). The thin biomolecular layer is then desorbed by a nanosecond laser pulse. The expanding noble gas jet entrains and cools the molecules before they can dissociate.

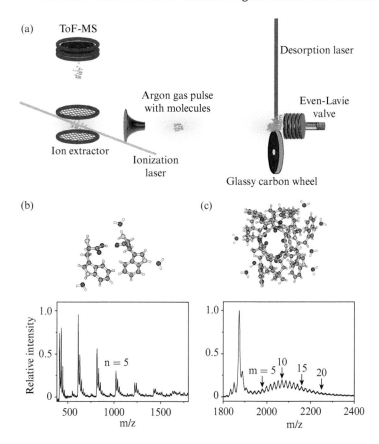

Fig. 10.10 (a) The pulsed nanosecond laser light desorbs molecules from a glassy carbon wheel which are immediately entrained into an expanding noble gas jet. The molecules are post-ionized with vacuum ultraviolet light (157 nm) and detected in a time-of-flight mass spectrometer (TOF-MS). This source generates hydrated cluster beams of tryptophan (b) as well as of polypeptides like gramicidin (c) or indolicidin. Here, n corresponds to the number of clustered amino acids and m to the number of attached water molecules.

This allows launching amino acid clusters (Fig. 10.10b), and peptides (Fig. 10.10c), also with a micro-hydration shell.

10.4.4 Laser-induced acoustic desorption

We may also ask whether it is possible to launch particles from a surface in vacuum without a direct exposure to the laser. This question has been explored by several groups in physical chemistry (Zinovev et al., 2011). *Laser-Induced Acoustic Desorption* (LIAD) has been successfully used to even load lowly charged individual viruses and bacterial cells into an ion trap (Peng et al., 2004). Our goal here is to prepare a beam of macromolecules, sufficiently intense and slow to comply with the coherence requirements in molecule interferometry. We performed demonstration experiments using the setup shown in Fig. 10.11. A nanosecond laser pulse (1 mJ) is directed onto

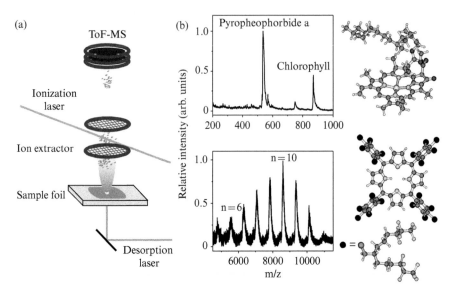

Fig. 10.11 (a) An energetic nanosecond laser pulse at 355 nm hits the back side of a thin titanium or tantalum foil. The ablation shock releases biomolecules from the front side of the metal sheet. The molecular velocity averages around 160 m/s for molecules in the mass range of porphyrins desorbed from tantalum. Surprisingly, we achieve velocities as low as 30 m/s when launching molecules from titanium foil. The internal molecular temperature has not been determined. For a given experimental setup the molecule velocity scales like $v \propto m^{-1/2}$. (b) The mass spectrum of chlorophyll shows that the fragile parent molecule is successfully released from the surface and detected after photo-ionization at 157 nm. The TPP derivatives represent an entire library of functionalized biochromophores which fly intact and slow using LIAD and can also be photo-ionized. Here, n is the number of attached side chains.

the back side of a thin metal sheet. This sheet is only 10–20 μm thick and covered on its front side by an even thinner layer of molecules. When the laser beam hits the back side of the foil, it generates a fast metal plume which imparts a strong recoil onto the sheet and also releases molecules from the front side.

In our experiment (see Fig. 10.11) we were able to prepare slow beams of neutral biomolecules and functionalized organic libraries (Sezer et al., 2015), which were post-ionized using vacuum ultraviolet light at 157 nm and mass-analyzed in time-of-flight mass spectroscopy. We observed neutral beams of chlorophyll and large functionalized porphyrins.

While intact chlorophyll has an average speed of 50 to 160 m/s depending on the foil material, tailored porphyrins with $m = 10^4$ amu were measured to have a mean velocity around $v = 20$ m/s. Their kinetic energy of 20 meV corresponds to $T \simeq 270\,^\circ$C assuming a thermal launching process. This is supported by the fact that the measured velocity scales like $v \propto m^{-1/2}$. Although the LIAD process is still not entirely understood it seems to have both a thermal and a mechanical contribution (Zinovev et al., 2011; Shea et al., 2007).

We also showed that LIAD can produce nanoparticles from pristine silicon wafers (Asenbaum et al., 2013) and nanostructured silicon templates with tailored nanorods

(Kuhn et al., 2016). Particles as massive as 10^{10} amu were broken off and launched at velocities between 0.2-30 m/s. The kinetic energy of such a particle flying at 10 m/s amounts to 5.2 keV. It seems that during the launching process energy stored in the deformation of the substrate is released. Even starting from such high energy, cooling in a high finesse cavity was observed to reduce the particle's transverse kinetic energy by about a factor of thirty (Asenbaum et al., 2013).

This observation is also a first step in a series of experiments towards preparing particles for quantum interferometry with masses around $10^6 - 10^7$ amu. This ongoing research work complements efforts in other places where active feedback cooling (Gieseler et al., 2012) or cavity cooling (Kiesel et al., 2013; Millen et al., 2015) of trapped and neutral particles is pursued.

10.4.5 Molecular ions as the basis for neutral molecular beams

Since the advent of *Matrix Assisted Laser Desorption Ionization* (MALDI) (Tanaka et al., 1988; Karas et al., 1989) and *Electrospray Ionization* (ESI) (Fenn et al., 1989) charged protein and DNA beams have been heavily studied using mass spectrometry. MALDI differs from laser desorption (LD) in that the analyte molecules (proteins, DNA, etc.) are embedded in an acidic molecular matrix on a substrate. The matrix molecules are optimized for high absorption around 330–350 nm which can be excited by the radiation of a nitrogen laser or a frequency-tripled Nd:YAG laser. Their wavelength is red-shifted with respect to the absorption spectrum of peptide bonds (~213 nm), aromatic amino acids, and nucleobases (260–290 nm). The photon energy is therefore dominantly deposited in the matrix molecules which evaporate and entrain the analyte particles. Since the matrix consists of proton donors it can also protonate and charge the analyte molecules. MALDI typically launches singly charged biomolecules of either polarity in high vacuum at moderate repetition rate and is well suited for short experiments with small samples. In contrast to that, electrospray ionization is advantageous for preparing steady continuous beams. Here analyte molecules are filled into a narrow capillary, which is placed at a distance to an opening in the vacuum machine. At high electric fields, the liquid forms a cone which breaks into a filament and further into droplets which partially dry in air and also eject further droplets, because of the growing imbalance between Coulomb repulsion and surface tension. A series of emission and evaporation processes leads finally to the release of individual analyte molecules. Hence, the mass-to-charge ratio in polypeptides and proteins usually ranges between 1000–2000 amu per elementary charge in ESI experiments. This is ideal for mass analysis in commercial quadrupole mass filters, but less advantageous for quantum experiments with neutral beams. Charge exchange techniques from aerosol science can reduce the charge per particle by collisions in bi-polar air (Bacher et al., 2001) before they are transferred into high vacuum through ion guides and a differential pumping system. The molecules are mass-selected in a quadrupole mass filter in high vacuum. A subsequent hexapole guide allows collisions with cryogenically cold (typically 60 K) neutral buffer gas (see Fig. 10.12).

ESI works over a very wide mass scale from amino acids to viruses. However, the manipulation, control and mass spectrometry needs to be adapted for a specific mass

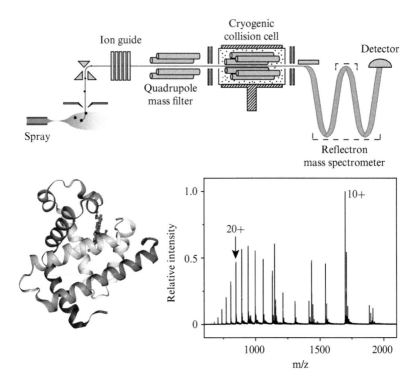

Fig. 10.12 Electrospray ionization yields highly charged biomolecules which are mass-selected and subsequently cooled in the presence of a cryogenic gas. The detection takes place in a high-resolution time-of-flight mass spectrometer. For proteins the size of myoglobin (153 amino acids, 17000 amu), ESI natively ejects highly-charged molecules.

scale. The setup in Vienna is currently optimized for a maximal mass-to-charge ratio of up to 30000 amu/e. The extraction of neutral particles by photo-depletion processes is an open challenge still under investigation (Sezer et al., 2011).

10.5 Conclusion

Quantum optics with complex molecules has many far-reaching goals and faces a plethora of intriguing challenges. The field shares many concepts and solutions with atom interferometry and atom optics, enriched by the complexity of strongly bound systems, composed of 1000 atoms at present and possibly 100 000 atoms in the future.

The branch of organic and biomolecular interferometry has been developing with new ideas for novel quantum and decoherence experiments, but also new hope for a radically different way of measuring biomolecular matter under controlled conditions. These experiments aim at single-molecule sensitivity, either isolated from the surroundings or hydrated with a controlled number of water molecules and also in interaction with single photons.

Matter-wave physics with the building blocks of life is still a young research activity which profits from vigorous and fruitful interactions with researchers in physical and computational chemistry, biochemistry, nanoimaging, and nanotechnology. Many quantum manipulation and interference techniques have already been developed. Further success is strongly tied with rapid development in molecular beam technologies.

Acknowledgments

This summary encompasses a selection of years of fruitful research in the Quantum Nanophysics Group in Vienna, including present and former members, as well as our collaboration partners around Ori Cheshnovsky (Tel Aviv University), Klaus Hornberger (University of Duisburg-Essen) and Marcel Mayor (University of Basel). MA is grateful to Anton Zeilinger (University of Vienna) for many years of fruitful collaboration on macromolecule interferometry. We thank them all for their inspiring and important input over many years that is only partially reflected in this lecture. We additionally thank Maxime Debiossac and Moritz Kriegleder for recording the mass spectrum of myoglobin. We are grateful to financial support through the European Research Council project 320694, the European Commission within project 304886 as well as the Austrian Science Funds project W1210-3.

References

Abfalterer, R. et al. (1997). Nanometer definition of atomic beams with masks of light. *Phys. Rev. A*, **56**, R4365–R4368.

Akhmetov, A. et al. (2010). Laser desorption postionization for imaging MS of biological material. *J. Mass Spectrom.*, **45**, 137–45.

Andrews, M. R. et al. (1997). Observation of interference between two Bose-Einstein condensates. *Science*, **275**, 637–41.

Arndt, M. et al. (2014). Matter-wave interferometry with composite quantum objects. In *Atom Interferometry - Proceedings of the International School of Physics "'Enrico Fermi"'*, *Vol. 188* (ed. G. M. Tino and M. A. Kasevich), pp. 89–141. IOSPress.

Arndt, M., and Hornberger, K. (2014). Testing the limits of quantum mechanical superpositions. *Nat. Phys.*, **10**, 271–7.

Arndt, M. et al. (1999). Wave-particle duality of C_{60} molecules. *Nature*, **401**, 680–2.

Asenbaum, P. et al. (2013). Cavity cooling of free silicon nanoparticles in high vacuum. *Nat. Commun.*, **4**, 2743.

Bacher, G. et al. (2001). Charge-reduced nano electrospray ionization combined with differential mobility analysis of peptides, proteins, glycoproteins, noncovalent protein complexes and viruses. *J. Mass Spectrom.*, **36**, 1038–52.

Bassi, A., Dürr, D., and Hinrichs, G. (2013a). Uniqueness of the equation for quantum state vector collapse. *Phys. Rev. Lett.*, **111**, 210401.

Bassi, A. et al. (2013b). Models of wave-function collapse, underlying theories, and experimental tests. *Rev. Mod. Phys*, **85**, 471–527.

Batelaan, H. (2007). Colloquium: illuminating the Kapitza-Dirac effect with electron matter optics. *Rev. Mod. Phys.*, **79**, 929–41.

Bateman, J. et al. (2014). Near-field interferometry of a free-falling nanoparticle from a point-like source. *Nat. Commun.*, **5**, 4788.

Becker, C. H. et al. (1995). On the photoionization of large molecules. *J. Am. Soc. Mass Spectrom.*, **6**, 883–8.

Berninger, M. et al. (2007). Polarizability measurements of a molecule via a near-field matter-wave interferometer. *Phys. Rev. A*, **76**, 013607.

Bloch, I., Hänsch, T., and Esslinger, T. (1999). Atom Laser with a cw output coupler. *Phys. Rev. Lett.*, **82**, 3008–11.

Bohr, N. (1949). Discussion with Einstein on epistemological problems in atomic physics. In *Albert Einstein, Philosopher-Scientist* (ed. P. A. Schilpp). Tudor, New York.

Bordé, Ch. J. (1989). Atomic interferometry with internal state labelling. *Phys. Lett. A*, **140**, 10–12.

Bouchendira, R. et al. (2011). New determination of the fine structure constant and test of the quantum electrodynamics. *Phys. Rev. Lett.*, **106**, 080801.

Brand, C. et al. (2015*a*). A Green's function approach to modeling molecular diffraction in the limit of ultra-thin gratings. *Ann. Phys. (Berlin)*, **527**, 580–91.

Brand, C. et al. (2015*b*). An atomically thin matter-wave beamsplitter. *Nat. Nanotechnol.*, **10**, 845–8.

Brezger, B., Arndt, M., and Zeilinger, A. (2003). Concepts for near-field interferometers with large molecules. *J. Opt. B*, **5**, S82–S89.

Brezger, B. et al. (2002). Matter-wave interferometer for large molecules. *Phys. Rev. Lett.*, **88**, 100404.

Cahn, S. B. et al. (1997). Time-domain de Broglie wave interferometry. *Phys. Rev. Lett.*, **79**, 784–7.

Carnal, O. et al. (1991). Imaging and focusing of atoms by a Fresnel zone plate. *Phys. Rev. Lett.*, **67**, 3231–4.

Chapman, M. et al. (1995). Optics and Interferometry with Na_2 Molecules. *Phys. Rev. Lett.*, **74**, 4783–6.

Chapman, M. S. et al. (1995). Photon scattering from atoms in an atom interferometer: coherence lost and regained. *Phys. Rev. Lett.*, **75**, 3783–7.

Clauser, J. F. (1997). De Broglie-wave interference of small rocks and live viruses. In *Exp. Metaphys.* (ed. R. S. Cohen, M. Horne, and J. Stachel), pp. 1–11. Kluwer Academic.

Clauser, J. F., and Li, S. (1994). Talbot-vonLau atom interferometry with cold slow potassium. *Phys. Rev. A*, **49**, R2213–R2216.

Cotter, J. P. et al. (2015). Coherence in the presence of absorption and heating in a molecule interferometer. *Nat. Commun.*, **6**, 7336.

Cronin, A. D., Schmiedmayer, J., and Pritchard, D. E. (2009). Optics and interferometry with atoms and molecules. *Rev. Mod. Phys.*, **81**, 1051–1129.

Davisson, C., and Germer, L. H. (1927). The scattering of electrons by a single crystal of nickel. *Nature*, **119**, 558–60.

de Vries, M. S., and Hobza, P. (2007). Gas-phase spectroscopy of biomolecular building blocks. *Annu. Rev. Phys. Chem.*, **58**, 585–612.

Delhuille, R. et al. (2001). High-contrast Mach-Zehnder lithium-atom interferometer in the Bragg regime. *Appl. Phys. B*, **74**, 489–93.

Denschlag, J. (2000). Generating solitons by phase engineering of a Bose-Einstein condensate. *Science*, **287**, 97–101.

Diósi, L. (1984). Gravitation and quantum-mechanical localization of macro-objects. *Phys. Lett. A*, **105**, 199–202.

Dörre, N. et al. (2014). Photofragmentation beam splitters for matter-wave interferometry. *Phys. Rev. Lett.*, **113**, 233001.

Dresselhaus, M. S., Dresselhaus, G., and Eklund, P. C. (1998). *Science of Fullerenes and Carbon Nanotubes* (2nd edn). Acad. Press, San Diego.

Eibenberger, S. et al. (2014). Absolute absorption cross sections from photon recoil in a matter-wave interferometer. *Phys. Rev. Lett.*, **112**, 250402.

Eibenberger, S. et al. (2013). Matter-wave interference of particles selected from a molecular library with masses exceeding 10,000 amu. *Phys. Chem. Chem. Phys.*, **15**, 14696–700.

Eibenberger, S. et al. (2011). Electric moments in molecule interferometry. *New J. Phys.*, **13**, 043033.

Estermann, I., and Stern, O. (1930). Diffraction of molecular beams. *Z. Phys.*, **61**, 95–125.

Fenn, J. B. et al. (1989). Electrospray ionization for mass spectrometry of large biomolecules. *Science*, **246**, 64–71.

Feynman, R., Leighton, R. B., and Sands, M. L. (1965). *The Feynman Lectures on Physics, Vol III, Quantum Mechanics*. Addison Wesley, Reading (Mass.).

Fielicke, A. et al. (2004). Structure determination of isolated metal clusters via far-infrared spectroscopy. *Phys. Rev. Lett.*, **93**, 023401.

Fray, S. et al. (2004). Atomic interferometer with amplitude gratings of light and its applications to atom based tests of the equivalence principle. *Phys. Rev. Lett.*, **93**, 240404.

Freimund, D. L., Aflatooni, K., and Batelaan, H. (2001). Observation of the Kapitza-Dirac effect. *Nature*, **413**(6852), 142–3.

Gahlmann, A., Park, S. T., and Zewail, A. H. (2008). Structure of isolated biomolecules by electron diffraction-laser desorption: Uracil and guanine. *J. Am. Chem. Soc.*, **131**, 2806–8.

Garcia-Sanchez, D. et al. (2008). Imaging mechanical vibrations in suspended graphene sheets. *Nano Lett.*, **8**, 1399–1403.

Gerlich, S. et al. (2008). Matter-wave metrology as a complementary tool for mass spectrometry. *Angew. Chem. - Int. Ed.*, **47**, 6195–8.

Gerlich, S. et al. (2007). A Kapitza–Dirac–Talbot–Lau interferometer for highly polarizable molecules. *Nat. Phys.*, **3**, 711–15.

Geyer, P. et al. (2016). Perspectives for quantum interference with biomolecules and biomolecular clusters. *Phys. Scr.*, **91**, 063007.

Ghirardi, G. C., Rimini, A., and Weber, T. (1986). Unified dynamics for microscopic and macroscopic systems. *Phys. Rev. D*, **34**, 470–91.

Gieseler, J. et al. (2012). Sub-Kelvin parametric feedback cooling of a laser-trapped nanoparticle. *Phys. Rev. Lett.*, **109**, 103603.

Gieseler, J., Novotny, L., and Quidant, R. (2013). Thermal nonlinearities in a nanomechanical oscillator. *Nat. Phys.*, **9**, 806–10.

Giltner, D. M., McGowan, R. W., and Lee, S. A. (1995). Theoretical and experimental study of the Bragg scattering of atoms from a standing light wave. *Phys. Rev. A*, **52**, 3966–72.

Giulini, D., and Großardt, A. (2011). Gravitationally induced inhibitions of dispersion according to the Schrödinger-Newton equation. *Class. Quantum Grav.*, **28**, 195026.

Gould, P. L., Ruff, G. A., and Pritchard, D. E. (1986). Diffraction of atoms by light: the near-resonant Kapitza-Dirac effect. *Phys. Rev. Lett.*, **56**, 827–30.

Gring, M. et al. (2010). Influence of conformational molecular dynamics on matter wave interferometry. *Phys. Rev. A*, **81**, 031604(R).

Grisenti, R. et al. (2000). He-atom diffraction from nanostructure transmission gratings: the role of imperfections. *Phys. Rev. A*, **61**, 033608.

Grisenti, R. E. et al. (1999). Determination of atom-surface van der Waals potentials from transmission-grating diffraction intensities. *Phys. Rev. Lett.*, **83**, 1755–8.

Hackermüller, L. et al. (2004). Decoherence of matter waves by thermal emission of radiation. *Nature*, **427**, 711–14.

Hackermüller, L. et al. (2007). Optical polarizabilities of large molecules measured in near-field interferometry. *Appl. Phys. B*, **89**, 469–73.

Hackermüller, L. et al. (2003). Wave nature of biomolecules and fluorofullerenes. *Phys. Rev. Lett.*, **91**, 090408.

Hagley, E. W. et al. (2001). The atom laser. *Opt. Photonics News*, **12**, 22–6.

Haslinger, P. et al. (2013). A universal matter-wave interferometer with optical ionization gratings in the time domain. *Nat. Phys.*, **9**, 144–8.

Hasselbach, F. (2010). Progress in electron- and ion-interferometry. *Rep. Prog. Phys.*, **73**, 016101.

Holmgren, W. F. et al. (2010). Absolute and ratio measurements of the polarizability of Na, K, and Rb with an atom interferometer. *Phys. Rev. A*, **81**, 053607.

Hornberger, K. et al. (2012). Colloquium: quantum interference of clusters and molecules. *Rev. Mod. Phys.*, **84**, 157–73.

Hornberger, K. et al. (2009). Theory and experimental verification of Kapitza-Dirac-Talbot-Lau interferometry. *New J. Phys.*, **11**, 043032.

Hornberger, K., Hackermüller, L., and Arndt, M. (2005). Influence of molecular temperature on the coherence of fullerenes in a near-field interferometer. *Phys. Rev. A*, **71**, 023601.

Jahns, J., and Lohmann, A. W. (1979). The Lau effect (a diffraction experiment with incoherent illumination). *Opt. Comm.*, **28**, 263–7.

Jönsson, C. (1961). Elektroneninterferenzen an mehreren künstlich hergestellten Feinspalten. *Z. Phys. A*, **161**, 454–74.

Juffmann, T. et al. (2012). Real-time single-molecule imaging of quantum interference. *Nat. Nanotechnol.*, **7**, 297–300.

Juffmann, T. et al. (2009). Wave and particle in molecular interference lithography. *Phys. Rev. Lett.*, **103**, 263601.

Kapitza, P. L., and Dirac, P. A. M. (1933). Reflection of electrons from standing light waves. *Proc. Cambridge Philos. Soc.*, **29**, 297–300.

Karas, M. et al. (1989). Laser desorption ionization mass-spectrometry of proteins of mass 100.000 to 250.000 Dalton. *Angew. Chem. Int. Ed.*, **28**, 760–1.

Kasevich, M. et al. (1991). Atomic velocity selection using stimulated Raman transitions. *Phys. Rev. Lett.*, **66**, 2297–2300.

Keith, D. et al. (1988). Diffraction of atoms by a transmission grating. *Phys. Rev. Lett.*, **61**, 1580–3.

Kiesel, N. et al. (2013). Cavity cooling of an optically levitated nanoparticle. *Proc. Natl. Acad. Sci. USA*, **110**, 14180–5.

Knobloch, C. et al. (2017). On the role of the electric dipole moment in the diffraction of biomolecules at nanomechanical gratings. *Fortschr. Phys.*, 65, 1600025.

Kokorowski, D. A. et al. (2001). From single- to multiple-photon decoherence in an atom interferometer. *Phys. Rev. Lett.*, **86**, 2191–5.

Kovachy, T. et al. (2015). Quantum superposition at the half-metre scale. *Nature*, **528**, 530–3.

Kuhn, S. et al. (2016). Full rotational control of levitated silicon nanorods. *arXiv*, 1608.07315v1.

Laogun, A. A., Sheppard, R. J., and Grant, E. H. (1984). Dielectric properties of insulin in solution. *Phys. Med. Biol.*, **29**, 519–24.

Lau, E. (1948). Beugungserscheinungen an Doppelrastern. *Ann. Phys.*, **437**, 417–23.

Mairhofer, L. et al. (2017). Quantum-assisted metrology of neutral vitamins in the gas phase. *Angew. Chem. Int. Ed.*, **56**, 10947-51.

Marksteiner, M. et al. (2008). Gas-phase formation of large neutral alkaline-earth metal tryptophan complexes. *J. Am. Soc. Mass Spectrom.*, **19**, 1021–6.

Millen, J. et al. (2015). Cavity cooling of a single levitated nanosphere. *Phys. Rev. Lett.*, **114**, 123602.

Millen, J., and Xuereb, A. (2016). Perspective on quantum thermodynamics. *New J. Phys.*, **18**, 011002.

Müller, H. et al. (2008). Atom interferometry with up to 24-photon-momentum-transfer beam splitters. *Phys. Rev. Lett.*, **100**, 180405.

Nairz, O., Arndt, M., and Zeilinger, A. (2003). Quantum interference experiments with large molecules. *Am. J. Phys.*, **71**, 319–25.

Nairz, O. et al. (2001). Diffraction of complex molecules by structures made of light. *Phys. Rev. Lett.*, **87**, 160401.

Nimmrichter, S. et al. (2011). Concept of an ionizing time-domain matter-wave interferometer. *New J. Phys.*, **13**, 075002.

Nimmrichter, S. et al. (2008). Absolute absorption spectroscopy based on molecule interferometry. *Phys. Rev. A*, **78**, 063607.

Patterson, D., and Doyle, J. M. (2015). A slow, continuous beam of cold benzonitrile. *Phys. Chem. Chem. Phys.*, **17**, 5372–5.

Peng, W.-P. et al. (2004). Measuring masses of single bacterial whole cells with a quadrupole ion trap. *J. Am. Chem. Soc.*, **126**, 11766–7.

Penrose, R. (1996). On gravity' s role in quantum state reduction. *Gen. Relativ. Gravit.*, **28**, 581–600.

Perreault, J. D., Cronin, A. D., and Savas, T. A. (2005). Using atomic diffraction of Na from material gratings to measure atom-surface interactions. *Phys. Rev. A*, **71**, 053612.

Purser, S. et al. (2008). Fluorine in medicinal chemistry. *Chem. Soc. Rev.*, **37**, 320–30.

Rasel, E. M. et al. (1995). Atom wave interferometry with diffraction gratings of light. *Phys. Rev. Lett.*, **75**, 2633–7.

Rashid, M. et al. (2016). Experimental realization of a thermal squeezed state of levitated optomechanics. *Phys. Rev. Lett.*, **117**(27), 273601.

Reiger, E. et al. (2006). Exploration of gold nanoparticle beams for matter wave interferometry. *Opt. Commun.*, **264**, 326–32.

Rodewald, J. et al. (2016). New avenues for matter-wave-enhanced spectroscopy. *Appl. Phys. B*, **123**, 3.

Sapmaz, S. et al. (2003). Carbon nanotubes as nanoelectromechanical systems. *Phys. Rev. B*, **67**, 235414.

Savas, T. A. et al. (1995). Achromatic interferometric lithography for 100-nm-period gratings and grids. *J. Vac. Sci. Technol. B*, **13**, 2732–5.

Schätti, J. et al. (2017). Tailoring the volatility and stability of oligopeptides. *J. Mass Spectrom.*, **52**, 550–6.

Schlag, E. W., Grotemeyer, J., and Levine, R. D. (1992). Do large molecules ionize? *Chem. Phys. Lett.*, **190**, 521–7.

Schöllkopf, W., and Toennies, J. P. (1994). Nondestructive mass selection of small van der Waals clusters. *Science*, **266**, 1345–8.

Schöllkopf, W., Grisenti, R. E., and Toennies, J. P. (2004). Time-of-flight resolved transmission-grating diffraction of molecular beams. *Eur. Phys. J. D*, **28**, 125–33.

Schrödinger, Erwin (1926). Über das Verhältnis der Heisenberg-Born-Jordanschen Quantenmechanik zu der meinen. *Ann. Phys.*, **384**, 734–56.

Schwing, K., and Gerhards, M. (2016). Investigations on isolated peptides by combined IR/UV spectroscopy in a molecular beam—structure, aggregation, solvation and molecular recognition. *Int. Rev. Phys. Chem.*, **35**, 569–677.

Sezer, U. et al. (2015). Laser-induced acoustic desorption of natural and functionalized Biochromophores. *Anal. Chem.*, **87**, 5611–19.

Sezer, U. et al. (2017). Selective photodissociation of tailored molecular tags as a tool for quantum optics. *Beilstein J. Nanotechnol.*, **8**, 325–33.

Shea, R. C. et al. (2007). Experimental investigations of the internal energy of molecules evaporated via laser-induced acoustic desorption into a Fourier transform ion cyclotron resonance mass spectrometer. *Anal. Chem.*, **79**, 1825–32.

Tanaka, K. et al. (1988). Protein and polymer analyses up to m/z 100 000 by laser ionization time-of-flight mass spectrometry. *Rapid Commun. Mass Spectrom.*, **2**, 151–3.

Tino, G. M., and Kasevich, M. A. (2014). *Atom Interferometry*. IOS Press, Amsterdam.

Toennies, J. P., Vilesov, A. F., and Whaley, K. B. (2001). Superfluid helium droplets: a ultracold nanolaboratory. *Phys. Today*, **54**, 31–7.

Thomson, G. P. (1927). The diffraction of cathode rays by thin films of platinum. *Nature*, **120**, 802.

Tonomura, A. (1999). *Electron holography.* Springer, New York.

Tüxen, J. et al. (2011). Highly fluorous porphyrins as model compounds for molecule interferometry. *Eur. J. Org. Chem.*, 4823–33.

Tüxen, J. et al. (2010). Quantum interference distinguishes between constitutional isomers. *Chem. Comm.*, **46**, 4145–7.

Walter, K., Nimmrichter, S., and Hornberger, K. (2016). Multiphoton absorption in optical gratings for matter waves. *Phys. Rev. A*, **94**, 043637.

Westbrook, N. et al. (1998). New physics with evanescent wave atomic mirrors: the van der Waals force and atomic diffraction. *Phys. Scr.*, **T78**, 7–12.

Yogev, S. et al. (2008). Charging of dielectrics under focused ion beam irradiation. *J. Appl. Phys.*, **103**, 64107.

Zeilinger, A. et al. (1988). Single- and double-slit diffraction of neutrons. *Rev. Mod. Phys.*, **60**, 1067–73.

Zinovev, A., Veryovkin, I., and Pellin, M. (2011). Molecular desorption by laser–driven acoustic waves: analytical applications and physical mechanisms. In *Acoustic Waves-From Microdevices to Helioseismology* (ed. M. G. Beghi), Chapter 16, pp. 343–68. InTech.

11
Schrödinger cat states
in circuit QED

STEVEN M. GIRVIN

Yale Quantum Institute
PO Box 208334
New Haven, CT 06520-8263 USA

Girvin, S. M. "Schrödinger cat states in circuit QED." In *Current Trends in Atomic Physics*.
Edited by Antoine Browaeys, Thierry Lahaye, Trey Porto, Charles S. Adams, Matthias
Weidemüller, and Leticia F. Cugliandolo. Oxford University Press (2019).
© Oxford University Press. DOI: 10.1093/oso/9780198837190.003.0011

Chapter Contents

The last fifteen years have seen spectacular experimental progress in our ability to create, control and measure the quantum states of superconducting 'artificial atoms' (qubits) and microwave photons stored in resonators. In addition to being a novel testbed for studying strong-coupling quantum electrodynamics in a radically new regime, 'circuit QED', defines a fundamental architecture for the creation of a quantum computer based on integrated circuits with semiconductors replaced by superconductors. The artificial atoms are based on the Josephson tunnel junction and their relatively large size (~mm) means that they couple extremely strongly to individual microwave photons. This strong coupling yields very powerful state-manipulation and measurement capabilities, including the ability to create extremely large (>100 photon) 'cat' states and easily measure novel quantities such as the photon number parity. These new capabilities have enabled a highly successful scheme for quantum error correction based on encoding quantum information in Schrödinger cat states of photons.

11.1   Introduction to Circuit QED

Circuit quantum electrodynamics ('circuit QED') (Blais et al., 2004; Wallraff et al., 2004; Blais et al., 2007; Schoelkopf and Girvin, 2008; Girvin, 2014; Devoret and Schoelkopf, 2013; Nigg et al., 2012; Solgun and DiVincenzo, 2015; Vool and Devoret, 2016; Wendin, 2017) describes the quantum mechanics and quantum field theory of superconducting electrical circuits operating in the microwave regime near absolute zero temperature. It is the analog of cavity QED in quantum optics with the role of the atoms being played by superconducting qubits. A detailed pedagogical introduction to the subject with many references is available in the author's lecture notes from the 2011 Les Houches School on Quantum Machines (Girvin, 2014). The present notes will therefore provide only a brief introductory review of the subject and will focus primarily on novel quantum states that can be produced using the strong coupling between the artificial atom and one or more cavities.

It is a basic fact of QED that each normal mode of the electromagnetic field is an independent harmonic oscillator with the quanta of each oscillator being photons in the corresponding mode. For an optical or microwave cavity, such modes are standing waves trapped inside the cavity. Cavities typically contain many modes, but for simplicity we will focus here on the case of a single mode, coupled to an artificial atom approximated as having only two-levels (and hence described as a pseudo spin-1/2). This leads to the Jaynes-Cummings model (Blais et al., 2004; Girvin, 2014)

$$H = \tilde{\omega}_c a^\dagger a + \frac{\tilde{\omega}_q}{2}\sigma^z + g[a\sigma^+ + a^\dagger\sigma^-] \tag{11.1}$$

describing the exchange of energy between the cavity and the atom via photon absorption and emission within the RWA (rotating wave approximation). Here $\tilde{\omega}_c$ is the (bare) cavity frequency, $\tilde{\omega}_q$ is the (bare) qubit transition frequency and g is the vacuum Rabi coupling which, because of the large qubit size and small cavity volume can be enormously large (~ 150 MHz which is several percent of the ~ 5GHz qubit frequency and several orders of magnitude larger than the cavity and qubit decay rates).

A 'phase diagram' (Schuster et al., 2007) for the Jaynes-Cummings model for different realizations of cavity and circuit QED is illustrated in Fig. 11.1. The vertical axis represents the strength g of the atom-photon coupling in the cavity and the horizontal axis represents the detuning $\Delta = \tilde{\omega}_q - \tilde{\omega}_c$ between the atomic transition and the cavity frequency. Both are expressed in units of $\Gamma = \max[\gamma, \kappa, 1/T]$, where γ and κ are the qubit and cavity decay rates respectively, and for the case of ('real') atoms, T is the transit time for the atoms passing through the cavity[1] In the region $g \geq \Delta$ the qubit and cavity are close to resonant and first-order degenerate perturbation theory applies. The lowest two excited eigenstates are the upper and lower 'polaritons' which are coherent superpositions of atom+photon excitations (Blais et al., 2004; Girvin, 2014). They are essentially the bonding and anti-bonding combinations of 'atom $\pm$ photon'. An excited state atom introduced into the cavity will 'Rabi flop' coherently between being an atomic excitation and a photon at the 'vacuum Rabi rate' $2g$ (which is the energy splitting between the upper and lower polariton). Within this degenerate region there is a 'strong-coupling' regime $g \gg \Gamma$, in which the cavity and qubit undergo multiple vacuum Rabi oscillations prior to decay.

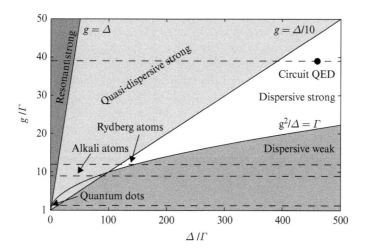

Fig. 11.1 A phase diagram for cavity QED. The parameter space is described by the atom-photon coupling strength, g, and the detuning Δ between the atom and cavity frequencies, normalized to the rates of decay represented by $\Gamma = \max[\gamma, \kappa, 1/T]$. Here γ and κ are the qubit and cavity decay rates respectively, and T is the transit time for atoms passing through the cavity. Different cavity QED systems, including Rydberg atoms, alkali atoms, quantum dots, and circuit QED, are represented by dashed horizontal lines. Since the time this graph was first constructed Γ has decreased dramatically putting circuit QED systems much deeper into the strong dispersive regime. Adapted from (Schuster et al., 2007).

---

[1] Another benefit of circuit QED is that our artificial atoms are 'glued down' and stay in the cavity indefinitely.

We will focus here on the case where the detuning Δ between the qubit and the cavity is large ($\Delta \gg g$). Because of the mismatch in frequencies, the qubit can only virtually exchange energy with the cavity and the vacuum Rabi coupling can be treated in second-order perturbation theory. Applying a unitary transformation which eliminates the first-order effects of g and keeping only terms up to second-order yields (Blais et al., 2004; Girvin, 2014)

$$H = \omega_c a^\dagger a + \omega_q |e\rangle\langle e| - \chi a^\dagger a |e\rangle\langle e|, \tag{11.2}$$

where we have shifted the overall energy by an irrelevant constant, ω_c and ω_q are renormalized cavity and qubit frequencies and $|e\rangle\langle e|$ is the projector onto the excited state of the qubit. The quantity $\chi \approx \frac{g^2}{\Delta} \sim 2\pi * (2 - 10\text{MHz})$ is the effective second-order coupling known as the dispersive shift[2]. In the dispersive regime, the qubit acts like a dielectric whose dielectric constant depends on the state of the qubit. This causes the cavity frequency to depend on the state of the qubit.

As illustrated in Fig. 11.2, when the qubit is in the excited state, the cavity frequency shifts by $-\chi$. As we will soon discuss, this dispersive shift has a number of useful applications, but the simplest is that it can be used to read out the state of the qubit (Blais et al., 2004). We can find the qubit state by measuring the cavity frequency by simply reflecting microwaves from it and measuring the resulting phase shift of the signal. Because the cavity frequency is very different from the qubit frequency, the photons in the readout do not excite or deexcite the qubit and the operation is non-destructive. Even though χ is a second-order effect in g it still can be several thousand times larger than the respective cavity and qubit decay rates, κ and γ. This is the so-called 'strong-dispersive' regime (Schuster et al., 2007; Girvin, 2014) of the phase diagram. The ability to easily enter this regime gives circuit QED great advantages over ordinary quantum optics and will prove highly advantageous for quantum state manipulation, control, and error correction.

The very same dispersive coupling term can also be viewed as producing a quantized 'light shift' of the qubit transition frequency by an amount $-\chi$ for each photon

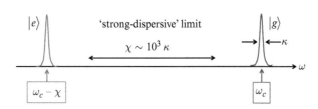

Fig. 11.2 Illustration of cavity response (susceptibility) depending on the state of the qubit. In the qubit ground state, the cavity resonance frequency is ω_c. When the qubit is in the excited state, the cavity frequency shifts by $-\chi$, which can be thousands of times larger than the cavity linewidth κ. This is the 'strong-dispersive' regime. When the dispersive shift is less than a linewidth of the cavity (or the qubit) we are in the 'weak-dispersive' regime.

---

[2] The simple expression given here for the dispersive shift is not accurate for the so-called transmon qubit because it is a weakly anharmonic oscillator and not necessarily well-approximated as a two-level system. For a quantitatively more accurate approach see (Nigg et al., 2012; Girvin, 2014).

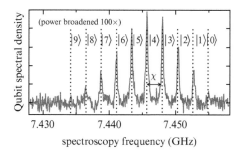

Fig. 11.3 Quantum jump spectroscopy of a transmon qubit dispersively coupled to a cavity illustrating the quantized light shift of the qubit transition frequency depending on how many photons are in the cavity. The weakly damped cavity is driven to produce a coherent state whose time-averaged photon number distribution is Poisson. A weak spectroscopy tone is used to excite the qubit at different frequencies. The spectral density of the qubit is determined by the probability that the spectroscopy tone excites the qubit. In this case a relatively strong spectroscopy tone was used resulting in a power-broadening of the spectral line by approximately 100x. The spectral peaks are actually about 10^3 times narrower than their separation. The symbol $|n\rangle$ denotes the number of photons in the cavity associated with each spectral peak of the qubit. The quantization of the light shift clearly shows that microwaves are particles! Courtesy R. Schoelkopf group.

that is added to the cavity. Fig. 11.3 shows quantum jump spectroscopy data illustrating this quantized light shift. In ordinary spectroscopy of a medium (a gas, say) one measures the spectrum by the absorption of the spectroscopy light at different frequencies. Here we instead use quantum jump spectroscopy in which we use the dispersive readout of the state of the qubit to determine when the qubit jumps to the excited state as a result of excitation by the spectroscopy tone.

The dispersive term commutes with both cavity photon number and qubit excitation number. It is thus 'doubly QND', and can be utilized to make quantum non-demolition measurements of the cavity (using the qubit) and the qubit (using the cavity). In the 'strong-dispersive' region $\chi \gg \Gamma$, quantum non-demolition measurements of photon number are possible. We will see later that in this regime it is even possible to make a QND measurement of the photon number parity without learning the value of the photon number!

In the 'weak-dispersive regime' $(\chi \ll \Gamma)$ of the phase diagram, the light shift per photon is too small to dispersively resolve individual photons, but a QND measurement of the qubit can still be realized by using many photons. The qubit-dependent cavity frequency shift is less than the linewidth of the cavity but with enough photons over a long enough period, this small frequency shift can be detected (using the small phase shift of the reflected signal, assuming that the qubit lifetime is longer than the required mesaurement time) (Clerk et al., 2010).

11.2   Measurement of photon number parity

In the strong-dispersive regime, if we measure the qubit transition frequency the state of the cavity collapses to a Fock state with definite photon number determined by the measured light-shift of the qubit transition frequency. Unlike a photomultiplier, this measurement is not only photon-number resolving, it is QND. The photon is not absorbed and we can repeat the measurement many times to overcome any imperfections in the measurement (improving the quantum efficiency and lowering the dark count). This novel feature has the potential to dramatically accelerate searches for axion dark matter particles which convert into microwave photons in the presence of a strong magnetic field (Zheng et al., 2016).

Another extremely powerful feature of the strong-dispersive regime is that it gives us the ability to measure the parity of the photon number without learning the photon number itself. Why is this interesting? It turns out that in quantum systems, what we do *not* measure is just as important as what we *do* measure! One way to measure the parity would be to measure the eigenvalue of the photon number operator $\hat{n}$. If the result is m, then we assign parity $\Pi = (-1)^m$. This process works, but it does a great deal of 'damage' to the state by collapsing it to the Fock state $|m\rangle$. This particular measurement has strong 'back action'. Let us formalize this by considering an arbitrary cavity state $|\psi\rangle$. The state of the system conditioned on the measurement result

$$|\psi\rangle_m = \frac{|m\rangle\langle m|\psi\rangle}{[\langle\psi|m\rangle\langle m|\psi\rangle]^{\frac{1}{2}}} = |m\rangle \tag{11.3}$$

is completely independent of the starting state. (Only the probability of obtaining the measurement outcome m depends on $|\psi\rangle$.)

If we somehow had a way to directly measure the eigenvalue of the photon number parity operator

$$\hat{\Pi} = e^{i\pi a^\dagger a} = (-1)^{\hat{n}}, \tag{11.4}$$

the back action would be much weaker. This is because there are only two measurement outcomes and the projection onto the even or odd subspaces only cuts out half of the overall Hilbert space (rather than all but one Fock state). Conditioned on the parity measurement being ± 1 we obtain

$$|\psi_\pm\rangle = \frac{\hat{\Pi}_\pm|\psi\rangle}{[\langle\psi|\hat{\Pi}_\pm|\psi\rangle]^{\frac{1}{2}}}, \tag{11.5}$$

where

$$\hat{\Pi}_+ = \sum_{j\in\text{even}} |j\rangle\langle j|, \tag{11.6}$$

$$\hat{\Pi}_- = \sum_{j\in\text{odd}} |j\rangle\langle j|. \tag{11.7}$$

Because we don't learn the value of the photon number, only its parity, the back action is minimized to that associated with the information we wanted to learn. In the first method we learned too much information (namely the value of the photon number not just its parity). The importance of what you do not measure will come to the fore when we discuss how Schrödinger cat states can be used for quantum error correction.

The parity operator in Eq. 11.4 is a non-trivial function of the number operator. At first sight it appears to be quite difficult to measure. It turns out however to be very straightforward to create a situation in which we entangle the state of the qubit with the parity of the cavity. Measuring the qubit then tells us the parity and nothing more. To see how this works, let us move to the interaction picture (i.e. go to an appropriate rotating frame) where the time evolution is governed solely by the dispersive coupling. Moving to a rotating frame via the time-dependent unitary

$$U(t) = e^{iH_0 t}, \qquad (11.8)$$

where

$$H_0 = \omega_c a^\dagger a + \omega_q |e\rangle\langle e|, \qquad (11.9)$$

the new Hamiltonian in the rotating frame becomes

$$V = UHU^\dagger - Ui\frac{d}{dt}U^\dagger = -\chi a^\dagger a |e\rangle\langle e|. \qquad (11.10)$$

Now consider time evolution under this Hamiltonian for a period $t = \pi/\chi$. The unitary evolution operator is

$$U_\pi = e^{+i\pi a^\dagger a |e\rangle\langle e|}. \qquad (11.11)$$

Using the fact that $|e\rangle\langle e|$ is a projector, we can reformulate this in terms of the photon number parity operator as

$$U_\pi = |g\rangle\langle g| + \hat{\Pi}|e\rangle\langle e| = \hat{\Pi}_+ \hat{I} + \hat{\Pi}_- \sigma^z, \qquad (11.12)$$

where $|g\rangle\langle g|$ is the projector onto the qubit ground state. Now sandwich this with Hadamard operators. These operators interchange the roles of the x and z components of the spin: $H|g\rangle = |+x\rangle$ and $H|e\rangle = |-x\rangle$). This yields

$$HU_\pi H = \hat{\Pi}_+ \hat{I} + \hat{\Pi}_- \sigma^x. \qquad (11.13)$$

Thus if the photon number is even, nothing happens to the qubit, whereas if the photon number is odd the qubit is flipped. Hence we have successfully entangled the parity with the qubit state. Starting with the qubit in the ground state, applying $HU_\pi H$ and then measuring the state of the qubit tells us the parity but not the photon number.

Essentially what we have done is the following. We use the Hadamard gate to put the spin in the $+x$ direction on the Bloch sphere. We then allow the qubit to precess around the z axis at a rate that depends on the quantized light shift. If the parity is even, the precession is through an even integer multiple of π radians and returns the qubit to the starting point (and erasing the information on the value of the even integer). If the parity is odd, the qubit precesses through an odd integer multiple of π radians and ends up pointing in the $-x$ direction on the Bloch sphere. The second Hadamard gate converts σ^x to σ^z which we then measure to determine the parity of the cavity photon number.

Provided that we are in the strong-dispersive limit (more precisely, provided that $\chi \gg \bar{n}\kappa$ where $\bar{n} = \langle \psi|\hat{n}|\psi \rangle$) then there is very little chance that an error will result from the cavity losing a photon during the course of the parity measurement. One could also imagine other errors which would make the measurement non-QND. The manipulations of the qubit could for example accidentally add a photon to the cavity. This is unlikely however because the qubit and cavity are strongly detuned from each other. Ref. (Ofek et al., 2016) were able to make parity measurements of a high-Q storage cavity that were 99.8% QND and hence could be repeated hundreds of times without extraneous damage to the state. (The fidelity with which the parity could be determined in a single measurement was 98.5% due to uncertainties in the readout, but these uncertainties were not associated with any adverse back action on the cavity.)

11.3   Application of parity measurements to state tomography

Quantum computation relies on the ability to create and control complex quantum states. The dimension of the Hilbert space grows exponentially with the number of qubits and the task of verifying the accuracy of a particular state (state tomography) or verifying a particular transformation of that state (process tomography) becomes exponentially difficult. To fully determine the state, one has to be able to measure not just the states of individual qubits, but also measure non-local multi-qubit correlators of arbitrary weight $\langle \sigma_i^z \sigma_j^z \sigma_k^x \sigma_l^y ... \sigma_n^z \rangle$, a task which is quite difficult.

For bosonic states of cavities, the quantum jump spectroscopy illustrated in Fig. 11.3 gives an excellent way to measure the photon number distribution in the state of the cavity. This however is far short of the information required to fully specify the quantum state. We need not just the probabilities of different photon numbers but the probability *amplitudes*. In general, the state need not be pure (and the processes we are studying need not be unitary) and so we need to be able to measure the full density matrix which has both diagonal and off-diagonal elements.

It turns out that the ability to make high-fidelity measurements of the photon number parity gives an extremely simple and powerful way to measure the Wigner function. As shown in App. 11.7, the Wigner function provides precisely the same information stored in the density matrix but displays it in a very convenient and intuitive format. The Wigner function $W(X, P)$ is a quasi-probablilty distribution in phase space and can be obtained by the following simple and direct recipe (Lutterbach and Davidovich, 1997): (1) displace the oscillator in phase space so that the point (X, P) moves to the origin; (2) then measure the value of the photon number parity. By repeating many times, one obtains the expectation value of this 'displaced parity'

$$W(X,P) = \frac{1}{\pi\hbar} \text{Tr} \left\{ \mathcal{D}(-X,-P)\rho \mathcal{D}^\dagger(-X,-P)\hat{\Pi} \right\}, \qquad (11.14)$$

where $\mathcal{D}$ is the displacement operator. As shown in App. 11.7, this is precisely the desired Wigner function which fully characterizes the (possibly mixed) state of the system. Related methods in which one measures not the parity but the full photon number distribution of the displaced state (using quantum jump spectroscopy) can in principle yield even more robust results in the presence of measurement noise (Shen et al., 2016).

The Wigner function is a quasi-probability distribution. Unlike wave functions it is guaranteed to be real, but unlike classical probabilities, it can be negative. In quantum optics, states with negative-valued Wigner functions are defined to be 'non-classical'. It should be noted however that the marginal distributions of momentum and position are always ordinary positive-valued probability distributions (much like the square of the wave function)

$$\rho_2(P) = \int_{-\infty}^{+\infty} dX\, W(X, P) \tag{11.15}$$

$$\rho_1(X) = \int_{-\infty}^{+\infty} dP\, W(X, P). \tag{11.16}$$

Exercise 11.1 From the definition of the Wigner function in App. 11.7 prove that

$$\iint dX dP W(X, P) = 1.$$

Since we are dealing with photons in a resonator, it is convenient to replace the position and momentum coordinates of phase space by a single dimensionless complex number β that expresses both position and momentum in units of 'square root of photons'. The displacement operator is then given by

$$\mathcal{D}(\beta) = e^{\beta a^\dagger - \beta^* a}. \tag{11.17}$$

We can check this expression by noting that

$$\mathcal{D}(-\beta) a \mathcal{D}(+\beta) = a + \beta \tag{11.18}$$

and that the displacement of the vacuum yields the corresponding coherent state

$$\mathcal{D}(\beta)|0\rangle = e^{\beta a^\dagger} e^{-\beta^* a} e^{-\frac{1}{2}[\beta a^\dagger, -\beta^* a]}|0\rangle = e^{-\frac{|\beta|^2}{2}} e^{\beta a^\dagger}|0\rangle = |\beta\rangle \tag{11.19}$$

with mean photon number $\bar{n} = |\beta|^2$. In these units, the (now-dimensionless) Wigner function is given by

$$W(\beta) = \frac{2}{\pi} \mathrm{Tr}\left\{\rho \mathcal{D}^\dagger(-\beta) \hat{\Pi} \mathcal{D}(-\beta)\right\}, \tag{11.20}$$

where we have used the fact that the Jacobian for the transformation from X, P to $\beta = \beta_R + i\beta_I$ obeys

$$dX\,dP = 2\hbar\,d\beta_R\,d\beta_I. \tag{11.21}$$

Exercise 11.2   Verify Eq. 11.17. Hint: differentiate both sides of the equation with respect to (the magnitude of) β.

Exercise 11.3   (a) Verify Eq. 11.21. (b) Verify the normalization $\iint d\beta_R\,d\beta_I\,W(\beta) = 1$ directly from Eq. 11.20.

Exercise 11.4   Using Eq. 11.17, show that the Wigner function of the coherent state $|\alpha\rangle$ is a gaussian centered at the point α and given by: $W(\beta) = \frac{2}{\pi}e^{-2|\alpha-\beta|^2}$.

As described in App. 11.7, the experimental procedure to measure the Wigner function is very simple. One applies a microwave tone resonant with the cavity and having the appropriate phase, amplitude, and duration to displace the cavity in phase space by the desired amount. One then uses the standard procedure described above to determine whether the parity is $+ 1$ or $- 1$. This procedure is repeated and the results averaged to obtain the mean value of the parity operator. This 'continuous variable' quantum tomography procedure is vastly simpler and more accurate than having to measure different combinations of multi-qubit correlators as required to do tomography in the discrete variable case. Furthermore the microwave cavity is simple (literally an 'empty box'), and the tomography can be performed with only a single ancilla transmon and cavity. This hardware efficiency is extremely powerful.

11.4   Creating cats

The Schrödinger cat paradox has a long and storied history in quantum mechanics. In circuit QED the cat being dead or alive is represented by a qubit being in the ground state $|g\rangle$ or the excited state $|e\rangle$. The role of the poison molecules in the air is played by a coherent state $|\alpha\rangle$ of photons in the cavity. The quantum state we want to create is

$$|\Psi\rangle = \frac{1}{\sqrt{2}}[|e\rangle|0\rangle \pm |g\rangle|\alpha\rangle. \tag{11.22}$$

This is a coherent superposition of 'cat alive, no poison in the air' and 'cat dead, poison in the air'. Notice that this is *not* a superpostion of 'cat dead' and 'cat alive'. (At this point the choice ± 1 of the phase of superposition is not particularly important, but we retain it for later use.) It is an *entangled* state between the cat and the poison. For large α the two states of the cavity are macroscopically distinct and orthogonal. Note that it is important that the cavity be long-lived because the (initial) rate of photons leaking out from the coherent state is $\kappa\bar{n} = \kappa|\alpha|^2$, where κ is the cavity damping rate. If a photon ever leaks out of the cavity into the environment, the environment immediately collapses the state to a product state in which the cat is dead. On the other hand, if no photons are detected after a long time, we grow more and more certain that they were never there in the first place and the second component of the state gradually damps out collapsing the system to $|e\rangle|0\rangle$. This quantum back action from *not* observing a photon entering the environment is quite a novel and subtle effect. See (Haroche and Raimond, 2006; Michael et al., 2016) for further discussion. See also the short story 'Silver Blaze', by Sir Arthur Conan Doyle in which Sherlock Holmes solves a crime by noting the curious fact that a dog did not bark in the night.

How do we create this Schrödinger cat state? It is surprisingly easy in the strong-dispersive coupling regime. The recipe is as follows. First apply a $\pi/2$ pulse to the qubit to put it in the state $|+x\rangle = \frac{1}{\sqrt{2}}[|g\rangle + |e\rangle]$. Then apply a drive tone to the cavity at frequency ω_c. As we can see from Fig. 11.2, this drive will be on resonance with the cavity (and hence able to displace the cavity state from $|0\rangle$ to $|\alpha\rangle$) if and only if the qubit is the ground state. If the qubit is in the excited state, the cavity remains in the vacuum state (to a very good approximation since the drive is thousands of linewidths off resonance in this case). Hence we immediately obtain the cat of Eq. 11.22. It is straightforward to produce very 'large' cats with $\bar{n} = |\alpha|^2$ corresponding to hundreds of photons but our ability to do high-fidelity state tomography begins to degrade above about 100 photons (Vlastakis et al., 2013a).

Another interesting state, confusingly referred to in the literature not as a 'Schrödinger cat' but as a 'cat state' of photons is given by

$$|\Psi_\pm\rangle = |g\rangle \frac{1}{\sqrt{2}}[|+\alpha\rangle \pm |-\alpha\rangle]. \tag{11.23}$$

This is a product state in which the qubit is not entangled with the cavity but the cavity is in a quantum superposition of two different (and for large α, macroscopically distinct) coherent states. (The normalization is approximate and only becomes exact in the asymptotic limit of large $|\alpha|$. We will ignore this detail throughout our discussion.) Because the qubit state factors out we will drop it from further discussion. The $\pm$ sign in the superposition labels the photon number parity of the state. That is, these are eigenstates of the parity operator:

$$\hat{\Pi}|\Psi_\pm\rangle = \pm|\Psi_\pm\rangle. \tag{11.24}$$

The parity of the state can be directly verified using the defining property of coherent states in Eq. 11.19 but is perhaps best visualzed in terms of the first-quantization wave function for the cat state. Since the ground state wave function is a gaussian, the cat state consists of a sum or difference of two displaced gaussians as illustrated in Fig. 11.4. We see that the spatial parity symmetry under coordinate reflection and the photon number parity are entirely equivalent. Mathematically this is because photon number parity reverses the position and momentum of the oscillator (i.e. inverts phase space), a fact which is readily verified using e.g. $\hat{\Pi}\hat{x}\hat{\Pi} \sim \hat{\Pi}(a+a^\dagger)\hat{\Pi} = -(a+a^\dagger)$.

The Wigner function of cat states of photons is very interesting. As one can see in Fig. 11.5, there are peaks of quasi-probability in phase space at $\pm\alpha$ as expected, but in addition the cat has 'whiskers'. These periodic oscillations are a kind of interference

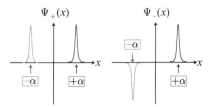

Fig. 11.4 First quantization wave functions for even and odd parity cat states.

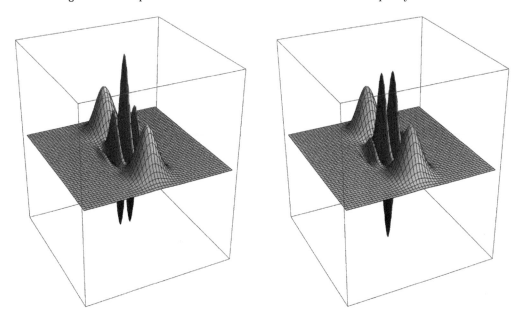

Fig. 11.5 Wigner functions for even- (left panel) and odd-parity (right panel) cat states of size $\alpha = 2.5$. The foreground and background peaks are associated with the coherent states $|\pm\, \alpha\rangle$. Note that the parity oscillations near the origin have opposite sign and represent interference fringes associated with the coherence of the superposition of the two distinct states.

pattern (much like a two-slit interference pattern between the two lobes) that are present for cat states but not for incoherent mixtures of $|+\alpha\rangle$ and $|-\alpha\rangle$.

Exercise 11.5   Using Eq. 11.20 to write down an (approximate) analytic expression for the Wigner function $W(\beta)$ of a cat state. Assume an even (odd) cat of the form in Eq. 11.23 with α real and positive, and sufficiently large that the normalization of the states is well-approximated by the $\sqrt{2}$ factor in Eq. 11.23. Show that in this limit

$$W(\beta) = \frac{2}{\pi} e^{-2|\beta|^2} \left\{ \pm \cos[4\alpha(\mathrm{Im}\beta)] + \cosh[4\alpha(\mathrm{Re}\beta)] e^{-2\alpha^2} \right\}. \qquad (11.25)$$

Using this result show that there are no interference fringes in the Wigner function of an incoherent mixture of an even and an odd cat.

There is a simple recipe for deterministically creating cat states of photons that begins with the Schrödinger cat of Eq. 11.22 except that the coherent state amplitude should be 2α instead of α. The trick is to figure out how to disentangle the qubit from the cavity. From Eq. 11.22 we see that we need to be able to flip the qubit if and only if the cavity is in the vacuum state. When the cavity is in the state $|2\alpha\rangle$ it has (for large $|\alpha|$) negligible amplitude to have zero photons. We can carry out this special disentangling operation again using the strong-dispersive coupling. We rely on the quantized light shift of the qubit transition frequency by applying a π pulse to the qubit at frequency ω_q which is the transition frequency of the qubit when there are zero photons in the cavity. This 'number selective π pulse' yields the product state

$$|\Psi'\rangle = |g\rangle \frac{1}{\sqrt{2}}[|0\rangle \pm |2\alpha\rangle].\tag{11.26}$$

The final step is simply to apply a drive to displace the cavity by a distance $-\alpha$ in phase space to produce the cat state shown in Eq. 11.23. The creation of this cat can be verified via quantum state tomography via measuring the Wigner function as described above. One can also use quantum jump spectroscopy to find the photon number distribution and see that even (odd) cats contain only even (odd) numbers of photons. For the case of circuit QED, both of these checks have been carried out in Ref. (Vlastakis et al., 2013b) with negative Wigner function fringes measured in states with size up to $d^2 = 111$ photons, where $d = 2\alpha$ is the distance between the two coherent states of the cat.

Large cat states can be readily produced for microwave photons in circuit QED (Vlastakis et al., 2013a) and cavity QED with Rydberg atoms (Haroche, 2013) and produced in the phonon modes of trapped ions using spin-dependent optical forces acting on the ion motional degree of freedom (Monroe et al., 1996; McDonnell et al., 2007; Poschinger et al., 2010; Wineland, 2013; Kienzler et al., 2016; Bermudez et al., 2007).

There is another interesting recipe for producing cat states. While simple, this recipe produces cats with non-deterministic parity. Start with a coherent state $|\alpha\rangle$ and write it as a coherent superposition of an even and an odd cat (again assuming $|\alpha|$ is large for simplicity).

$$|\alpha\rangle = \frac{1}{\sqrt{2}}[|\Psi_+\rangle + |\Psi_-\rangle],\tag{11.27}$$

where $|\Psi_\pm\rangle$ is given by Eq. 11.23 (except we continue to drop the qubit state since it factors out). Now simply follow the parity measurement protocol described above. This collapses the state onto definite (but random) parity (Brune et al., 1992; Sun et al., 2014) and hence the measurement back action creates a cat state! Fig. 11.6 shows the Wigner function for this process conditioned on the state of the qubit used to perform the parity measurement (Sun et al., 2014).

Exercise 11.6 For large $|\alpha|$ the non-deterministic procedure above produces even and odd cats randomly but with equal probability. Show that for small $|\alpha|$ there is a bias towards producing even cats more frequently than odd cats. Compute the probability. (Hint: in the limit $|\alpha| \to 0$, we obtain the vacuum state which is definitely even parity.)

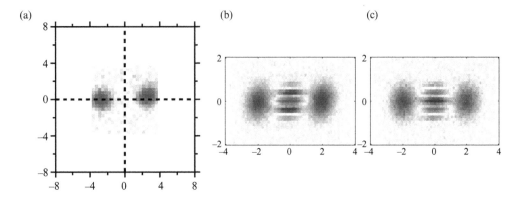

Fig. 11.6 Non-deterministic production of a cat state via the back action of a parity measurement of a coherent state with amplitude $\alpha = 2$ ($\bar{n} = 4$). (a) Wigner function of the incoherent mixture state resulting from tracing over the outcome of the parity measurement; (b) Wigner function when the outcome of the parity measurement is odd; (c) Wigner function when the outcome of the parity measurement is even. The phase of the fringe oscillations is opposite in the even and odd cats. Adapted from Sun et al. (2014).

11.5   Decoherence of cat states of photons

For the case of physical objects like trapped ions, the cat state splits the ion wave packet into two positions which can be separated by an amount much larger than their zero-point position uncertainty. If environmental degrees of freedom are able to gain information about the position of the ions then of course the superposition collapses to a single coherent state. For example, a stray gas atom might bounce off one of the trapped ions or the electric field from the moving ions may induce damping via energy transfer into nearby metallic structures. For the electromagnetic oscillator the 'position' of the 'object' that is oscillating is the electric field strength (say) at some selected point inside the cavity. (That point can be selected arbitrarily as long as it is not a nodal point of the mode.) Because the frequency of the cavity is set by geometry it is very stable and there is very little dephasing of the electromagnetic oscillations. The primary interaction with the environment is through weak damping of the resonator via the port that brings in the drive tones. When a photon leaks out of the cavity the parity of the cat must of course change. Equivalently this follows from the property of coherent states that they are eigenstates of the destruction operator:

$$a|+\alpha\rangle = (+\alpha)|+\alpha\rangle \tag{11.28}$$

$$a|-\alpha\rangle = (-\alpha)|-\alpha\rangle \tag{11.29}$$

Interestingly this means that photon loss does not dephase a coherent state (indeed it does nothing to a coherent state!) but photon loss is a dephasing error on cat states. Since the rate of photon loss $\kappa|\alpha|^2$ grows with the 'size' of the cat, macroscopic cats dephase quickly and become classical mixtures of two coherent states.

If photon loss does nothing to a coherent state how does its energy decay? This comes from the back action associated with the intervals in which photons are not observed leaking out that was mentioned above. This leads to a deterministic decay of the amplitude

$$|\alpha(t)\rangle = |e^{-\frac{\kappa}{2}t}\alpha(0)\rangle. \tag{11.30}$$

One remarkable feature of this is that even though the cavity is emitting photons into the environment, it remains in a pure state and does *not* become entangled with the environment. This is a unique feature of coherent states in simple harmonic oscillators (Haroche and Raimond, 2006).

Let us now move beyond simple coherent states to more general (and possibly mixed) states of the damped oscillator. The time evolution of the density matrix can be solved exactly by considering all the possible quantum trajectories specified by the random instants in time that photons leak out of the cavity (Haroche and Raimond, 2006; Michael et al., 2016). The so-called Kraus operators that describe the CPTP (completely positive trace-preserving) map can be organized according to the total number of photons lost in time t. In the absence of any Kerr non-linearities, the actual time that the photons are lost has no effect on the time evolution:

$$\rho(t) = \sum_{\ell=0}^{\infty} \hat{E}_\ell \rho(0) \hat{E}_\ell^\dagger, \tag{11.31}$$

where the ℓth Kraus operator describing the loss of ℓ photons is

$$\hat{E}_\ell(t) = \sqrt{\frac{(1 - e^{-\kappa t})^\ell}{\ell!}} e^{-\frac{\kappa t}{2}\hat{n}} a^\ell. \tag{11.32}$$

Exercise 11.7 Prove that the CPTP in Eq. 11.31 is in fact trace-preserving by showing that

$$\sum_{\ell=0}^{\infty} \hat{E}_\ell^\dagger \hat{E}_\ell = \hat{I},$$

where $\hat{I}$ is the identity operator. Hint: prove the identity is true for an arbitrary Fock state $|n\rangle$ and use the fact that the Fock states are complete.

Exercise 11.8 Compute the expectation value of the parity over time as a damped oscillator decays starting from: (a) an initial even cat state, and (b) an initial odd cat state. The expectation value of the parity will begin at ± 1, then (for large cats) decay close to zero and then end up at $+1$ because the vacuum is even parity.

Exercise 11.9 Compute the Wigner function over time for a damped oscillator decaying from: (a) an initial even cat state, and (b) an initial odd cat state.

11.5.1 Quantum Error Correction Using Cat States

Cat states are well-known to be delicate and their phase coherence is notoriously subject to rapid decay due to photon loss. It therefore seems like a very bad idea to try to use them to store quantum information. In fact the opposite is true. Ref. (Mirrahimi et al., 2014) proposes and Ref. (Ofek et al., 2016) demonstrates a clever scheme for encoding quantum information in two (nearly) orthogonal code words consisting of even-parity cat states

$$|W_1\rangle = \frac{1}{\sqrt{2}}[|+\alpha\rangle + |-\alpha\rangle] \tag{11.33}$$

$$|W_2\rangle = \frac{1}{\sqrt{2}}[|+i\alpha\rangle + |-i\alpha\rangle], \tag{11.34}$$

where (without loss of generality) α is real and positive. (We assume α is large enough that the states are nearly orthonormal.) These states are indeed highly sensitive to photon loss and quickly dephase. However as noted above, the source of the incoherence is the loss of parity information when we do not know how many photons have leaked out of the resonator. Fortunately, this problem can be overcome because, as noted above, we have the ability to make rapid, high fidelity, and highly QND measurements of the photon number parity that can be repeated hundreds of times without back action damage to the state.

Let us now examine in detail how by (semi-) continuously monitoring the parity jumps in the system, we can recover the stored quantum information. Under photon loss the two code words obey the following relations

$$a|W_1\rangle \to \frac{1}{\sqrt{2}}[|+\alpha\rangle - |-\alpha\rangle] \tag{11.35}$$

$$a^2|W_1\rangle \to +|W_1\rangle \tag{11.36}$$

$$a|W_2\rangle \to i\frac{1}{\sqrt{2}}[|+i\alpha\rangle - |-i\alpha\rangle] \tag{11.37}$$

$$a^2|W_2\rangle \to -|W_2\rangle. \tag{11.38}$$

After the loss of two photons the parity has returned to being even but we have a phase flip error. Thus it takes the loss of four photons for an arbitrary superposition of the two code words to return to the original state. We do not need to correct immediately for each photon loss. We need only keep track of the total number lost modulo 4 and then conditioned on that, apply one of four unitaries to restore the state. This is extremely powerful and extremely simple and has allowed circuit QED to be the first technology to reach the break-even point for quantum error correction in which the lifetime of the quantum information exceeds that of the best single component of the system (Ofek et al., 2016). Assuming perfect parity tracking (and no dephasing of the cavity) the only source of error from parity jumps would be the possibility that the parity jumps more than once in the short interval Δt between measurements. Two jumps would leave the parity untouched and the error monitoring would miss this fact. If photon loss occurs at rate $\Gamma \sim \kappa \bar{n}$, then parity monitoring at intervals Δt would reduce the error rate from first order Γ to second order $\mathcal{O}(\Gamma^2 \Delta t)$.

As noted above, in addition to the photon number jumps which occur at random times, the amplitude of the coherent state decays at rate $\kappa/2$. Because this is fully deterministic, the logical qubit decoding circuit can simply take this into account. The code fails at long times because the amplitude $\alpha e^{-\frac{\kappa}{2}t}$ becomes sufficiently small that the two code words are no longer orthonormal. Novel ideas for 'cat pumping' to keep the coherent states energized indefinitely have already been demonstrated experimentally for two-legged cats (Leghtas et al., 2015; Touzard et al., 2017). Bosonic codes for quantum error correction are now an active area of research (Gottesman et al., 2001; Terhal and Weigand, 2016; Chuang et al., 1997; Michael et al., 2016; Li et al., 2017; Albert et al., 2017) and will be discussed in detail at a future Les Houches School.

11.6 Conclusions and outlook

Circuit QED offers access to strong-coupling cavity QED with artificial atoms based on Josephson junctions coupled via antenna elements to microwave photons. In the dispersive regime where the atom and cavity are strongly detuned from each other, the dispersive coupling between atom and cavity can be several orders of magnitude larger than dissipation rates. This 'strong-dispersive' regime allows robust universal quantum control of the coupled system and permits creation of novel entangled Schrödinger cats as well as cat states of photons. In this regime it is easy to make high-fidelity and highly QND measurements of photon number parity. This allows production of cat states by measurement back action on coherent states and also allows repeated measurements of the primary error syndrome for cavity decay. This in turn permits highly hardware-efficient quantum error correction protocols and very simple quantum state tomography through measurement of the Wigner function of the cavity.

In addition to yielding a wonderful new regime to explore cavity QED, these capabilities will be key to a novel quantum computer architecture in which the logical qubits are stored in microwave resonators using bosonic codes and controlled by Josephson-junction based non-linear elements. Rapid progress towards this goal is being made as evidenced by recent demonstration of a universal gate set for a logical qubit encoded in a cavity (Heeres et al., 2017), entangling photon states (Wang et al., 2011) and cat states (Wang et al., 2016) between separate cavities, implementation of a CNOT gate between bosonic codes words stored in separate cavities (Rosenblum et al., 2018) and development of a 'catapult' to launch cat states stored in cavity into flying modes (Pfaff et al., 2017) for use in error correction in quantum communication (Li et al., 2017) and remote entanglement.

11.7   Appendix: the Wigner function and displaced parity measurements

The density matrix for a quantum system is defined by

$$\rho \equiv \sum_j |\psi_j\rangle p_j \langle\psi_j| \tag{11.39}$$

where where p_j is the statistical probability that the system is found in state $|\psi_j\rangle$. As we will see further below, this is a useful quantity because it provides all the information needed to calculate the expectation value of any quantum observable $\mathcal{O}$. In thermal equilibrium, $|\psi_j\rangle$ is the jth energy eigenstate with eigenvalue ϵ_j and $p_j = \frac{1}{Z} e^{-\beta\epsilon_j}$ is the corresponding Boltzmann weight. In this case the states are all naturally orthogonal, $\langle\psi_j|\psi_k\rangle = \delta_{jk}$. It is important to note however that in general the only constraint on the probabilities is that they are non-negative and sum to unity. Furthermore there is *no* requirement that the states be orthogonal (or complete), only that they be normalized.

The expectation value of an observable $\mathcal{O}$ is given by

$$\langle\langle\mathcal{O}\rangle\rangle = \mathrm{Tr}\,\mathcal{O}\rho \tag{11.40}$$

where the double brackets indicate both quantum and statistical ensemble averages. To prove this result, let us evaluate the trace in the complete orthonormal set of eigenstates of $\mathcal{O}$ obeying $\mathcal{O}|m\rangle = O_m|m\rangle$. In this basis the observable has the representation

$$\mathcal{O} = \sum_m |m\rangle O_m \langle m| \tag{11.41}$$

and thus we can write

$$\begin{aligned}
\mathrm{Tr}\,\mathcal{O}\rho &= \sum_m \langle m|\mathcal{O}\rho|m\rangle \\
&= \sum_m O_m \sum_j \langle m|\psi_j\rangle p_j \langle\psi_j|m\rangle \\
&= \sum_j p_j \sum_m \langle\psi_j|m\rangle O_m \langle m|\psi_j\rangle \\
&= \sum_j p_j \langle\psi_j|\mathcal{O}|\psi_j\rangle \equiv \langle\langle\mathcal{O}\rangle\rangle.
\end{aligned} \tag{11.42}$$

The considerations above are quite general. We now specialize to the case of a single-particle system in one spatial dimension. A useful example is the harmonic oscillator model which might represent a mechanical oscillator or the electromagnetic oscillations of a particular mode of a microwave or optical cavity. Our first task is to understand the relationship between the quantum density matrix and the classical phase space distribution.

In classical statistical mechanics we are used to thinking about the probability density $P(x, p)$ of finding a particle at a certain point in phase space. For example, in thermal equilibrium the phase space distribution is simply

$$P(x,p)\frac{dxdp}{2\pi\hbar} = \frac{1}{Z}e^{-\beta H(x,p)}\frac{dxdp}{2\pi\hbar}, \tag{11.43}$$

where the partition function is given by

$$Z = \int\int\frac{dxdp}{2\pi\hbar}e^{-\beta H(x,p)}, \tag{11.44}$$

and where for convenience (and planning ahead for the quantum case) we have made the phase space measure dimensionless by inserting the factor of Planck's constant. The marginal distributions for position and momentum are found by

$$\rho_1(x) = \frac{1}{2\pi\hbar}\int_{-\infty}^{+\infty}dp\,P(x,p) \tag{11.45}$$

$$\rho_2(p) = \frac{1}{2\pi\hbar}\int_{-\infty}^{+\infty}dx\,P(x,p). \tag{11.46}$$

Things are more complex in quantum mechanics because the observables $\hat{x}$ and $\hat{p}$ do not commute and hence cannot be simultaneously measured because of Heisenberg uncertainty. To try to make contact with the classical phase space distribution it is useful to study the quantum density matrix in the position representation given by

$$\rho(x,x') = \langle x|\rho|x'\rangle = \sum_j p_j\psi_j(x)\psi_j^*(x'), \tag{11.47}$$

where the wave functions are given by $\psi_j(x) = \langle x|\psi_j\rangle$. It is clear from the Born rule that the marginal distribution for position can be found from the diagonal element of the density matrix

$$\rho_1(x) = \rho(x,x) = \sum_j p_j|\psi_j(x)|^2. \tag{11.48}$$

Likewise the marginal distribution for momentum is given by the diagonal element of the density matrix in the momentum representation

$$\rho_2(p) = \langle p|\rho|p\rangle. \tag{11.49}$$

We can relate this to the position representation by inserting resolutions of the identity in terms of complete sets of position eigenstates[3]

---

[3] Note that we are using unnormalized momentum eigenstates $\langle x|p\rangle = e^{\imath px/\hbar}$. Correspondingly we are not using a factor of the system size L in the integration measure ('density of states in k space') for momentum.

$$\tilde{\rho}(p,p') = \frac{1}{2\pi\hbar} \int_{-\infty}^{+\infty} dx dx' \, \langle p|x\rangle \langle x|\rho|x'\rangle \langle x'|p'\rangle$$

$$= \frac{1}{2\pi\hbar} \int_{-\infty}^{+\infty} dx dx' \, e^{-ipx/\hbar} \rho(x,x') e^{+ip'x'/\hbar}. \tag{11.50}$$

Thus the momentum representation of the density matrix is given by the Fourier transform of the position representation. We see also that the marginal distribution for the momentum involves the off-diagonal elements of the real-space density matrix in an essential way

$$\rho_2(p) = \frac{1}{2\pi\hbar} \int_{-\infty}^{+\infty} dx dx' e^{-ip(x-x')/\hbar} \rho(x,x'). \tag{11.51}$$

For later purposes it will be convenient to define 'center of mass' and 'relative' coordinates $y = \frac{x+x'}{2}$ and $\xi = x - x'$ and reexpress this integral as

$$\rho_2(p) = \frac{1}{2\pi\hbar} \int_{-\infty}^{+\infty} dy \int_{-\infty}^{+\infty} d\xi \, e^{-ip\xi/\hbar} \rho(y+\xi/2, y-\xi/2). \tag{11.52}$$

Very early in the history of quantum mechanics, Wigner noticed from the above expression that one could write down a quantity which is a natural extension of the phase space density. The so-called Wigner 'quasi-probability distribution' is defined by

$$W(x,p) \equiv \frac{1}{2\pi\hbar} \int_{-\infty}^{+\infty} d\xi \, e^{-ip\xi/\hbar} \rho(x+\xi/2, x-\xi/2). \tag{11.53}$$

Using this, Eq. 11.52 becomes (changing the dummy variable y back to x for notational clarity)

$$\rho_2(p) = \int_{-\infty}^{+\infty} dx \, W(x,p). \tag{11.54}$$

Similarly, by using Eq. 11.53, we can write Eq. 11.48 as

$$\rho_1(x) = \int_{-\infty}^{+\infty} dp \, W(x,p). \tag{11.55}$$

These equations are analogous to Eqs. 11.45, 11.46 and show that the Wigner distribution is analogous to the classical phase space density $P(x, p)$. However the fact that position and momentum do not commute turns out to mean that the Wigner distribution need not be positive. In fact, in quantum optics one often takes the defining characteristic of non-classical states of light to be that they have Wigner distributions which are negative in some regions of phase space.

The Wigner function is extremely useful in quantum optics because, like the density matrix, it contains complete information about the quantum state of an electromagnetic oscillator mode, but (at least in circuit QED) is much easier to measure. Through a remarkable mathematical identity (Lutterbach and Davidovich, 1997) we can relate the Wigner function to the expectation value of the photon number parity, something that can be measured (Bertet et al., 2002) and is especially easy to measure (Vlastakis et al. 2013b) in the strong-coupling regime of circuit QED (a regime not easy to reach in ordinary quantum optics).

We are used to thinking of the photon number parity operator in its second quantized form

$$\hat{\Pi} = e^{i\pi a^\dagger a} \tag{11.56}$$

in which its effect on photon Fock states is clear

$$\hat{\Pi}|n\rangle = (-1)^n|n\rangle, \tag{11.57}$$

and indeed it is in this form that it is easiest to understand how to realize the operation experimentally using time evolution under the cQED qubit-cavity coupling $\chi\sigma^z a^\dagger a$ in the strong-dispersive limit of large χ relative to dissipation. However because the Wigner function has been defined in a first quantization representation in terms of wave functions, it is better here to think about the parity operator in its first-quantized form. Recalling that the wave functions of the simple harmonic oscillator energy eigenstates alternate in spatial reflection parity as one moves up the ladder, it is clear that photon number parity and spatial reflection parity are one and the same. That is, if $|x\rangle$ is a position eigensate

$$\hat{\Pi}|x\rangle = |-x\rangle, \tag{11.58}$$

or equivalently in terms of the wave function

$$\hat{\Pi}\psi(x) = \langle x|\hat{\Pi}|\psi\rangle = \psi(-x). \tag{11.59}$$

To further cement the connection, we note that since the position operator is linear in the ladder operators, it is straightforward to verify from the second-quantized representations that

$$\hat{\Pi}\hat{x} = -\hat{x}\hat{\Pi}. \tag{11.60}$$

We now want to show that we can measure the Wigner function $W(X, P)$ by the following simple and direct recipe (Lutterbach and Davidovich, 1997): (1) displace the oscillator in phase space so that the point (X, P) moves to the origin; (2) then measure the expectation value of the photon number parity

$$W(X,P) = \frac{1}{\pi h}\mathrm{Tr}\left\{\mathcal{D}(-X,-P)\rho\mathcal{D}^\dagger(-X,-P)\hat{\Pi}\right\}, \tag{11.61}$$

where $\mathcal{D}$ is the displacement operator. Related methods in which one measures not the parity but the full photon number distribution of the displaced state can in principle yield even more robust results in the presence of measurement noise (Shen et al., 2016).

Typically in experiment one would make a single 'straight-line' displacement. Taking advantage of the fact that $\hat{p}$ is the generator of displacements in position and $\hat{x}$ is the generator of displacements in momentum, the 'straight-line' displacement operator is given by

$$\mathcal{D}(-X,-P) = e^{-\frac{i}{\hbar}(P\hat{x}-X\hat{p})}. \tag{11.62}$$

In experiment, this displacement operation is readily carried out by simply applying a pulse at the cavity resonance frequency with appropriately chosen amplitude, duration and phase. In the frame rotating at the cavity frequency the drive corresponds to the following term in the Hamiltonian

$$V(t) = i[\epsilon(t)a^{\dagger} - \epsilon^*(t)a] \tag{11.63}$$

where $\epsilon(t)$ is a complex function of time describing the two quadratures of the drive pulse. The Heisenberg equation of motion

$$\frac{d}{dt}a = i[V(t),a] = \epsilon(t), \tag{11.64}$$

has solution

$$a(t) = a(0) + \int_{-\infty}^{t} d\tau\, \epsilon(\tau), \tag{11.65}$$

showing that the cavity is simply displaced in phase space by the drive. For the 'straight-line' displacement discussed above, $\epsilon(t)$ has fixed phase and only the magnitude varies with time.

Exercise 11.10   Find an expression for $\epsilon(t)$ such that time evolution under the drive in Eq. 11.63 will reproduce Eq. 11.62. Ignore cavity damping (an assumption which is valid if the pulse duration is short enough).

For theoretical convenience in the present calculation, we will carry out the displacement in two steps by using the Feynman disentangling theorem

$$e^{\hat{A}+\hat{B}} = e^{\hat{A}}e^{\hat{B}}e^{\frac{1}{2}[\hat{B},\hat{A}]} \tag{11.66}$$

(which is valid if $[\hat{B},\hat{A}]$ itself commutes with both $\hat{A}$ and $\hat{B}$) to write

$$\mathcal{D}(-X,-P) = e^{i\theta}\mathcal{D}(0,-P)\mathcal{D}(-X,0) = e^{i\theta}e^{-\frac{i}{\hbar}P\hat{x}}e^{+\frac{i}{\hbar}X\hat{p}}, \tag{11.67}$$

where $\theta \equiv \frac{i}{2\hbar} X P$. This form of the expression represents a move of the phase space point (X, P) to the origin by first displacing the system in position by $-X$ and then in momentum by $-P$. This yields the same final state as the straightline displacement except for an overall phase θ which arises from the fact that displacements in phase space do not commute. For present purposes this overall phase drops out and we will ignore it henceforth.

Under this pair of transformations the wave function becomes

$$\psi(x) \to \psi(x+X) \to e^{-iPx/\hbar}\psi(x+X). \tag{11.68}$$

More formally, we have two results which will be useful in evaluating Eq. 11.61

$$\langle \xi | \mathcal{D}(0,-P)\mathcal{D}(-X,0)|\psi\rangle = e^{-iP\xi/\hbar}\psi(\xi+X) \tag{11.69}$$

$$\langle \psi | \mathcal{D}^\dagger(-X,0)\mathcal{D}^\dagger(0,-P)|\xi\rangle = e^{+iP\xi/\hbar}\psi^*(\xi+X). \tag{11.70}$$

Taking the trace in Eq. 11.61 in the position basis yields

$$
\begin{aligned}
W(X,P) &= \frac{1}{\pi\hbar}\sum_j p_j \int_{-\infty}^{+\infty} d\xi\, \langle \xi | \mathcal{D}(-X,-P)|\psi_j\rangle\langle\psi_j|\mathcal{D}^\dagger(-X,-P)\hat{\Pi}|\xi\rangle \\
&= \frac{1}{\pi\hbar}\sum_j p_j \int_{-\infty}^{+\infty} d\xi\, \langle \xi | \mathcal{D}(-X,-P)|\psi_j\rangle\langle\psi_j|\mathcal{D}^\dagger(-X,-P)|-\xi\rangle \\
&= \frac{1}{\pi\hbar}\sum_j p_j \int_{-\infty}^{+\infty} d\xi\, e^{-iP2\xi/\hbar}\psi_j(\xi+X)\psi_j^*(-\xi+X) \\
&= \frac{1}{2\pi\hbar}\sum_j p_j \int_{-\infty}^{+\infty} d\xi\, e^{-iP\xi/\hbar}\psi_j(X+\xi/2)\psi_j^*(X-\xi/2), \tag{11.71}
\end{aligned}
$$

which proves that the displaced parity is indeed precisely the Wigner function.

Acknowledgments

The ideas described here represent the collaborative efforts of many students, postdocs, and faculty colleagues who have been members of the Yale quantum information team over the past fifteen years. The author is especially grateful for the opportunities he has had to collaborate with long-time friends and colleagues Michel Devoret, Leonid Glazman, Liang Jiang, and Rob Schoelkopf as well as frequent visitor, Mazyar Mirrahimi. Most of the ideas presented in these notes originated primarily with them and not the author. This work was supported by the National Science Foundation through grant DMR-1609326, by the Army Research Office and the Laboratory of Physical Sciences through grant ARO W911NF1410011, and by the Yale Center for Research Computing and the Yale Quantum Institute.

References

Albert, Victor V. et al. (2018). Performance and structure of single-mode bosonic codes. *Phys. Rev. A*, **97**, 032346.

Bermudez, A., Martin-Delgado, M. A., and Solano, E. (2007). Mesoscopic superposition states in relativistic Landau levels. *Phys. Rev. Lett.*, **99**, 123602.

Bertet, P. et al. (2002). Direct measurement of the Wigner function of a one-photon Fock state in a cavity. *Phys. Rev. Lett.*, **89**, 200402.

Blais, A. et al. (2007). Quantum-information processing with circuit quantum electrodynamics. *Phys. Rev. A*, **75**, 032329.

Blais, A. et al. (2004). Cavity quantum electrodynamics for superconducting electrical circuits: an architecture for quantum computation. *Phys. Rev. A*, **69**, 062320.

Brune, M. et al. (1992). Manipulation of photons in a cavity by dispersive atom-field coupling: quantum-nondemolition measurements and generation of 'Schrödinger cat' states. *Phys. Rev. A*, **45**, 5193–214.

Chuang, I. L., Leung, D. W., and Yamamoto, Y. (1997). Bosonic quantum codes for amplitude damping. *Phys. Rev. A*, **56**, 1114.

Clerk, A. A. et al. (2010). Introduction to quantum noise, measurement and amplification. *Rev. Mod. Phys.*, **82**, 1155–1208. (Longer version with pedagogical appendices available at: arXiv.org:0810.4729).

Devoret, M. H., and Schoelkopf, R. J. (2013). Superconducting circuits for quantum information: an outlook. *Science*, **339**, 1169–74.

Girvin, S. M. (2014). *Proceedings of the 2011 Les Houches Summer School on Quantum Machines*, Chapter Circuit QED: Superconducting Qubits Coupled to Microwave Photons. Oxford University Press, Oxford.

Gottesman, D., Yu. Kitaev, A., and Preskill, J. (2001). Encoding a qubit in an oscillator. *Phys. Rev. A*, **64**, 012310.

Haroche, S. (2013). Nobel lecture: controlling photons in a box and exploring the quantum to classical boundary. *Rev. Mod. Phys.*, **85**, 1083–1102.

Haroche, S., and Raimond, J.-M. (2006). *Exploring the Quantum: Atoms, Cavities and Photons*. Oxford University Press, Oxford.

Heeres, R. W. et al. (2017). Implementing a universal gate set on a logical qubit encoded in an oscillator. *Nat Commun.*, **8**, 94.

Kienzler, D. et al. (2016). Observation of quantum interference between separated mechanical oscillator wave packets. *Phys. Rev. Lett.*, **116**, 140402.

Leghtas, Z. et al. (2015). Confining the state of light to a quantum manifold by engineered two-photon loss. *Science*, **347**, 853–7.

Li, L. et al. (2017, Jul.). Cat codes with optimal decoherence suppression for a lossy bosonic channel. *Phys. Rev. Lett.*, **119**, 030502.

Lutterbach, L. G., and Davidovich, L. (1997). Method for direct measurement of the Wigner function in cavity qed and ion traps. *Phys. Rev. Lett.*, **78**, 2547–50.

McDonnell, M. J. et al. (2007). Long-lived mesoscopic entanglement outside the Lamb-Dicke regime. *Phys. Rev. Lett.*, **98**, 063603.

Michael, M. H. et al. (2016). New class of quantum error-correcting codes for a bosonic mode. *Phys. Rev. X*, **6**, 031006.

Mirrahimi, M. et al. (2014). Dynamically protected cat-qubits: a new paradigm for universal quantum computation. *New J. Phys.*, **16**, 045014.

Monroe, C. et al. (1996). A "Schrödinger cat" superposition state of an atom. *Science*, **272**, 1131–6.

Nigg, S. E. et al. (2012). Black-box superconducting circuit quantization. *Phys. Rev. Lett.*, **108**, 240502.

Ofek, N. et al. (2016). Extending the lifetime of a quantum bit with error correction in superconducting circuits. *Nature*, **536**, 441.

Pfaff, W. et al. (2017). Controlled release of multiphoton quantum states from a microwave cavity memory. *Nature Physics*, **13**, 882–7.

Poschinger, U. et al. (2010). Observing the phase space trajectory of an entangled matter wave packet. *Phys. Rev. Lett.*, **105**, 263602.

Rosenblum, S. et al. (2018). A CNOT gate between multiphoton qubits encoded in two cavities. Nature Communications, **9**, 652.

Schoelkopf, R. J., and Girvin, S. M. (2008). Wiring up quantum systems. *Nature*, **451**, 664.

Schuster, D. I. et al. (2007). Resolving photon number states in a superconducting circuit. *Nature*, **445**, 515–18.

Shen, C. et al. (2016). Optimized tomography of continuous variable systems using excitation counting. *Phys. Rev. A*, **94**, 052327.

Solgun, F., and DiVincenzo, D. P. (2015). Multiport impedance quantization. *Annals of Physics*, **361**, 605–69.

Sun, L. et al. (2014). Tracking photon jumps with repeated quantum non-demolition parity measurements. *Nature*, **511**, 444–8.

Terhal, B. M., and Weigand, D. (2016). Encoding a qubit into a cavity mode in circuit QED using phase estimation. *Phys. Rev. A*, **93**, 012315.

Touzard, S. et al. (2018). Coherent oscillations inside a quantum manifold stabilized by dissipation. *Phys. Rev. X* **8**, 021005.

Vlastakis, B. et al. (2013). Deterministically encoding quantum information using 100-photon Schrödinger cat states. *Science*, **342**, 607–10.

Vool, U., and Devoret, M. H. (2017). Introduction to quantum electromagnetic circuits. *International Journal of Circuit Theory and Applications*, **45**, 897.

Wallraff, A. et al. (2004). Circuit quantum electrodynamics: coherent coupling of a single photon to a Cooper pair box. *Nature*, **431**, 162–7.

Wang, C. et al. (2016). A Schrödinger cat living in two boxes. *Science*, **352**, 1087–91.

Wang, H. et al. (2011). Deterministic entanglement of photons in two superconducting microwave resonators. *Phys. Rev. Lett.*, **106**, 060401.

Wendin, G. (2017). Quantum information processing with superconducting circuits: a review. *Rep. Prog. Phys.*, **80**, 106001.

Wineland, David J. (2013). Nobel lecture: superposition, entanglement, and raising Schrödinger's cat. *Rev. Mod. Phys.*, **85**, 1103–14.

Zheng, H. et al. (2016). Accelerating dark-matter axion searches with quantum measurement technology. *arXiv:1607.02529.*

12

Hanbury Brown and Twiss, Hong Ou and Mandel effects and other landmarks in quantum optics: from photons to atoms

ALAIN ASPECT

Institut d'Optique Graduate School, Université Paris Saclay
91120 Palaiseau, France

Aspect, A. "Hanbury Brown and Twiss, Hong Ou and Mandel effects and other landmarks in quantum optics: from photons to atoms." In *Current Trends in Atomic Physics*. Edited by Antoine Browaeys, Thierry Lahaye, Trey Porto, Charles S. Adams, Matthias Weidemüller, and Leticia F. Cugliandolo. Oxford University Press (2019). ⓒ Oxford University Press. DOI: 10.1093/oso/9780198837190.003.0012

Chapter Contents

For a long time, the name "quantum optics" referred mostly to the physics of lasers. In fact, in a laser, what is fully quantum is the amplifying medium itself, with the quantized levels of the emitter (atoms, molecules, ions), but light can be perfectly described as a classical electromagnetic wave, as it is done in the Lamb theory of the laser (Lamb, 1964). In fact, in 1960, there was only one phenomenon involving visible light that demanded the use of the quantum theory of light: it was spontaneous emission. Absorption or stimulated emission could be described by the semi-classical classical theory of matter-light interaction [see for instance (Grymberg, Aspect and Fabre, 2010)], in which matter only is quantized. And when it came to freely propagating light, the description as classical electromagnetic waves was found perfectly adequate, provided that one used the semiclassical model of photo detection, and that a statistical description were used to describe incoherent light, such as thermal radiation, or light emitted by discharge lamps.

It is only in the 1960s that emerged the idea that it might be important to describe freely propagating light as a quantized system, involving in particular the notion of photon. When Roy Glauber took the challenge of using the quantum formalism to describe the Hanbury Brown and Twiss (HBT) effect, he had to develop a formalism, which not only allowed him to describe the HBT effect, but also was available, from then on, to allow physicists to render an account of new genuine quantum optics effects that were discovered in the next decades. Among these effects, whose description and understanding demand quantization of the free electromagnetic field, one must cite the violations of Bell's inequalities with pairs of entangled of photons, observed in the 1970s and early 1980s [see references in (Aspect, 1999; Aspect, 2015)], and the Hong Ou and Mandel (HOM) effect (Hong, Ou and Mandel, 1987), which also involves pairs of entangled photons. There were other quantum effects, related to properties of single photons (Aspect, 2017), such as photon anti-bunching in resonance fluorescence (Kimble, Dagenais and Mandel, 1977), or anticorrelation for a single photon on a beam splitter (Grangier, Roger and Aspect, 1986). But in this lecture I will put the emphasis on the HBT and the HOM effects, which have been recently revisited in our laboratory with atoms replacing photons. These effects are remarkable landmarks in quantum optics since their description demands to use the notion of two photon amplitudes interference, which is a major ingredient[1] of the second quantum revolution (Dowling and Milburn, 2003; Aspect, 2004).

Today, I will first present my views on the second vs the first quantum revolution, then describe the Hanbury Brown and Twiss effect with photons, and indicate why it was so important in the development of modern quantum optics. The presentation of our experiments on the HBT effect with atoms will allow me to emphasize the analogies but also the increased richness of the effect when going from photons to atoms. I will similarly describe the HOM effect for photons and its significance, and then present the analogous experiment with atoms. In conclusion, I will put these two effects in the long list of landmarks in the development of quantum optics, and indicate what has been done and what remains to be done with atoms in lieu of photons.

---

[1] The other ingredient of the second quantum revolution is the experimental ability to observe and manipulate individual quantum objects, and the Quantum Monte-Carlo methods that suggest clear intuitive images for the evolution of these individual quantum objects. See (Dowling and Milburn, 2003; Aspect, 2004).

12.1 Two great quantum mysteries

In the early 1960s, in chapter 1 of volume 3 of his famous lectures on physics (Feynman, 1963), Feynman described wave particle duality as "the only quantum mystery". Two decades later, however, in a paper that is considered the founding paper of quantum information (Feynman, 1982), he recognized that there was another great mystery, entanglement. Why are these two extraordinary features of quantum mechanics different in nature?

Wave particle duality refers to a single quantum particle, which can be described both as a wave and a particle. Each of these descriptions involves a classical notion: a wave propagating in the ordinary space-time, or a particle whose trajectory is developed in the ordinary space-time. What is quantum, and hard to swallow, is that these two descriptions are used for the same object, which belongs a priori to one of the two categories: an electron, a neutron, is a priori a particle, but we must also think of it as a wave; light is a priori a wave, but we must also think of it as composed of particles, the photons. But each of these behaviors can be described without any problem, in the usual ordinary space-time.

In contrast, entanglement between several particles must be described in an abstract Hilbert space, which is the tensor product of the spaces of each of the entangled objects. A problem may arise, then, when one wants to give an image of what happens in our ordinary space-time. For instance, for two maximally entangled particles separated in space, such as the photons entangled in polarization used for most Bell's inequalities tests (Aspect, 1999; Aspect, 2015), any image in our ordinary space-time involves either negative probabilities, or some-degree of non-locality, i.e., a contradiction with the notion that nothing can propagate faster than light (Aspect, 2002). Both sides of the alternative are very hard to swallow, as stressed by Feynman (Feynman, 1982): "I've entertained myself always by squeezing the difficulty of quantum mechanics into a smaller and smaller place, so as to get more and more worried about this particular item. It seems to be almost ridiculous that you can squeeze it to a numerical question that one thing is bigger than another. But there you are—it is bigger than any logical argument can produce, if you have this kind of logic." It could be tempting to content oneself with the observation that there is no non-locality in the Hilbert space where the two entangled particles are described. But as emphasized by Asher Peres, a famous quantum optics theorist: "Quantum phenomena do not occur in a Hilbert space. They occur in a laboratory"[page 373 in (Peres, 1995)].

Because it is difficult to renounce locality or positiveness of probabilities, phenomena based on entanglement are much more difficult to swallow than the ones based on wave particle duality. This is why, in this lecture, I will focus on two quantum optics landmarks that we have recently revisited with atoms, in which entanglement must be invoked to give a consistent quantum description.

12.2 The Hanbury Brown and Twiss effect for photons

The experiment reported by R. Hanbury Brown and R. Q. Twiss in 1956 (Hanbury Brown and Twiss, 1956) is considered the landmark signaling the beginning of modern quantum optics. It is indeed for giving a fully consistent description of that experiment that R. Glauber developed the Quantum Optics formalism that we still use today.[2]

12.2.1   Experimental observation

Fig. 12.1 describes the original experiment, which was a study of the intensity fluctuations of light emitted by an incoherent source (laser had not yet been invented). Two photomultipliers, almost images of each other by reflection in a beam splitter, allowed one to monitor the correlation function of the photocurrents associated with light detection at $(\mathbf{r}_1, t)$ and $(\mathbf{r}_2, t + \tau)$, with $(\mathbf{r}_1, t)$ and $(\mathbf{r}_2, t + \tau)$ as close to each other as one wants. One also monitors the average photocurrent at each detector.

According to the semi-classical theory of quantum optics, the photocurrent is proportional to the light intensity, i.e., the squared modulus of the classical complex electric field

$$i(\mathbf{r},t) \propto I(\mathbf{r},t) = \left| E^{(+)}(\mathbf{r},t) \right|^2 = E^{(-)}(\mathbf{r},t) E^{(+)}(\mathbf{r},t), \tag{12.1}$$

so that the normalized current correlation function is equal to the normalized correlation function of the light intensity:

$$g^{(2)}(\mathbf{r}_1,\mathbf{r}_2;\tau) = \frac{\langle i(\mathbf{r}_1,t) i(\mathbf{r}_2,t+\tau) \rangle}{\langle i(\mathbf{r}_1,t) \rangle \langle i(\mathbf{r}_2,t) \rangle} = \frac{\langle I(\mathbf{r}_1,t) I(\mathbf{r}_2,t+\tau) \rangle}{\langle I(\mathbf{r}_1,t) \rangle \langle I(\mathbf{r}_2,t) \rangle}. \tag{12.2}$$

Note that in spite of a somewhat misleading, but traditional, notation, $g^{(2)}$ is in fact a fourth order correlation function of the classical complex electric field

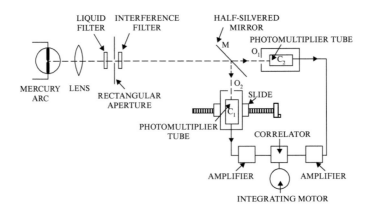

Fig. 12.1 Schematic view of the original HBT experiment.

---

[2] According to Claude Cohen-Tannoudji (private communication) the interpretation of the Forrester et al. experiment (Forrester, Gudmundsen) on quantum beats in light emitted by a spectral lamp had provoked intense discussions, which had prepared the minds to the necessity of a fully consistent quantum description of such phenomena.

$$g^{(2)}(\mathbf{r_1},\mathbf{r_2};\tau) = \frac{\langle E^{(-)}(\mathbf{r_1},t)E^{(+)}(\mathbf{r_1},t)E^{(-)}(\mathbf{r_2},t+\tau)E^{(+)}(\mathbf{r_2},t+\tau)\rangle}{\langle E^{(-)}(\mathbf{r_1},t)E^{(+)}(\mathbf{r_1},t)\rangle\langle E^{(-)}(\mathbf{r_2},t+\tau)E^{(+)}(\mathbf{r_2},t+\tau)\rangle}. \tag{12.3}$$

Fig. 12.2 shows the results of the original HBT experiment. At zero distance and time delay, the normalized correlation function is nothing else than the average of the square of the intensity. The value greater than 1 for $\mathbf{r_1} = \mathbf{r_2}$ and $\tau = 0$ means that light intensity fluctuates. More precisely, the value of 2 indicates that the variance $\langle I^2\rangle - \langle I\rangle^2$ is equal to the squared average intensity $\langle I\rangle^2$. At long distance ($|\mathbf{r_1} - \mathbf{r_2}| \gg L_c$) and/or large delay ($\tau \gg \tau_c$), the autocorrelation function drops to 1, which means no correlation between the fluctuations.

The goal of HBT was to perform such a measurement on the light emitted by a star, in order to determine its angular diameter α. The reason is that the correlation length L_c is linked to the angular diameter under which one sees the star (Fig. 12.3) by the relation

$$L_c = \frac{\lambda}{\alpha} \tag{12.4}$$

where λ is the wavelength of the light. This would allow them to measure angular diameters of stars, a measurement made impossible with standard astronomical methods by the atmospheric fluctuations.

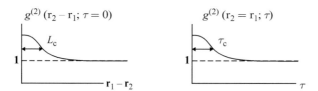

Fig. 12.2 Results of the original HBT experiment. The normalized correlation function is maximum, with a value of 2, at zero distance and delay, where it characterizes intensity fluctuations. It drops to the value of 1, which means no correlation, for a delay larger than the correlation time τ_c, or a distance larger than the correlation length L_c.

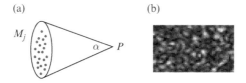

Fig. 12.3 Intensity pattern produced by an incoherent source: (a) the source, composed of many independent emitters, is seen from the detection point under an angular diameter α; (b) at a given time, the intensity pattern is a speckle pattern, whose "grains" have a characteristic size of the order of $L_c = \lambda/\alpha$; this pattern is a random process, which evolves with a characteristic time τ_c.

12.2.2 Semi-classical interpretation

We describe classically the field emitted by the source as the sum of many contributions issued from independent emitters j in the source. The complex electric field at P (Fig. 12.3) is thus

$$E^{(+)}(P,t) = \sum_j a_j \exp\left\{ \phi_j + \frac{\omega_j}{c} M_j P - \omega_j t \right\} \qquad (12.5)$$

where the ϕ_j are independent random variables. The field $E^{(+)}(P,\ t)$ is a sum of many independent random variables with the same statistical properties. It is thus a Gaussian random process, as a consequence of the Central Limit Theorem. We can then use the Wick theorem to express $g^{(2)}(\mathbf{r}_1, \mathbf{r}_2; \tau)$, which is a fourth order moment of the complex electric field (Eq. 12.3), as

$$g^{(2)}(\mathbf{r}_1, \mathbf{r}_2; \tau) = 1 + \left| g^{(1)}(\mathbf{r}_1, \mathbf{r}_2; \tau) \right|^2 \qquad (12.6)$$

where

$$g^{(1)}(\mathbf{r}_1, \mathbf{r}_2; \tau) = \frac{\left\langle E^{(-)}(\mathbf{r}_1, t) E^{(+)}(\mathbf{r}_2, t+\tau) \right\rangle}{\left\langle E^{(-)}(\mathbf{r}_1, t) E^{(+)}(\mathbf{r}_1, t) \right\rangle^{1/2} \left\langle E^{(-)}(\mathbf{r}_2, t+\tau) E^{(+)}(\mathbf{r}_2, t+\tau) \right\rangle^{1/2}} \qquad (12.7)$$

is the second order moment of the complex electric field. In fact, $g^{(1)}$ is the so-called first order coherence function, whose spatial and temporal widths are respectively the coherence length L_c and the coherence time τ_c. Within a factor of the order of 1, the functions $g^{(1)}(\mathbf{r}_1, \mathbf{r}_2; \tau)$ and $g^{(2)}(\mathbf{r}_1, \mathbf{r}_2; \tau)$ have thus the same widths. Since $g^{(1)}(\mathbf{r}_1 - \mathbf{r}_2 = 0; \tau = 0) = 1$, one has $g^{(2)}(\mathbf{r}_1 - \mathbf{r}_2 = 0; \tau = 0) = 2$. This factor of 2 is characteristic of a Gaussian process.

Note in passing an illuminating interpretation of the width of $g^{(2)}(\mathbf{r}_1, \mathbf{r}_2; \tau)$. Let us think of the intensity pattern produced around P by the source of Fig. 12.3. At any given time, it is a speckle pattern, whose "grains" have a characteristic size of the order of $L_c = \lambda/\alpha$; this pattern fluctuates with a characteristic time τ_c. If two detectors are separated by less than the grain size, the detected intensities are correlated fluctuating quantities. For a larger separation, the fluctuations of the detected intensities are uncorrelated.

12.2.3 A hot debate

When HBT proposed to build what they called "an intensity interferometer" to measure $g^{(2)}$ in order to deduce stars' angular diameters, their application to get funding was rejected (Hanburry Brown, 1974), based on the following argument.

Let us think of the experiment with the two detectors working in the photon counting mode. The correlation function $g^{(2)}(\mathbf{r}_1, \mathbf{r}_2; \tau)$ is then expressed as a function of single and joint detection probabilities

$$g^{(2)}(\mathbf{r}_1, \mathbf{r}_2; \tau) = \frac{\pi^{(2)}(\mathbf{r}_1, t; \mathbf{r}_2, t+\tau)}{\pi^{(1)}(\mathbf{r}_1, t) \cdot \pi^{(1)}(\mathbf{r}_2, t+\tau)}. \qquad (12.8)$$

A value of 2 for $g^{(2)}$ would then mean that the photons "come in pairs", a totally unacceptable hypothesis, according to the referees, since photons emitted at different, possibly very distant, points of a star, are obviously independent. In spite of their efforts to argue with the referees, including the realization of the table top experiment of reference (Hanburry Brown and Twiss, 1956), shown on Fig. 12.1, they had to move to Australia, to find support, and build an observatory in the desert of Narrabri,

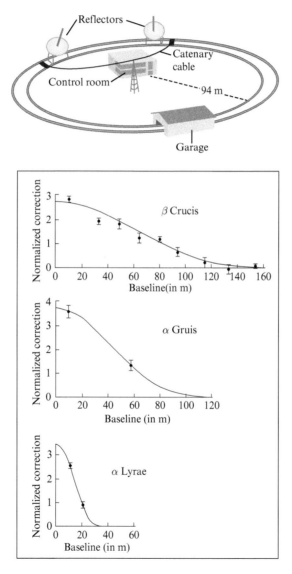

Fig. 12.4 The intensity interferometer built in Australia by R. Hanbury Brown et al. The intensity correlation function could be measured up to separations of 188 m. The right panel shows some examples, which allowed them to determine the angular diameter of stars of the southern hemisphere.

where they measured the angular diameter of several stars of the southern hemisphere (Fig. 12.4).

Beyond its interest in astronomy, the HBT experiment is now celebrated as the landmark whose quantum interpretation prompted the development of the modern quantum optics formalism, by Roy Glauber (Glauber, 1963a; Glauber, 1963b; Glauber, 1963c; Glauber, 1965).

12.2.4   Quantum interpretation

From a quantum point of view, the HBT effect is related to the quantum statistics of bosons, which tend to be detected in pairs if they cannot be distinguished. The quantum statistics is automatically taken into account in the formalism of Glauber, which is a version of second quantization well adapted to the case of photons. I will not recall here the full treatment of reference (Glauber, 1963a), and will only emphasize the role of two-photon amplitudes interference, or equivalently, in that case[3], of entanglement, in the quantum description of the HBT correlations. This can be done using a toy model introduced by Glauber in his Les Houches course of 1964 (Glauber, 1965), and shown in Fig. 12.5.

In this model, two one-photon wave-packets, emitted by two independent excited atoms, will overlap and be detected by two detectors. The full process consists of an evolution from an initial state to a final state, shown on Fig. 12.5. One can see by simple inspection that there are two paths to go from the initial state to the final state. These two paths are sketched on the right panel of Fig. 12.5. In order to calculate the probability of a joint detection at D_1 and D_2, one must add the amplitudes of these paths, before taking the squared modulus of the sum:

$$\pi^{(2)}(\mathbf{r}_1;\mathbf{r}_2) = |\langle D_1|U|E_1\rangle\langle D_2|U|E_2\rangle + \langle D_2|U|E_1\rangle\langle D_1|U|E_2\rangle|^2. \qquad (12.9)$$

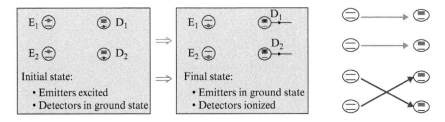

Fig. 12.5 Toy model to understand how two photons amplitudes interference plays a role in the joint detections at D_1 and D_2. One must add the amplitudes associated with the two processes sketched on the right panel, which correspond to the same initial and final states of the emitters and detectors. If the detectors are close enough to each other that the two photons wave packets are indistinguishable, the two amplitudes have almost the same phase factor and the interference is constructive.

---

[3] It must be recalled that while indistinguishability of quantum particles leads, in a first quantization point of view, to entangled states, the reciprocal is not true: entanglement can also happen between fully distinguishable particles.

This is an example of a two-photon amplitudes interference effect. It is deeply linked to the notion of entanglement since the state of the photons between the emitters and the detectors is

$$|\Psi\rangle = \frac{1}{\sqrt{2}} \left(|1_{E_1 D_1}, 1_{E_2 D_2}\rangle + |1_{E_1 D_2}, 1_{E_2 D_1}\rangle \right) \tag{12.10}$$

where $|1_{E_1 D_1}\rangle$ refers to one photon travelling from E_1 to D_1, etc....

When one considers many different pairs of emitters, the phases of the two terms that interfere in Eq. 12.9 differ by a random quantity, and the result is a Gaussian random process, with fluctuations such that relation (12.6) holds. From a mathematical point of view, the bump may seem due to the same effect as in the classical interpretation. But in the quantum point of view, the quantities that interfere are abstract two photon quantum amplitudes, while in the classical description the quantities that interfere are classical electromagnetic fields propagating in our ordinary space-time.

12.2.5 A paradoxical situation

As shown in Sec. 12.2.3, and emphasized as early as 1956 by E. Purcell (Purcell, 1994), the HBT effect can fully be described by a semi-clasical model in which light is not quantized. Moreover, one must admit that the quantum description is more involved than the semi-classical one. It is thus remarkable that, in order to answer a question which could have been considered a simple curiosity rather than a necessity, R. Glauber developed a full fledged formalism, which would turn out to be necessary to interpret and analyze the genuine quantum effects that would appear later, in particular the violations of Bell's inequalities, or the HOM effect.

12.3 The Hanbury Brown and Twiss effect for atoms

12.3.1 From light to atoms

In Sec. 12.2, we have seen that the interpretation of the HBT effect demands the quantum notion of two-photons amplitudes interference, and entanglement, if light is considered as made of photons. It is therefore interesting to consider the HBT effect for other kinds of particles. As a matter of fact, as described in (Baym, 1998), the HBT effect has been observed with nuclear particles, and used in order to determine collision cross sections between these particles. At the other end of the energy scale, ultra-cold atoms offer nowadays exquisite experimental methods allowing physicists to revisit the photon quantum optics experiments. Following a pioneering experiment, which demonstrated the effect using metastable Neon atoms (Yasuda and Shimizu, 1996), we decided to start a systematic program of study of quantum effects related to atoms entanglement, using metastable helium atoms, the workhorse of programs of Quantum Atom Optics in our group or in the group at ANU (see for instance (Hodgman et al., 2011)).

12.3.2   Metastable helium: the workhorse of Quantum Atom Optics

As explained in the caption of Fig. 12.6, helium atoms in a triplet state can be manipulated with light, and thus laser cooled and trapped. When they are released on the MCP, they can be detected individually, with the position and time of detection of each atom recorded. Since all the free falling atoms arrive on the MCP with almost the same velocity, time can be converted into a vertical position in the free falling cloud, and the 3D ensemble of positions in an individual cloud can be reconstructed.

The emergence of modern Quantum Optics had been permitted by the development, after World War Two, of photon counting techniques, which allowed pioneers to measure correlation functions $g^{(2)}$ in light. MCP with He* offers similar possibilities; better in fact, since our system with delay lines is equivalent to 10^5 independent detectors (the MCP has a diameter of 70 mm, and the resolution is 0.2 mm), while landmark Quantum Optics experiments were performed with two detectors only. We show now how that system was used to study atomic HBT.

12.3.3   Atomic HBT

Fig. 12.7 shows the result of the experiment reported in (Schellekens et al., 2005), whose ingredients have been sketched in Sec. 12.3.2.

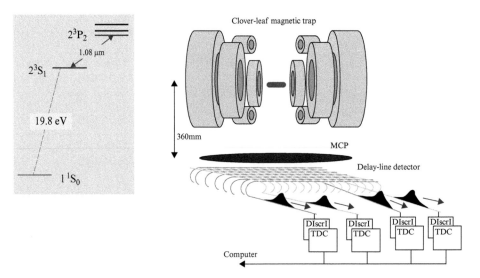

Fig. 12.6 Metastable helium He*. Left panel: radiative transitions from triplet levels $2^3 P_2$ or $2^3 S_1$ to the singlet ground state $1^1 S_0$ are forbidden. Level $2^3 S_1$, which is the lowest triplet state, is thus a metastable state, which plays the role of an effective ground state for atoms in any triplet state interacting with light, including when they emit spontaneous photons. The transition at 1.08 μm can thus be used for cooling and trapping He* atoms. Right panel: when a He* atom in a triplet state falls on the Micro Channel Plate (MCP), a transition to the ground state $1^1 S_0$ happens, and at least 19.8 eV of energy is released. This is more than enough to extract an electron from the upper face of the MCP. After multiplication in the MCP, a macroscopic electric pulse emerges on the lower face of the MCP, and is divided in four pulses propagating along delay lines, so that one can register the time and the location of the detection of each atom.

A thermal cloud of ultracold atoms is dropped onto the detector, and we register the 3D positions of the atoms. More precisely, we define 3D pixels centered around positions $\mathbf{r}_i$ and count how many atoms we find in each pixel (actually the number is most of the time 0 and sometimes 1). We can then determine the probability to have pairs of atoms separated by $\Delta \mathbf{r} = \mathbf{r}_i - \mathbf{r}_j$, in the whole cloud. Dividing by the product of the probabilities of having one atom in each pixel, we obtain the correlation function for one cloud

$$g^{(2)}_{1\text{cloud}}(\Delta \mathbf{r}) = \frac{\pi^{(2)}(\Delta \mathbf{r})}{[\pi^{(1)}]^2}. \tag{12.11}$$

where the probabilities are defined for 1 cloud.

The result is usually quite noisy, but averaging $g^{(2)}_{1\text{cloud}}(\Delta \mathbf{r})$ over many clouds yields the correlation function $g^{(2)}(\Delta \mathbf{r})$ with a good signal-to-noise ratio, as shown on Fig. 12.7, which is extracted from (Schellekens et al., 2005) where one can find more details.

In that reference, we also show that, when the temperature of the initial cloud is lowered below the transition temperature, one obtains a Bose Einstein Condensate (BEC), for which the correlation function is found flat, as it was the case for laser light (Arecchi, Gatti, and Sona, 1966). A similar result has been obtained with rubidium atoms extracted from a BEC (Ottl et al., 2005), while (Hodgman et al., 2011) reports on measurements of third order correlation functions in a BEC of He* atoms.

Beyond these proofs of principle, the atomic HBT effect can be used as a tool to probe many-body states of ultracold atoms (Altman, Demler and Lukin, 2004).

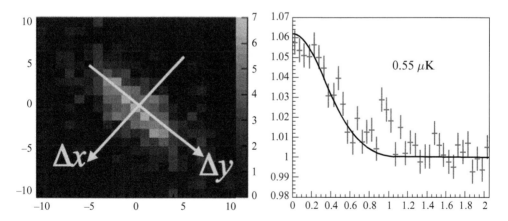

Fig. 12.7 Correlation function $g^{(2)}(\Delta \mathbf{r})$ for an initial thermal cloud at 0.55μ K, with a cigar shape elongated along x. The $g^{(2)}(\Delta \mathbf{r})$ function is found symmetrical by rotation around the x axis, as expected, and its shape, which is shown on the cut in the (x, y) plane, corresponds to the Fourier transform of the shape of the initial cloud. The maximum value of 1.06 rather than 2 is a consequence of the finite resolution of the detector, which has a point spread function wider than the atoms distribution along x but narrower than the distribution along y.

12.3.4   Fermionic HBT effect

While photons are bosons, atoms can come either in bosonic or in fermionic forms. The experiment described in Sec. 12.3.3 was performed with ^{4}He atoms, which are bosons, but we have also performed a similar experiment with ^{3}He atoms, which are fermions. The result is shown on Fig. 12.8, extracted from (Jeltes et al., 2007).

In that experiment, carried out in collaboration with colleagues at the VU of Amsterdam, we were able to realize a direct comparison of the effects for ^{4}He and ^{3}He atoms, initially held in the same trap at the same temperature, i.e. in clouds with identical shapes and widths. We could choose at will to drop either of the two isotopes. The density was small enough that interaction energy was negligible, and the observations the result of quantum statistical effects only. One clearly sees that in the case of fermions one has a dip around zero rather than a bump. This is easily understood by referring to Fig. 12.5. In the case of fermions, the (entangled) state describing the two particles propagating from source to detectors must be antisymmetrized rather than symmetrized, so the amplitudes associated with the two diagrams of the right panel of Fig. 12.5 must be added with opposite signs. For detectors close enough to each other, and atoms with almost equal velocities so that they are undistinguishable, it results in a null probability of joint detection, in agreement with the Pauli principle. Analogous results have been reported in (Rom et al., 2006) for ^{40}K atoms, and, with a lower visibility, for electrons in solids (see (Kiesel, Renz and Hasselbach, 2002) with references).

In conclusion of that section, it must emphasized that there is absolutely no classical interpretation for the HBT effect with fermions. This is in contrast with the case of light, for which the HBT can be understood classically as a consequence

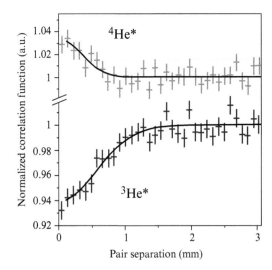

Fig. 12.8 Comparison of the HBT effect with bosons and fermions. Bosons tend to bunch while fermions tend to antibunch. The differences in width and amplitude of the dip vs. the bump are related to the difference in de Broglie wavelengths, which are in ratio 4 to 3 for atoms with the same velocity.

of the Cauchy-Schwarz inequality for light intensity

$$\langle I^2 \rangle \geq \langle I \rangle^2. \tag{12.12}$$

In analogy, the HBT effect for fermions would be associated with the following inequality for atomic density

$$\langle n^2 \rangle < \langle n \rangle^2. \tag{12.13}$$

This is mathematically impossible if n is a classical quantity, but it becomes possible in the framework of second quantization, where n is considered an operator expressed as a function of the creation and annihilation operators for fermions.

12.4 The Hong Ou and Mandel effect for photons

12.4.1 A two photon interference effect

The Hong Ou and Mandel (HOM) effect was first described in a paper (Hong, Ou, and Mandel, 1987) emphasizing its use to determine with a high resolution the "simultaneity" of the two photons (Fig. 12.9) of pairs emitted in parametric down conversion from a CW laser beam, another landmark in quantum optics (Burnham and Weinberg, 1970). Nowadays, the HOM paper is mostly cited as an emblematic example of a two photon interference effect, as shown on Fig. 12.10. Indeed, a joint detection at D_3 and D_4 corresponds to two possible processes, which will interfere if the two photons are indistinguishable, i.e. if the two wave packets exactly overlap at the beam splitter. A careful examination of the situation shows that for a balanced splitter the two two-photon amplitudes are opposite, so that the interference yields a null probability of joint detections. The opposite signs are related to the unitarity of the matrix describing the effect of the beam splitter. More precisely, if we choose the phase references such that all amplitude reflection and transmission coefficients

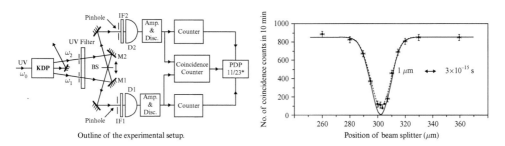

Outline of the experimental setup.

Fig. 12.9 Observation of the HOM dip in the joint detection of two photons emitted in parametric down conversion and recombined on a beam splitter. The probability of joint detection drops to zero when the two photon wave-packets arrive exactly at the same time on the beam splitter. The width of the dip, of about 50 fs, indicates a simultaneity at that scale, a time resolution not previously heard of.

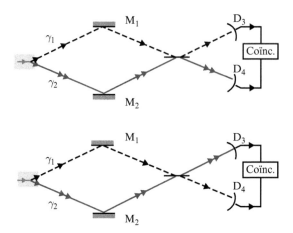

Fig. 12.10 The HOM effect: an emblematic two photon interference effect. If the two photons are undistinguishable and exactly overlap at the balanced beam splitter, the two processes sketched on the two panels are undistinguishable, and their amplitudes must be added. It turns out that these amplitudes have the same modulus and opposite signs, so the interference is destructive, and the probability of a joint detection is null. One can equivalently understand the phenomenon by writing the state of the two photons as $|\Psi\rangle = 1/\sqrt{2}\,[|2_3, 0_4\rangle + |0_3, 2_4\rangle]$, which is a (NOON) entangled state ($|2_3, 0_4\rangle$ means 2 photons in mode 3 and 0 photon in mode 4, propagating respectively from the beam splitter to detector D_3 or D_4, etc....).

be real, the transmission coefficients involved in the lowest panel are equal, but the two reflection coefficients involved in the upper panel have opposite signs.

12.4.2 A fully quantum effect

The HOM effect is the observation that both photons are always detected in the same channel, either the upper one or the lower one, and never one in one channel and one in the other channel. This is an intriguing effect. If we thought of photons as classical particles with equal probabilities to be transmitted or reflected, the probability to observe a joint detection would be $1/2$, while the probability of detecting both photons in the upper channel would be $1/4$ and similarly for the probability to detect both photons in the lower channel.

By analogy with the HBT effect, one might think that a semi-classical model involving classical waves could render an account of the situation. Let us indeed consider two classical waves with the same frequency entering in the two inputs of the balanced beam splitter. If their phase difference is ϕ, their interference leads to rates of single detections in the output channels, $w^{(1)}(D_3)$ and $w^{(1)}(D_4)$, respectively proportional to $\sin^2\phi$ and $\cos^2\phi$. The rate of joint detections $w^{(2)}(D_3; D_4)$ is thus proportional to $\sin^2\phi\cos^2\phi = 1/4\sin^2 2\phi$. In order to render an account of the random character of the detections in either channel, we take ϕ a random variable uniformly distributed over an interval of 2π. The average rates of single detection are then $1/2$ each, while the average rate of joint detection is $1/8$, i.e., $w^{(2)}(D_3; D_4) = 1/2 w^{(1)}(D_3) \cdot w^{(1)}(D_4)$. There is thus a suppression of the joint detection in D_3 and D_4, since the interference favors double detections in the same channel. That suppression, however, is limited to a factor $1/2$, while the quantum calculation leads to a total suppression, in agreement with the observation. A quantum calculation of the shape of the dip obtained with parametric down conversion pairs can be found, for instance, in Sec. 7.4.6 of (Grynberg, Aspect, and Fabre, 2010).

It is remarkable that the effect can be generalized to the case where the two indistinguishable photon wave packets come from different sources. For instance, it has been observed with two spontaneous photons emitted by two different atoms, and terminating by chance in two modes of the electromagnetic field exact images of each other in the beam splitter (Beugnon et al., 2006). A calculation of the dip obtained with two independent one-photon wave packets can be found in Complement 5B of (Grymberg, Aspect, and Fabre, 2010).

12.5   The Hong Ou and Mandel effect for atoms

In order to revisit, with atoms rather than photons, quantum optics landmarks that are based on pairs of photons, we have developed a versatile source of pairs of ^{4}He* atoms. This source is somewhat analogous to the sources of pairs of photons based on parametric down conversion of laser photons in a non-linear crystal (Burnham and Weinberg, 1970). We start with a Bose Einstein Condensate of ^{4}He*, dense enough that interaction energy between atoms plays a role analogous to a $\chi^{(3)}$ non-linearity for light. According to a suggestion of (Molmer, 2006), first demonstrated in (Campbell et al., 2006), we apply a moving laser standing wave on a 1D interacting BEC in order to favor the emission of pairs with well defined velocities (Bonneau et al., 2013). The phenomenon favoring this velocities selection is conservation of energy and quasi-momentum in the periodic potential provided by the standing wave. It is analogous to phase matching in non-linear optics in non-linear crystals. The non-linear process responsible for the emission of pairs is a dynamical instability associated with a repulsive interaction between atoms.

With this source, we have implemented the experiment described in Fig. 12.11. The atoms are allowed to circulate along the vertical z axis, and held on that axis by a far-off red detuned laser beam propagating along z. At time t_0, we apply the standing wave entailing the emission of a pair of atoms with controlled vertical velocities, as explained above. Their motion in gravity is represented as parabolas in the z, t diagram (panel b of Fig. 12.11).

If now we consider the equivalent diagram in a frame of reference falling freely as the center of mass of the two atoms, the motion is represented by two symmetrical strait lines (panel c of Fig. 12.11). At time t_1, we apply another laser standing wave, stationary in the free falling frame of reference thanks to the chirp $\Delta\omega(t)$ shown in the figure. That second standing wave realizes a Bragg diffraction of each atom, whose velocities are reverted if the standing wave is applied during a time and with an amplitude corresponding to a π pulse. At time t_2 such that $t_2 - t_1 = t_1 - t_0$, we apply again the second laser standing wave, still stationary in the free falling frame of reference, but for half the time only, realizing a $\pi/2$ pulse, which is equivalent to a balanced beam splitter.

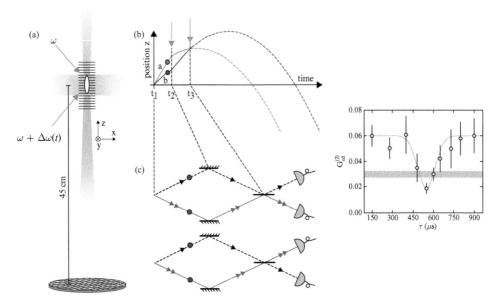

Fig. 12.11 Atomic HOM effect.

As shown in the panel c of Fig. 12.11, when viewed in the free falling frame of reference, the process is equivalent to the one shown for photons in Fig. 12.10, and if the atoms are in indistinguishable modes of matter-waves, we observe the HOM dip. This is shown in the right panel of Fig. 12.11, where the time delay is controlled by changing the time t_2 around the value given by $t_2 - t_1 = t_1 - t_0$.

Note that in the experiment with atoms, we have a bunch of atoms submitted to the process, and we do not have two detectors, but we have the equivalent of many detectors monitoring all the atoms (cf. Sec. 12.3.2 and Fig. 12.6). Among all the registered detections, we then look a posteriori for pairs of atoms corresponding to modes symmetrical in the final beam splitter. Compared to the case of photons, where corresponding modes are selected a priori with pinholes placed before the mirrors and beam splitter, our selection is done a posteriori, which is possible thanks to our many pixels detectors.

The fact that the dip does not go to zero is fully accounted for by the fact that in the pair creation process there is some amplitude to have 2 atoms rather than 1 in an elementary mode. The amount of that contamination can be determined by using the stored data to calculate the $g^{(2)}$ function in an elementary mode. For one atom only, that function should be zero, but we find it different from zero and infer the amplitude for 2 atoms in the mode. Even with that imperfection of our experiment, the visibility of the dip is larger than $1/2$, which means that the observed effect could not be explained by "ordinary" interferences of atomic matter-waves in the ordinary space-time, and demands an interpretation in terms of two atom amplitudes.

12.6 Outlook: towards Bell's inequalities test with atoms

Experimenting on light has played a major role in the development of both the first and second quantum revolution. Wave particle duality for light was recognized by Einstein as early as 1909 (Einstein, 1909), while it was only in 1923 that Louis de Broglie proposed that wave-particle duality should also apply to material particles. When it comes to the second quantum revolution, of which entanglement and two particles interference effects are key ingredients, light is again far ahead, with first violations of Bell's inequalities reported in the early 1970s, and the conflict with relativistic locality demonstrated in the early 1980s (Aspect, 1999), with polarizers varied during the flight of photons (Fig. 12.12). A new generation of experiments, started in the late 1990s, has led to improved tests, culminating in 2015 with almost perfect, so-called loophole-free, experiments (Aspect, 2015). In contrast, no Bell's inequalities tests have been performed on the external degrees of freedom (position or momentum) of material particles[4].

Such tests would be highly desirable, not only because they would complete the long series of correspondance between landmarks of photon optics and of atom optics [Table12.1], but also because they may open the way to experiments that could shed some light on the elusive frontier between quantum physics and relativity.

One may wonder why the HOM scheme is not sufficient to carry out such investigations. The reason is that the HOM effect does not address the question of non-locality, i.e. the tension between relativity and quantum mechanics, which was a major element in the EPR argument against the completeness of Quantum Mechanics. This is because two modes only (3 and 4) are involved in the entangled state of the two HOM particles

$$|\Psi\rangle = \frac{1}{\sqrt{2}} \left[|2_3, 0_4\rangle + |0_3, 2_4\rangle \right]. \tag{12.14}$$

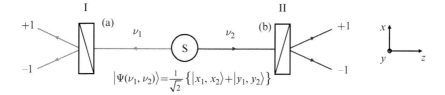

Fig. 12.12 Bell's inequalities test with well separated photons. Having two modes ($|x_i\rangle$ and $|y_i\rangle$) for each separated photon ($i = 1$ or $i = 2$) allows one to choose at will, for each photon, between different directions of polarization measurement. If the polarizers are far enough from each other, and can be adjusted fast enough, the choice can be made while the photons are in flight, enforcing locality, i.e. relativistic separation between the measurements. A test of Bell's inequalities in such a configuration allows one to decide between Einstein's Local Realism and Quantum Mechanics.

---

[4] Tests of Bell's inequalities with ions of (Rowe et al., 2001) bore on internal degrees of freedom.

Table 12.1 Photon vs. Atom quantum optics: some landmarks.

Photon Optics	Date	Atom Optics	Date
Interference, diffraction	1800s	Interference, diffraction	1990s
Single photons	1974, 1985	Single atoms	2002
Photon correlations (HBT)	1955	Atom correlations (HBT)	2005
$\chi^{(2)}$ photon pairs	1970s	$\chi^{(3)}$ atom pairs	2007
Beyond SQL: squeezing	1985	Beyond SQL: squeezing	2010
Bell tests spontaneous photons	1972, 1982	Bell tests molecules dissoc.	?
HOM with $\chi^{(2)}$ pairs	1987	HOM with $\chi^{(3)}$ pairs	2014
Bell tests with $\chi^{(2)}$ pairs	1989–98	Bell tests with $\chi^{(3)}$ pairs	?

"SQL": Standard Quantum Limits. "?": not (yet) done.

while Bell's inequalities tests demand to have an entangled state of two particles in four modes, such as

$$|\Psi\rangle = \frac{1}{\sqrt{2}} \left[|x_1, x_2\rangle + |y_1, y_2\rangle \right], \tag{12.15}$$

(see Fig. 12.12). More precisely in order to test locality, one must be able to choose between two non-commuting observables for each of two spatially separated particles, and this demands a two-dimensional space on each side, for each particle.[5]

As a first step towards a Bell's test with massive particles entangled in momentum, the experiment described in Sec. 12.5 has allowed us to find evidence of entanglement of two atoms in four modes associated with different momenta (Dussarat et al., 2017). This is hopefully the last step towards a genuine test of Bell's inequalities with a pair of material particles entangled in momentum, following a scheme in the spirit of the experiment of (Rarity and Tapster, 1990), sketched in Fig. 12.13. It would complete

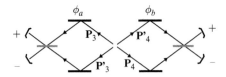

Fig. 12.13 Proposed configuration to test Bell's inequalities with two atoms in the momentum entangled state $|\Psi\rangle = \frac{1}{\sqrt{2}} \left[|p_3, p_4\rangle + |p_3', p_4'\rangle \right]$

---

[5] One can note that in contrast to Bell's inequalities tests, the quantum behavior in the HOM experiment can be mimicked by a local hidden variable theory where the two photons are determined, from the moment of the emission, to both go either on one side or the other side.

the series of landmark quantum optics experiments revisited with atoms, and initiate a series of such experiments with heavier particles, allowing one to address the interface between quantum physics and relativity.

References

Altman, E., Demler, E., and Lukin, M. D. (2004). "Probing many-body states of ultracold atoms via noise correlations," *Physical Review A*, vol. 70, no. 1.

Arecchi, F., Gatti, E., and Sona, A. (1966). "Time distribution of photons from coherent and Gaussian sources," *Physics Letters*, vol. 20, no. 1, pp. 27–9.

Aspect, A. (1999). "Bell's inequality test: more ideal than ever," *Nature*, vol. 398, no. 6724, pp. 189–90.

Aspect, A. (2002). "Bell's theorem: the naive view of an experimentalist," in *Quantum (un)Speakables: From Bell to Quantum Information*, Springer, New York; https://arxiv.org/abs/quant-ph/0402001.

Aspect, A. (2004). *Introduction: John Bell and the Second Quantum Revolution*, pp. xvii–xl. Cambridge: Cambridge University Press, 2nd edn.

Aspect, A. (2015). "Viewpoint: closing the door on Einstein and Bohr's quantum debate," *Physics*, vol. 8, p. 123.

Aspect, A. (2017). "Mooc: Quantum optics 1: single photons," https://www.coursera.org/learn/quantum-optics-single-photon.

Baym, G. (1998). "The physics of Hanbury Brown–Twiss intensity interferometry: from stars to nuclear collisions," *arXiv preprint nucl-th/9804026*.

Beugnon, J. et al. (2006). "Quantum interference between two single photons emitted by independently trapped atoms," *Nature*, vol. 440, no. 7085, pp. 779–82.

Bonneau, M. et al. (2013). "Tunable source of correlated atom beams," *Physical Review A*, vol. 87, no. 6, p. 061603.

Brown, R. H., and Twiss, R. (1956). "Correlation between photons in two coherent beams of light," *Nature*, vol. 177, no. 4497, pp. 27–9.

Brown, R. H. (1974). "The intensity interferometer: its application to astronomy," Research supported by the Department of Scientific and Industrial Research, Australian Research Grants Committee, US Air Force, et al. London, Taylor and Francis, Ltd.; New York, Halsted Press, 1974. 199 p.

Burnham, D. C., and Weinberg, D. L. (1970). "Observation of simultaneity in parametric production of optical photon pairs," *Physical Review Letters*, vol. 25, no. 2, pp. 84–7.

Campbell, G. K. et al. (2006). "Parametric amplification of scattered atom pairs," *Physical review letters*, vol. 96, no. 2, p. 020406.

Dowling, J. P., and Milburn, G. J. (2003). "Quantum technology: the second quantum revolution," *Philosophical Transactions of the Royal Society of London A: Mathematical, Physical and Engineering Sciences*, vol. 361, no. 1809, pp. 1655–74.

Dussarrat, P. et al. (2017). "Two-particle four-mode interferometer for atoms," *Physical Review Letters*, vol. 119, no. 17, p. 173202.

Einstein, A. (1909). "On the evolution of our vision on the nature and constitution of radiation," *Physikalische Zeitschrift*, vol. 10, pp. 817–26.

Feynman, R. P. (1963). *Lectures on Physics*. Addison-Wesley.

Feynman, R. P. (1982). "Simulating physics with computers," *International Journal of Theoretical Physics*, vol. 21, no. 6–7, pp. 467–88.

Forrester, A. T., Gudmundsen, R. A., and Johnson, P. O. (1955). "Photoelectric mixing of incoherent light," *Physical Review*, vol. 99, no. 6, p. 1691.

Glauber, R. J. (1963a). "Photon correlations," *Physical Review Letters*, vol. 10, no. 3, pp. 84–6.

Glauber, R. J. (1963b). "Coherent and incoherent states of the radiation field," *Physical Review*, vol. 131, no. 6, p. 2766.

Glauber, R. J. (1963c). "The quantum theory of optical coherence," *Physical Review*, vol. 130, no. 6, p. 2529.

Glauber, R. (1965). "Les houches lecture notes, 1964 (quantum optics and electronics, ed. C. de Witt et al., Gordon and Breach, NY, 1965).'2) g," *Lachs: Phys. Rev*, vol. 138, p. B1012.

Grangier, P., Roger, G., and Aspect, A. (1986). "Experimental-evidence for a photon anticorrelation effect on a beam splitter—a new light on single-photon interferences," *Europhysics Letters*, vol. 1, no. 4, pp. 173–9.

Grynberg, G., Aspect, A., and Fabre, C. (2010). *Introduction to Quantum Optics: From the Semi-Classical Approach to Quantized Light*. Cambridge University Press.

Hodgman, S. et al. (2011). "Direct measurement of long-range third-order coherence in Bose-Einstein condensates," *Science*, vol. 331, no. 6020, pp. 1046–9.

Hong, C. K., Ou, Z. Y., and Mandel, L. (1987). "Measurement of subpicosecond time intervals between 2 photons by interference," *Physical Review Letters*, vol. 59, no. 18, pp. 2044–6.

Jeltes, T. et al. (2007). "Comparison of the Hanbury Brown-Twiss effect for bosons and fermions," *Nature*, vol. 445, no. 7126, pp. 402–5.

Kiesel, H., Renz, A., and Hasselbach, F. (2002). "Observation of Hanbury Brown–Twiss anticorrelations for free electrons," *Nature*, vol. 418, no. 6896, pp. 392–4.

Kimble, H. J., Dagenais, M., and Mandel, L. (1977). "Photon anti-bunching in resonance fluorescence," *Physical Review Letters*, vol. 39, no. 11, pp. 691–5.

Lamb, Jr, W. E. (1964). "Theory of an optical maser," *Physical Review*, vol. 134, no. 6A, p. A1429.

Molmer, K. (2006). "Phase-matched matter wave collisions in periodic potentials," *New Journal of Physics*, vol. 8.

Ottl, A. et al. (2005). "Correlations and counting statistics of an atom laser," *Physical Review Letters*, vol. 95, no. 9.

Peres, A. (1995). *Quantum Theory: Concepts and Methods*, vol. 57. Springer, New York.

Purcell, E. (1994). "Reproduced from nature (1956) 17 8, 1449–50," *Journal of Astrophysics and Astronomy*, vol. 15, no. 1, pp. 27–32.

Rarity, J. G., and Tapster, P. R. (1990). "Experimental violation of Bell inequality based on phase and momentum," *Physical Review Letters*, vol. 64, no. 21, pp. 2495–8.

Rom, T. et al. (2006). "Free fermion antibunching in a degenerate atomic fermi gas released from an optical lattice," *Nature*, vol. 444, no. 7120, pp. 733–6.

Rowe, M. A. et al. (2001). "Experimental violation of a Bell's inequality with efficient detection," *Nature*, vol. 409, no. 6822, pp. 791–4.

Schellekens, M., Hoppeler, R., Perrin, A., Gomes, J. V., Boiron, D., Aspect, A., and Westbrook, C. I. (2005). "Hanbury brown twiss effect for ultracold quantum gases," *Science*, vol. 310, no. 5748, pp. 648–51.

Yasuda, M., and Shimizu, F. (1996). "Observation of two-atom correlation of an ultracold neon atomic beam," *Physical Review Letters*, vol. 77, no. 15, pp. 3090–3.